G. Campardo · R. Micheloni · D. Novosel

VLSI-Design of Non-Volatile Memories

Giovanni Campardo · Rino Micheloni · David Novosel

VLSI-Design of Non-Volatile Memories

With 568 Figures

Springer

Giovanni Campardo
Rino Micheloni

STMicroelectronics Srl
Memory Product Groups
Flash Division
Via C. Olivetti 2
20041 Agrate Brianza (MI)
Italy

David Novosel

IMD Intelligent
Micro Design, Inc.
Mercer-West Middlesex Road 2456
West Middlesex, PA 16159
U.S.A.

ISBN 3-540-20198-X **Springer Berlin Heidelberg New York**

Library of Congress Control Number: 2004116726

Springer is a part of Springer Science+Business Media

springeronline.com

Typesetting: Data conversion by the author.
Final processing by PTP-Berlin Protago-TEX-Production GmbH, Germany
Cover-Design: medionet AG, Berlin
Printed on acid-free paper 62/3141/Yu - 5 4 3 2 1 0

Preface

The electronics and information technology revolution continues, but it is a critical time in the development of technology. Once again, we stand on the brink of a new era where emerging research will yield exciting applications and products destined to transform and enrich our daily lives! The potential is staggering and the ultimate impact is unimaginable, considering the continuing marriage of technology with fields such as medicine, communications and entertainment, to name only a few.

But who will actually be responsible for transforming these potential new products into reality? The answer, of course, is today's (and tomorrow's) design engineers!

The design of integrated circuits today remains an essential discipline in support of technological progress, and the authors of this book have taken a giant step forward in the development of a practice-oriented treatise for design engineers who are interested in the practical, industry-driven world of integrated circuit design.

The authors, Giovanni Campardo and Rino Micheloni, are very well qualified to effectively address this challenging objective. Both have a solid track record of leading design activities at the STMicroelectronics Flash Division. I should probably mention at this point my association with and knowledge of the accomplishments of these authors. In April 2003, they published a unique Special Issue on the subject of Flash Memories for the Proceedings of the IEEE, the journal for which I am Managing Editor. Therefore, I have firsthand knowledge of their approach to the development of very well crafted technical material.

In addition, a third member of the author team, David Novosel, has provided invaluable assistance, particularly in the translation efforts on this material. David is President and founder of Intelligent Micro Design, Inc. which specializes in the design of custom memories and analog circuits, which is located in Pittsburgh, PA.

This book is intended for Electrical Engineering graduates who want to enter into the integrated circuit design world. Nonvolatile memories, in many cases, are treated as an example to explain general design concepts (basic circuits, layout, design flow, etc.). Practical illustrative examples of nonvolatile memories, including flash types, are showcased to give insightful examples of the design approaches that are discussed. The authors introduce the key nonvolatile memories design issues in the first section and they discuss the various functions and capabilities of these memories; much of these discussions are based on their recently published Special Issue on the subject of Flash memories.

After a complete review of the general design issues, the authors begin to focus on the important concepts that should be understood. For example, they introduce MOS Process and briefly review the CMOS building blocks and the four available components. In the next section the authors describe Memory cell operations including Read and Erase operations, as well as coverage of software and program issues.

Comprehensive sections on Design Building Blocks, Integrated circuits layout, and Matrix architecture are included. They explore in some depth Input buffers, Output Buffers Decoders and Program and Erase operations circuitry. They discuss the subjects of Timers and Read operation sensing techniques.

Two sections deal with quality and dependability-related issues, and cover the subjects of Redundancy and Test Modes. Coverage is also provided on ESD & latch-up issues as well as Embedded Flash algorithms.

A collection of photos is included to make the reader familiar with silicon aspects.

Throughout all parts of this book, the authors have taken a practical and applications-driven point of view, providing a comprehensive and easily understood approach to all the concepts discussed.

I greatly appreciate the very kind invitation of the authors to submit these words of introduction for their exciting new book. I wish them continued success in their latest publishing enterprise, but more importantly I hope that all who are reading these words will derive knowledge and understanding from the sincere and dedicated efforts of the authors.

Jim Calder
Managing Editor
PROCEEDINGS OF THE IEEE

Acknowledgements

Introducing this book, we want to acknowledge the contribution of all those people who greatly helped us in the preparation of the manuscript.

Our colleague Roberto Ravasio, who took care of the sections related to the logic, the burst, the description of the algorithms executed by the Flash and related implementation, either using PLA methodology or the sequencer.

Our colleagues Miriam Sangalli and Ilaria Motta for the section related to charge pumps and regulators, and again Ilaria Motta for the High Voltage Management section.

We also wish to thank Paolo Cappelletti, ST Non Volatile Memory Technology Development Director in Agrate, who wrote the foreword for this book. Paolo, who has always supported the various initiatives that we have undertaken in these years, has shown once again his availability and his passion for this kind of activities.

We also want to thank Jim Calder, Managing Editor for the Proceedings of the IEEE: the publication of the Proceedings allowed us to contact the Editor for this book. Jim has written the preface to the text.

We have to thank Dieter Merkle for giving us the possibility of publishing this work together with Gaby Maas who greatly helped us as editor. Special thanks to Adalberto Micheloni for drafts review.

Last but not least, we wish to thank our colleagues Marcello Pesare and Stefano Commodaro, who have taken care of the translation of the text from Italian to English.

Our thanks go to all these people and to many others who have indirectly contributed to this book with their job work and dedication.

Gianni, Rino and Dave

Contents

Foreword: Non-Volatile Memory Technology Evolution

Memories represent a significant portion of the semiconductor market and they are key components of all electronic systems.

From a system view point, semiconductor memories can be divided into two major categories: the RAM's (Random Access Memories), whose content can be changed in a short time and for a virtually unlimited number of times, and the ROM's (Read Only Memories), whose content cannot be changed, or at least not in a time comparable with the clock period of the system. But there is another very important feature which differentiates the two families: RAM's loose their content when the power is switch-off while ROM's retain their content virtually for ever.

The ideal memory would combine the writing properties of RAM's with the data retention properties of ROM's: increasing efforts have been devoted to find a viable technology for the "universal" memory and some very promising candidates have been recently identified, but none of them is ready for volume production yet.

So far, among the semiconductor memories that have reached the industrial maturity, the one that has got closest to the goal belongs to the category of non-volatile memories. That category includes all the memories whose content can be changed electrically but it is retained even when the power supply is removed. Those memories are still ROM's from a system view point because the time to change their content is too long with respect to the clock period, but they are significantly more flexible than the masked ROM's, whose content is defined during their fabrication process and can never be changed.

The history of non-volatile memories started in the early 70's, with the introduction in the market of the first EPROM (Erasable Programmable Read Only Memory). Since then, non-volatile memories have been always considered one of the most important families of semiconductor memory. However, until mid 90's, the relevance of this kind of memories was related more to the key role they play in most electronic systems and to the scientific interest for their memory cell concepts, than to the economical size of their market segment. The dramatic growth of the non-volatile memory market, which started in 1995, has been fuelled by two major events: the first was the introduction of Flash memories and the second was the development of battery-supplied electronic appliances, mainly the mobile phones but also PDA's, MP3 players, digital still cameras and so on. This preliminary chapter will shortly review the technology evolution of non-volatile memories.

Systems Needs for Non-Volatile Storage

Almost in every electronic system, some pieces of information must be stored in a permanent way, i.e. they must be retained even when the system is not powered.

Program codes for microcontrollers are probably the most popular example: any system based on microcontrollers needs the permanent storage of the set of instructions to be executed by the processors to perform the different tasks required for a specific application.

A similar example is given by the parameters for DSP's (Digital Signal Processors); those also are pieces of information to be stored in a non-volatile memory. In general, any programmable system requires a set of instructions to work; those instructions, often called "the firmware", cannot be lost when the power supply is switched off. But solid state non-volatile memories are widely used not only for the firmware.

In most systems there are data which are set either by the system manufacturer, or by the distributors, or by the end users, and those data must be retained at power-off. The examples are many and for a variety of different functions: identification and security codes, trimming of analog functions, setting of system parameters, system self-diagnostic, end-user programmable options and data, in-system data acquisition and many others. Due to their pervasiveness, non-volatile memories have penetrated all electronic market segments: industrial, consumer, telecommunication, automotive and computer peripherals (Fig. 1).

	Cell Phone	Consumer	Automotive	Computer & Communication
EPROM	Analog, Residential	Games, Set Top Box	Engine Mgt	HDD, Copiers, Fax, Switching
FLASH	Digital (GSM)	Set Top Box PDA	All Power Train Car Navigation, ABS, GPS	HDD, PC Bios, CDROM
EEPROM	Digital (GSM)	Audio, Video	All Car Body	PC SPD, Graphic boards, Printers

Fig. 1. Main applications of Non-Volatile Memories

Even in personal computers, that host a magnetic memory media (the hard disk) for mass storage and RAM's as working memory, there are solid state non-volatile memories: the system boot, which tells the system what to do at power-up, before the operating system is downloaded from the hard disk into the RAM, is stored in a non-volatile memory. Moreover, in the hard disk drive itself, which is a micro-controller based system, there is a non-volatile memory.

To cover such a variety of application needs, non-volatile memories are available in a wide range of capacities, from few Kbits to hundreds of Mbits, and with a number of different product specifications. Moreover, thanks to the evolution of integration technology, non-volatile memories can also be embedded into the processor chip; today a significant portion of them, mainly in the small and medium size range, is not sold as stand-alone memory but integrated with other logic functions. This trend, initiated by microcontrollers, has been extended to a wider range of products and to higher complexity towards the integration of complete systems in a single chip (SoC's: System-on-a-Chip); non-volatile memory has been an enabling technology for this evolution to happen and smart cards are just an example of a popular single chip electronic systems that could not exist without that kind of technology.

Until the introduction of Flash memory, there were two different categories of electrically programmable non-volatile memories: the EPROM's and the EEPROM's (Electrically Erasable PROM). EPROM's have a one-transistor memory cell and therefore can provide high density and cost effectiveness, but they can only be erased by exposure to UV light. Mounted in very expensive ceramic packages with a transparent window, EPROM's were used for system debugging, to be substituted in volume production by either masked ROM's or by OTP (One Time Programmable) memories, i.e. the same EPROM chips in more cost effective plastic packages.

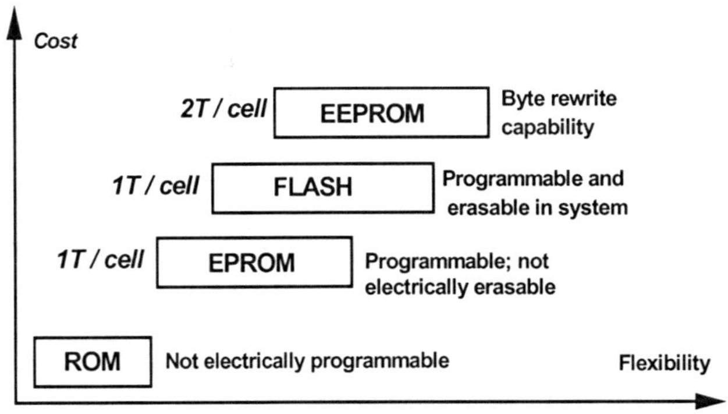

Fig. 2. Comparison of Non-Volatile Memories; the common feature of all types is that they retain the stored data even without power supply

EEPROM's feature the electrical erase capability, with a fine granularity (even single byte) and a pretty good endurance (over 1 million program/erase cycles); however, because of the complex structure of their memory cell (Fig. 2), they are quite expensive, they offer a much lower density than EPROM's at same technology node and their cell size cannot be scaled in proportion to the lithography feature. As a consequence of their different cost/performance trade-offs, EPROM's have been mostly used for code storage while EEPROM's have been used to store parameters and user's data.

Offering the electrical erase capability, traditionally featured by the expensive EEPROM's, at cost and density comparable to EPROM's, Flash memories not only have taken a big portion of their progenitor's markets, but in addition they have greatly expanded the fields of application of non-volatile memories.

This impressive growth has been also associated to the development and the diffusion of personal portable electronic appliances. Systems like PDA's and mobile phones cannot use magnetic disks because of size and power consumption; therefore in these systems, besides the usual requirements for non-volatile storage of codes and parameters, there is a demand for mass storage (operating system, application programs, user's files) that must be covered by semiconductor memories.

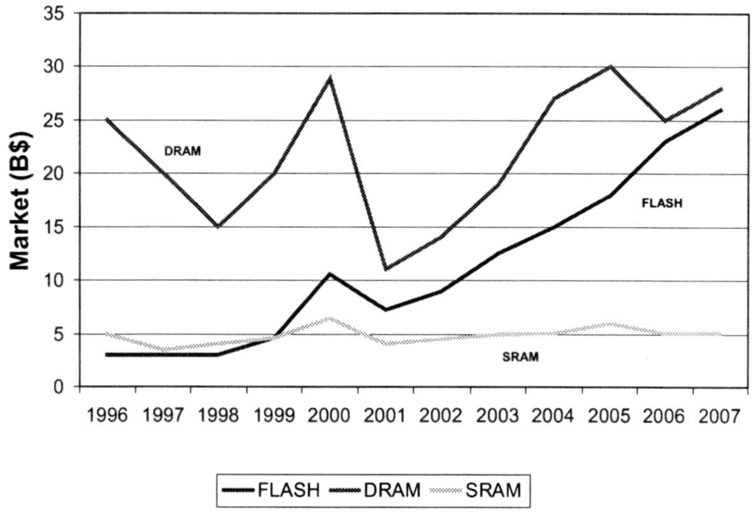

* Forecast of Web Feet

Fig. 3. Semiconductor memory market

Moreover the development of multimedia applications and the convergence of personal consumer appliances towards portable systems that manage data, communication, images and music, is dramatically increasing the demand for a friendly way of storing and moving large files: memory cards, in different formats, are the rising segment that is further fuelling the growth of Flash memory market. Indeed, the dramatic increase of flash memory has been the most relevant event in the semiconductor memory market in the last decade (Fig. 3). In 1999 flash memory revenues passed the ones of SRAM's, making flash the second largest memory family after DRAM's; today flash market is not far from the DRAM size and this trend is foreseen to continue in the future.

Among the many different Flash technologies that have been conceived and the less that have been developed to volume production, we can identify two dominant ones:

- NOR Flash, which is the mainstream technology for the applications that requires the storage of codes and parameters, and more generally for embedded memories (system-embedded and chip-embedded) that has to provide random memory access
- NAND Flash, which provide only serial access, but higher density and lower cost than NOR, and it is therefore the dominant technology for data storage and memory cards

NOR Flash Memory

NOR Flash memory was born in mid 80's and it was introduced at the end of that decade as EPROM replacement. The *first generation* products were actually looking like erasable EPROM's, because they required an external 12 V supply for program and erase, they only offered a bulk erase capability (all memory content erased at once) and they required the time-consuming erase procedure to be managed by an external machine (programmer or system microcontroller).

In mid 90's, there was a *second generation* of NOR Flash memory; those new products stated to be significantly different from EPROM, mainly in the direction of being more flexible a and better suited for in-system reprogramming. The most important new features offered by second generation Flash memories were:

- single power supply: the high programming voltage was generated on-chip by a charge pump from the standard 5 V or 3 V external power supply, removing the quite troublesome requirement of a second power line on the application board
- sector erase: the memory array was divided in sectors, of equal (64 KB) or different sizes (8 KB - 64 KB), to allow the modification of a portion of the memory while keeping the information stored in the rest of it
- embedded algorithms: a state machine was provided on-chip to run the erase algorithms locally, without keeping the system busy for all the time needed to complete the operation

The explosion of mobile phone market, which has really been the killer application for Flash memory, has pushed the development of a *third generation* of products, specifically designed for that application, whose main new features are:

- very low power supply voltage (1.8 V), to minimize power consumption both in reading and writing
- different power supply pins, one for programming voltage, one for the chip main supply voltage and one for input/output circuitry, to allow the maximum flexibility for power management at system level
- different memory banks, to allow reading one portion of the memory while writing another one (read-while-write feature)

- fast read modes (burst and page) to enhance the data throughput between the processor and the memory, which the bottleneck for system performance

Indeed, the advanced architecture of latest generation NOR Flash memory is effectively conceived to meet the requirements of mobile phones: it optimizes the trade-off between speed and power consumption, and it gives the possibility of using a single chip to store both code and data, through the read-while-write feature.

NOR Flash memories are all but commodities: they are available in a variety of densities (from 1Mbit to 512 Mbit), of voltages (from 1.8 V to 5 V and with single or triple voltage supply), of read parallelism (serial or random x8, x16 and x32, burst or page access), of memory partitioning (bulk erase or sector erase, equal sectors or boot block sector scheme, single bank, dual banks or multiple banks). All the different product specifications are meant to fit the different needs of specific applications.

The variety of products is the best demonstration of the versatility of NOR Flash technology, which together with its excellent cost/performance trade-off and its superior reliability have been key success factors of this technology.

All the nice features of NOR Flash products are inherently related to the memory cell concept and the memory array organization (Fig. 4).

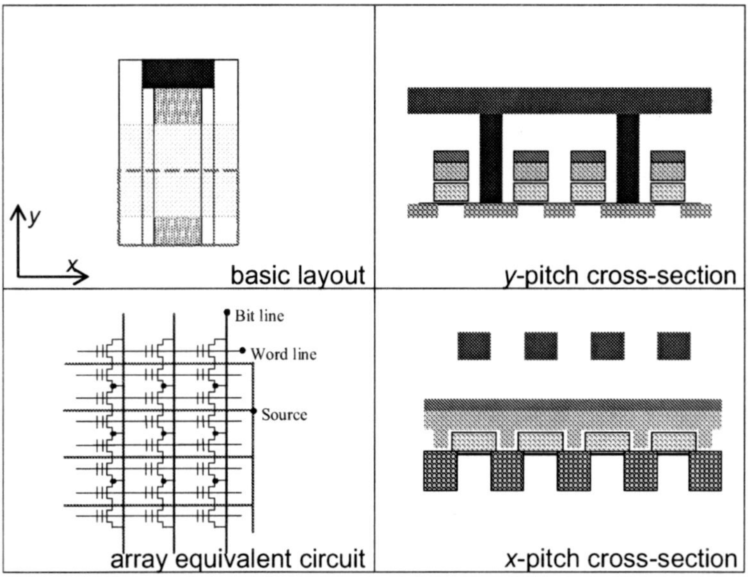

Fig. 4. NOR flash cell structure and array organization

The cell, as described in the following chapters, is a one-transistor cell made of a stacked-double-poly floating-gate MOS device, which is programmed by channel-hot-electron (CHE) injection and erased by Fowler-Nordheim (FN) tunneling.

The memory cells are arranged in a NOR type array organization, which means that all the cells are parallel connected with a common ground node and the bit lines are directly connected to the drains of memory cells.

If we exclude space applications and related cosmic ray effects, charge storing in a floating gate is the most reliable mechanism employed in programmable memory technology, as fare as data retention, thanks to the very high (3.2 eV) energy barrier electrons have to overtake for escaping from the floating gate. The channel-hot-electron programming mechanism is the best for immunity to program disturbs and it does not require scaling tunnel oxide to reduce the memory cell channel length, allowing to preserve a good data retention while scaling cell size.

The NOR array organization is best for high speed and noise immunity, because of the direct access to the memory cell.

The combination of NOR array and CHE programming make this Flash technology the most suitable for multilevel storage, which helps to boost the density for very cost sensitive applications; high density NOR memories that store two bits in each cell, allowing 30-40% cost saving versus traditional one-bit-per-cell memories of the same capacity in the same technology node, are currently available in volumes.

Last but not least, this technology has been proven to be well compatible with advanced logic processes and it is widely used for embedded memory in SOC's (system-on-a-chip).

There are issues to be addressed to keep on scaling NOR memory cell; however, we believe it can be scaled down to 45 nm technology node, and maybe further if no competitive technology will materialize by that time.

NAND Flash Memory

NAND Flash has basically the same memory cell structure as NOR, but it has a totally different array organization (Fig. 5) and it employs a different programming mechanism. The memory array is organized in NAND arrangement, i.e. a number (16 or 32) of cells are connected in series between ground and the bit line contact. That allows increasing the density vs. NOR, which instead requires a ground line and a bit line contact every two cells, but it dramatically affects speed. In fact every cell must be read through a number (15 or 31) other cells, strongly reducing read current; that results in much longer access time (microseconds compared with the tens of nanoseconds of NOR) and it practically prevents the usage of this technology for random access memories and restricts it to serial memories only. More over, the read-through mechanism make this memory type much more noise and pattern sensitive than NOR; therefore, implementing multilevel storage in NAND Flash more difficult and, although two-bit-per-cell products are available now, the mainstream for NAND is still one-bit-per-cell.

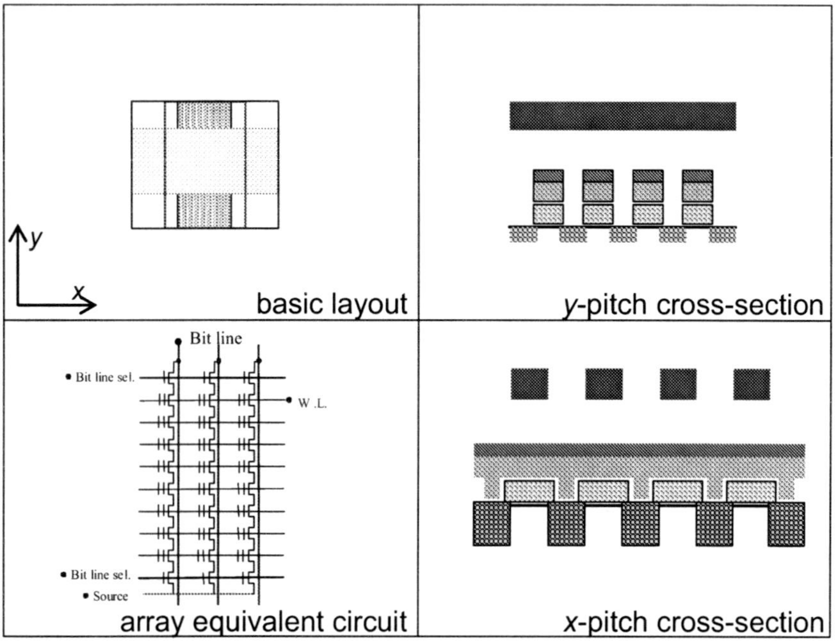

Fig. 5. NAND flash cell structure and array organization

The programming mechanism utilized by NAND Flash is Fowler-Nordheim tunneling. This mechanism is less reliable than CHE because it requires a thinner tunnel oxide; that is taken care in NAND memory by error correction techniques, which would strongly penalize random access memories but they are more compatible with serial memories. On the other hand, being a programming mechanism that requires very low current, FN tunneling allows a very high on-chip parallelism for programming and, as a consequence, a very high writing throughput, which is a key feature for mass storage.

The higher density and the higher programming throughput make NAND the dominant Flash technology for memory cards (Fig. 6); as the data storage is the fastest growing application for flash memories, the portion of market served by NAND technology is increasing (Fig. 7).

This technology is believed to face the same scaling issues than NOR; still the effort to push it down to 45 nm and beyond will be maintained, unless an alternative technology will show better cost/performance combination.

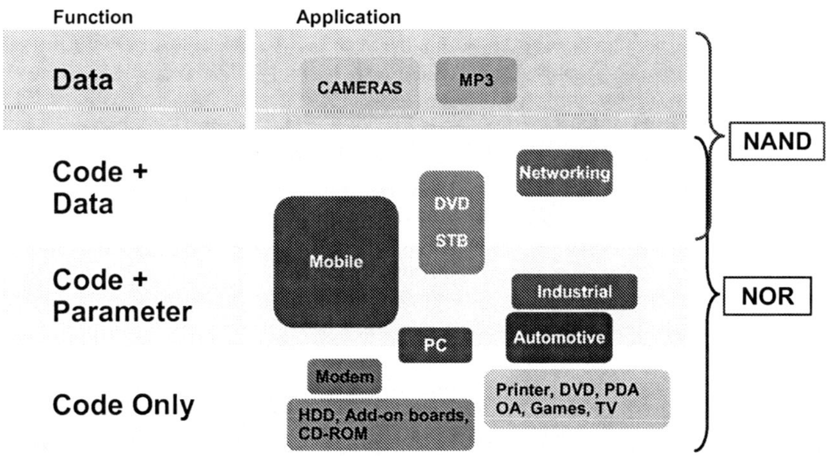

Fig. 6. Main applications of flash memories

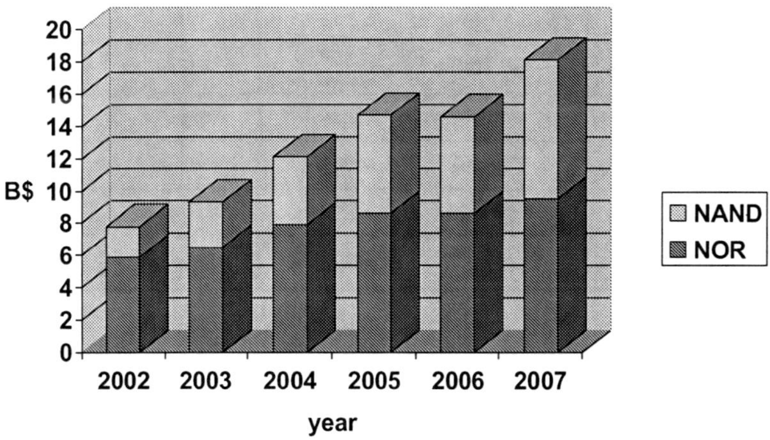

Fig. 7. Flash memory market sharing by technology (source: Web Feet Inc.)

New Memory Concepts

Different alternative memory concepts have been explored in the last twenty years, aiming to overtake the major limitations of existing semiconductor memories, i.e. the volatility of RAM's and the slow programming and limited endurance of Flash memories. The ultimate goal of this search is the "universal memory", a dense memory that can be written at the speed of a RAM for a virtually unlimited number of times and retains data like a Flash memory. The major efforts have been focused on three concepts: Ferroelectric RAM (FRAM), Magnetic RAM (MRAM) and PCM (Ovonics Universal Memory). The benchmark of these three technologies (Fig. 8) shows that each of them has its own peculiar characteristic, for which it is better than the other two, but none of them is best in all aspects. FRAM is the lowest power but is the most difficult to scale, MRAM is the fastest but it is the one requiring most current for programming, PCM has the smallest and most scalable cell but is the most critical as far as endurance. Hence, the "universal memory" is probably still to be found; nevertheless all of the three types of memory will find their own field of application provided that they reach the maturity of viable industrial technologies, which is not the case yet.

	Flash	FRAM (ferroelectric)	MRAM (magnetic)	PCM (phase change)
Relative bit size*	0.25 - 1	3 - 10	1 - 3	0.5 - 2
Relatv. mask count	1.1	1	1	1
Scalability	Fair	Poor	Poor	Good
Endurance	10^5	10^{10} (destructive read)	$>10^{14}$	10^{12}
Data retention	> 10years	> 10years	> 10years	> 10years
Write time	µs/ms	< 100ns	< 100ns	< 100ns
Write power/B (VxI)	5V x 1mA	3Vx100µA	1.8Vx10mA	3Vx1mA
Maturity	Volume prod.	Limited prod.	Test chips	Test chips

* 1= NOR flash cell size

Fig. 8. Benchmark of emerging NVM technologies

Among the three, the most interesting technology as potential successor of Flash memory, which, as we have seen, will start to face some scaling difficulties in a few technology generations from now, is the PCM, mainly because there are no evident physical limitations that prevents scaling down its memory cell.

The physical mechanism at the base of PCM concept (Fig. 9) is the change of conductivity associated to a phase change, from polycrystalline to amorphous, exhibited by a certain class of chalcogenide compounds as a stable and reversible modification induced by proper current pulses. The memory element is the portion of the chalcogenide material (the most popular is $Ge_2Sb_2Te_5$) in contact with a "heater", a structure properly designed to generated an hot spot under a fairly limited current flow; the memory cell is completed by an access device, either a MOS transistor or a diode, which is the preferred solution for a high density array. The current required to set the device in the crystalline state is almost half of the one necessary to re-set it into the amorphous phase; both current are below 1 mA. It takes 50-100 ns to set the device, while re-setting it is much faster. Individual cells have been proven to withstand up to 10^{12} program/erase cycles, while ensuring the same level of endurance on large arrays is still one of the most critical issues at the present stage of maturity of this technology.

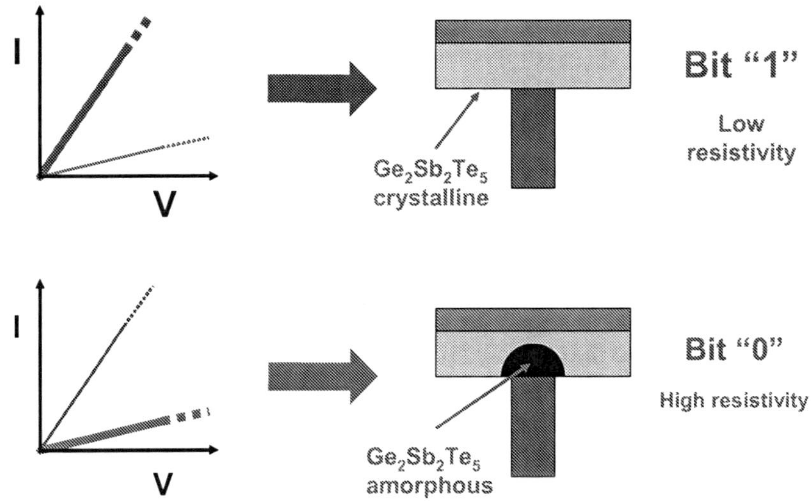

Fig. 9. Basic concepts of phase change memory

Compared to Flash, PCM, which by nature offers a single-bit erase granularity, can feature a random access time similar to NOR, a byte programming 10-100 times faster than NOR and a writing throughput comparable to NAND, an endurance orders of magnitude better that Flash; it requires a programming current similar to NOR and a much lower programming voltage (abut 3 V, compared to 9 V for NOR and 20 V for NAND) and that makes it much more compatible with scaled CMOS technologies. The memory cell size at present technology nodes is smaller than NOR and it is comparable to NAND, but the projections to future technology generation (Fig. 10) show the potential of PCM to became smaller than any Flash cell, thanks to its superior scalability.

PCM is not yet proven to be an industrial technology and it has still a learning curve to climb before catching-up with the well established Flash technologies; however, this new technology not only promises better performances than NOR Flash but it has also the potential to eventually become cost competitive even with NAND Flash.

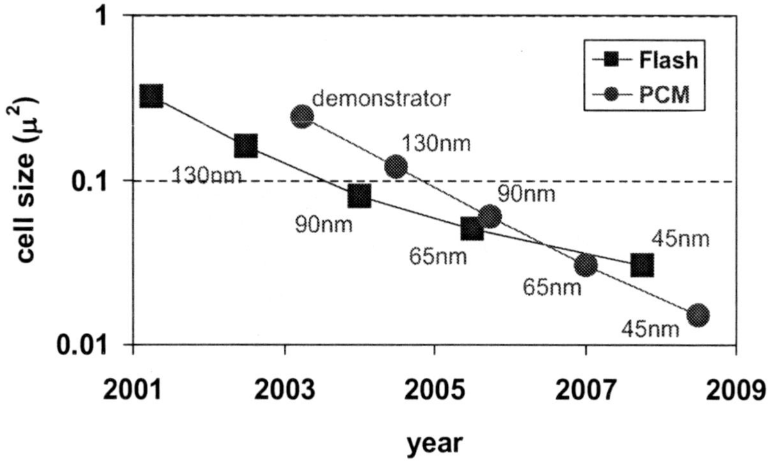

Fig. 10. Perspective road map of phase change memory

Conclusions

Flash memories have been the fastest growing among the different semiconductor memory families; their market has reached a size comparable to the DRAM one and it is expected to keep growing in the coming years. The fantastic success of Flash memories is related to the key role they play as storing media for both code and data in battery-supplied electronic systems. Moreover the diffusion of multimedia consumer products will drive the increasing demand of Flash memory cards. NOR and NAND Flash technologies will continue to dominate in their respective field for at least five more years, following the lithography scaling for two or three generations beyond the 130 nm node, before facing severe scaling limitation. PCM is the best candidate, among the different emerging memory concepts, to take by that time the baton and to lead the non-volatile memory technology race down to the 10 nm frontier.

Paolo Cappelletti
Non Volatile Memory Technology Development Director
STMicroelectronics, Agrate Brianza (MI), Italy

1 Non-Volatile Memory Design

This first chapter gives an overview of the main items that will be dealt with in the rest of the book.

Operating modes of the memory devices are introduced, namely Read, Write and Erase, together with their basic concepts and related issues, aimed at defining the general scenario. Understanding of this chapter requires the knowledge of the basic principle behind MOS transistor theory.

1.1 Introduction

Aim of this book is to explain how to design an integrated circuit. Since our design experience is mainly related to Non-Volatile Memories, both EPROM and Flash[1], the guiding thread will be the study of circuital solutions applied, in particular, to Non-Volatile Memory design, i.e. devices capable of keeping the stored information even without external power supply.

This type of device is widely used in almost all electronic applications; for instance, wherever a microprocessor is present, a bank of Non-Volatile Memory is required, to store the boot code.

Throughout the book, we try to depict both basic circuits, which are essential for design and which are – undoubtedly – part of the fundamental knowledge of any designer, and the more complex circuit solutions, which are mainly implemented in memory devices, to show how to connect basic blocks together. The knowledge of the fundamental notions of electronics is a requirement for comprehending the book. We try to express ourselves in the same way as we do when we discuss with our colleagues to implement a new project or to solve an issue.

Math is therefore reduced to a minimal extent, and it is used when it is indispensable to summarize a concept, while the discussion is aimed at the analysis of the physical behavior of the system, whose ultimate purpose is the knowledge of:

- the behavior of the voltage at circuit nodes;
- the current flowing in a circuit net;
- input or output impedance of a circuit;

---

[1] Flash derives from the fact that this type of memory has been conceived upon a specific request from the military; to prevent the enemy from reading the codes of fallen aircraft or unexploded missiles, they specifically asked for a EPROM which could be electrically erased in case of need.

- frequency response of a circuit;
- the influence of temperature, supply voltage, process, etc. over the above mentioned.

These are the issues that an electronic designer should consider during the design phase.

Several (unresolved) problems are posed throughout the book. Our intent is to stimulate discussion and brainstorming, and a strict numerical resolution is not required.

Let's start with an overview of the main issues related to Non-Volatile Memories.

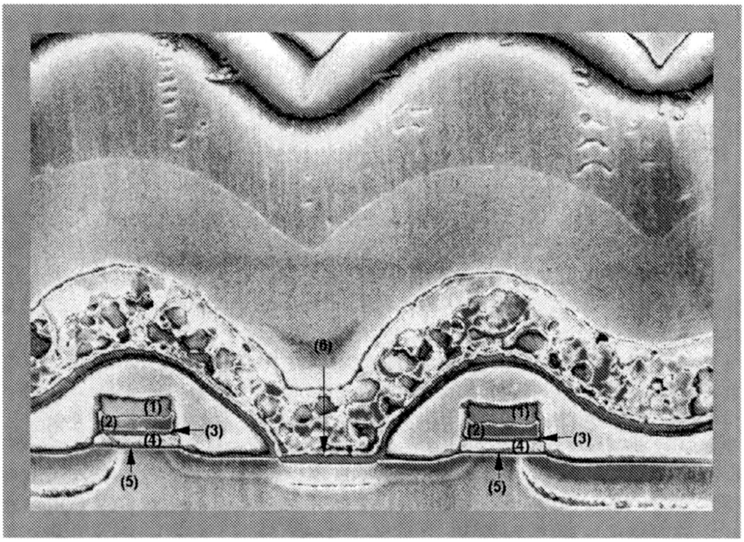

Fig. 1.1. Cross-section of a Flash Memory device; silicide (1) and polysilicon (2) which constitute the control gate, the interpoly oxide layer (3), the floating gate (4), the thin oxide (5), drain contact, shared by the two cells and connected to Metal 1 (the granular wire)(6)) and source implant, deeper than the drain implant

1.2 Main Features of Non-Volatile Memories

The basic principle of the majority of Solid State Non-Volatile Memories currently in use is the concept of a floating gate. A MOS transistor is designed, where two overlapping (but separate) gates are present: the former is completely isolated in the Silicon Oxide, the latter is capacitively coupled to the floating and becomes the gate terminal.

The floating gate is a perfect "trap" for the electrons, where the charge ca be retained for more than 10 years (Fig 1.1).

The operations that allow trapping electrons on to and to remove electrons from the floating gate are called Program and Erase[2]. The macroscopic variable that is modified by these processes is the memory cell threshold voltage.

1.3 Program

Figure 1.2 represents a cross-section of the Flash memory, where the two gates are clearly shown.

During the Write operation, the control gate and drain are biased at "high" voltage[3]; i.e.12 V for the gate and 5 V for the drain, while the source is kept at ground.

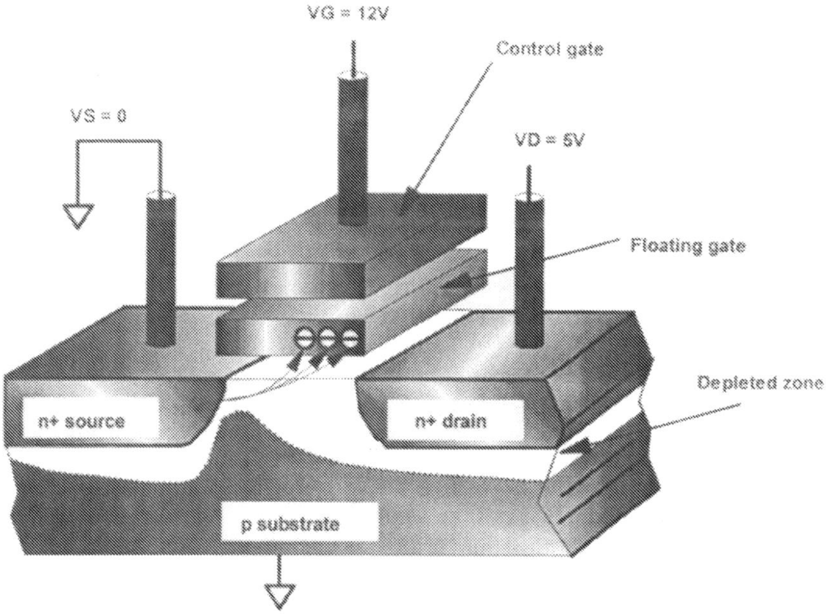

Fig. 1.2. Cross-section of a Flash cell in program mode: depleted zone is highlighted, and typical voltage values are shown

[2] By convention, the operation which is executed over a set of cells is called Erase. With Flash and EPROM memories, Erase operation results in cells with low threshold voltage; with EEPROM, it results in cells with high threshold voltage.

[3] It would be more precise to say high values of the electric field: in fact, 10 V over a thickness of 100 Å mean 10 MV/cm!!; for devices working at 5 V, a 12 V potential can be considered high.

Under these conditions, with a given oxide thickness of 200 Å between the two gates and 120 Å between the floating gate and channel, there are very strong electric fields which let the electrons pass from the channel to the floating gate, overcoming the potential barrier represented by the oxide. This mechanism is known as Hot Electrons Injection. Due to the high voltage on the drain, the electrons flowing from the source towards the drain gain energy from the orthogonal electric field while losing energy in the interactions with the lattice (acoustic and optical phonons). A dynamic equilibrium situation holds for values up to 100 KV/cm. In the instance of stronger electric fields, electron energy starts to increase relative to the conduction band. Electrons are heated by the high orthogonal electric field, some can gain enough energy to overcome the barrier between the oxide and the silicon conduction band. Hot electrons must overcome the barrier in the right direction to be collected in the floating gate.

The electrons trapped in the floating gate cause the cell threshold voltage (V_T). to rise. Therefore, when a Read operation occurs, the cell will appear in the switched off state, since it cannot conduct current due to its high V_T. Writing data at a given address brings the cells from a neutral (erased) state, which is typically a logic state "1", to a logic state "0", i.e. charge is trapped in the floating gate.

The time required by this process is usually microseconds.

1.4 Erase

The advantage of Flash memories over EPROM is the electrical erase capability. In the EPROM, the depletion of the floating gate is achieved by exposing the memory to UV light. This operation requires the removal of the device from the motherboard.

In Flash memories, a positive voltage is applied between the control gate and source node. This condition can be achieved by grounding the control gate and raising the source node to a value of about 12 V or by lowering the control gate down to –8 V and raising the source node to about 5 V.

The former is referred to as positive source Erase, the latter as negative gate erase (Fig. 1.3 and Fig. 1.4). The drain terminal, in either case, is left floating. The erase mode uses the mechanism known as Fowler-Nordheim (FN) tunneling, and the execution time is hundreds of milliseconds.

1.5 Distributions and Cycles

Simultaneous erase of several bytes grouped into sectors is a feature particular to Flash technology. In comparison, EPROM memories erase simultaneously all the cells inside the memory bank, while EEPROM memories, which implement a selection transistor, erase the content byte by byte.

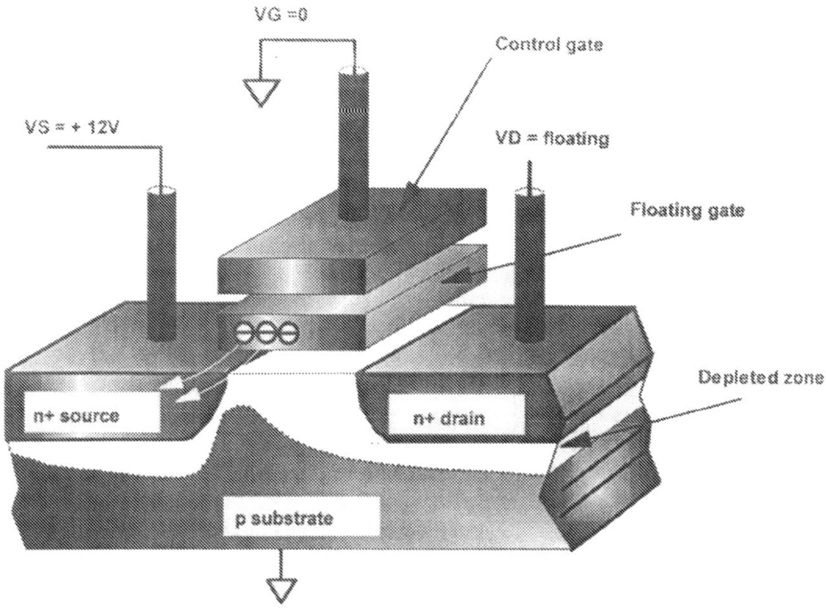

Fig. 1.3. Cross-section of a Flash cell in positive source erase mode

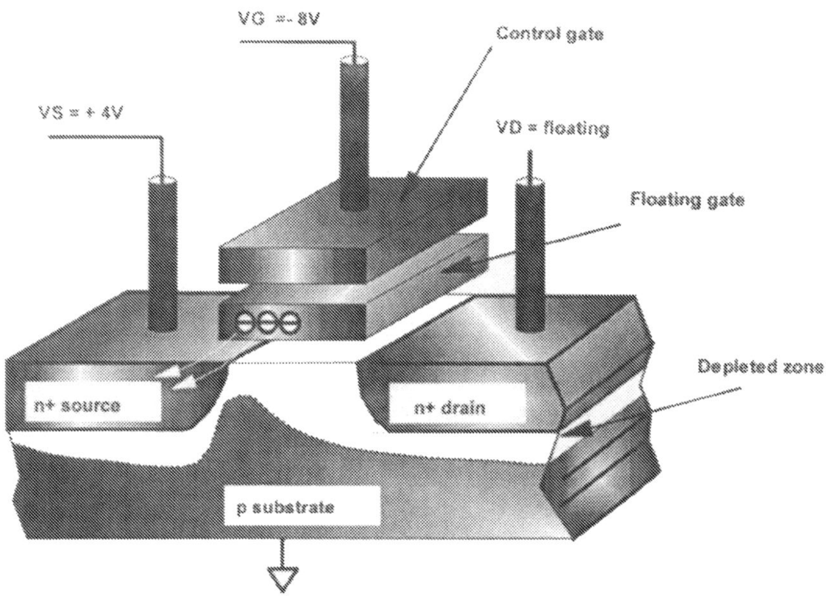

Fig. 1.4. Cross-section of a Flash cell in negative gate erase mode

Flash memories can achieve a higher density than EEPROMs since a single transistor cell is used, while the drawback is that several cells must be erased at the same time. This set of cells is referred to as a *sector*. Every erase operation is preceded by a program operation to bring all the cells of the same sector to the same state, so that the cells share a common history and, therefore, a common behavior. Cycling, and the ability to carry out a very high number of erase and program operations, is undoubtedly one of the key requirements for a Flash memory.

Currently manufactured Flash memories can guarantee up to 100,000 cycles. An EEPROM memory can achieve 10 million cycles, while volatile memories, like static and dynamic RAM, have practically no limit to the number of possible cycles.

Now the concept of threshold voltage distribution should be considered. It has been shown that the storage of charge on the floating gate creates a negative electrical potential. This becomes a higher threshold voltage, V_T, of the equivalent transistor that constitutes the cell. A non-programmed cell has a threshold $V_T < V_{TE}$, while V_T is greater than V_{TP} for the programmed one. Unfortunately the cells that compose the memory matrix are not uniform. Small geometrical asymmetries, process variation, etc. result in a distribution of the output characteristic of the cells, including the threshold values. Figure 1.5 shows typical V_T distributions for both erased and programmed cells.

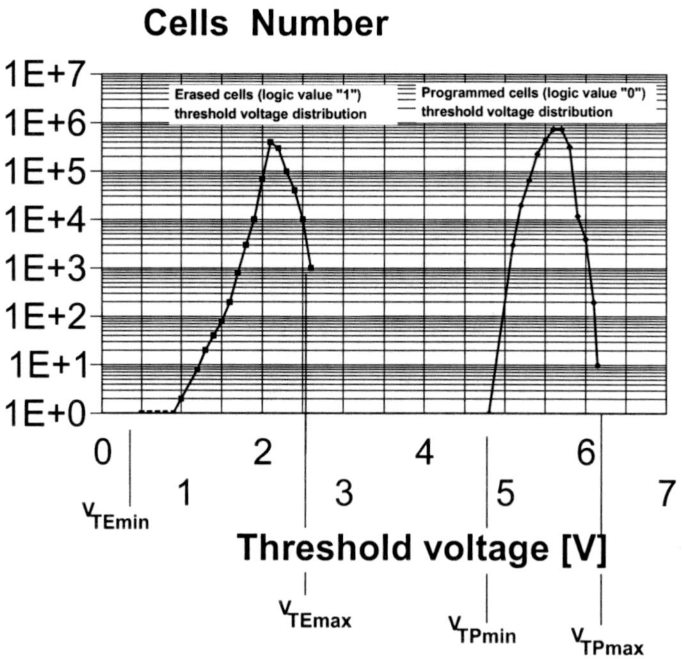

Fig. 1.5. Threshold distributions for both erased and programmed cells

Such distributions create problems for the read operation of the device, since a cell must be read as erased if its threshold falls in the interval $V_{TEmin} < V_T < V_{TEmax}$ and this interval is greater than 1 V. During the read operation, the proper voltage which permits the erased cells to sink current, is applied to the control gate. Simultaneously the source node is grounded and the drain node is biased to a proper voltage[4]. In this way, a written cell (programmed) sinks little or no current at all. Cell current is brought via the column contact (*bit line*), to a circuit that converts the current into a voltage, which is provided as an input to a comparator. This voltage is compared against a reference voltage to determine the logic value to be placed on the memory outputs.

1.6 Read Mode Architecture

Figure 1.6 schematically represents Read mode, where the connection to the single cell is highlighted. The resistor R provides the current to the cells and such current is converted into a voltage (i.e. the voltage drop across R itself). The connection of the cell to the resistor is achieved by the column decoding. This combined with row decoding (which provides the biasing voltage to the gate of the selected cell) constitutes the address of the data. The combination of the current to voltage converter and the comparator is called *sense amplifier.*

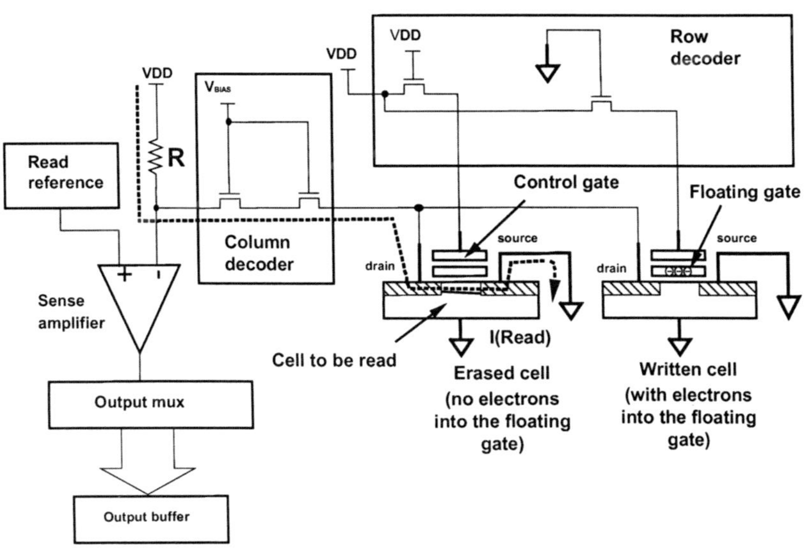

Fig. 1.6. Read Mode Architecture

4 "Proper" is not a scientifically correct term. Explanations must be gradual; every value will eventually be quantified.

1.7 Write Mode Architecture

The Write (Program) operation (see Fig. 1.7) is completely handled by the circuitry contained inside the memory device, which generates the proper voltages and correct signal timings required by the write procedures. The circuitry that handles Program and Erase is controlled by a control unit called FSM (Finite State Machine). The FSM controls the data to be programmed, which are provided on the same data pins used during read. These data are input to the circuitry which turns on the transistors to pass the VD voltage to the drains of the cells. This biasing causes the current to flow and generate the hot electrons, thus causing the writing of the cell. The FSM generates the right timing for programming pulse, and subsequently verifies, with a specific read, the proper result of the operation.

The unit called CUI (Command User Interface) is the command interpreter, whose task is to translate the user request to the FSM. The output of the sense amplifier is compared with the original data to be written, and the result of this operation is provided to the FSM.

If needed, more than a pulse can be used to write the single cell. The FSM applies multiple pulses to accomplish the write operation, up to a predetermined maximum, until the intended has been successfully written. At the end of the Program operation, the FSM communicates the success or failure of the operation to the user, by a defined protocol.

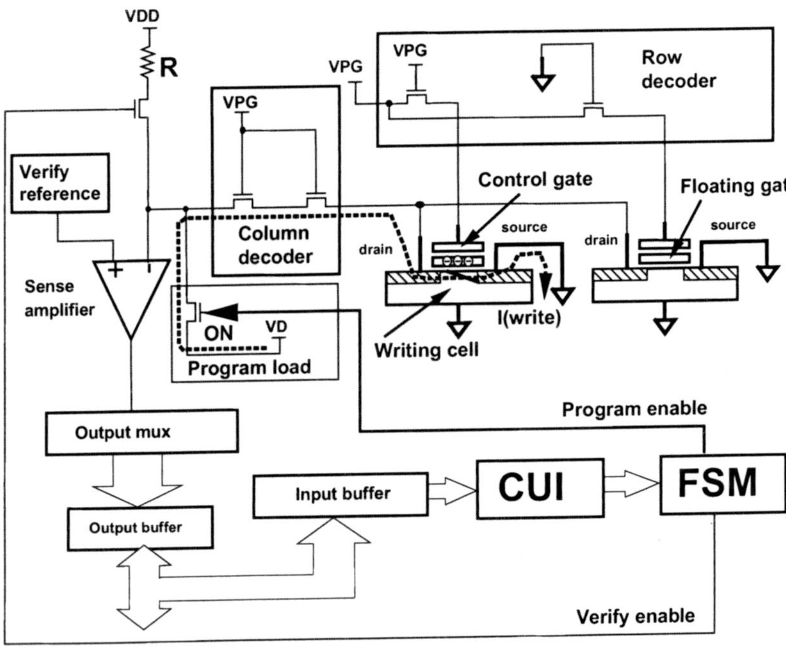

Fig. 1.7. Write Mode Architecture. The addressed cell, whose gate is VPG > VDD, sinks current from the generator VD > VDD. Electrons are therefore stored into the floating gate

1.8 Erase Mode Architecture

Erase operation forces all the memory cells of the selected sector to a logical value of "1". The FN tunneling mechanism is used; erase voltages are applied on the gate and source, while the drain is left floating (Fig. 1.8). Inside a biased cell, the electrons stored in the floating gate are attracted towards the source contact, carrying out the tunneling through the gate oxide, which produces the desired change in the threshold voltage of the cell.

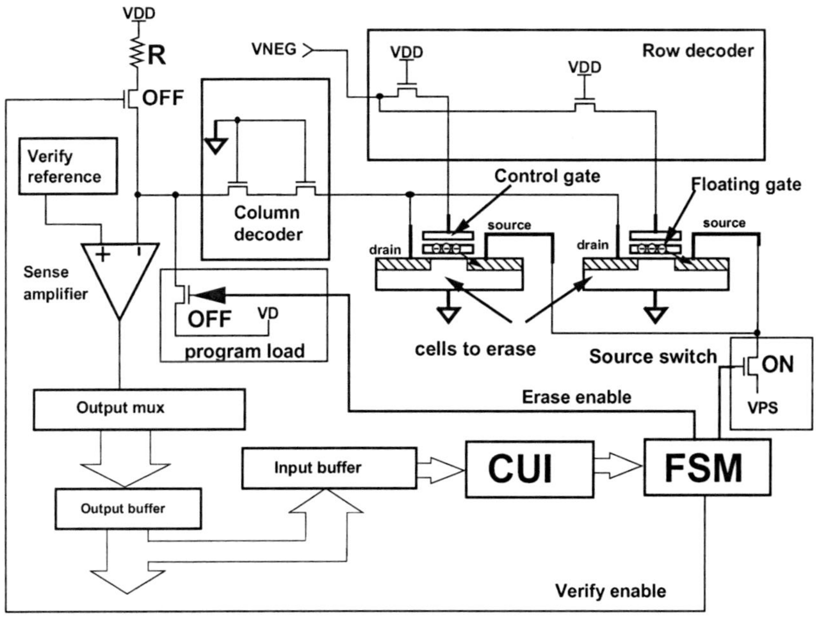

Fig. 1.8. Erase Mode Architecture (Negative Gate Erase). The cells inside the same sector have the gate biased to –8 V, source to 5 V and the drain floating. The electrons move from the floating gate to the substrate through source diffusion. The handling of the Erase is governed by the FSM and instructions are communicated from the user to the CUI

To avoid reliability issues, the internal erase algorithm (handled by the FSM as well) performs a *preconditioning* operation before erase, i.e. it programs all the cells of the sector under erase. Erase, as well as Program, is performed by means of consecutive pulses, whose typical duration in this case is 10 ms, for a maximum of some hundreds of milliseconds After every erase pulse, the result of the operation is verified by the FSM for every byte/word with a proper read that verifies the achieved erase state. An internal counter is used to scan the addresses inside the sector. As is the case with the Program, the FSM communicates to the user either the successful completion or failure of the Erase operation.

1.9 Elements of Reliability

A cycle is composed of an Erase and a Program operation; the number of cycles that can be executed on a memory cell is one of the main quality parameters for a Flash memory device. Cycling causes several failure mechanisms due, for instance, to a charge trap or spurious program/erase that cause erased cells to be written and vice versa. There are also some complex and subtle effects that will be described later.

Program and Erase operations generate cell threshold distributions that change cycle after cycle with a Flash device. The Erase operation is a "social" phenomenon and therefore it produces undesired tails in the distributions, which might cause the device to not operate properly. The design must carefully evaluate all these issues and operate in order to reduce any possible marginality. The presented analyses should indicate that designing a Non-Volatile Memory device is very challenging. It is a sort of "living being", which changes over time, and sometimes, like any other being, becomes "ill". Thus some of its cells might not work properly, and they need to be "healed" by specific temporary recovery algorithms. On the contrary, a device composed of pure logic, e.g. a microprocessor, is more complex from a system architecture point of view, but it is "frozen" in time once operational, and therefore, more easily reproduced on an industrial scale.

1.10 Influence of Temperature and Supply Voltage

Flash memory devices are used extensively, from the BIOS for the personal computers to mobile phones, cars, etc. Therefore, required specifications are very general: supply voltage range is very wide (either 1.8 V to 2.7 V or 2.7 V to 3.6 V), and acceptable temperature range can be as large as 165 °C, from –40 °C to +125 °C (automotive range). Characteristic features of the cell, which we learned in the capability of programming and erasing within the required timings and threshold distributions, greatly suffer from both temperature and supply voltage variations. The design must be able to anticipate, control, and overcome the previously described issues to meet the lifetime requirement according to the specifications.

Temperature, for instance, influences write speed; a higher temperature reduces the number of hot electrons available for injection into the floating gate, hence retarding the programming characteristics. Erase worsens at low temperature, because the value of breakdown voltage at the cell source decreases, and therefore the electrical field that can be used for erase decreases as well.

The aim of the design is to operate on the voltages used during Program and Erase to balance parasitic effects, and lead the timings within the target values of the specification. In the same way, VDD supply voltage variations heavily affect not only Read, but also Program and Erase. Thus it is mandatory to have a bandgap reference circuit to generate stable and accurate voltages within the entire supply voltage and temperature ranges. Inside the latest generation of Flash devices, all the required voltages, both positive and negative, are generated inside the device by means of charge pumps using the external supply voltage. Finally, process variations make the design activity more complex. A principle that the designer should follow is:

> Every circuit must work in simulation under dimensional, process, temperature and supply variations of ± 30%! If it not functional in simulation over these ranges, then the device will not yield functional devices in silicon.

The following example can be used as a basis for discussion:
Our goal is to design and sell 20 million parts a year, for at least 10 years, of a device containing 8 Megabits of cells. Since we are going to realize 8,388,608 cells inside every device, we are going to sell nothing less than $1.67 \cdot 10^{15}$ cells in this 10 year period ! The target is that, after erase in the entire allowed temperature and VDD ranges, the cells must all fall in a 1.5 V wide distribution.

The floating gate of the memory cell has a typical value of equivalent capacitance of 0.7 fF. Considering a threshold delta (ΔV_T) of 3 V for the written cell, we are going to have a charge on the floating gate:

$$Q = C_T \cdot \Delta V_T = 0.7\,fF \cdot 3V = 2.1\,fC \tag{1.1}$$

In terms of the number of electrons we have:

$$2.1\,fC \cong 2.1 \cdot 10^{-15} \cdot 1.6 \cdot 10^{19} = 33{,}600\ electrons \tag{1.2}$$

It is as if we wanted to place the same number of grains of sand, 33,600, in $1.67 \cdot 10^{15}$ boxes using a bulldozer.

Furthermore, we want to repeat this task 100,000 times, emptying the boxes with the same bulldozer, without breaking them, while moving from the Equator to the North Pole over a 10 years period!

1.11 Lab Activities

Several complex activities must be carried out in order to lead a device to maturity, making it producible, i.e. reproducible in large volume.

The first activity is *characterization*, i.e. the analysis of the operation of the integrated circuit under all the different modes; voltages on the various internal nodes are measured, in order to highlight not only failures, but also marginalities that might prevent the device from working over the entire required parameter range. Characterization for a Non-Volatile Memory can be divided into two different steps: the first step covers all the circuitry, while the second step involves the matrix. This is the chronological sequence of events.

After wafer manufacturing is complete, some of the wafers are taken out of the process prior to the deposition of the passivation so that the upper metal layer is exposed and accessible during the analysis. Before being delivered, these unpassivated wafers undergo an electric test, in order to check the values of the parameters of both the transistors and the memory cells. This test is performed on sample structures that have been inserted around the devices along the line where the individual chips will be cut out of the wafer so that they can be packaged.

Unpassivated wafers are then erased to remove any electric charge that may have built up during the manufacturing process; this is done by exposition to Ultra Violet (UV) rays. Finally, the wafers are submitted to a rough EWS[5] that tests their functionality. Then several fully functional chips are assembled and delivered to the lab.

This begins the "investigation" task on the device. The end of this phase might results in requesting new masks to correct the errors found during the analysis. The second phase, devoted to the matrix, starts. It is important to analyze the behavior of the matrix cells with respect to their geometrical position, to verify if the distance from ground, the length of drain connection (in metal 1), or its position with respect to matrix border have an influence on Program, Erase or Read

Then there are cycles, with their infinite complications, caused by oxide damage, which progressively ages both cells and transistors, modifying both their threshold voltage and transconductance (gain).

1.12 Working Tools

Once assembled in a package, our parts are placed under an optical microscope and connected to a test equipment which stimulates both input and control pins to reproduce every working condition.

The voltages on the nodes internal to the device are checked by probes, whose diameter is a few microns, which are pressed on to specific metal structures, drawn on the layout to facilitate these measurements. Available probes are controlled by micromanipulators and there are two types: passive, i.e. just a "wire" which forces or senses voltage while presenting a capacitive load of some pico-Farad; active, i.e. a MOS gate that is able to sense voltages up to a maximum of about 5÷6 V, capacitively loading the nodes with some tens of fF. Debug, i.e. the search for errors, or bugs, is then performed as if the circuitry was on a conventional printed circuit board.

Besides the classical instrumentation, like oscilloscopes, voltage generators, logic state analyzers, etc., other more capable tools are used to investigate and debug the internal circuitry. One powerful tool is the Electron Beam Tester (EBT), a Scanning Electron Microscope (SEM) that can produce a beam of electrons which can be directed on the metal connections of the device. The fraction of electronic beam reflected by the electric field that is present on the selected metal is used to reconstruct the shape of the originating signal. This reconstructed waveform is ultimately displayed on a PC screen. Figure 1.9 shows the analysis performed using an EBT of a Flash memory device, supplied at 1.8 V, at room temperature, with a load capacitance of 100 pF. As the addresses vary, the ATD signal is generated; the signal triggers the read and starts the boost of the row, which is brought to the supply voltage value at the end of the read, by means of the ENDREAD signal.

[5] EWS is the acronym for Electrical Wafer Sort: it is the test performed on silicon wafers

The meaning of the various signals will become clear throughout the text of this book.

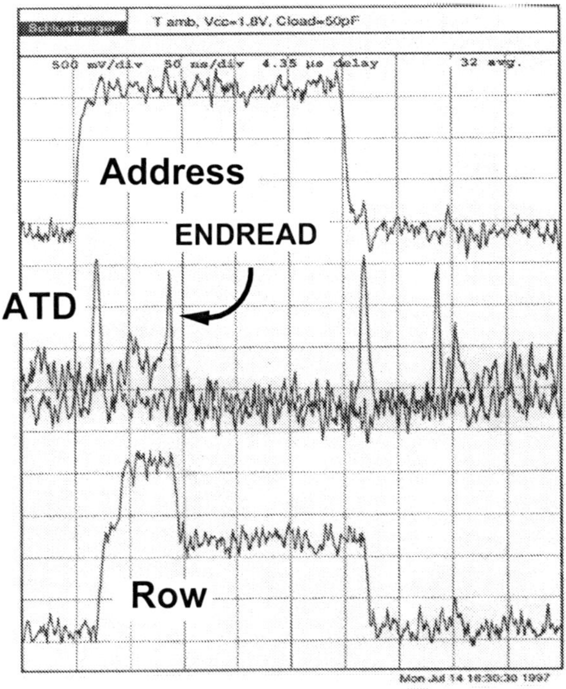

Fig. 1.9. A measurement performed in the lab using an EBT

The FIB (Focused Ion Beam) is another widely used tool; it is very similar to a SEM, except that the incident beam is composed of ions (usually Ga^{+}) and not electrons.

When the ion beam hits the sample, a local removal of atoms takes place (sputtering process). The small spot size of the incident beam allows cutting of electrical connections in metal on a device, even passing through a dielectric layer. Furthermore, the FIB can depose either conductive or insulating material with a high spatial resolution. In this way it is possible to modify internal connections of the device without changing the mask set of the entire integrated circuit, resulting in saving both time and mask costs.

As the integration level grows, failure analysis becomes more and more difficult, forcing the designer to become equipped with a wide range of tools. In CMOS technology, most of the failure situations derive from a high absorption (current absorbed from VDD, input leakage, output leakage). This increase in the dissipated power causes a heat (and even light) generation that, if localized, allows a fast detection of the defect. Heat revelation techniques and emission microscopy are the tools that are normally used to localize the spots where leakage phenomena occur.

Liquid crystals represent the most popular technique for heat revelation. The device that is anomalously consuming is covered with crystals and placed on a thermo chuck, so that its temperature can be precisely regulated to a value just below transition point of the crystals. Once the device is powered up, the region where an anomalous consumption occurs gets hotter, modifying the color of the crystals in an easily visible way. Ease of use, a good sensitivity and low cost are the main features of this technique.

Emission microscopy allows the revelation of the issues that cause light emission. Main emission processes are electron-hole recombination; electron (holes) transition between two states in the conduction (valence) band and Fowler-Nordheim tunneling. The radiation emitted by the device is focused on a photocathode. The electrons produced in the process are amplified and then sent to a fluorescent screen, where they produce an intensified image that is then collected by a CCD. Latch-up, snap-back, gate-oxide defects and direct-biased junctions are the typical phenomena that are investigated using this technique.

Analysis continues in a similar fashion, solving one problem after another, using the state-of-the-art tools and techniques It is electronics reproducing itself.

1.13 Shmoo Plots

Shmoo plots are a commonly used method to characterize the behavior of a part. These are tri-dimensional plots represented on a plane[6].

For instance, access time is evaluated as a function of supply voltage. The shmoo will have access time (in nanoseconds) on the X-axis, and supply voltage (in volt) on the Y-axis. The presence of an asterisk represents a failed read operation.

Figure 1.10 is the access time in the case where the device is continuously selected and the addresses are toggled. Read is done using a strobe for each read cycle, at a different timing value, until the whole written pattern is correctly read.

For each pair of values, the plot shows the read of the slowest word of the whole array.

[6] Shmoo are creatures invented in 1948 by Al Capp, the famous American cartoonist. They are lovely pets, whose only concern is to make men happy. They were able to instantaneously produce eggs, milk, top-quality exotic fruits, etc. If needed, they merrily died to provide starving people an exquisite meat. Shmoo represented Utopia on hearth, the return to Nature, and as such they represented a threat to the System, to the consumer society, to the full employment and to all the others taboos of the Industrial Society. Therefore they were fought against and destroyed because " shmoo are evil because they are so good!". *Linus, May 1965.*

```
ADDRESS ACCESS TIME

TADD.(nS)+50    +75    +100   +125   +150   +175   +200   +225   +250   +275   +300
VCC   (mV)
  1200    |****|****|****|****|****|****|****|****|****|****|****|****|**
  1300    |****|****|****|****|****|****|****|****|****|****|****|****|**
  1400    |****|****|****|****|****|****|****|****|****|****|****|****|**
  1500    |****|****|****|****|****|****|****|****|****|****|****|****|**
  1600    |****|****|****|****|****|****|****|****|****|****|****|****|*.
  1700    |****|****|****|****|****|****|****|****|****|****|**..|....|..
  1800    |****|****|****|****|****|****|****|****|**..|....|....|....|..
  1900    |****|****|****|****|****|****|****|**..|....|....|....|....|..
  2000    |****|****|****|****|****|****|*...|....|....|....|....|....|..
  2100    |****|****|****|****|****|....|....|....|....|....|....|....|..
  2200    |****|****|****|****|....|....|....|....|....|....|....|....|..
  2300    |****|****|****|**..|....|....|....|....|....|....|....|....|..
  2400    |****|****|***.|....|....|....|....|....|....|....|....|....|..
  2500    |****|****|**..|....|....|....|....|....|....|....|....|....|..
  2600    |****|****|*...|....|....|....|....|....|....|....|....|....|..
  2700    |****|****|....|....|....|....|....|....|....|....|....|....|..
  2800    |****|****|....|....|....|....|....|....|....|....|....|....|..
  2900    |****|****|....|....|....|....|....|....|....|....|....|....|..
  3000    |****|***.|....|....|....|....|....|....|....|....|....|....|..
  3100    |****|***.|....|....|....|....|....|....|....|....|....|....|..
  3200    |****|***.|....|....|....|....|....|....|....|....|....|....|..
  3300    |****|**..|....|....|....|....|....|....|....|....|....|....|..
  3400    |****|**..|....|....|....|....|....|....|....|....|....|....|..
  3500    |****|**..|....|....|....|....|....|....|....|....|....|....|..
  3600    |****|**..|....|....|....|....|....|....|....|....|....|....|..
  3700    |****|*...|....|....|....|....|....|....|....|....|....|....|..
  3800    |****|*...|....|....|....|....|....|....|....|....|....|....|..
  3900    |****|*...|....|....|....|....|....|....|....|....|....|....|..
  4000    |****|*...|....|....|....|....|....|....|....|....|....|....|..
  4100    |****|*...|....|....|....|....|....|....|....|....|....|....|..
  4200    |****|....|....|....|....|....|....|....|....|....|....|....|..
  4300    |****|....|....|....|....|....|....|....|....|....|....|....|..
  4400    |****|....|....|....|....|....|....|....|....|....|....|....|..
  4500    |****|....|....|....|....|....|....|....|....|....|....|....|..
  4600    |****|....|....|....|....|....|....|....|....|....|....|....|..
  4700    |****|....|....|....|....|....|....|....|....|....|....|....|..
  4800    |****|....|....|....|....|....|....|....|....|....|....|....|..
  4900    |****|....|....|....|....|....|....|....|....|....|....|....|..
  5000    |***.|....|....|....|....|....|....|....|....|....|....|....|..
  5100    |***.|....|....|....|....|....|....|....|....|....|....|....|..
  5200    |***.|....|....|....|....|....|....|....|....|....|....|....|..
  5300    |***.|....|....|....|....|....|....|....|....|....|....|....|..
  5400    |***.|....|....|....|....|....|....|....|....|....|....|....|..
  5500    |****|....|....|....|....|....|....|....|....|....|....|....|..
  5600    |****|....|....|....|....|....|....|....|....|....|....|....|..
  5700    |****|....|....|....|....|....|....|....|....|....|....|....|..
  5800    |****|....|....|....|....|....|....|....|....|....|....|....|..
  5900    |****|....|....|....|....|....|....|....|....|....|....|....|..
  6000    |****|....|....|....|....|....|....|....|....|....|....|....|..
```

Fig. 1.10. Shmoo plot for the access time to the memory as a function of the supply voltage

1.14 Testing

Once debug and characterization (are they one and the same?) phase is complete, production starts and testing helps the non-infallibility of circuit and component designers. Flash memory testing is the more severe version of the "Tarpea crag" for the devices. Most subtle and selective tests are invented, and such tests must be run on all the devices that will be sold, so that a selection is achieved, and only perfect devices are acceptable. This kind of selection requires electrical tests to be run both directly on the wafers and on the packaged component. When wafers come out of the fab, they undergo two very demanding EWS.

The matrix for each device on the wafer is verified in order to find out defects and marginalities in Read, Program and Erase. All possible behaviors that might lead to improper future functioning are checked as well, e.g. the bits, if any, that are too slow to either program or erase.
Together with the matrix, the circuitry that allows the execution of all the operations is checked. The first EWS finishes with a Program on all the cells. Between the two EWS, a Bake operation is performed, i.e. the wafers are placed in an oven at high temperature to verify the quality of charge retention. Then all the devices that passed both EWS are cut from the wafer and closed in the container for the customer to use on his board.

After this operation, which is called assembly, the devices undergo a further Final Test (FT verifies the quality of the assembly). Defects can be introduced by both the thermal and mechanical stresses induced by wire bond soldering and fusion of the resin packing operation (performed at 150 °C). Then some good parts are cycled up to the specification limits in order to verify reliability issues.

There are additional tests executed on samples, i.e. ElectroStatic Discharge (ESD) tests, verifying the circuitry robustness by applying discharges up to 4KV on each pin; the pressure pot test, to verify the impermeability of the package to humidity, mechanical shocks; and tests for rejection to noise, on both the supply lines (i.e. causing perturbation on VDD and GND). The global yield for all the various steps is about 70%. To produce our 20 million pieces per year for 10 years we need to produce at least 28 million parts per year.

It's worth contemplating the extensive amount of work required to produce, test, understand the failures (and corrections to eliminate them), assembly, deliver, and application in various electronics designs… but this is another book.

1.15 Memory Pins Description

We would like to close this introductory chapter describing the pins that are used in a typical Flash memory. Figure 1.11 shows the pinout of a hypothetical 128 Mbit memory. In reality, it is advisable to read the datasheet of the product to know exactly which pins and functionality are available.

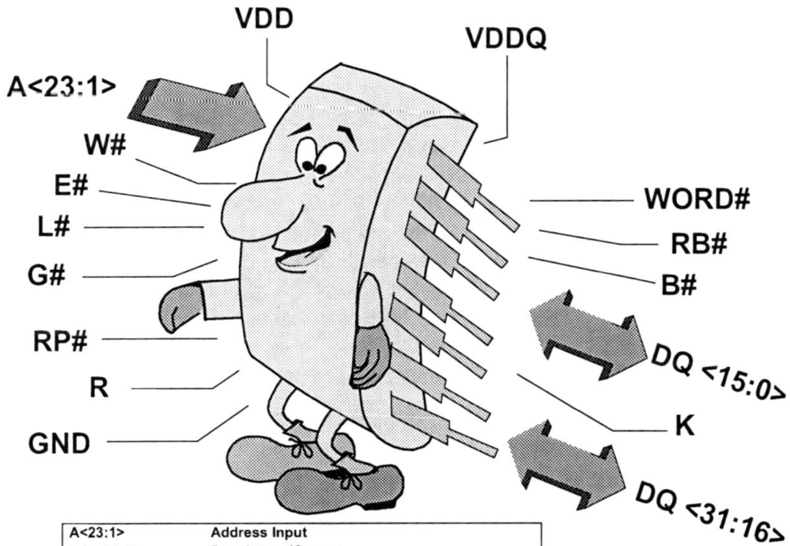

A<23:1>	Address Input
DQ <15:0>	Data Inputs/Output
DQ <31:16>	Data Inputs/Output (x32 bus width only)
B#	Burst Address Advance
E#	Chip Enable
G#	Output Enable
K	Clock
L#	Latch Enable
R	Valid Data Ready
RB#	Ready/Busy
RP#	Reset/Power-Down
W#	Write Enable
WORD#	Word Organisation
VDD	Supply Voltage
VDDQ	Input/Ouput Supply Voltage
GND	Ground

Fig. 1.11. Example of pinout of a 128 Mbit Flash memory

Address Inputs A<23:1>. The Address Inputs are used to select the cells to access in the memory array either to read or to program data to. During write operations they control the commands sent to the Command Interface of the internal state machine. Chip Enable must be low when selecting the addresses. The address inputs can be latched on the rising edge of Chip Enable, Write Enable or Latch Enable depending on the specifications. The address latch is transparent when Latch Enable is low, V_{IL}. The address is internally latched in a program or erase operation.

With a x32 bus width, WORD# = V_{IH}, address input A<1> is ignored; the Least Significant Word (LSW) is output on DQ<15:0> and the Most Significant Word (MSW) is output on DQ<31:16>. With a x16 Bus width, WORD# = V_{IL}, the LSW is output on DQ<15:0> when A<1> is low, and the MSW is output on DQ<15:0> when A<1> is high.

Same considerations can be done considering a memory organized x8/x16: in this case the least significant address is A<0> and BYTE# pin is used to switch between x8 and x16 outputs configuration.

Data Inputs/Outputs DQ<31:0>. The Data Inputs/Outputs pins output the data stored at the selected address during a read operation, or are used to input the data during a program operation. During write operations they represent the commands sent to the Command Interface of the device. When used to input data or write commands they can be latched on the rising (falling) edge of Chip Enable or Write Enable depending on the specifications.

When Chip Enable and Output Enable are both low, the data bus outputs data from the memory array. The data bus is high impedance when the chip is deselected, Output Enable is high or the Reset/Power-Down signal is low. With a x16 bus width, WORD# is low, DQ<31:16> are not used and are high impedance.

Chip Enable (CE# or E#). The Chip Enable input activates the memory control logic, input buffers, decoders and sense amplifiers. CE# high deselects the memory and reduces the power consumption to the standby level.

Output Enable (G# or OE#). The Output Enable gates the outputs through the data output buffers during a read operation. When G# is High the outputs are high impedance. G# can be used to inhibit the data output during a burst read operation.

Write Enable (W# or WE#). The Write Enable input controls writing to the Command Interface, input address and data latches. Both addresses and data can be latched on the rising (falling) edge of WE#.

Reset/Power-Down (RP#). The RP# pin can be used to apply a Hardware Reset to the memory or, in some devices, to temporarily unprotect all blocks that have been protected. A hardware reset is achieved by holding RP# low for a minimum time. The device is deselected and outputs are high impedance. If RP# goes low during program or erase operations, the operation is aborted and the data may be corrupted. After RP# goes high, the memory will be ready for read and write operations after a minimum time.

Latch Enable (L#). Address Inputs can be latched on the rising edge of L# input pin. In synchronous operations the address is latched on the active edge of the Clock when L# is low. Once latched, the addresses may change without affecting the address used by the memory. When L# is low, the address latch is transparent.

Clock (K). The Clock is used to synchronize the memory with the external bus during synchronous operations. The K pin can be configured to have an active rising or falling edge. Bus signals are latched on the active edge of the Clock during synchronous operations. In burst read mode, the address is latched on the first active clock edge when L# is low or on the rising edge of L#, whichever occurs first. During asynchronous operations the Clock is not used. See Chap. 11 for more details.

Burst Address Advance (B# or BAA#). B# input controls the advancing of the address by the internal address counter during synchronous operations. B# is only sampled on the active K edge when X or Y latency time has expired. If B# is low, the internal address counter advances; if B# is V_{IH}, the same data remains on the data Input/Output pins because the internal address counter does not change.

Valid Data Ready (R). The Valid Data Ready output is an open drain output that can be used to identify if the memory is ready to output data or not. The R pin is only active during burst read and it can be configured to be active on the clock edge of the invalid data read cycle or one cycle before. R pin low, V_{OL}, indicates that the data is not, or will not be valid. Valid Data Ready pin may be tied to other components with the same functionality to create a unique System Ready signal. Usually an external pull-up resistor is used to meet the external timing requirements for R rising.

Word Organization (WORD#). The word organization input selects the x16 or x32 bus width. When WORD# is V_{IL}, x16 bus is set. Data are read and written to DQ<15:0>; DQ<31:16> are at high impedance and A<1> is the LSB of the address bus.

Ready/Busy (RB#). The RB# output is an open drain output that can be used to identify if the internal program/erase controller is currently active. Ready/Busy is low during write or program operations. When the device is busy, it will not accept any additional program or erase commands except program/erase suspend. When the program/erase controller is idle, or suspended, RB can be driven high through a pull-up resistor. The use of an open drain output allows RB pins from several memories to be connected to a single pull-up resistor. A low signal will then indicate that one, or more, of the memories are busy.

VDD supply voltage. The supply voltage VDD is the core power supply. All internal circuits draw their current from the VDD pin. Usually an external capacitor should be connected between VDD and GND.

VDDQ. VDDQ is the input/output buffers power supply. Depending on the specifications, VDDQ can be lower than VDD in order to decrease power dissipation.

GND. Ground, or GND, is the reference for all core power supply voltages.

Bibliography

S. Aritome, "Advance Flash memory technology and trends for file storage application", in IEDM Tech. Dig., pp. 763-766, (2000).

H.P. Belgal et al., "A new reliability model for post-cycling charge retention of Flash memories", Proc. IRPS, pp. 7-20, (2002).

R. Bez, "Introduction to Flash Memory", IEEE Proceeding of the, Vol. 91, No. 4, pp. 489-502, (April 2003).

W.D. Brown and J. E. Brewer, eds., Nonvolatile Semiconductor Memory Technology. New York, NY: IEEE Press, (1998).

C. Calligaro et al., Proc. 3rd IEEE Int. Conf. on Electronics Circuits and Systems, pp.1005, (1996).

G. Campardo, R. Micheloni, "Scanning the special issue on Flash Memory technology", IEEE Proceeding of the, Vol. 91, No. 4, pp. 483-488, (April 2003).
P. Cappelletti, R. Bez, D. Cantarelli and L. Fratin, "Failure mechanisms of Flash cell in program/erase cycling", IEDM Tech. Dig., pp. 291-264, (1994).
P. Cappelletti, A. Modelli, "Flash memory reliability", in Flash memory, P. Cappelletti et al., Ed Norwell, Ma: Kluwer, (1999).
A. Chimenton , P. Pellati , P. Olivo , "Analysis of Erratic Bits in FLASH Memories", Proc. IRPS, 17-22, (2001).
A. Conci, et al., Current criticalities and innovation perspective in Flash memory design automation", IEEE Proceeding of the, Vol. 91, No. 4, pp. 581-593, (April 2003).
G. Crisenza, C. Clementi, G, Ghidini and M. Tosi, "Floating gate memories", Qual. Reliab. Eng. Int., vol. 8, pp.177-187, (1992).
G. Crisenza, G. Ghidini, S. Manzini, A. Modelli, M. Tosi, "charge loss in EPROM due to ion generation and transport in interlevel dielectrics", IEDM Tech. Dig., pp. 107-110, (1990).
D. Ielmini, A.S. Spinelli, A.L. Lacaita, L. Confalonieri and A. Visconti, "New technique for fast characterisation of SILC distribution in Flash arrays", Proc. IRPS, 73-80, (2001).
D. Ielmini, A.S. Spinelli, A.L. Lacaita, R. Leone and A. Visconti, "Localisation of SILC in Flash memories after program/erase cycling", Proc. IRPS, 1-6, (2002).
D. Ielmini, A..S. Spinelli, A..L. Lacaita, A. Modelli, "Statistical Model of reliability and scaling projections for Flash memories", IEDM Tech. Dig., (2001).
International Technology Roadmap for Semiconductors, (2001).
S. Keeney, "A 130nm generation high-density ETOX Flash memory technology", IEDM Tech. Dig., p.41, (2001)
V.N. Kynett, A. Baker, M. Fandrich, G. Hoekdtrsa, O. Jungroth, J. Kreifels and S. Well, "An in-system reprogrammable 256 K CMOS Flash memory", ISSCC, Conf. Proc., pp. 132-133, (1988).
S. Lai, "Flash memories: Where we were and where we are going", IEDM Tech. Dig., pp. 971-973, (1998).
F. Masuoka, M.Momodomi, Y. Iwata and R. Shirota, "New ultra high density EPROM and Flash with NAND structure cell", IEDM Tech. Dig., pp. 552-555, (1987).
A. Modelli, "Reliability of thin dielectrics for non-volatile applications", Microelectronic Engineering, 48, 403 (1999).
S. Mukherjee, T. Chang, R. Pang, M. Knecht and D. Hu, "A single transistor EEPROM cell and its implementation in a 512K CMOS EEPROM", IEDM Tech. Dig., pp. 616-619, (1958).
T.C. Ong et al., VLSI Symp. on Tech., 7A-2, p.83, (1993)
P. Pavan and R. Bez, " The industrial standard Flash memory cell", in Flash memory, P. Cappelletti et al., Ed Norwell, Ma: Kluwer, (1999).
P. Pavan, R. Bez, P. Olivo and E. Zanoni, "Flash memory cells-An overview", Proc. IEEE, vol. 85, pp. 1248-1271, (Aug. 1977).
B. Ricco et al., Proc. IEEE, vol. 86, pp. 2399, (1998).
L. Selmi and C. Fiegna, "Physical aspects of cell operation and reliability", in Flash memory, P. Cappelletti et al., Ed Norwell, Ma: Kluwer, (1999).
Webfeet Inc., "Semiconductor industry outlook", presented at the 2002 Non-Volatile memory Conference, Santa Clara, CA.

2 Process Aspects

Before starting the design of the device, it is important to know the basic components that are at the designer's disposal to realize the circuitry, i.e. transistors, capacitors, resistors, diodes, bipolars and memory cells.

The characteristics of such components are defined by the process that, in the case of Flash Memories, is CMOS with two layers of poly and two or more of metal, with the recent addition of the triple well. After the process and the components have been defined, the circuit is designed to maximize the electrical performance, while the layout utilizing the available layers is realized in parallel.

The success of a design depends on the correct combination of process, design, and layout.

Before analyzing the components of the circuits, it is worth giving a brief overview of the different steps of fabrication of a typical CMOS process.

2.1 Introduction

The fabrication of an integrated circuit resembles the preparation of a *sandwich* in which, on a good slice of bread as a substrate, the sauces diffuse and several layers are harmoniously placed, one on top of the other, to satisfy the customer's taste. All is sealed with the final slice of bread that keeps the whole product together if the underlying layers are planar.

The technological process for the fabrication of integrated circuits is similar to the process described above.

2.2 Main Steps of Fabrication for a CMOS Process

The main steps that allow the fabrication of a device in CMOS technology will be described hereafter, with particular attention to the aspects that have the most impact on the design. The following description will show the sequence of process steps for an NMOS and a PMOS *low and high voltage* transistor, and two memory cells. The definition of the different layers is obtained by means of a number of photomasks, chrome images on glass plates that reproduce the polygons of the layout, defining different areas on the wafer through a photolithographic process.

The fabrication starts from a wafer of *p*-type silicon that forms the substrate; during the first operations, two twin-tubs, *n-well* and *p-well*, are realized.

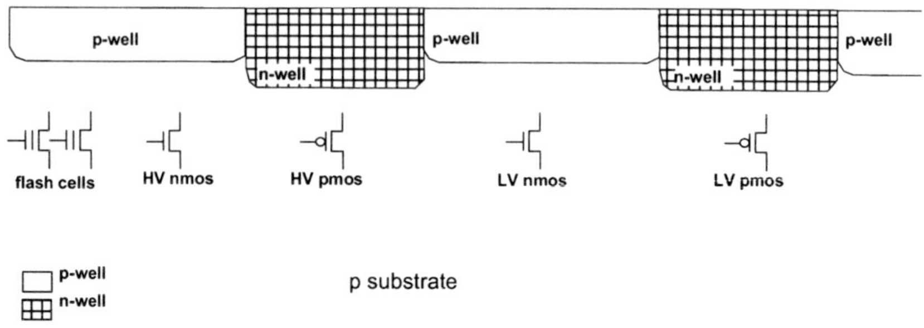

Fig. 2.1. Twin tub diffusion

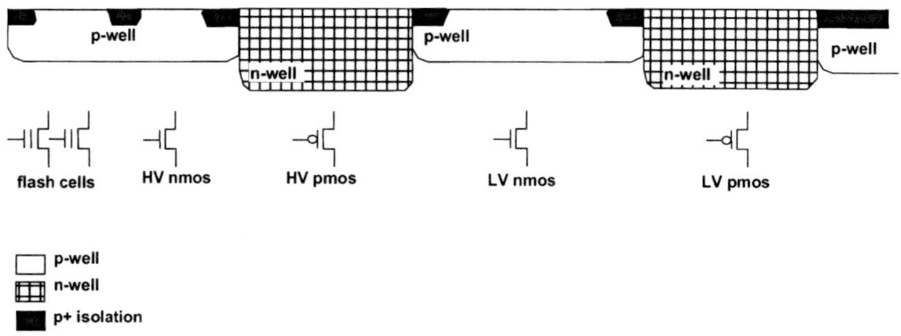

Fig. 2.2. Creation of the p+ insulation areas

The n-well in the *p*-substrate is necessary to fabricate *p*-type transistors, whereas the p-well is necessary to create an *ad-hoc* substrate for the *n*-type transistors (Fig. 2.1).

Subsequently, the insulating diffusions are created, so as to increase the threshold voltage of the parasitic elements that form between the junctions of two contiguous transistors. These parasitic devices are located beneath the thick oxide, also called *field oxide* (around 5000 Å of thickness) that constitutes an insulating barrier between active components (Fig. 2.2). The insulating diffusion, like any other selective step that acts in specific areas of the wafer, is realized by suitably masking the areas that must not be involved in that process step.

Subsequently, the thick oxide is grown on the insulating junctions (Fig. 2.3). The oxide for *high voltage* (*HV*) and *low voltage* (*LV*) *p* and *n* transistors are realized (typically around 250 Å and 120 Å, respectively) by means of two masks (Fig. 2.4).

The existence of HV and LV transistors is a characteristic of the memory devices that require voltages higher than VDD to program and erase.

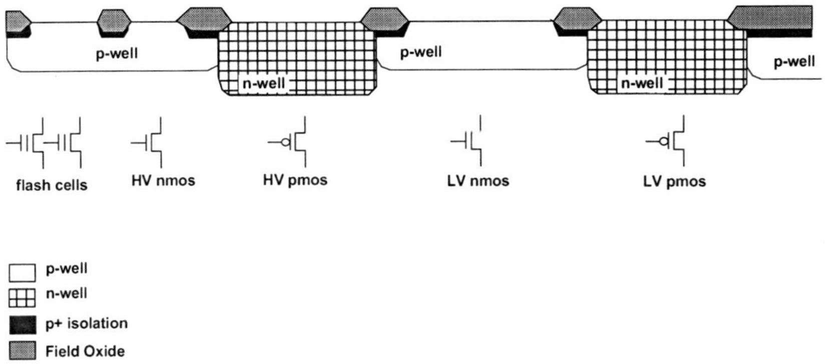

Fig. 2.3. Thick oxide growth

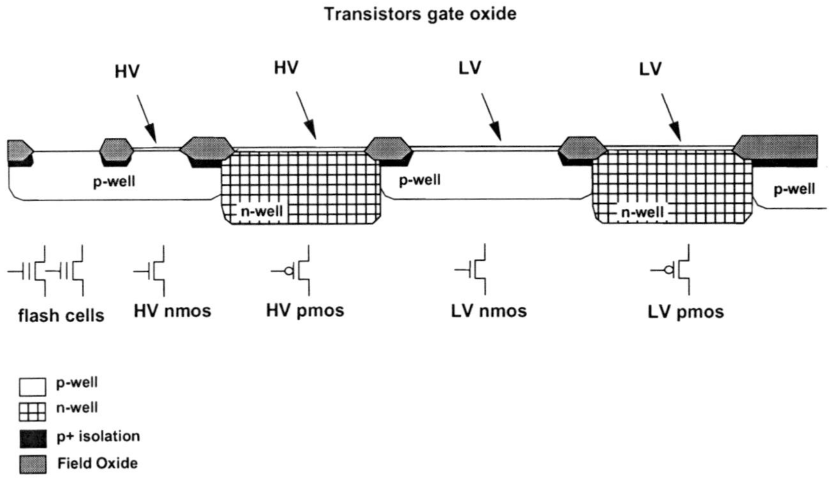

Fig. 2.4. Transistor gate oxide growth

The presence of two oxides allows optimizing the low voltage logic circuits, which have to be fast, and the high voltage circuits that, on the other hand, have to be robust. This also improves the functional aspects of the device such as reading speed, standby consumption, etc., since thick oxides limit the electric field at their side to about 4 MV/cm[1] at high voltages. Subsequently, the EPM[2] area is realized so as to implant the channel of the cells of the array. Such implants differentiate the cells from the typical n-channel transistors by setting the threshold voltage and

[1] This value is suggested by experience and reliability measurements.

[2] EPM is the acronym for Enhanced Program Mask.

optimizes the electrical characteristics since the cells are to be programmed and erased (Fig. 2.5).

Afterward, the tunnel oxide (100 Å) is grown directly on the EPM implant that delimits the area of the array. This step is perhaps the most critical since the quality of the tunnel oxide is essential to the device performance and reliability. (Fig. 2.6). The first polysilicon layer, poly1, which will form the floating gate of the cells of the array, is now defined. Typically this poly layer is deposited on all the transistors with the second poly deposited directly on poly1 to form the gate of the standard transistors.

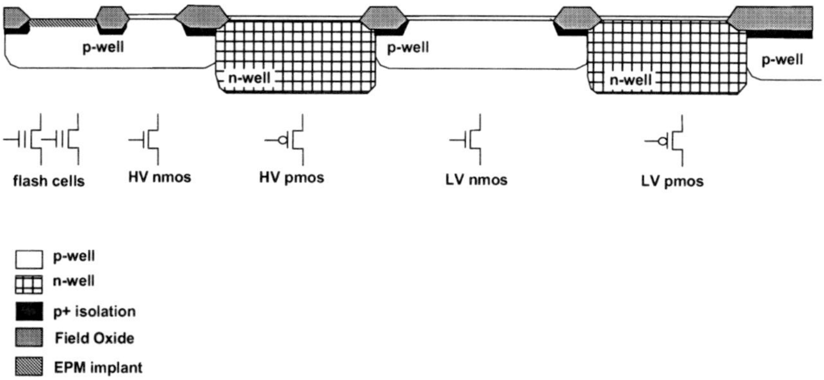

Fig. 2.5. Channel diffusion for memory cells

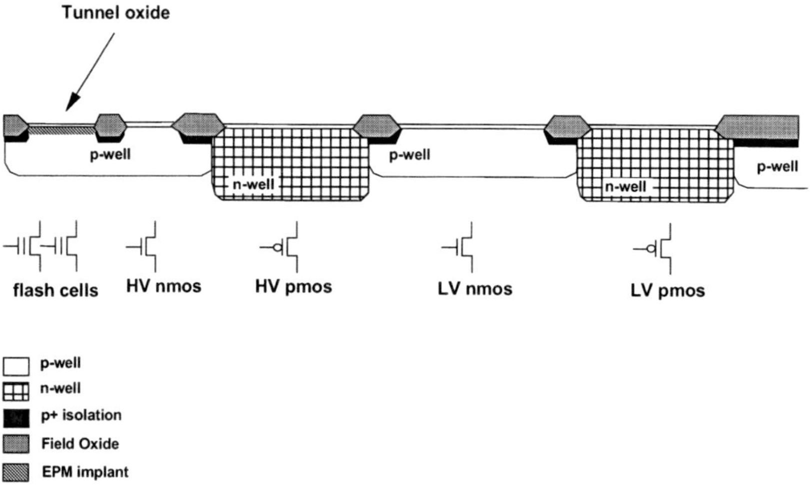

Fig. 2.6. Tunnel oxide growth

In this case, the process is called DSCP (Double Short-Circuited Poly) (Fig. 2.7). The layer that separates poly1 from poly2 is called ONO, which is the acronym for the composing materials, i.e. Oxide-Nitride-Oxide. This stacked composition improves the overall quality of the insulation between floating gate and control gate (Fig. 2.8).

Subsequently, the implants for low and high voltage transistors can be defined with another group of masks, to selectively differentiate the threshold voltages of the transistors and, then, deposit the poly2 layer that constitutes the control gate of transistors and cells (Fig. 2.9).

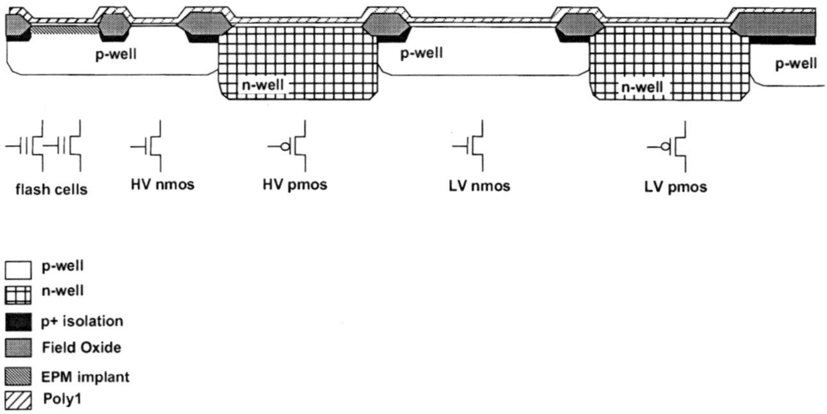

Fig. 2.7. Poly1 deposition

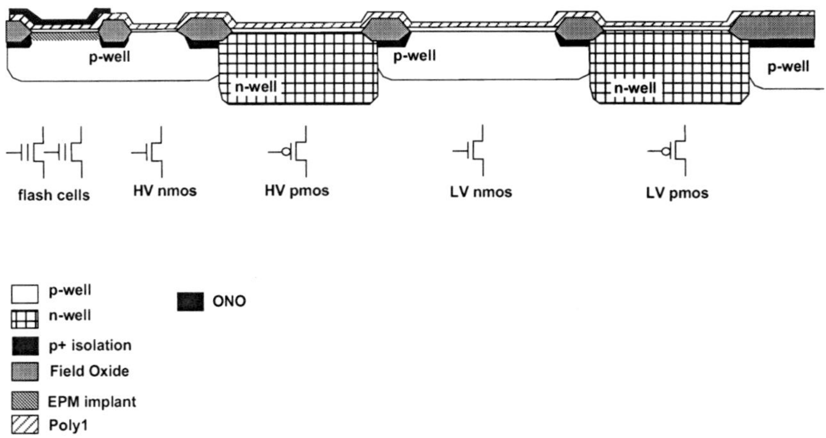

Fig. 2.8. Realization of the insulation between poly1 and poly2 (ONO)

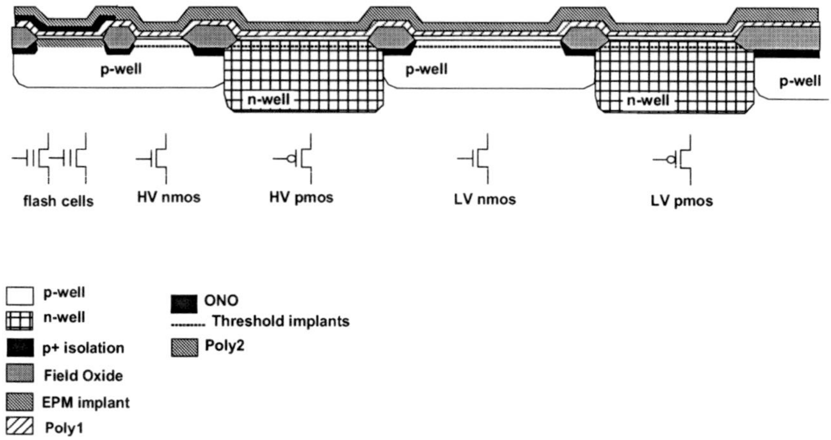

Fig. 2.9. Diffusions to define the MOS threshold voltages and deposition of poly2

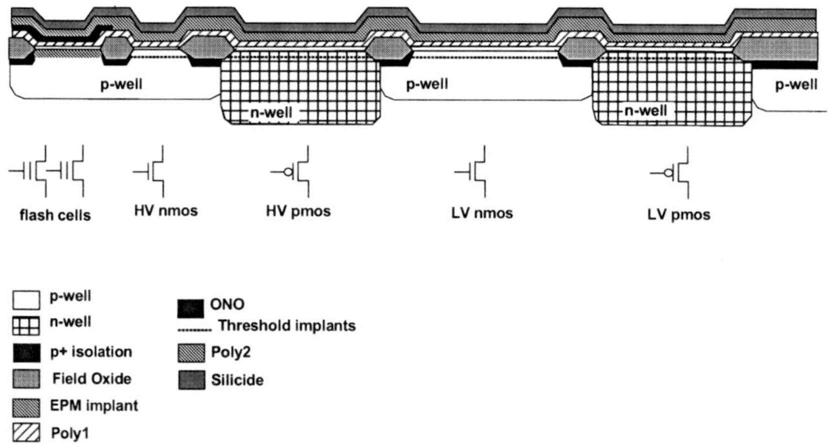

Fig. 2.10. Silicide deposition

The implants for the thresholds are realized through the poly1 already deposited.

On top of the poly2 layer, a metallic compound WSi_2, called silicide, is deposited, which decreases the resistivity of poly2 from around 50 Ω/□ to 5 Ω/□. This step is very important in determining the access time because the rows of the cells of the array are realized with poly2 as a conductor and, hence, their resistivity determines also the RC time constant of the rows (Fig. 2.10).

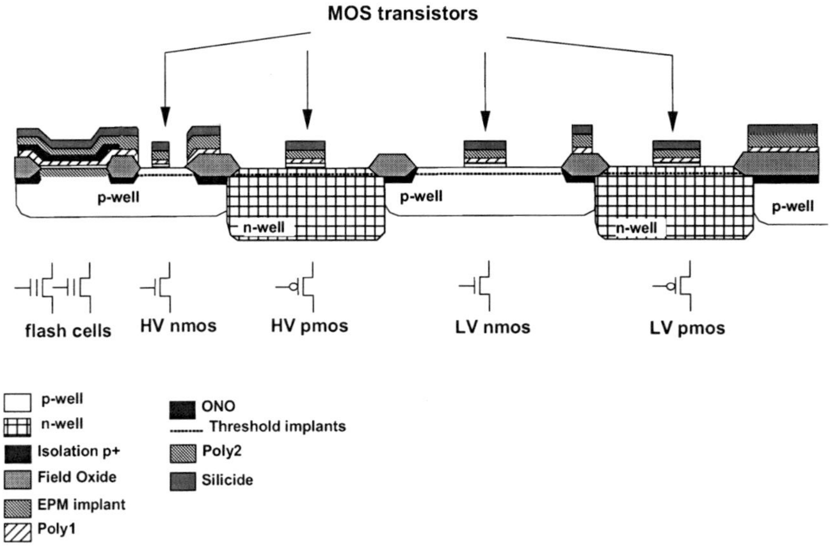

Fig. 2.11. Removal of the layers to define the transistors

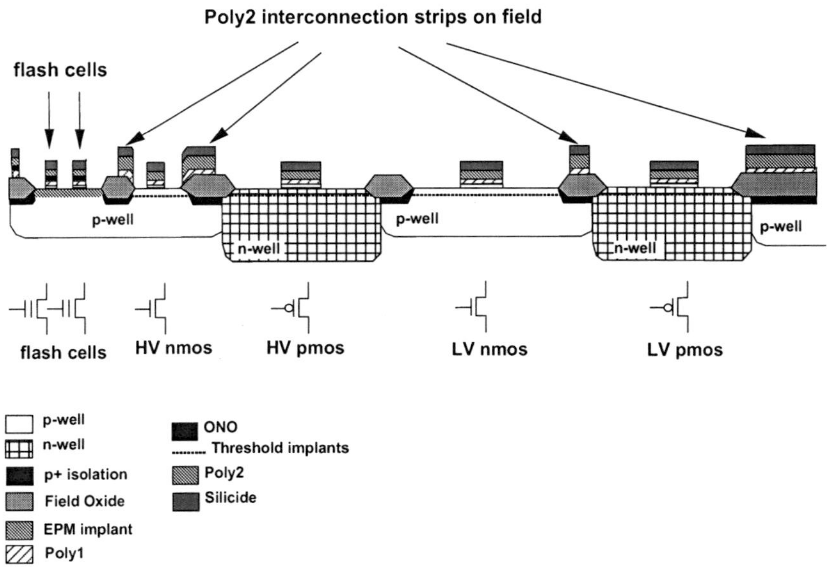

Fig. 2.12. Flash cell definition

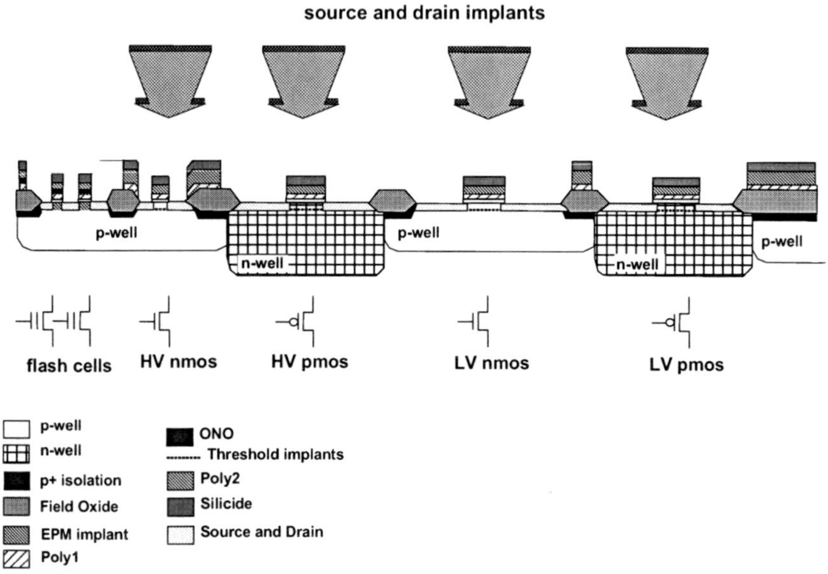

Fig. 2.13. Source and drain implants

After the basic layers have been fabricated, it is possible to etch them simultaneously, so as to obtain the structure of MOS transistors and cells as single element.

Transistors and cells are fabricated with different operations. In Fig. 2.11, the transistors have been obtained and, subsequently, the Flash cells are defined (Fig. 2.12). The portions of polysilicon that remain on the field oxide represent the strips of poly2 used as interconnect and are not the gates of the transistors. The source and drain areas for transistors and cells are diffused using other masks, as shown in Fig. 2.13.

The aluminum-silicon alloy connection layer called *Metal* is then deposited. Before that, however, the "holes" that allow the contacts between poly2, active areas, and the different layers of metal, are created. The first step is the insulation of the surface with an intermediate layer of oxide (Fig. 2.14). Afterward, the "holes", called *contacts* in this case, are cut into the insulator so as to allow the connection between *metal1*, i.e. the first metal layer, and either poly2 or active area (Fig. 2.15).

The *metal1* layer is deposited so as to penetrate into the contacts, thus forming the electrical connection. Metal1 is deposited on the whole wafer (Fig. 2.16), and removed from the areas where it is not needed by means of a masking process. The surface is then completely covered with dielectric materials having the function of both insulating metal1 from the layers that will be deposed thereafter, and planarizing the surface itself. Such materials are called TEOS[3] and SOG[4].

[3] TEOS is the acronym for TetraEthylOrho Silicate.

[4] SOG is the acronym for Spin-On-Glass.

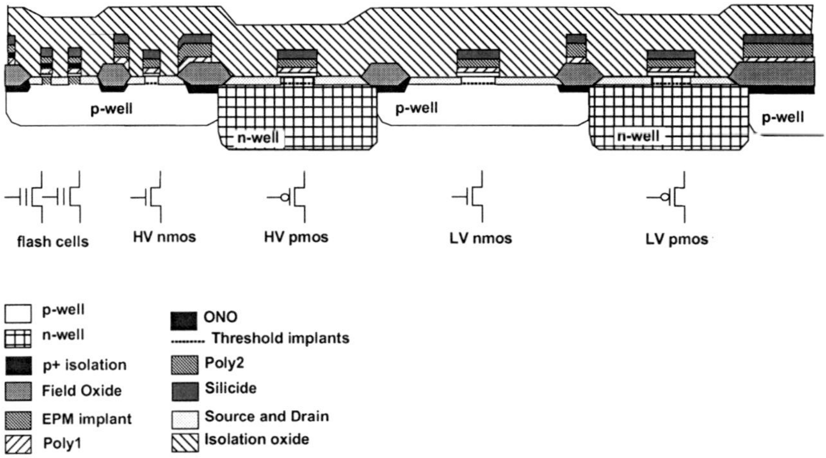

Fig. 2.14. Intermediate insulating oxide

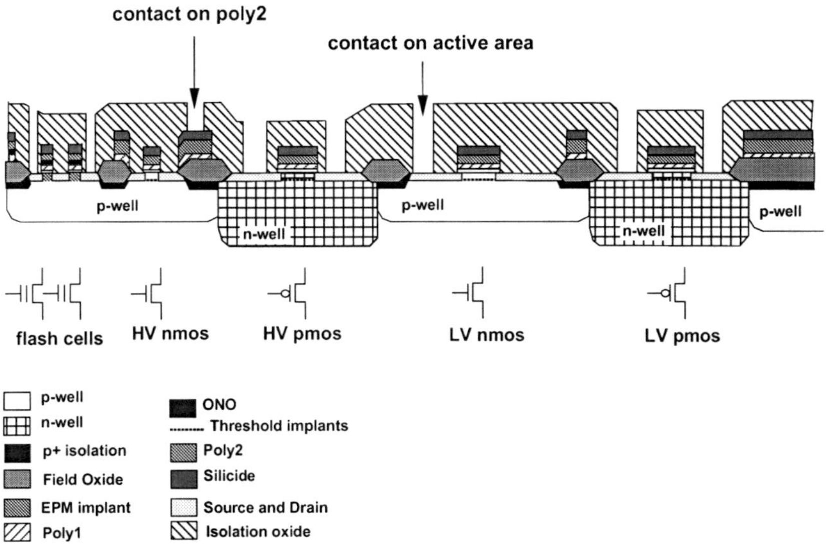

Fig. 2.15. Contact definition

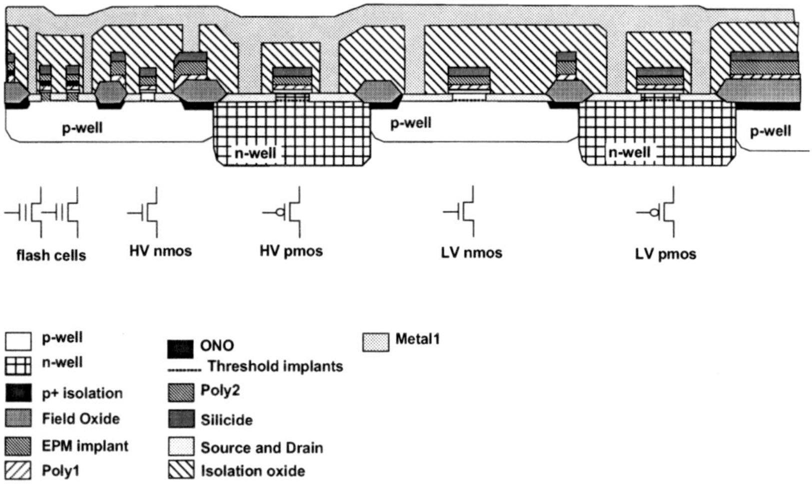

Fig. 2.16. Metal1 deposition

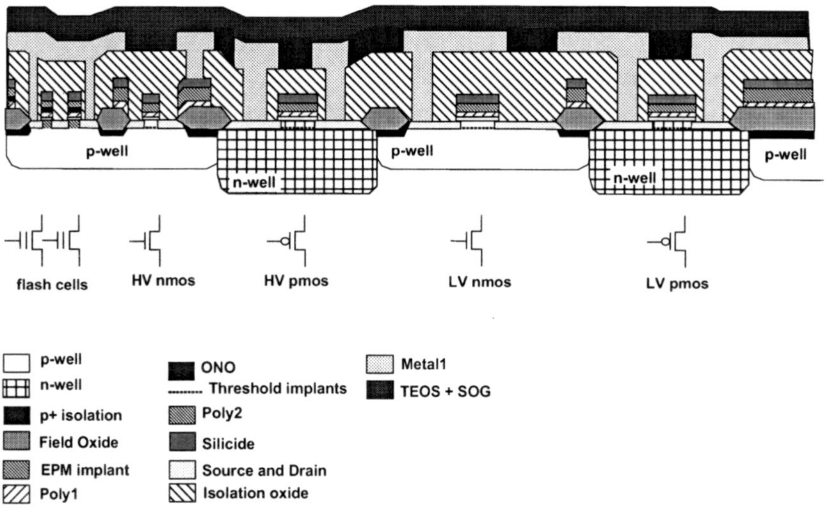

Fig. 2.17. Metal1 definition and intermetal dielectric deposition

Planarity enables the deposition and etching to be more precise and, therefore, smaller dimensions can be defined (Fig. 2.17). Afterward the planarization, the "holes" that allow the connection between metal1 and *metal2*, called *vias*, are opened (Fig. 2.18). The metal2 is deposed with the same technique used for metal1 and then it is etched and removed from the areas where it is not necessary (Fig. 2.19). Finally, the entire area of the surface is covered with a vitreous layer, or passivation, having the function of sealing and protecting from the external environment.

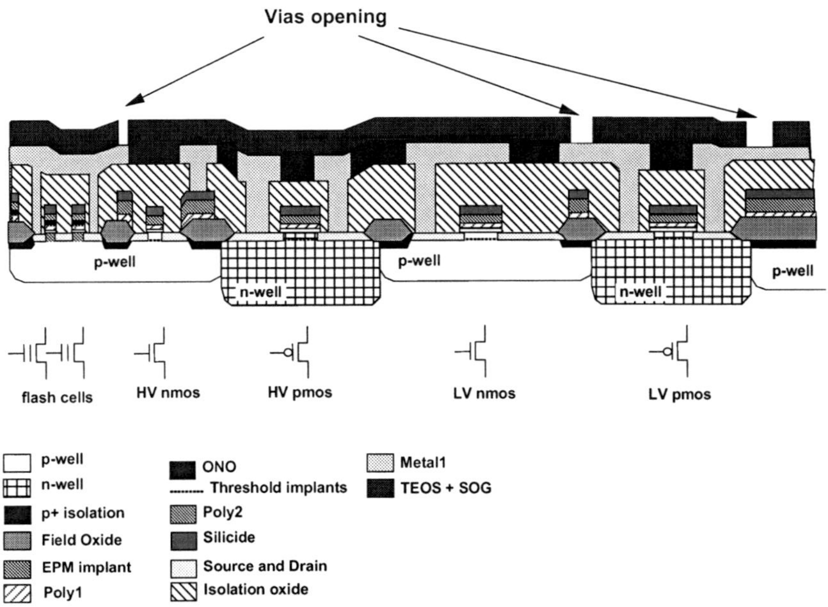

Fig. 2.18. Vias opening

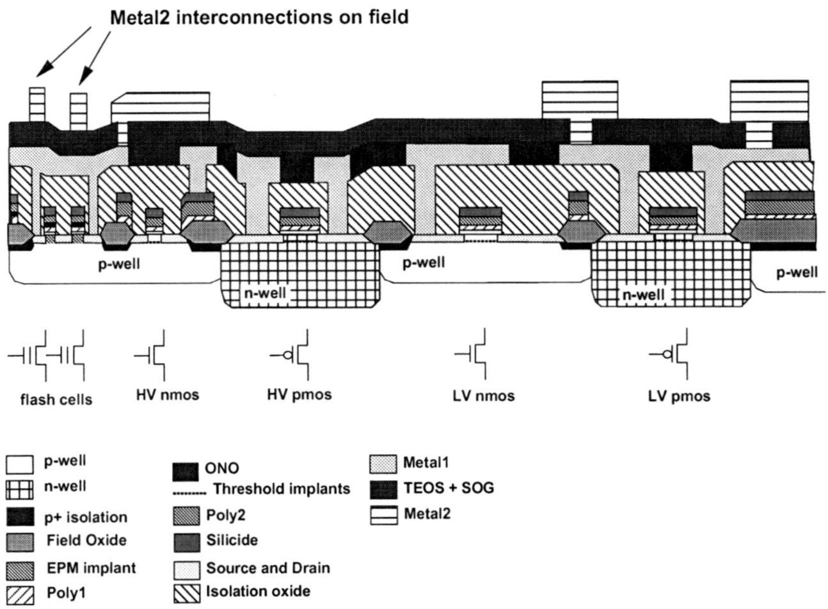

Fig. 2.19. Metal2 deposition

The last operation is the opening of the pad areas in the passivation layer, where the wires are bonded to make connection from the integrated circuit to the pins of the package (Fig. 2.20).

Subsequently, the wafer is lapped, i.e. it is thinned by means of chemical abrasion so as to make it easier to assemble it into the package and, moreover, improve the conduction of heat. The device is then submitted to electrical and functional tests.

The most important steps of a typical CMOS process with two layers of poly and metal for non-volatile cells have been presented. The process is actually constituted of many more steps that can reach 300, the majority of which are cleaning steps. For each major operation, the recipes are defined in terms of temperature environment, gas, atomic species etc. and only the steps that are directly involved in the circuit design have been described in this chapter.

Another characteristic of the process is the *triple well*, which facilitates a simpler solution to erase the Flash memory.

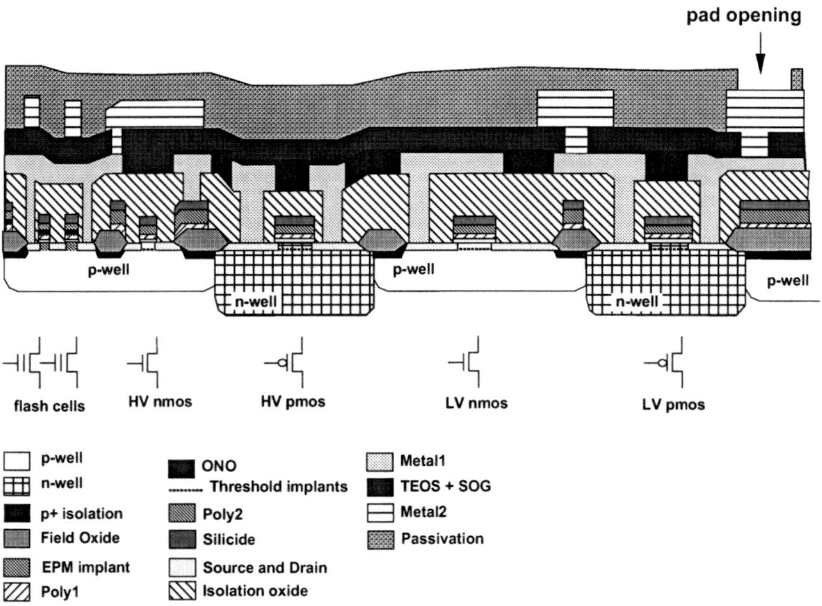

Fig. 2.20. Passivation and pad opening

In order to solve the problem of the erasing Flash cells by use of a negative voltage on the row without increasing the complexity of the circuit, it is useful to create n-channel transistors having substrate that can be biased independently. Areas with n-wells are fabricated through a doped buried layer, thus realizing a p-type tub insulated from the substrate, which is also p-type (Fig. 2.21).

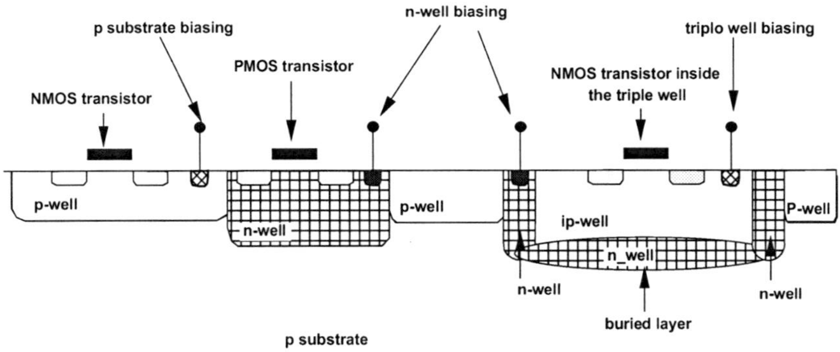

Fig. 2.21. Device section that highlights the triple-well transistors with their bias

Bibliography

C. Auricchio, R. Bez, A. Lo savio, A. Maurelli, C. Sala, P. Zabberoni G. Baccarani and M. Rudan, "A triple-weel architecture for low voltage operation in submicron CMOS devices", (Eds.), Proc. ESSDERC 96, Bologna, Italy, p. 613, (1996).

T.M. Bloomstein, M. Rothshild, R.R. Kunz, D.E. Hardy, T.B. Goodman,and S.T. Palmacci, "Critical issues in 157 nm lithography", J. Vac. Sci. Thecnol. B, vol.16, pp. 3154, (1998).

J.O. Borland, R. Koelsch, "MeV implantation technology: next generation manufacturing with current generation equipment", Solid State Technology, P.1 (1993).

P. Cappelletti, A. Tutolo, L. Frattin, L. Ravazzi, C. Riva, "The Flash E2PROM cell with Boron P-pocket architecture: advantages and limitations", Proc. Nonvolatile Semiconductor Memory Workshop, (1995).C.T. Gabriel, and J.P. McVittie, "How Plasma Etching Damages Thin Gate Oxides", Sol. State Technology, 81, (June 1992).

T.Y. Chan, P.K. Ko and C. Hu "Dependence of channel electric field on device scaling" IEEE Electron Dev. Lett., EDL-6, p.551, (Oct. 1985).

M.L. Chen et al. "Suppression of Hot-Carrier Effects in Submicron CMOS Technology" IEEE Trans. Electron Dev., ED-35, p2210, (Dec. 1988).

E.G. Cromer, "Mask aligners and stepper for precision microlithography, Solid State Technology, 36, (April 1993).

J.D. Cuthbert, "Optical projection printing", Solid State Technology, 20, 59, August (1977).

F.H. Dill, W.P. Hornberger, P.S. Hauge, J.M. Shaw, "Characterization of positive resist", IEEE Trans. Electron. Dev., ED22, 445, (1975).

G. Dunn, S.S. Scott, "Channel hot-carrier stressing of reoxidizied nitrided silicon dioxide", IEEE Trans. On Electron Device, 37, 7, p. 1719, (1990).

L. Forester, A.L. Butler and G.Schets, "SOG planarization for Polysilicon and First Metal Interconnect in a One Micron CMOS Process" Proceedings of the 6th International IEEE VLSI Multilevel Interconnection Conference, p.72, (1989).

Fowler-Nordheim tunneling on source doping concentration of an n-MOSFET", IEEE Electron Device Letters, 17,11, p.525, (1996).

G. Ginami. et al., "Survey on Flash technology with specific attention to the critical process parameters related to manufacturing", IEEE Proceeding of the, Vol. 91, No. 4, pp. 503-522, April (2003).

C. Hill, S. Jones and D. Boys "Rapid Thermal Annealing – Theory and Practice" in R.A. Levy Ed., Reduced Thermal Processing for ULSI, Plenum Press, New York, p. 143, (1989).

IEEE Standard Department, "IEEE P1005 draft standard for definitions, symbols, and characteristics of floating gate memory arrays", approved 1998.

E. Kooi, "The invention of LOCOS", IEEE Press, NY, (1991).

M.D. Levenson, N.S. Viswanathan, R.A. Simpson, "Improving resolution in photolithography with a phase-shifting mask", IEEE Trans. Electron. Dev, ED29, 1828, (1982).

B.J. Lin, "The optimum numerical aperture for optical projection microlithography", Proc. SPIE, 1463, 42, (1991).

R.A. Morgan, "Plasma etching in semiconductor fabrication", Elsevier, NY, (1985).

S. Mori, E. Sakagami, H. Araki, Y. Kaneko, K. Marita, Y. Ohshima, N. Arai, K. Yoshikawa, "ONO interpoly dielectric scaling for non-volatile memories applications", IEEE Trans. On Electron Devices, 38, 2, p. 386-391, (1991).

E.G. Moore, "Cramming more components onto integrated circuits", Electronics Magazine, vol. 8, pp. 114-117, (April 1965). M. Born , E. Wolf "Principles of optics", Pergamon Press

A. Nakae, K. Kamon, T. Hanawa, H. Tanabe, "Improvement in optical proximity correction by optimizing second illumination source", Jpn. J. Appl. Phys., Pt.1, vol.35, pp. 6396, (1996).

J.K. Roberts "Heat and Thermodinamic" 4th ed. Blackie and Son, Ltd., Glasgow, (1995).

K.C. Saraswat, D.L. Brors, J.A. Fair, K.A. Monnig and R.Beyers "Properties os Low Pressure CVD Tungsten Silicide for MOS VLSI Interconnections" IEEE Trans. Electron Dev., ED-30, 1497, (1983).

K.C. Saraswat, F. Mohammadi and J.D. Meindl "WSi_2 Gate MOS Device" Tech. Dig. IEDM, 462, (1979).

Semiconductors Industry Association, "International Roadmap for Semiconductors: 2002".

K. Shibahara et al., "Trench Isolation with V-shaped Buried Oxide for 256-Mbit DRAMs" Tech Dig. IEDM, p. 275, (1992).

Y. Tang, J. Chen, C. Chang, D. Liu, S. Haddad, Y. Sun, A. Wang, M. Ramskey, M. Kwong, H. Kinoshita, W. Chan, J. Lien, "Different dependence of band-to-band and H. Tsai, C.L. Yu, C.Y. Wu, "A bird's beak technique for LOCOS in VLSI fabrication", IEEE Electron Device Letters, 7, 2, pp. 122-123, (1986).

K. Tsukamoto et al. "High Energy Ion Implantation for ULSI: Well Engineering and Gettering", Solid State Technology, p.49, (June 1992).

N.A.H. Wils, P.A. van der Plas, A.H. Montree, "Dimensional characterization of poly buffer LOCOS in comparison with suppressed LOCOS", ESSDERC 90, pp. 535-538, (1990).

3 The MOSFET Transistor and the Memory Cell

Although this book addresses topics that require previous basic knowledge of the MOS transistor, an overview of the principles of such component will not be omitted. For a rigorous and complete exposition on the MOS transistor behavior, the reader should refer to specialized literature. This book will focus on certain aspects of the MOS transistor characteristics, and examples are given from the standpoint of the designer, who tends to reason in terms of potential, current, and impedance. Furthermore, the main characteristics of the Flash memory cell will be presented in read, program, and erase mode. Also in this case, the dissertation is reduced to the essential notions to understand the problems of the design. A brief history of the electrical erasing will be outlined in the case of Flash Memories, from the first generation of devices with double bias, to the current devices having a single VDD.

3.1 The MOSFET Transistor

The MOSFET transistor (or simply MOS) is defined as a voltage controlled current source. When the behavior of the device with respect to small signals is to be studied, which means small deviations with respect to a fixed steady condition (called bias or operating point), the equations of the representing model can be linearized[1].

The equivalent circuit of the MOS is represented in Fig. 3.1, in which the gate terminal G is insulated, since it is supposed that no current flows into the gate, and the current generator is controlled only by means of the voltage.

The source terminal S has lower potential than the drain D in case of n-channel transistor.

---

[1] There are several ways to define what is meant by "small signal". For a MOS transistor, for example, the ratio between bias current in saturation region and signal current can be considered. The small signal condition is then $v_{gs} \ll (V_{GS} - V_T)$, where the voltage in lowercase refers to the signal, whereas the bias quantities are indicated in uppercase. Owing to the notion of small signal, it is possible to consider only the variations and, hence, in the analysis of the equivalent circuit, both VDD and ground are considered to be at the same potential, being stationary with respect to the signal variations.

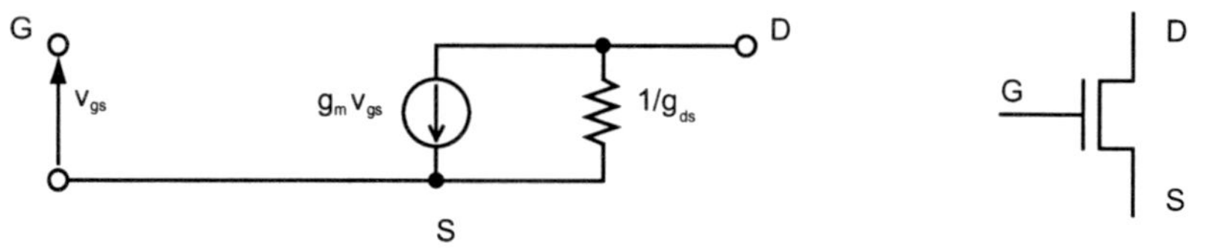

Fig. 3.1. Equivalent circuit and graphic symbol of an NMOS transistor

In Fig. 3.1 g_{ds} is the output conductance, g_m is the transconductance since it links the variation of an input quantity, the voltage V_{GS}, to an output quantity, the drain current I_{DS}.

$$g_m = \frac{\partial I_{DS}}{\partial V_{GS}} \tag{3.1}$$

$$g_{ds} = \frac{\partial I_{DS}}{\partial V_{DS}} \tag{3.2}$$

The parameters described are rarely used when designing from scratch, the time when the ideas are converted into circuits. Only the simulation programs can manage them so as to solve the systems of equations and obtain the potentials of all the nodes, and the currents of all the branches. The relevant aspect is that, as it will be shown, g_m is proportional to the aspect ratio (W/L), which is the only parameter that the designer can control.

What is the equivalent resistance R_{eq} of a PMOS transistor connected to the supply voltage, as shown in Fig. 3.2? If we suppose that the load resistance R is large enough so that the current drawn from the transistor is less than the capability of supplying current of the transistor itself, the voltage V_{OUT} is very close to VDD.

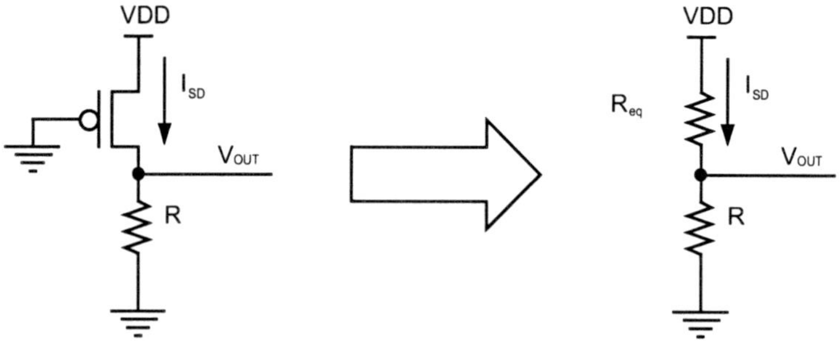

Fig. 3.2. Equivalent resistance for a PMOS transistor working as a current source

For example, if V_{OUT} is 100 mV lower than VDD, R_{eq} can be calculated by means of the Ohm's law:

$$R_{eq} = \frac{100mV}{I_{SD}} \tag{3.3}$$

For example, with I_{DS} of 300 μA, R_{eq} of 333 Ω is obtained.

A more accurate model that takes into account the effect of bias of the substrate (body effect) is shown in Fig. 3.3.

In this case g_{mb} is defined as:

$$g_{mb} = \frac{\partial I_{DS}}{\partial V_{BS}} \tag{3.4}$$

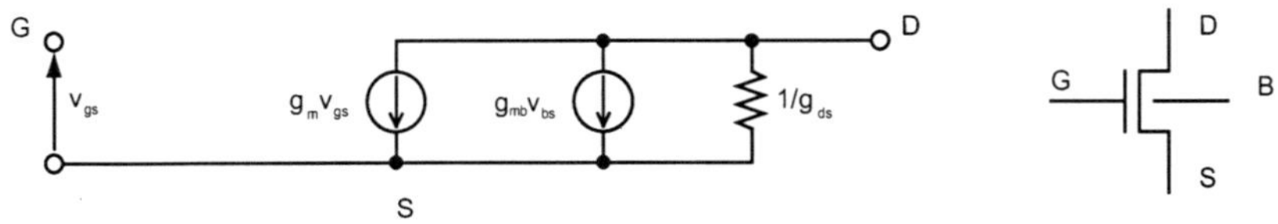

Fig. 3.3. Equivalent circuit and graphic symbol for an NMOS transistor where the substrate bias source V_{BS} is highlighted

The effect of the bias between bulk B and source S for a MOS transistor is quantified by means of the expression of the threshold voltage V_T. Conventionally, it is assumed that the conduction in a MOS transistor starts when the difference of potential between gate and source exceeds the value V_T that, in case of n- and p-channel, can be calculated through the following formulae:

$$V_{T,n} = V_{T0} + \gamma \cdot \left(\sqrt{2 \cdot |\Phi_p| + |V_{SB}|} - \sqrt{2 \cdot |\Phi_p|} \right) \tag{3.5}$$

$$V_{T,p} = V_{T0} - \gamma \cdot \left(\sqrt{2 \cdot |\Phi_n| + |V_{SB}|} - \sqrt{2 \cdot |\Phi_n|} \right) \tag{3.6}$$

$$\Phi_p = -\frac{kT}{q} \ln\left(\frac{N_a}{n_i}\right) \tag{3.7}$$

$$\Phi_n = -\frac{kT}{q} \ln\left(\frac{N_d}{n_i}\right) \tag{3.8}$$

$$\gamma = \frac{\sqrt{2 \cdot q \cdot \varepsilon_0 \cdot \varepsilon_{Si} \cdot N_a}}{C_{ox}} \tag{3.9}$$

q	$= 1.602 \cdot 10^{-19}$ [C]	electronic charge
k	$= 8.62 \cdot 10^{-5}$ [eV/K]	Boltzmann's constant
T	= [K]	absolute temperature
e_0	$= 8.854 \cdot 10^{-14}$ [F/cm]	permittivity of vacuum
e_{Si}	= 11.7	relative permittivity of Si
e_{ox}	= 3.9	relative permittivity of SiO_2
N_a	= [atoms/cm^3]	doping density of the p-type substrate
N_d	= [atoms/cm^3]	doping density of the n-type substrate
Ni	$= 1.45 \cdot 10^{10}$ [cm^{-3}]	intrinsic carrier concentration
C_{ox}	$= \varepsilon_0 \cdot \varepsilon_{ox} / t_{ox}$	oxide capacitance per unit area
t_{ox}	= [nm]	oxide thickness

Φ_p is the p-type substrate potential, about 0.6 V, whereas V_{T0} is the threshold without the contribution of the body (i.e. with $V_{SB} = 0$). For a rough estimation, it is useful to consider the variation of the threshold as proportional to the square root of the voltage between bulk and source multiplied by the body effect coefficient γ. Such parameter ranges between 0.3 and 1.0 $V^{1/2}$ in a typical CMOS process.

Problem 3.1: Identify some circuits in which the body effect is disadvantageous and others in which, on the contrary, such effect is useful. Discuss the circuits identified.

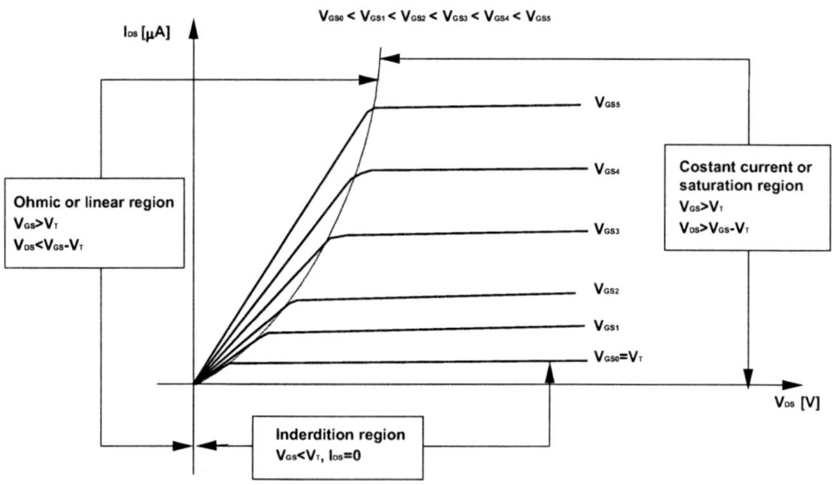

Fig. 3.4. Output characteristic for an NMOS transistor

In Fig. 3.4 the output characteristic of an NMOS transistor is reported. In the linear region (or triode) the following relationships exist:

$$V_{DS} < V_{GS} - V_T \tag{3.10}$$

$$I_{DS} = \frac{1}{2} \cdot \mu_n \cdot C_{ox} \cdot \left(\frac{W}{L}\right) \cdot \left[2 \cdot (V_{GS} - V_T) \cdot V_{DS} - V_{DS}^2\right] \tag{3.11}$$

The symbol μ_n is the electron mobility that, in silicon, is 1417 $cm^2V^{-1}s^{-1}$. The W/L ratio is called aspect ratio of the transistor, where W is the channel width and L is the channel length.

In the saturation region in which $V_{DS} \geq (V_{GS} - V_T)$, Eq. (3.11) is no longer valid and the drain current can be expressed as:

$$I_{DS} = \frac{1}{2} \cdot \mu_n \cdot C_{ox} \cdot (V_{GS} - V_T)^2 \cdot (1 + \lambda \cdot V_{DS}) \tag{3.12}$$

The term $(1 + \lambda V_{DS})$ accounts for the fact that in saturation region the transistor characteristic is not parallel to the x-axis but, on the contrary, there is a certain slope (the value of λ ranges between 0.03 and 0.005 V^{-1}). The parameter λ is usually referred to as channel modulation parameter. Finally, the important parameters for frequency analysis are the gate, gate-to-source, and gate-to-drain parasitic capacitance.

At this point, the transconductance and output conductance of the small signal model g_m and g_{ds} can be calculated in the two working regions.

In the linear region, using Eq. (3.11) we obtain:

$$g_m = \mu_n \cdot C_{ox} \cdot \left(\frac{W}{L}\right) \cdot V_{DS} \tag{3.13}$$

$$g_{ds} = \mu_n \cdot C_{ox} \cdot \left(\frac{W}{L}\right) \cdot (V_{GS} - V_T - V_{DS}) \tag{3.14}$$

Starting from Eq. (3.12), the transconductance in saturation region can be calculated:

$$g_m = \mu_n \cdot C_{ox} \cdot \left(\frac{W}{L}\right) \cdot (V_{GS} - V_T) \cdot (1 + \lambda \cdot V_{DS}) = \frac{2 \cdot I_{DS}}{(V_{GS} - V_T)} \tag{3.15}$$

$$g_{ds} = \frac{\lambda \cdot I_{DS}}{(1 + \lambda \cdot V_{DS})} \tag{3.16}$$

Similar expressions can be obtained for p-channel transistors, keeping in mind that $V_{T,p}$ is negative and that in silicon μ_p equals 471 $cm^2V^{-1}s^{-1}$.

Problem 3.2: The name of the working regions (linear and saturation) for MOS transistors is inverted with respect to bipolar transistors: find out why.

3.2 Transistors Available

The transistors that the designer has at his disposal depend on the kind of process and on the performance required to realize the memory device. For a device operating at low supply voltage, for example, it is important the use of transistors having different thresholds.

Hereafter a description of the most used types of transistors will be given: p-channel transistors with thin and thick oxide, n-channel transistors fabricated in both common substrate and local triple well p-type tubs, called ip-well for sake of clarity. N-channel transistors can be LVS type (Light Voltage Shift) or NAT (Natural) depending on the kind of threshold required, and, moreover, the LVS can be either HV or LV, depending on the oxide thickness. The transistors with thin oxide have the advantage of a lower threshold voltage than those with thick oxide, but they are not able to manage voltages above VDD, and, typically, they also have shorter minimum channel length. N-channel natural transistors (NAT) have threshold voltage than is even lower than LVS, and, generally, only the low voltage version is fabricated. The HV, LV and NAT versions are also available for p-channel transistors. Table 3.1 summarizes the possible types of transistors with the corresponding symbol that will be found in the schematics of the following chapters. In the table, the values of the threshold voltage at room temperature, the oxide thickness for a typical process, the value of the body effect coefficient γ, and, finally, the value of the current that a square of transistor is able to source or sink in saturation with V_{GS} and V_{DS} equals to VDD are reported.[2]

Table 3.1. Typical transistors available in a CMOS process and main useful parameters

Name	Symbol outside and inside triple well	Threshold typical @ 27 °C [V]	T_{ox} thickness gate oxide [Å]	Body effect coefficent $[V]^{1/2}$	I_{DS} current/square [μA/square]
NMOS LVS LV	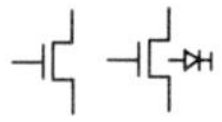	0.6	120	0.95	360
NMOS LVS HV	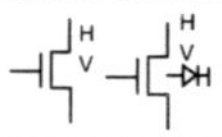	0.8	250	0.96	300
NMOS NAT LV	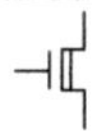	-0.1	120	0.41	380
PMOS LVS LV	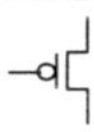	-0.6	120	0.42	150
PMOS LVS HV	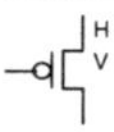	-0.8	250	0.6	130
PMOS NAT LV		-1.5	120	0.42	250
NMOS DEP LV	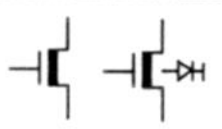	-3.5	120	0.9	360

[2] Such parameters are specific to each process and in this case must be considered merely as an example, to provide some figures to discuss.

The circuit design must be based on actual measurements of the transistors from recently processed wafers, to insure the correct dimensioning of the circuitry.

It is important to state that the transistors do not act as ideal current sources. As we know, two well defined operating regions exist, the saturation region, in which the current supplied by the transistor I_{DS} is nearly constant with respect to the voltage V_{DS}, and the linear (Ohmic) region, in which the behavior of the transistor is resistive.

Let's consider a circuit like the one in Fig. 3.5 in which M1 is used to charge the capacitor C_{LOAD}, with V_{IN} = VDD; if M1 works as an ideal current source having value of I_L, we obtain:

$$I_{LOAD} = C_{LOAD} \cdot \frac{dV_{OUT}}{dt} \tag{3.17}$$

$$\Delta V_{OUT} = \frac{I_{LOAD}}{C_{LOAD}} \cdot \Delta t \tag{3.18}$$

with the initial condition of V_{OUT} = 0V and t = 0. Thus, the relationship between the voltage of the output node and the charging time is linear.

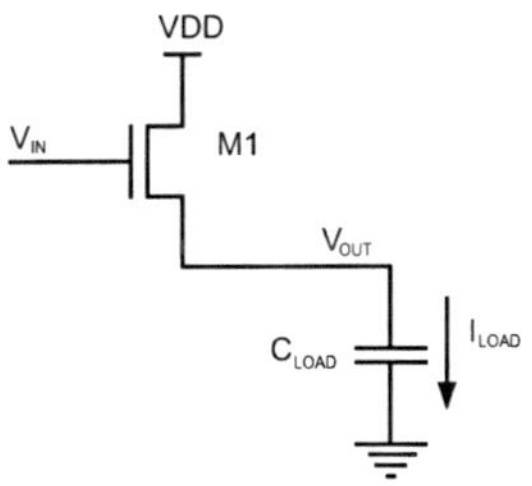

Fig. 3.5. N-channel transistor used to charge a capacitor

In reality, the capacitor is initially uncharged, the current starts flowing and the output voltage increases, varying V_{GS} and V_{DS} of the transistor according to the load characteristic shown in Fig. 3.6. The charge finishes when M1 switches off, i.e. when V_{OUT} reaches the value of VDD minus the threshold voltage of M1.

Moreover, the body effect has to be taken into account, since it causes the increase of the threshold voltage of M1 as the output voltage V_{OUT} increases, slowing down the charging of the output node and stopping it at a value lower than VDD – V_{T0}.

Problem 3.3: Determine the value of V_{OUT} taking into account the body effect.

Let's analyze how the charge of C_{LOAD} occurs through a p-channel transistor M2. In this case, V_{IN} = GND, V_{GS} is fixed because it depends on VDD, not on V_{OUT} like in the case of the n-channel transistor. The capacitor is charged with a constant current until V_{DS} of M2 is in saturation region.

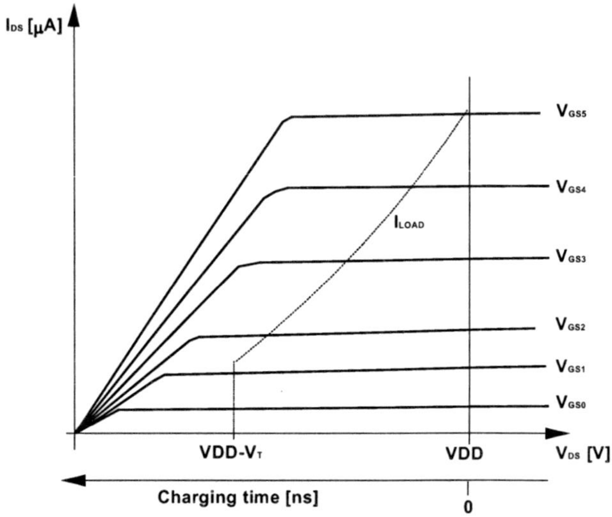

Fig. 3.6. M1 current during C_{LOAD} charge

After that, in the linear region, the output voltage transient is of exponential type, considering that the transistor dynamic impedance varies for each value of V_{DS} of M2. In this case, I_{LOAD} characteristic is represented by the curve named V_{GS5} in Fig. 3.7.

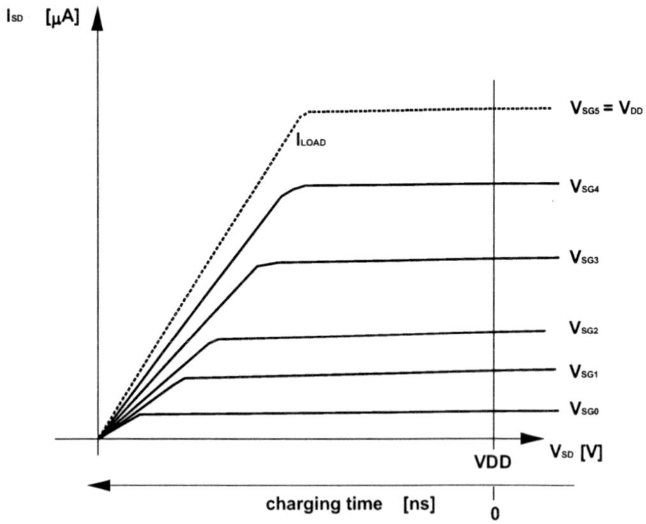

Fig. 3.7. Charging of a capacitor through a PMOS

Let's give another example, considering the circuit in Fig. 3.8, used as voltage reference with $V_{OUT} < V_{IN}$.

For M1, the biasing conditions are $V_{GS} = V_{IN}$ and $V_{DS} = V_{OUT}$, while for M2, $V_{GS} = (V_{IN} - V_{OUT})$ and $V_{DS} = (VDD - V_{OUT})$. M1 is in linear region, being:

$$V_{DS} = V_{OUT} < V_{IN} = V_{GS} \tag{3.19}$$

whereas M2 is in saturation

$$V_{DS} = VDD - V_{OUT}; V_{GS} = V_{IN} - V_{OUT}; VDD \geq V_{IN} \tag{3.20}$$

Recalling Eqs. (3.11) and (3.12) the currents that flow in M1 and M2 can be calculated.

$$I_{D2} = \frac{\beta_2}{2} \cdot \left(V_{IN} - V_{OUT} - V_{T,nat}\right)^2 \tag{3.21}$$

$$I_{D1} = \beta_1 \cdot \left[\left(V_{IN} - V_{T,nat}\right) \cdot V_{OUT} - \frac{1}{2} \cdot V_{OUT}^2\right] \tag{3.22}$$

$$\beta = \mu \cdot C_{ox} \cdot \left(\frac{W}{L}\right) \tag{3.23}$$

Where $V_{T,nat}$ is the threshold voltage of the natural transistor considered. Equaling the two currents and assuming $V_{T,nat} = 0$, after some manipulations the following relation is obtained:

$$V_{OUT} = V_{IN} \cdot \left(1 - \frac{1}{\sqrt{\rho + 1}}\right); \rho = \frac{\beta_2}{\beta_1} \tag{3.24}$$

Table 3.2 reports simulated and calculated values for V_{OUT} at different values of ρ.

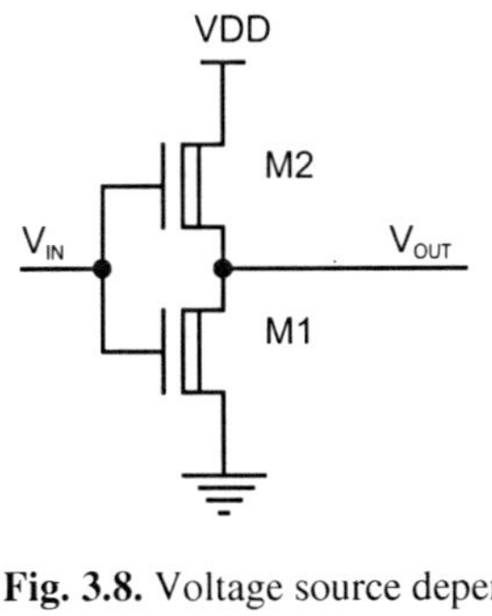

Fig. 3.8. Voltage source dependent on V_{IN} and on the size ratio between M1 and M2

Problem 3.4: Remake all the calculations of the theoretical V_{OUT} taking into account the body effect of M2.

Table 3.2. Simulated and calculated values of V_{OUT} as the size ratio of transistors M1 and M2 of Fig. 3.8 varies.

W/L M2 [μm]	W/L M1 [μm]	ρ	V_{IN} [V]	V_{OUT} theoretical [V]	V_{OUT} simulated [V]	VDD [V]
20/3	20/3	1	5	1.46	1.3	5
10/3	20/3	0.5	5	0.91	0.8	5
20/3	10/3	2	5	2.1	2.1	5
3/20	3/20	1	5	1.46	1.1	5
3/40	3/5	0.125	4	0.2	0.1	5
3/40	3/3	0.075	3	0.1	0.1	5
40/3	3/5	22	5	4	4.1	5

3.3 The Memory Cell

The memory cell we will deal with is the non-volatile cell, which is able to retain the information even without supply voltage. This is possible due to the insulated gate. Several versions of the non-volatile cell, based on the principle of the floating gate, exist. We will deal in detail with the cell named "T" because of the geometric shape, which constitutes the basic element of the array organization of the so-called NOR-type non-volatile memory. The picture or, better, the layout of the cell is shown in Fig. 3.9 with its basic features. It stores the single bit of information, the reading of several bits in parallel allows obtaining the bytes and the words.

The single cell structure is repeated to create the array that constitutes the bank of memory. Figure 3.10 shows an array of eight cells. As it can be noted, the contact is shared between two cells, which allows reducing the overall area by diminishing the number of contacts and, moreover, increasing the reliability of the device since the contact are critical in the fabrication process. The consequent parallelism of cells is often paid from the electrical point of view as we will see later on.

The cells that are on the same row of poly2 have also the same source junction, which is also shared with the cells of the following row. A source contact at intervals[3] guarantees the connection of the source to ground or to a fixed potential.

[3] Generally every 16 cells.

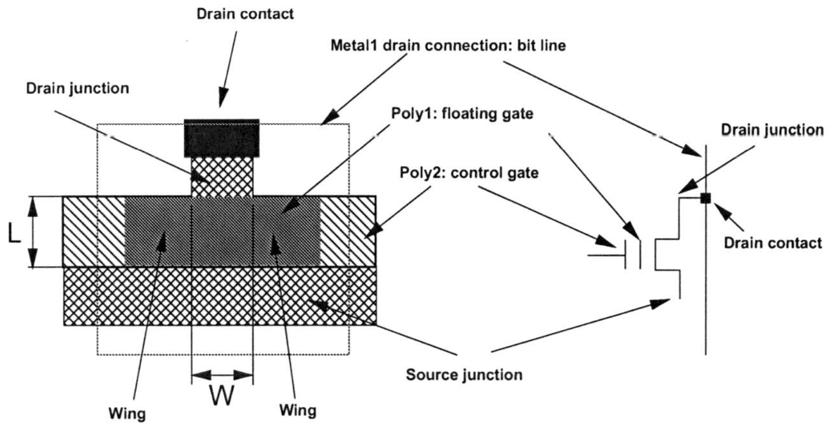

Fig. 3.9. Layout of the non-volatile cell and corresponding schematic. The size of L and W are presently less than one micron. The contacts are drawn as square even though they actually result in a round shape on silicon.

Particular attention is then paid to the side cells, the cells that constitutes the rows and columns placed at the edge of the array. As in any repetitive structure, breaking the symmetry at the end of the framework causes non-uniformity that results in memory cells that are different in terms of size and characteristics from those inside the array. Some dummy rows and columns are present but not electrically connected, with the function of reducing the edge effect for the used cells.

Problem 3.5: How should side rows and columns be fabricated in an EPROM cell array and in a Flash cell array?

The starting point for the analysis of a memory cell with floating gate is the equivalent capacitance one-dimensional model depicted in Fig. 3.11. In this model, merely electrostatic, the four electric terminals are coupled to the floating one by means of capacitors that can be derived assuming flat and parallel planes. Therefore, the capacitance depends on the thickness of the dielectric and on the area of overlap between floating gate and electrodes.

Considering that a certain charge Q is present on the floating gate, it is possible to use the relationship among capacitance, charge, and difference of potential at the sides of a capacitor to calculate the potential of the floating gate V_{FG} with reference to the external potentials:

$$\begin{aligned} Q = C_{FC} \cdot (V_{FG} - V_{CG}) + C_S \cdot (V_{FG} - V_S) + \\ + C_D \cdot (V_{FG} - V_D) + C_B \cdot (V_{FG} - V_B) \end{aligned} \tag{3.25}$$

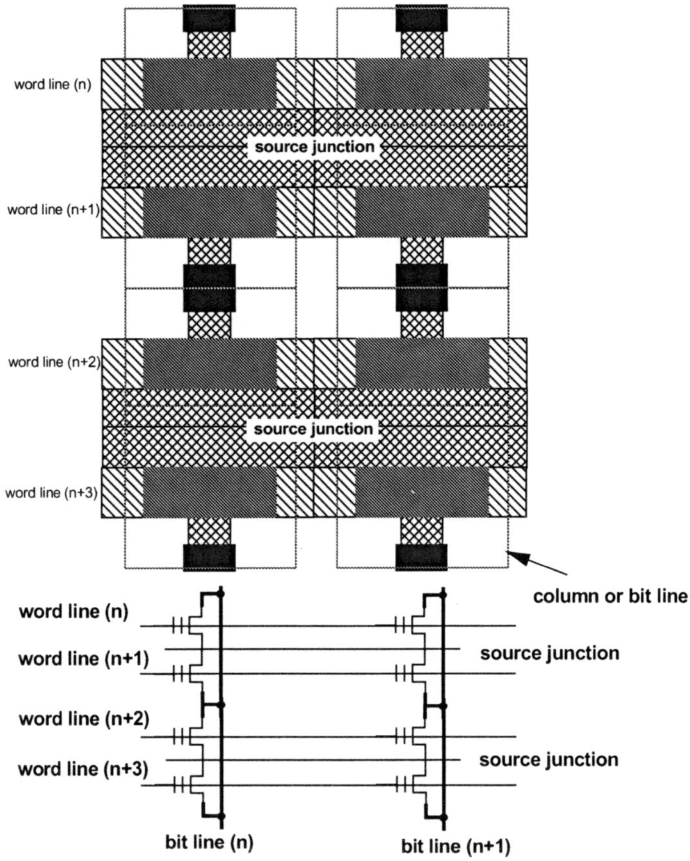

Fig. 3.10. Layout of eight cells connected so as to form an NOR-type array, with the source shared between two rows of cells and the drain contact shared between two single cells

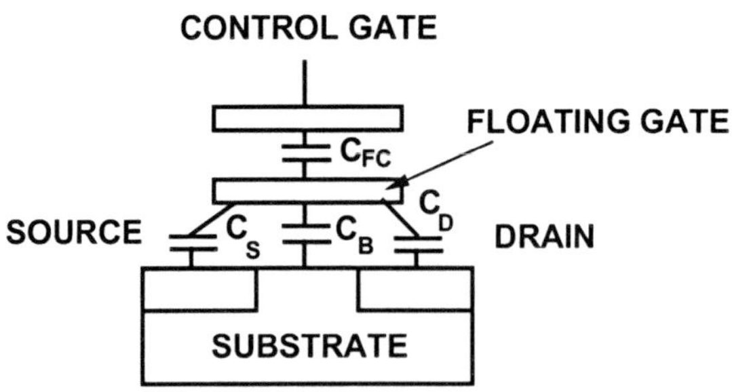

Fig. 3.11. One-dimensional model for floating gate transistor

At this point, we can use the following definitions:

$$C_T = C_{FC} + C_D + C_S + C_B \tag{3.26}$$

$$\alpha_S - \frac{C_S}{C_T}; \alpha_D = \frac{C_D}{C_T}; \alpha_B = \frac{C_B}{C_T}; \alpha_G = \frac{C_{FC}}{C_T} \tag{3.27}$$

to obtain:

$$V_{FG} = \alpha_S \cdot V_S + \alpha_D \cdot V_D + \alpha_B \cdot V_B + \alpha_G \cdot V_{CG} + \frac{Q}{C_T} \tag{3.28}$$

It is interesting to point out that the behavior of the cell is identical to the n-channel transistor having the gate terminal at V_{FG}, i.e. the polarization of the electrodes. This observation is the essential principle of operation of the Flash cell.

Let's analyze the behavior of the cell in the linear region. Substituting the voltage of the floating gate in Eq. (3.11), we get:

$$I_D = \beta^{FG} \cdot \left[\left(V_{FG} - V_T^{FG} \right) \cdot V_D - \frac{V_D^2}{2} \right] \tag{3.29}$$

The superscript FG indicates quantities that refer to the floating gate. With the source and bulk at ground potential, it is possible to calculate the drain current of the cell with reference to the external potentials only. Substituting Eq. (3.28) into Eq. (3.29) we have:

$$I_D = \beta^{FG} \cdot \left[\left(\alpha_D \cdot V_D + \alpha_G \cdot V_{CG} + \frac{Q}{C_T} - V_T^{FG} \right) \cdot V_D - \frac{V_D^2}{2} \right] \tag{3.30}$$

Now it is possible to define the parameters β and V_T with respect to the control gate, i.e.:

$$\beta^{CG} = \alpha_G \cdot \beta^{FG}; V_T^{CG} = \frac{1}{\alpha_G} \cdot \left(V_T^{FG} - \frac{Q}{C_T} \right) \tag{3.31}$$

In the case Q = 0, it is simple to give a circuit representation of the definition of the threshold voltage with respect to the control gate, by considering the capacitive divider formed by C_{FC} and C_T and imposing that the voltage of the floating gate equals the threshold voltage. It is now possible to rewrite the drain current of the cell in linear region only:

$$I_D = \beta^{CG} \cdot \left[\left(V_{CG} - V_T^{CG} \right) \cdot V_D + \frac{1}{\alpha_G} \left(\alpha_D - \frac{1}{2} \right) \cdot V_D^2 \right] \tag{3.32}$$

Problem 3.6: Write the working equations of the floating gate cell in saturation region. Can the cell switch on also for gate voltage lower than V_T?

3.4. Reading Characteristics

Due to the array organization, the cells that are on the same column share the bias contact of the drain, whereas the cells placed on the same row share the same gate contact, as depicted in Fig. 3.12[4].

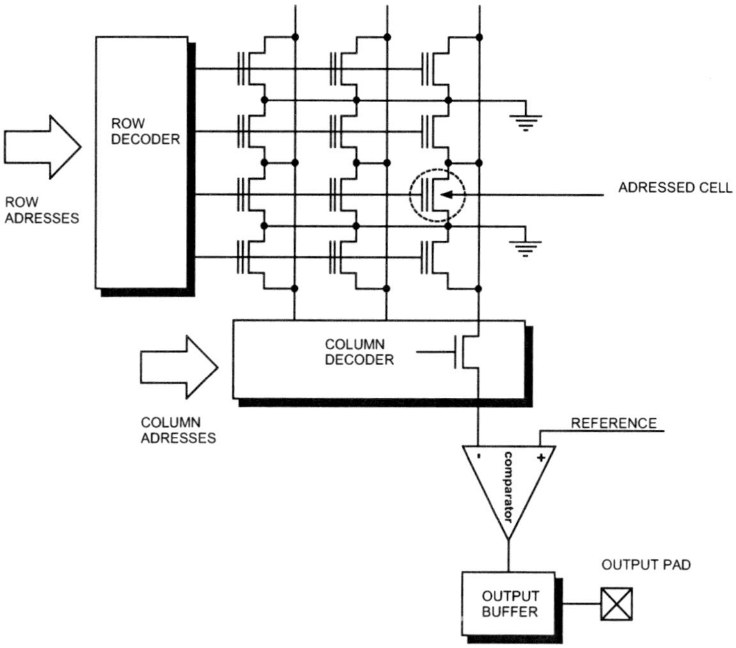

Fig. 3.12. Array organization: the cells are connected in the x-direction, i.e. the row, by the gate and in the y-direction, i.e. the column, by the drain contact. The decoding univocally allows addressing each single cell of the array.

Thus, in order to program the cell shown in the figure, the gate must be biased at about 12 V and the drain at about 5 V. Obviously, in such a condition, all the cells on the same row have the gate at 12 V and all those on the same column have the drain at 5 V. Therefore, a cell having gate at ground but belonging to the same column of the cell that is being programmed has the drain at 5 V and suffers the effect of drain stress that tends to erase it. In Fig. 3.13 the impact of such effect on programmed cells for a typical process is shown.

If the columns contain many cells and all of them are to be programmed, in the worst case the maximum programming time allowed, say 250 μs, will be neces-

[4] The figure anticipates many issues, but what is important at this point is the organization of the array that allows many cells to be connected to the same drain contact.

sary for each cell. Assuming a column of 1024 cells, the time of drain stress t_{ds} applied to the first cell of the column equals:

$$t_{ds} = 1023 \cdot 250\mu s \cong 0.256s \tag{3.33}$$

This kind of stress should not lead to any problem but, in case the drain voltage were 5 V also during the reading phase there would be a loss of intrinsic charge due not only to the tunnel oxide retention but also to the electric field applied. The characteristic shown in Fig. 3.13 refers to a typical "good" cell, but the situation may be worse by several orders of magnitude for defective cells in terms of threshold shift.

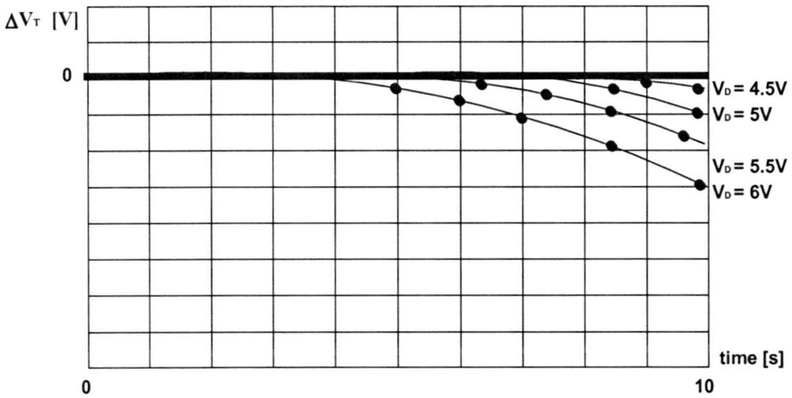

Fig. 3.13. Impact of the drain stress on a programmed cell. The deviation from the curve corresponding to zero volts indicates the percentage of charge loss. The family of curves has the drain voltage as parameter.

The effect of the drain stress can potentially occur even at low drain potentials. Thus, it is important to choose values of the drain potential that do not induce stress to the cells, which would lead to loss of information. In practice, the voltage of the drain terminal is driven to 1 V, to the maximum[5]. Considering that the voltage applied between control gate and source during the reading is generally greater than 4 V, this causes the cell to operate in the linear region. It is now possible to calculate the transconductance of the cell in linear region (with constant V_D), basing on Eq. (3.32):

$$g_m = \frac{\partial I_{DS}}{\partial V_{CG}} = \beta^{CG} \cdot V_D \tag{3.34}$$

It is possible to observe that the cell transconductance is independent from the content of charge of the floating gate, resulting in the same value for erased or programmed cells.

[5] What is stated here will become more evident after studying the chapter that deals with the techniques to read the cells.

Let's define V_{T0} as the threshold of the cell with Q = 0. The step of voltage ΔV_T with respect to the control gate, which differentiates the erased cell (logic "1") from the programmed cell (logic "0"), can be calculated starting from Eq. (3.31):

$$\Delta V_T = V_T^{CG} - V_{T0} = -\frac{Q}{C_{FC}} \tag{3.35}$$

Therefore, the drain current of a cell in linear region results to be:

$$I_D = \beta^{CG} \cdot \left[\left(V_{CG} - V_{T0} - \Delta V_T \right) \cdot V_D + \frac{1}{\alpha_G} \cdot \left(\alpha_D - \frac{1}{2} \right) \cdot V_D^2 \right] \tag{3.36}$$

Once the drain voltage used during reading has been fixed, the characteristics of the "1" and "0" cells result to be parallel and separated by a fixed quantity equal to ΔV_T. The diagram of the Fig. 3.14 reports the typical quantities for the input characteristics on the axes. The axis of abscissas can be regarded as both V_{GS} and bias voltage VDD, since it is suitable to drive the gate terminal to the maximum voltage available, which is generally the bias voltage.

For sake of simplicity, hereafter the superscript CG will be omitted and V_{CG} will be referred to as V_{GS} considering the source at ground.

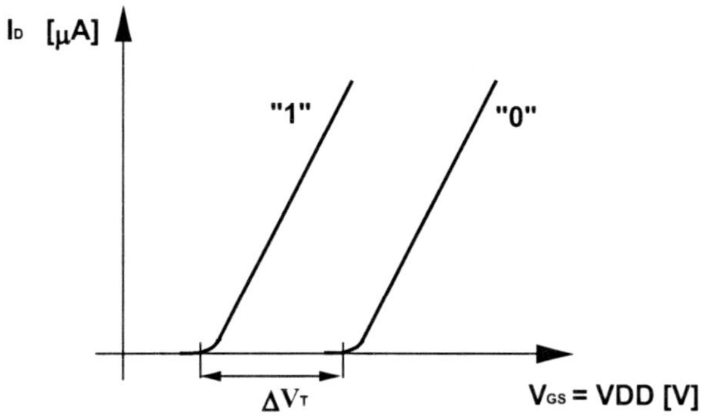

Fig. 3.14. The characteristics of programmed or erased cells result to be parallel during the read operation

3.5 Programming

The writing operation, or programming for a Flash cell is carried out by means of the so-called hot electrons that are able to overcome the energy barrier corresponding to the thin oxide between the drain area and the insulated gate. Thus, the two terminals of the cell, source and drain, are used in different ways during the two operations: during the erase the electrons flow from the insulated gate to the source, whereas, during the program, the electrons flow from drain to gate. Such specialization of the two terminals allows optimizing the device for the specialization of the two terminals allows optimizing the device for the two different operations[6]. Let's now examine the program mechanism.

Let's consider an NMOS transistor (the Flash cell is an n-channel transistor) biased with a difference of potential between the gate and source terminals greater than its threshold voltage. In such a condition, minority carriers (electrons) are present all along the channel. If the drain junction is reverse-biased, some minority charges are drawn from the nearby region of channel and, thus, a reduction in the electron density along the channel from source to drain is determined. We define pinch-off voltage the V_{DS} bias for which complete depletion of the channel is achieved at the drain junction. If the V_{DS} voltage is further increased, the depleted region expands toward the source and its extremity is called pinch-off point. The presence of such a point defines the saturation condition for a transistor. In the specific case of the program operation, the cell is always in saturation region.

The longitudinal component of the electric field, always pointing from the source to the drain node, has a very intense gradient in the depleted region of the drain. Due to this unhomogeneity, the profile of the energy distribution of the electrons varies significantly along the direction parallel to the channel.

In general, conduction electrons tend to maintain thermal equilibrium with the lattice, yielding the energy that they acquire from the electric field through collisions with acoustic phonons, impurity, or other electrons. As a consequence, their average energy keeps around (3/2) KT, where K is the Boltzmann constant and T is the temperature of the lattice, while their speed varies linearly with the applied electric field. However, increasing the intensity of the field beyond a given limit, some electrons acquire more energy from the field than they can lose (no scatter mechanism is effective). In practice, these carriers are no longer in thermal equilibrium with the lattice and are called hot electrons.

In order to describe the kinetic of the hot electrons, the Fermi-Dirac energy distribution is introduced, similarly to what happens to the charge in thermal equilibrium, but with a specific associated temperature $T_e > T$. Silicon lattice, having atomic density of $5 \cdot 10^{22}$ atoms/cm^3, is characterized by electrons with covalent bonds (Fig. 3.15). Thermal energy allows some of these electrons to break the atomic bonds and, hence, they are free to move throughout the lattice, passing in conduction band. In a semiconductor, together with the electrons, also holes exists, i.e. absence of electrons, which can be represented as positive charge carriers, that move in the opposite way in the valence band. An important property of the electrons in the lattice is their distribution with respect to the allowed energy states, in the condition of thermal equilibrium. The Fermi-Dirac distribution function, f_{FD}, provides the probability that an energy status, E, is occupied by an electron:

$$f_{FD}(E) = \frac{1}{1 + e^{(E-E_f)/kT}} \tag{3.37}$$

where E_F is the Fermi energy (or level), corresponding to a 50% probability that an electron occupies the related energy status. In an intrinsic semiconductor, i.e. not doped, the number of carriers in the conduction band (electrons) equals the num-

[6] With the introduction of the array in the triple well, the erasing can take place along the entire channel length, making the specialization of the source no longer necessary.

ber of carriers in the valence band (holes), and the Fermi-Dirac function is symmetric with respect to that level. In a doped semiconductor, instead, the Fermi level shifts as a function of the type and quantity of doping present.

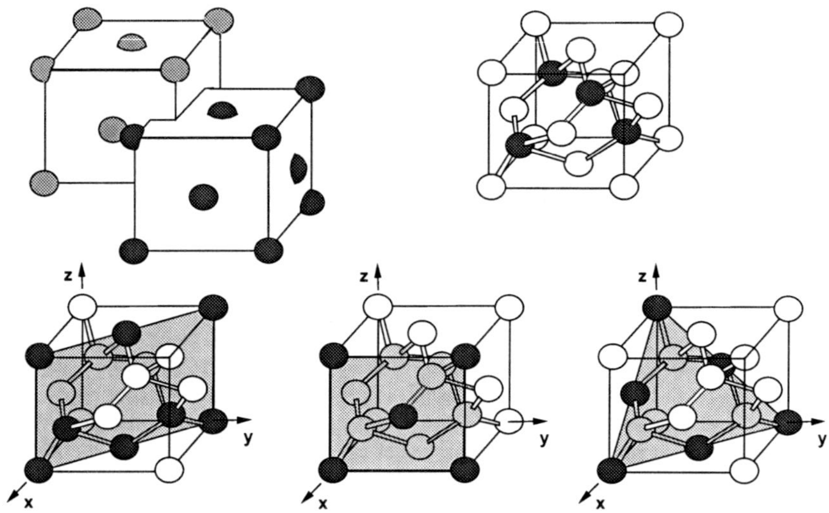

Fig. 3.15. Silicon crystallizes in a diamond structure, which is composed of two cubic lattices with centered sides, shifted with respect to each other by a quarter of diagonal. The different planes of the crystal and the respective Miller indexes are reported

The idea of attributing an electronic temperature, T_e, greater than the lattice one so as to characterize the condition of the hot electrons is not completely correct, since temperature is a concept typical of the thermodynamic equilibrium. However, it is useful to understand the drift of the carriers in presence of strong electric fields.

Hot electrons are responsible for a series of mechanisms that, in various ways, feed the channel, substrate, and gate currents. In Fig. 3.16, the effects that contribute to the gate current are highlighted. Let's now analyze in detail the phenomena of multiplication of the hot electrons in the channel.

Let's considering the phenomenon of the generation of electron-hole pairs due to impact ionization. Impact ionization is a process of coulombian interaction among electrons that activates in presence of high electric field. In practice, it happens that a very energetic electron, impacting against an electron that is in the valence band, yields enough energy for it to pass in the conduction band. At the end of the interaction, both the electrons are in the conduction band while a hole has been generated in the valence band. The energy threshold for which such a process takes place is nearly equal to 3/2 of the silicon energy gap[7]. Considering that the gate current is around some nanoAmperes and the substrate current around microAmperes, it descends that, for the channel electrons, the ionization phenomenon is favored from the energy point of view with respect to their passage across the tunnel oxide. The negative charges, e_1, that originate due to primary multiplication, increase the population of electrons in the channel, phenomenon also referred to as C.H.E.I.A. (Channel Hot Electron Induced Avalanche). On the other hand, the ionization process provokes loss of energy. The electrons produced are re-accelerated by the longitudinal electric field and their probability of crossing the oxide voltage gap is concentrated in the final part of the channel. In the region of channel where the transverse electric field is direct toward the floating gate, there is a non-null probability that the holes produced by ionizing impacts may cross the oxide, providing a negative contribution to the gate current. It is necessary to bear in mind that they have to pass an energy barrier of 4.7 V, instead of the 3.2 V that is necessary for the electron injection and, furthermore, their effective mass is higher.

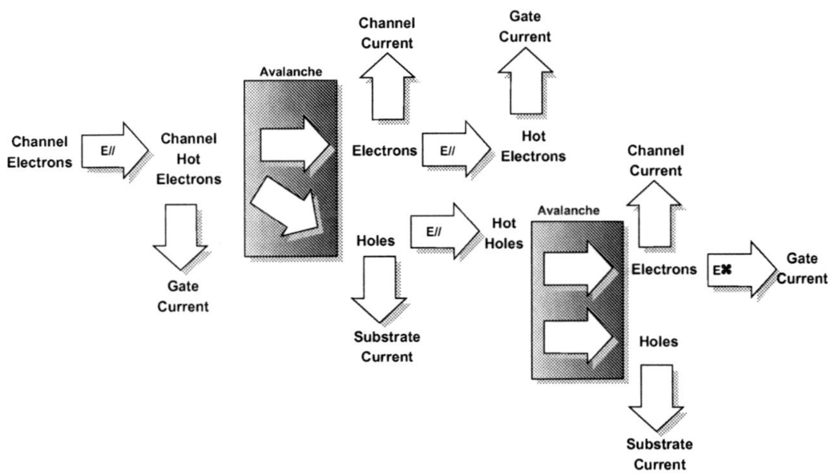

Fig. 3.16. Mechanism of charge injection onto the floating gate

As a consequence of the analyzed mechanisms, the amount of electrons (primary or generated by ionization) that are able to cross the tunnel oxide concentrates close to the drain junction. Recently, a physical mechanism known as C.H.I.S.E.L. (Channel Initiated Secondary Electron Injection) has been identified, which is able to produce an electron flow toward the gate, and whose peak of intensity is located far from the drain junction. The holes produced by the first impact ionization acquire enough energy from the electric field direct toward the substrate to start a new multiplication phenomenon. Due to the presence of a suit-

[7] This derives from the momentum and energy conservation equations, assuming that all the carriers have the same final speed and the same mass.

able electric field, the electron generated by this mechanism are heated and attracted by the channel where they can be injected onto the floating gate. Obviously, in the first phase of the programming, this contribution is negligible, since it depends upon the probability that multiplication phenomena occur. However, as the programming operation goes on, the transverse electric field strongly opposes against the electron injection and, thus, the injection efficiency due to the C.H.E. and C.H.E.I.A. loses the most consistent component. The flow of the electrons produced through C.H.I.S.E.L. is reduced by a lower amount, since it is located in a portion of channel closer to the source, where the direction of the electric field is favorable. In this situation, the contribution due to the secondary electrons becomes more important in the expression of the overall gate current. The phenomenon described above can be regulated by means of the substrate bias. Increasing the difference of potential applied to the drain-substrate junction, the intensity of the transverse field below the channel is also increased, which is the fundamental element to the effectiveness of the injection mechanism of secondary electrons.

In conclusion, the ionization phenomenon in the channel causes an increase in the amount of conduction electrons and in the decrease of their average energy.

The tail of hot electrons is hence remarkably increased nearby the drain junction and, as a consequence, the charge flow toward the floating gate. During program, when the floating gate voltage reaches the imposed value, V_{DS}, the transverse electric field dramatically reduces the charge injection in the channel region close to the drain, causing a quick reduction in the gate current.

It is evident that the charge flow across the oxide tightly depends upon the profile of the energy distribution of the hot electrons. Despite the fact that it is very complicated to express the relationship with the voltage applied, it is evident that an increase in the drain voltage acts to increase the percentage of the electrons having energy greater than the oxide energy gap.

From the operating point of view, the minimum speed with which the device is due to accomplish the program operation is determined, and this also defines the voltage that must be applied. Supposing that the threshold voltage is increased by 3 V from an initial V_T of 2.5 V, the bias configurations that fulfills the requirements can be found. The choice will select the lowest voltages, thus minimizing power consumption and electric stress.

The flow of secondary electrons across the tunnel oxide induced by the negative bias of the substrate is one of the most popular approaches to reduce the voltages applied to the cell terminals. Of course, the technological process must account for the possibility of fabricating the array in triple-well as highlighted in Fig. 3.17. By reducing the voltage of the control gate and the drain, it is possible to reduce the intensity of the electric field across the tunnel oxide, with the consequent improvement in terms of gate and drain stress (see Chap. 20).

In practice, the negative substrate bias reduces the voltage of the pinch-off point, increasing the longitudinal component of the electric field. Moreover, owing to the effect on the threshold voltage (the Flash cell is an NMOS transistor), a decrease in the substrate voltage greatly reduces the overall drain current during program, with the consequent power saving.

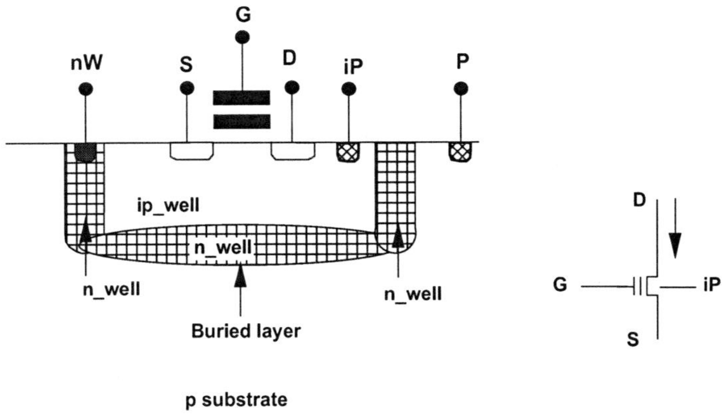

G	D	S	iP	nW	P
12V	5V	0V	-5V	5V	0V

Fig. 3.17. The substrate of the array (ip-well) is insulated from the substrate of the rest of the circuitry by means of the n-well tub. In this way, it can be biased with a negative voltage during program to increase the writing effectiveness and diminish the drain current of the cell.

In Fig. 3.18 the characteristic of the threshold voltage step, ΔV_T, is reported versus the programming time in the case of cell with grounded substrate: this is the so-called program curve. We recall that the threshold voltage of the cells is varied by the gate current by means of charge accumulation on the floating gate.

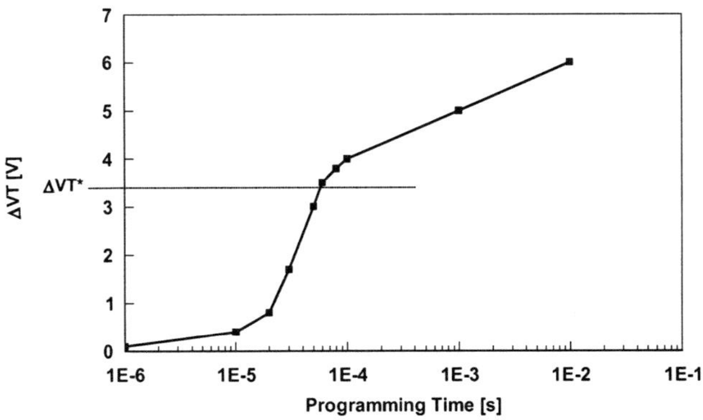

Fig. 3.18. Program curve of a Flash cell

In Fig. 3.18 two different zones can clearly be distinguished. The first in which there is a fast increase in V_T in a short time; the second in which there is a slower increase and a logarithmic function of the time. The point of separation between the two zones corresponds to the condition in which V_{FG} equals V_{DS}, i.e. the voltage of the floating gate equals the voltage of the drain, and, on top of that, it corresponds to the threshold voltage step also known as ΔV_T^*.

Let's suppose to apply a V_{GS1} gate voltage and a V_D drain voltage to an erased cell. The writing by means of hot electrons will continue until the voltage of the floating gate equals the drain voltage. If, on the contrary, the gate voltage applied is $V_{GS2} > V_{GS1}$, we obtain a higher initial gate voltage of the floating gate. As a consequence, in order to reach the ΔV_T^* condition, we have to inject more electrons onto the floating gate, which results in a higher threshold voltage of the written cell.

In the case in which $V_{FG} > V_D$, the conditions favor the charge injection onto the floating gate, whereas if $V_{FG} < V_D$, the charge that has already been injected provokes a reduction in the potential of the floating gate below the value imposed by the drain. The part of the channel closer to the drain diffusion has higher voltage than the floating gate, which does not allow the majority of the hot electrons (generated in this zone of high longitudinal electric field) to cross the oxide[8]. Thus, the slope of the program curve in the zone where $V_{FG} < V_D$ changes, and the programming speed quickly diminishes. In terms of gate current, we obtain that I_G is nearly constant when $V_{FG} > V_D$, whereas the gate current diminishes exponentially in the region where $V_{FG} < V_D$.

Substituting Eq. (3.35) in Eq. (3.28), the potential of the insulated gate can be expressed as:

$$V_{FG} = \alpha_G (V_{CG} - \Delta V_T) + \alpha_D V_D + \alpha_S V_S + \alpha_B V_B \tag{3.38}$$

If we want to calculate the value of ΔV_T^*, i.e. the intrinsic voltage step, we have to insert the value of the potentials in the expression above, i.e. $V_B = V_S = 0$, and express ΔV_T in the hypothesis that $V_{FG} = V_D$

$$\Delta V_T^* = V_{CG} - \frac{1-\alpha_D}{\alpha_G} V_D \tag{3.39}$$

The value of ΔV_T^* is an important indicator of the programming speed[9]. If the minimum threshold voltage to regard the cell as programmed is greater than ΔV_T^*, the cell will follow the knee of the program curve, going in a region where the gate current is very low, with a penalty in terms of programming speed. The problem of the maximization of ΔV_T^* through the variation of the physical parameters that impact on the capacitive ratios poses. What we do is increasing the overlap area between control and floating gate by means of the so-called wings. As it can

---

[8] Near the beginning of the injection process, the inversion layer extends almost all the way to the drain, and the field in the oxide is attractive except for a small portion very near the drain. Current begins to flow through the oxide at the point where the electrons are the hottest and where the oxide field is most favorable. As the floating gate charges up, the floating gate-to-source voltage drops and the drain pinch-off region moves toward the source.

[9] In the cells of the latest generation, such a sharp distinction between the two programming zones no longer exists.

be seen in Fig. 3.9, while the capacitive coupling between control and floating gate increases because of the wings, the coupling with the substrate is nearly constant due to the high thickness of the field oxide underneath. In this way, the α_G coefficient increases and, taking into account Eq. (3.39), also the value of ΔV_T^*.

When operations of modification of the charge content of the floating gate are examined, it is necessary to account for the topological distribution of the cells within the array. For example, the resistive paths that connect the source of the single cells to the ground contact present resistance ranging from a few to several thousands Ohms, depending on the size and the kind of insulation used. This resistance worsens the programming characteristics and further widens the cell distribution. Finally, temperature variations and misalignments of the masks during fabrication worsen the problem.

3.6 Program Algorithm

The circuitry necessary to accomplish the programming is very complex. The advent of technologies with more and more reduced dimensions imposes precise control on the potentials to apply to the cells and to timings. In the case of devices with single supply, another problem for the designer is the cell current consumption during this phase. The control circuitry that governs the operation is realized in the same way also in the case of double supply devices. Major attention is paid to the way in which the program voltages are applied. First, the gate voltage must go high in order for the channel to form, and then the drain node must be pulled up with the program pulse. If the drain went up before the gate, unpleasant accidents might occur to all the cells connected to the column, such as spurious erase or program, *snap-back* phenomena (Fig. 3.19) and degradation of the cell performances. Thus, the control circuitry verifies the correct timing application of the signal, their duration and amplitude.

The instant of application of the drain pulse, with the gate already up, causes the cell to sink very high current, a part of which charges the parasitic capacitance of the bit line, whereas the other part crosses the channel producing the necessary hot electrons. This current peak can be estimated around 1 mA in the worst case.

Single supply devices cannot afford charge pumps that, in the hypothesis that we want to program an entire word, are able to source 16 mA during program with a 5 V output voltage. The first step is the adoption of the program on a byte basis instead of on a word basis, even though this increases the programming time[10].

[10] The programming time directly impacts on the cost of the final system since a part of the memory is usually programmed by the manufacturer. The typical programming time ranges from 5 to 10 μs to program a word.

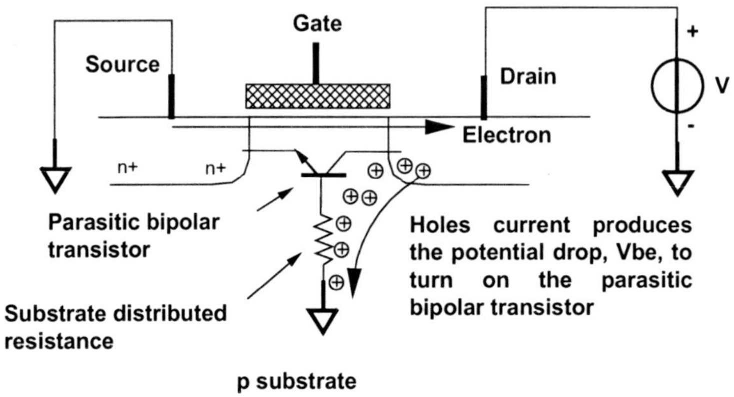

Fig. 3.19. Snap-back effect due to the switching on of the parasitic bipolar associated to each MOS. The potential applied to the drain may cause the breakdown of the junction, injecting holes into the substrate and pulling up the voltage of the base of the *npn* transistor indicated. When the bipolar switches on, the current does not flow along the surface, but is sunk into the ground contact of the substrate. Thus, the potential of the drain node decreases, which switches off the bipolar, and so forth. An oscillation is triggered on the drain node

The introduction of a control algorithm that distinguishes the two bytes that compose the word only in the case that the overall number of non-programmed bits is greater that eight may help. Even in this case, the problem of the power consumption is still present. In order to try to reduce its impact, new circuit configurations may be introduced. The first consist in programming by means of a ramp on the gate, as shown in Fig. 3.20.

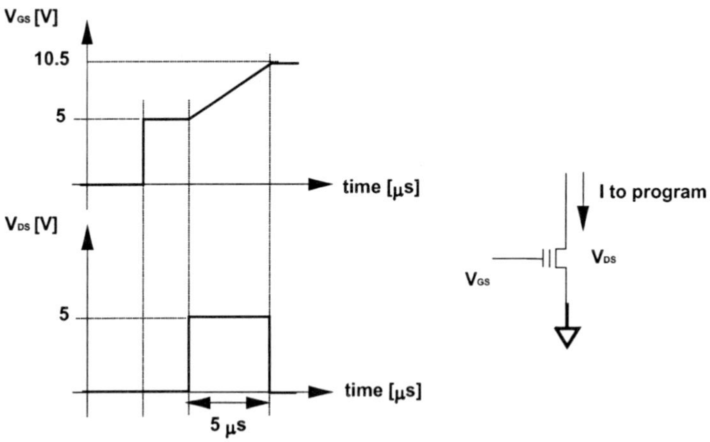

Fig. 3.20. Application of the program pulse during the gate ramp to limit the current consumption

During the first phase, the gate is biased at 5 V, which is enough to create the channel and, subsequently it is pulled to the final value of about 10 V by means of a ramp. During the application of the ramp, the drain pulse is applied that, in the case of the example shown, has a minimum duration of 5 μs. In this way, the initial peak of current is reduced by acting on the value of the potential applied.

As already discussed, the placement of the array in triple-well allows biasing the cell substrate, improving the program effectiveness and, thus, reducing the value of the current sunk to about 100 μA for each cell.

3.7 Erase Operation

Let's now examine the electrical erase operation in detail. Let's start by considering a gated diode, which is a MOS structure where only the diffusion area beneath the gate is active (Fig. 3.21). In our case, the gated diode represents a system composed of the isolated gate and the cell source junction, where the erase phenomenon takes place.

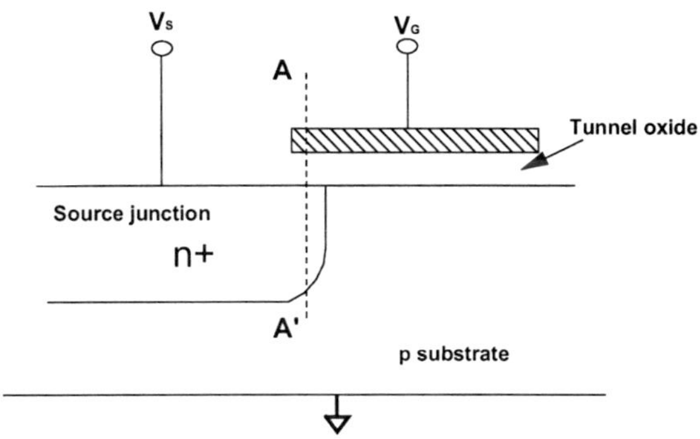

Fig. 3.21. Gate diode structure to study the erase operation

If we consider the cross section along the A-A' dashed line and draw the energy band diagram during erase, we obtain Fig. 3.22. The source has positive potential, the gate has negative potential, whereas the body is grounded. Moreover, the cell is programmed, i.e. the floating gate is charged with electrons. Such electrons have a finite probability of crossing the energy barrier related to the gate oxide, due to the Fowler-Nordheim (FN) tunneling. The gate current associated with the FN tunneling phenomenon can be expressed as:

$$I_G = A_{FN} \cdot E_{ox}^2 \cdot \exp\left(-\frac{B_{FN}}{E_{ox}}\right) \tag{3.40}$$

where E_{ox} is the electric field across the gate oxide (between source and floating gate), while A_{FN} and B_{FN} are constants.

Bands' bending is due to the applied potentials and implies that in the valence band there are electrons having the same energy as those in the conduction band (point A in Fig. 3.23).

The possibility of such an interband electron flow depends on the donor concentration in the silicon that is responsible for the band curvature. In the examined case, the n^+ doping concentration of the cell source is indicated in Fig. 3.23.

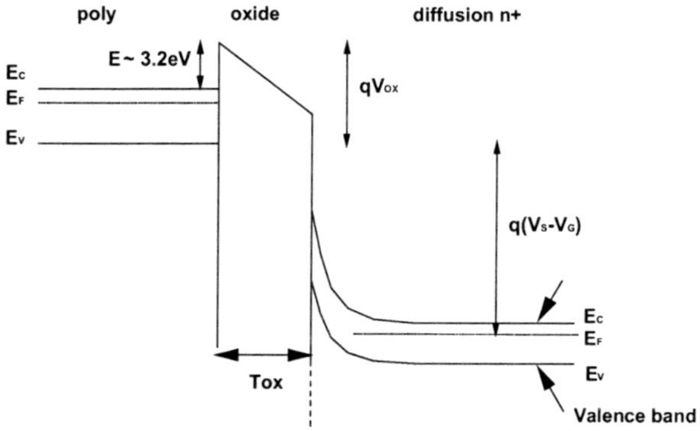

Fig. 3.22. Energy band structure with the applied erase potentials

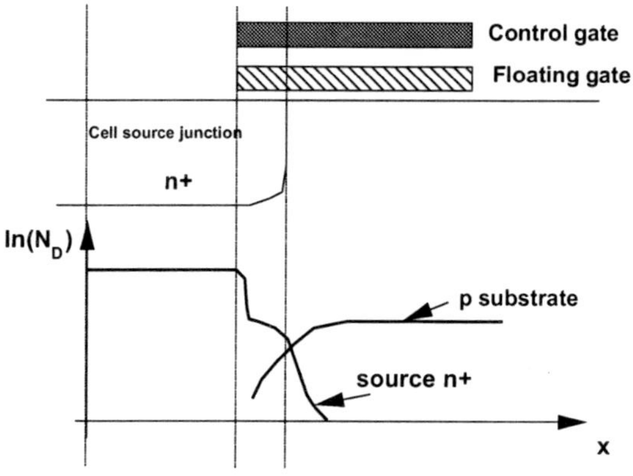

Fig. 3.23. Variation of the n^+ doping concentration along the source-to-gate direction

The voltage drop in the silicon increases as the doping concentration decreases and, thus, an intense electric field is present in the oxide beneath the gate while, on the opposite side, at the source junction end, all the potential drop is located in the silicon. In Fig. 3.24, the region where the interband tunneling phenomenon due to the electrons that leave holes in the valence band can take place is shown. Such holes flow through the depleted region of the substrate into the ground terminal and are responsible for the current called I_{BBT} (Band-to-Band Tunnel). The gate current is around some picoAmperes, while the BBT current is three orders of magnitude larger, i.e. around some nanoAmperes per cell. Also I_{BBT} can be expressed through an expression similar to the one used for the gate current:

$$I_{BBT} = I_S = A_{BB} \cdot E_{Si}^2 \cdot \exp\left(\frac{-B_{BB}}{E_{Si}}\right) \tag{3.41}$$

where A_{BB} and B_{BB} are constant and E_{Si} is the electric field at the silicon surface in the source region underneath the gate. E_{ox} and E_{Si} are bound together by the continuity of the displacement vector at the surface.

$$\varepsilon_{ox} \cdot E_{ox} = \varepsilon_{Si} \cdot E_{Si} \tag{3.42}$$

$$E_{ox} = 3 \cdot E_{Si} \tag{3.43}$$

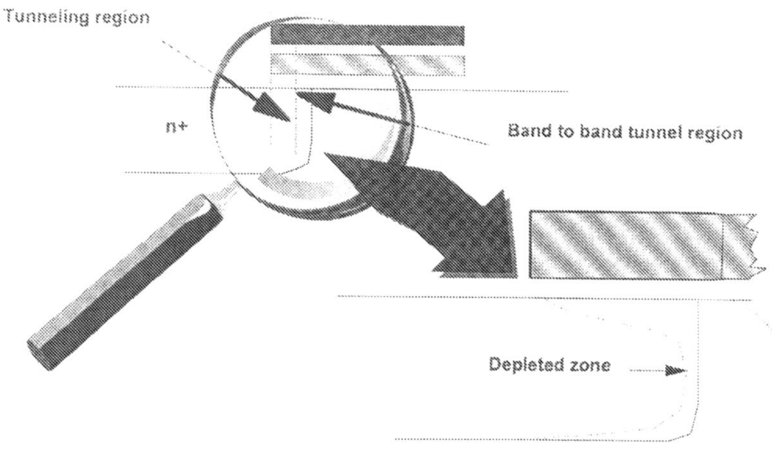

Fig. 3.24. The region underneath the gate where the tunneling across the oxide and the interband tunneling can take place

If we draw the logarithmic diagram of the BBT current as a function of the source voltage and regard the gate voltage as a parameter, we obtain Fig. 3.25. By substituting the difference of potential between the source and the gate nodes for the source voltage on the abscissa axis in Fig. 3.25, we obtain the diagram in Fig. 3.26 where the characteristics of the BBT current are overlapped.

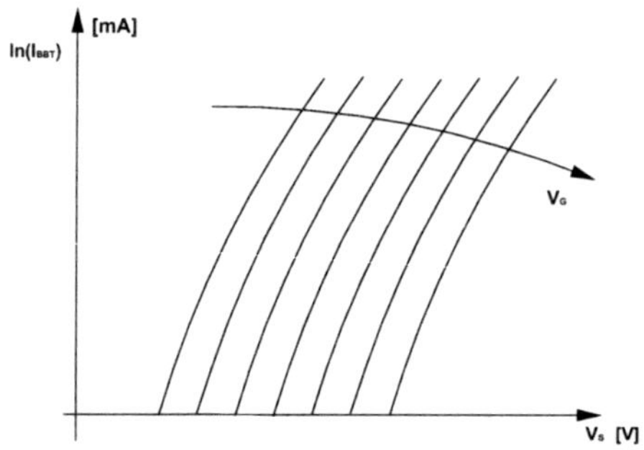

Fig. 3.25. Band-to-Band Tunnel current as a function of the source and isolated gate voltage

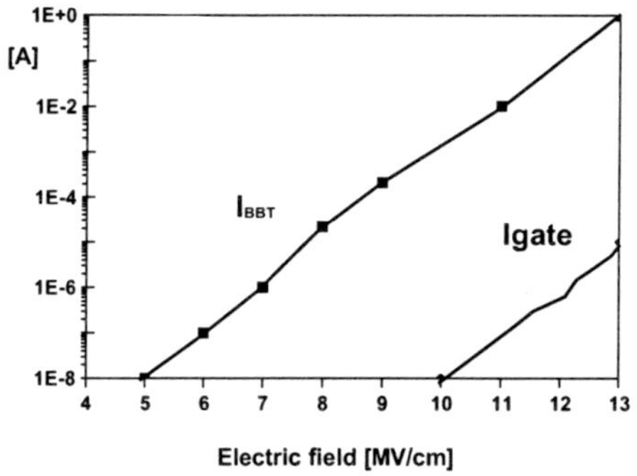

Fig. 3.26. BBT and gate current as a function of the difference of potential between the source node and the isolated gate

V_S-V_G is the transverse component of the electric field and I_{BBT} does not depend on V_S. Therefore, the current due to the interband tunneling depends only on the transverse component of the electric field.

Let's now consider the memory cell taking into account Eq. (3.25) that associates the voltage of the isolated gate with the source and gate node potentials, the charge trapped, and the capacitive ratios. The drain does not have any impact (at least at a rough estimate) since it is left floating during the erase operation while the substrate is grounded.

$$V_{FG} = \alpha_S \cdot V_S + \alpha_G \cdot V_{CG} + \frac{Q}{C_T} \tag{3.44}$$

If we neglect the voltage drop in the floating gate and in the silicon, the electric field across the tunnel oxide can be calculated as follows:

$$E_{ox} = \frac{|V_{FG} - V_S|}{t_{ox}} = \frac{\left|(\alpha_S - 1)\cdot V_S + \alpha_G \cdot V_{CG} + \frac{Q}{C_T}\right|}{t_{ox}} \tag{3.45}$$

3.7.1 Erasing at Constant Voltage

In this case the erase operation is carried out by leaving the drain floating, whereas the gate is grounded and the source is pulled up to 10–12 V. The necessary electric field is therefore obtained by controlling only one of the cell terminals. This mode is implemented in flash memories that, besides VDD, have a second bias voltage dedicated to program and erase operations, named VPP. At the initial time, the electric field across the thin oxide depends only upon the charge stored on the floating gate, Q. The erase operation is carried out by discharging the isolated gate and, at the same time, the electric field and the current diminish in an exponential fashion with the electric field.

The problem is that the initial peak of the electric field, caused by the application of the source voltage, may provoke the breakdown of the source junction. The solution that prevents the junction damage and the cell degradation consists in applying the erase voltage to the source by means of a voltage ramp instead of applying it directly. A simple alternative is the insertion of a resistor, R, to limit the voltage (Fig. 3.27).

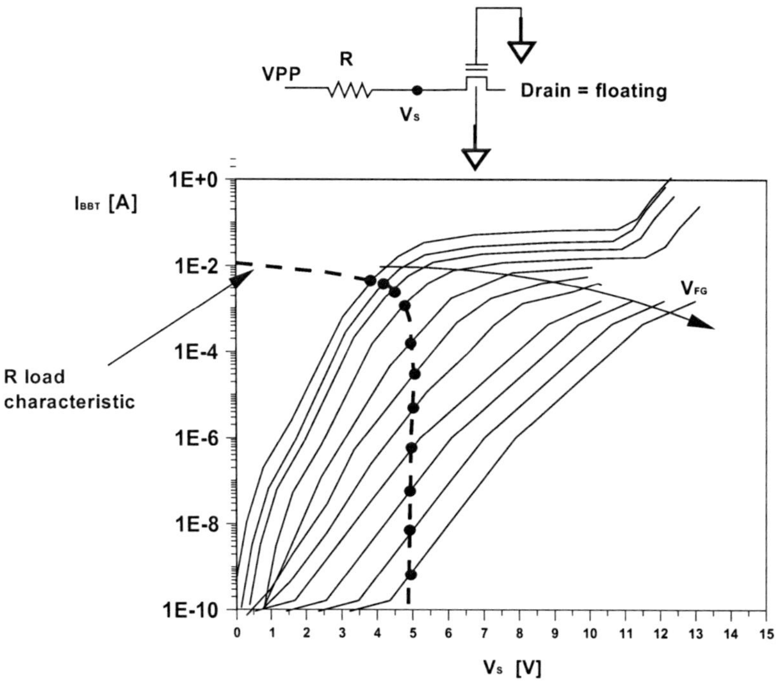

Fig. 3.27. Circuit configuration and voltage of the source node in the case of constant erase voltage

The load characteristic of the resistor is overlapped to the I_{BBT} characteristics. With the resistor we do not erase through a ramp since the horizontal part of the characteristic, which becomes vertical later, is covered in a short time. Erase can be performed because the voltage drop on the resistor is small and allows obtaining the necessary electric field due to the fact that the gate current is much less than I_{BBT}.

The resistor value is determined so as to pull the source potential to a value definitely smaller than the breakdown potential at the beginning of the erase phase, when the BBT current is maximum.

As for program, we can define the erase characteristic as the curve that expresses the threshold voltage shift as a function of time. Figure 3.28 shows as the time that is necessary to reach the $V_{T,end}$ threshold voltage is independent of the V_T voltage of the programmed cells at the beginning of the erase phase. Of course, this is true for a given tunnel oxide thickness.

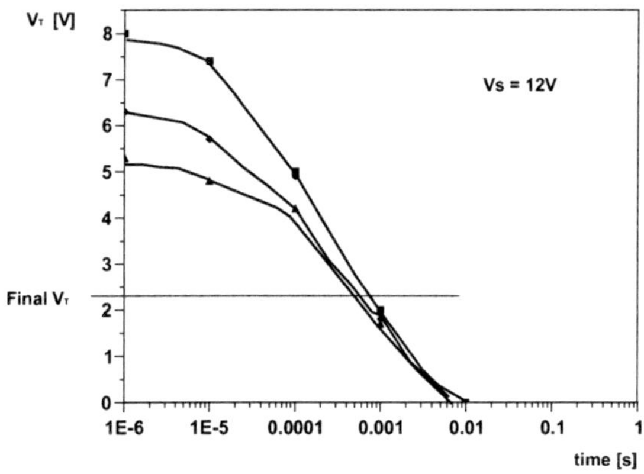

Fig. 3.28. Threshold voltage as a function of time

In order to better understand this behavior, let's erase two cells, geometrically and electrically identical, that have different initial charge on the floating gate (Fig. 3.29), i.e. different V_T. During erase we suppose that the voltage applied to the terminals of the two cells is the same, which means that the two control gates are equipotential as well as the two sources since the local currents are so small that they cannot modify the voltage drop.

At the beginning, the cell having more charge and, therefore, greater threshold voltage, has a more intense electric field and is erased more quickly. The faster cell reaches the amount of charge of the slower cell and, from that moment on, they are erased together since they undergo the same electric field.

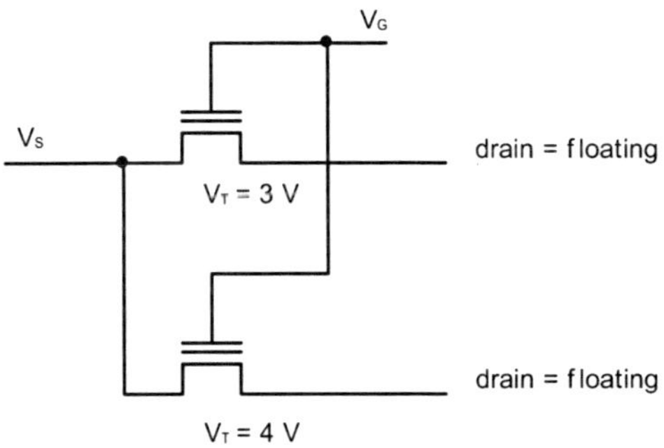

Fig. 3.29. Two identical cells with different initial amount of charge are erased simultaneously since they belong to the same sector. The slower cell waits for the faster cell so that they have the same electric field applied

A clarification is now necessary. The tunneling phenomenon that causes the erasing is located in a very narrow region, at least in the channel direction. The polysilicon grains that make up the isolated gate have the same size of the erase region. Besides the geometrical differences, also the possible charge trapped, due to the various process steps that eventually modifies the band structure at the interface[11], must be taken into account. All the variations can be regarded as an equivalent variation of the oxide thickness. It is important to notice that the shape of the threshold voltage distribution of the erased cells does not depend on the shape of the corresponding distribution of the programmed cells (before erase). It is also possible to regard such a distribution as the overlap of a gaussian distribution and a poissonian tail.

The adoption of the mixed erase instead of the source erase increases the breakdown margin that, we recall, may be a problem only if it involves few cells that undergo source junction degradation caused by the high current flow.

In single bias source devices, high voltages are internally generated by means of charge pumps (see Chap. 15). The limited availability of current sourced by such circuits has led to the negative gate erase.

Taking into account Eq. (3.45) and supposing that:

$$\alpha_S = 0.15; \Delta V_T = 3V; T_{ox} = 10nm; E_{ox} = 10MV/cm \tag{3.46}$$

we obtain

$$V_G = -8V; V_S = 5V \tag{3.47}$$

Notice that, in this case, also the gate capacitive coupling takes place. The advantage in terms of reduction of the voltage applied to the cell source is evident.

[11] The trapped charge is responsible for the phenomenon known as "erratic bit". During the write/erase cycles, some cells can be found whose V_T might be negative after the *n-th* erase cycle, and might become positive after *m* more cycles and eventually negative again after *n+m* further cycles.

3.7.2 Constant Current Erase

The main drawback of the standard erase method with constant voltages applied is the strong dependence of the erase time on V_S (typically VPP or VDD) and temperature. Moreover, this approach implies a complete dependence of the electric field on the cell process variation, giving rise to a variable peak of both source and gate current, occurring at the beginning of the erase operation.

On the contrary, the constant current erase method consists in keeping the gate current constant during the whole erase operation. This is obtained by forcing a constant current in the source node in such a way to maintain the electric field always constant. It is based on the observation that, given a certain electric field, the ratio between gate and source current is constant. In fact, as clearly shown in Fig. 3.30, given a determined I_S and considering Eq. (3.40) and Eq. (3.41), E_{Si} is fixed and also I_G is known. The I_G/I_S ratio is independent of the bias conditions and depends only on the electric field. Moreover, since the gate current is constant, the threshold voltage shift of the cell, proportional to the integral of I_G, follows a linear time law, whose slope is a function of the chosen source current. The value of the source current can be determined according to the defined erase time and maximum electrical field, chosen according to reliability considerations.

In fact, the charge trapping/generation in the oxide is a strong function of the charge flowing through it and the electric field. The higher the electric field, the greater the oxide degradation. Hence, it is possible to control the erase time degradation, which depends on the oxide damage, by controlling the maximum electric field.

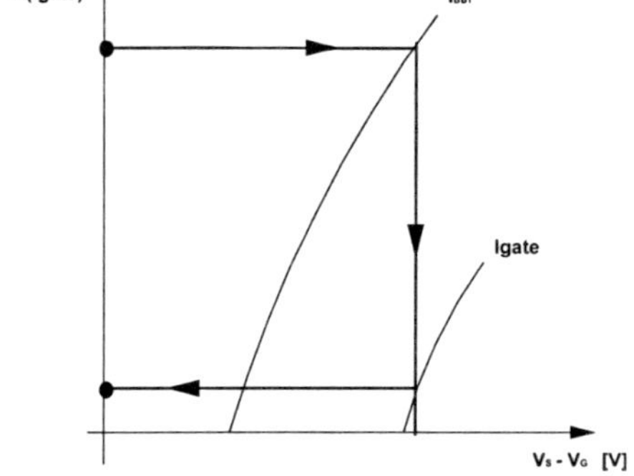

Fig. 3.30. Erase speed defined through the design of the BBT and gate currents

Therefore, using this kind of approach, the threshold voltage variation is constant with time, differently from all the other cases in which the variation is fast at the beginning and slows down progressively. On the other hand, what happens if

ΔV_T is constant and two cells have different amount of charge? Is the distribution of the erased cells similar to the distribution of the programmed cells?

The answer is negative since the cells tend to behave uniformly, i.e. those that have less charge "wait" for those that are erased more quickly because they have more charge at the beginning. The choice of the current determines the electric field so that the erase speed is the same also in the case of different devices. The source voltage will be tuned in order for the electric field to fulfill the design value.

The main drawback is the precision with which erase is stopped. In the previous cases, the voltage step obtained though a constant pulse at the end of the erasing diminishes and, hence, the precision of the final value of the threshold voltage increases. In this case, instead, the variation of the threshold voltage is the same during all the pulses and, thus, the error with respect to the desired threshold voltage is larger.

3.7.3 Erasing at Negative Gate and Triple-Well Array

The fabrication of the cell array in triple-well (Fig. 3.31) allows biasing the substrate with positive voltage, eliminating the voltage drop responsible for the curvature of the energy bands that causes the spurious band-to-band tunneling current. This result has a fundamental importance for the present memories that work with a single bias voltage that is smaller than the program and erase voltages. The high voltages are generated by means of charge pumps (see Chap. 15) that typically have a limited capability to source current. Increasing the available current means increasing the size of the capacitors of the pumps, with evident repercussions on the overall device area.

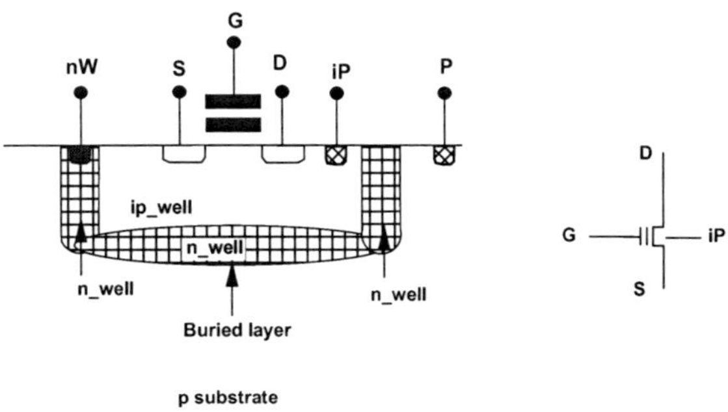

G	D	S	iP	nW	P
-8V	floating	5V	5V	5V	0V

Fig. 3.31. Memory array in triple-well. The array substrate (ip-well) is isolated from the rest of the circuits by an n-well tub

3.8 Erase Algorithm

After recognizing the code that the user communicates to the CUI[12], i.e. the internal memory controller, to erase a sector, the preconditioning operation starts, that is all the cells are programmed. Such a preliminary operation is executed to guarantee uniform aging of the cell population so as to limit the width of the erase distribution. If we erased a sector containing cells with low V_T, such cells would be over-erased and probable depleted to negative V_T while decreasing the V_T of the programmed cells. Figure 3.32 illustrates the failure mode induced by a depleted cell.

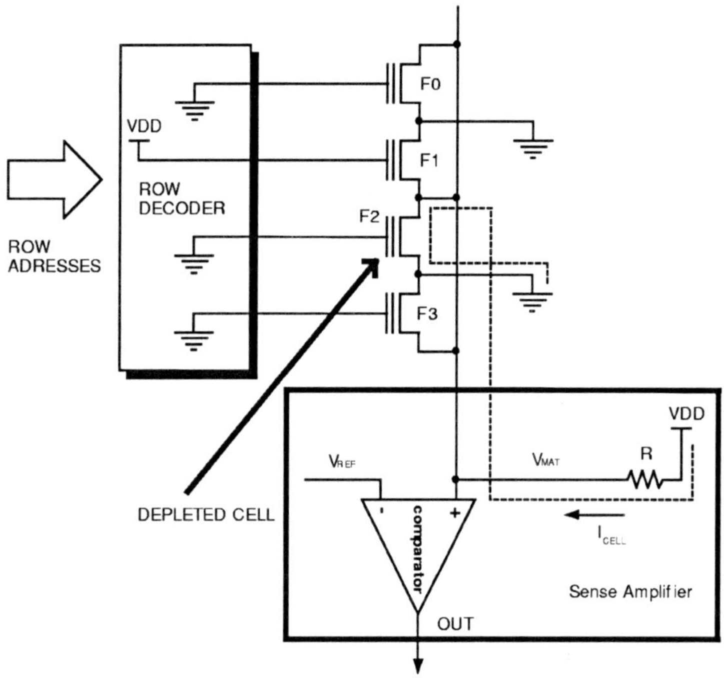

Fig. 3.32. Effect of a depleted cell on the reading of a written cell belonging to the same column. The current sunk by the depleted cell simulates, on the sense amplifier point of view, the reading of an erased cell

The row decoder selects the written cell, F1, based on the address applied by the user. This cell does not sink current even when V_{GS} equals VDD. The sense amplifier (see Chap. 12) recognizes the cell as written since the potential of node MAT equals VDD. The presence of a depleted cell, F2, with threshold voltage

[12] It is the acronym for Command User Interface. It is a finite state machine used to decode the commands the user applies to the memory.

equal to -2 V, determines a current flow along the bit line. The potential of node MAT decreases and the reading is not correct.

Let's go back to the erase algorithm. At this point, we have a group of programmed cells with gaussian distribution of V_T. During the actual electrical erase phase, the source and gate voltages of the cells are forced to the proper values with respect to the adopted erase technique (see previous section). The time counter limits the erase pulse to the typical duration of 10 ms. At the end of the pulse, the erase verify phase is carried out. Such an operation is performed through a margined read that guarantees the correct cell recognition in normal read mode.

The erase pulses continue until the correct verification of all the cells of the sector has been completed. At this point it is necessary to verify that there are no depleted cells that may induce errors during read. It is difficult to detect the depleted cells since the presence of a cell with negative V_T implies that the reading of all the other cells belonging to the same columns is dominated by the current of the depleted one. Therefore, a search algorithm is applied to verify the presence of leakage current on the columns when all the rows are grounded. When a column with such an anomaly is found, the first cell is addressed and a program pulse is applied with low gate voltage, so as to slightly increase its V_T, without overcoming the limit used during the erase verify phase. Subsequently, the same algorithm is applied to the second cell of the same column. If no current is drawn, it means that the depleted cell was the previous one that was already recovered with the soft programming pulse; otherwise a pulse is applied to the present cell and so forth, until the end of the column. At the end of the column, the verify phase is repeated and, in the case of presence of current, the procedure is repeated with increased gate voltage, since it means that the program pulse applied to all the cells of the column was not sufficient to recover the depleted cells.

In order to simplify the algorithm controlled by the internal finite state machine, the gate voltage is used as a parameter, whereas the drain voltage is kept to the value used during the normal program phase and the same holds for the pulse duration. As we will see in Chap. 14, the gate voltage control limits the maximum threshold voltage step for a fixed time duration of the pulse.

The erase procedure has also an important feature called *erase suspend* that allows the user to suspend the erase operation for an indefinite time, storing the current state of the procedure, so as to be able to restart after a proper command called erase resume. Such a feature allows the external microprocessor to access a sector of the array to retrieve data, apart from the one that is undergoing erase. In this way, the penalization due to the long erase duration (around 1 s) can be partially overcome.

Bibliography

S. Aritome, R. Shirota, G. Hemink, T. Endoh, and F. Masoka, "Reliability issues of Flash memory cells," Proc. IEEE, vol. 81, no. 5, pp. 776-788, (May 1993).

R. Bez et al., "Depletion Mechanism of Flash cell induced by parasitic drain stress contidion", VLSI Technology Symposium, (1994).

J. D. Bude, "Gate current by impact ionization feedback in sub-micron MOSFET technologies", in 1995 Symposium VLSI Technology Dig. Tech. Pap., pp. 101–102, (June 1995).

J. D. Bude, M.R. Pinto and R. K. Smith, "Monte Carlo Simulation of the CHISEL Flash Memory Cell," IEEE Tran. Electron Devices, vol. 47, pp. 1873-1881, (Oct. 2000).

E. Burstein, S. Lundqvist, "Tunneling Phenomena in Solida", Plenum Press, New-York, (1969).

E. Camerlenghi, P. Caprara, and G. Crisenza: "A 18 μm^2 cell for megabit CMOS EPROM", in Proc. 17th European Solid State Device Research Conf., pp. 765–768, (Sept. 1987).

John Y. Chen, CMOS devices and technology for VLSI, Prentice Hall, (1990).

A. Chimenton, P. Pellati, and P. Olivo, "Constant Charge Erasing Scheme for Flash Memories," IEEE Tran. Electron Devices, vol. 49, pp. 613-618, (Apr. 2002).

A. Chimenton, et al., "Overerase Phenomena: An insight Into Flash Memory reliability", IEEE Proceeding of the, Vol. 91, No. 4, pp. 617-626, (April 2003).

C. Dunn, C. Kaya, T. Lewis, T. Strauss, J. Schreck, P. Hefley, M. Middendorf, and T. San, "Flash EEPROM disturb mechanisms," in Proc. Int. Rel. Phys. Symp., pp. 299-308, (April 1994).

B. Eitan and D. Frohman-Bentchkowski, "Hot electron injection into the oxide in n-channel MOS devices", IEEE Trans. Electron Devices, vol. ED-28, pp. 328–340, (March 1981).

B. Eitan, R. Kazerounian, A. Roy, G. Crisenza, P. Cappelletti, and A. Modelli, "Multilevel Flash cells and their trade-offs", in 1996 IEDM Tech. Dig., pp. 169-172, (Dec. 1996).

Leo Esaki, "Long Journey into Tunneling", Proceedings of IEEE, vol 62, No 6, pp 825-835, (June 1974).

D. Frohman-Bentchkowsi, "Memory behavior in a floating gate avalanche-injection MOS (FAMOS) structure", Appl. Phys. Lett., vol. 18, pp. 332-334, (1971).

D. Frohman-Bentchkowsi, "FAMOS-A new semiconductor charge storage device", Solid State Electron, vol. 17, pp. 517-520, (1974).

C. Hu, "Lucky-electron model for channel hot-electron emission", 1979 IEDM Tech. Dig., pp. 22–25, (Dec. 1979).

C. Hu, "Future CMOS scaling and reliability", Proc. IEEE, vol. 81, pp. 682–689, (May 1993).

Y. Igura et al., "New Device Degradation Due to "Cold" Carriers Created by Band-to Band Tunneling", IEEE Electro Device Letters, VOL. 10, NO. 5, MAY (1989).

C. Kittel, Introduction to Solid State Physic, John Wiley & Sons, New York, (1966).

M. Lenzlinger and E. H. Snow, "Fowler-Nordheim tunneling into thermally grown SiO_2", J. of Applied Physics, vol. 40, pp. 273-283, (Jan. 1969).

S. Mahapatra, S. Shukuri, and J. Bude, "CHISEL flash EEPROM–Part I: performance and scaling", IEEE Trans. Electron Devices, vol. ED-49, pp. 1296–1301, (July 2002).

S. Mahapatra, S. Shukuri, and J. Bude, "CHISEL flash EEPROM–Part I: reliability", IEEE Trans. Electron Devices, vol. ED-49, pp. 1302–1307, (July 2002).

Yohsuka Mochizucki, "Read-disturb Failure in Flash Memory at low field", Intel reports, Nikkei Electronics Asia, pp. 35-36, (May 1993).

J. Van Houdt, et al., "The HIMOS Flash technology: The alternative solution for low-cost embedded Mmeory", IEEE Proceeding of the, Vol. 91, No. 4, pp. 627-635, (April 2003).

Samuel Tuan Wang, "On the I-V characteristics of Floating-Gate Mos transistors", IEEE Transaction on electron devices, Vol ED-26, No 9, September (1979).

4 Passive Components

In this chapter, a brief summary of the main characteristics of the passive components that are essential to the design of any integrated circuits will be given.

4.1 MOS Capacitors

The analysis of the MOS capacitor is generally the first step in the study of the MOSFET transistor. As a matter of principle, we can regard it as a capacitor having plane and parallel plates, the polysilicon gate and the doped silicon, with the gate oxide as insulator.

In practice, it is possible to fabricate a capacitor by contacting the source and drain junctions of a transistor, as depicted in Fig. 4.1.

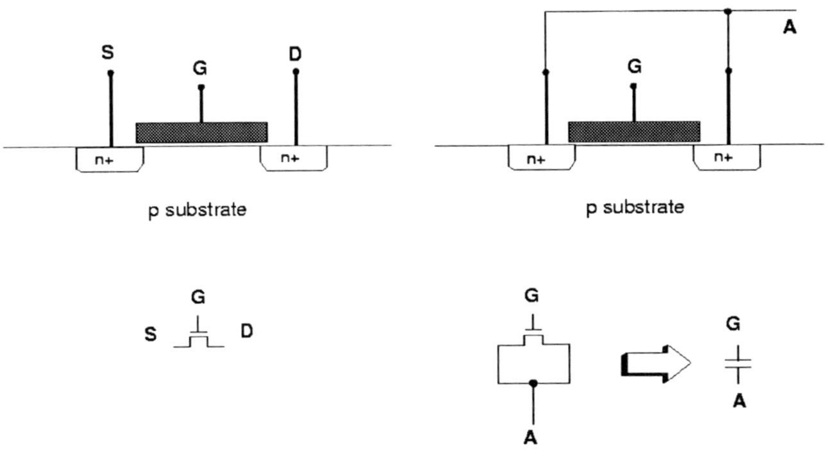

Fig. 4.1. Fabrication of a capacitor from a MOS transistor

For the simplest configurations to fabricate a capacitor realized by directly contacting the p substrate (Fig. 4.2), three different configurations should be distinguished. Let's consider the p-type substrate connected to ground.

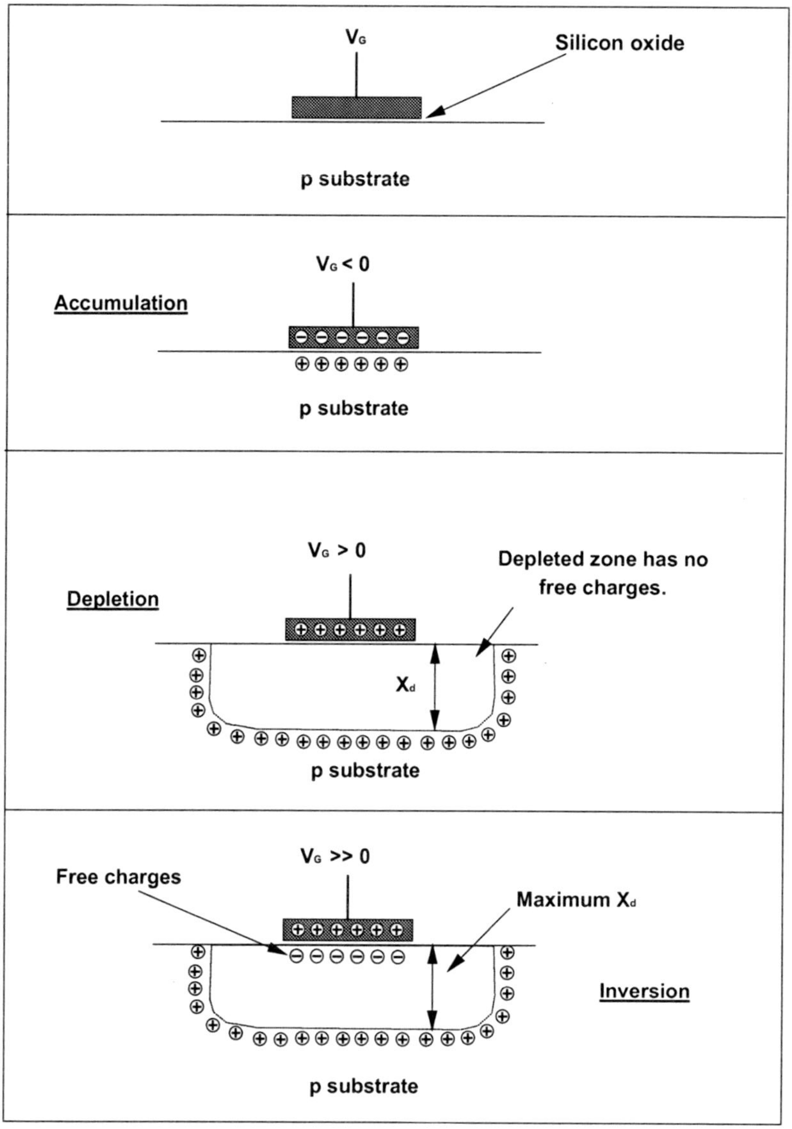

Fig. 4.2. Operating regions of a MOS capacitor: accumulation, depletion, and inversion

- *Accumulation.* Applying a negative voltage to the gate ($V_G < 0$), a layer of holes is induced at the oxide-semiconductor interface. The net charge of the semiconductor is, hence, positive, owing to the accumulation of the holes in excess with respect to the equilibrium.

The value of the capacitance is then $C = \varepsilon/t_{ox}$, where ε is the silicon oxide permittivity and t_{ox} is the oxide thickness. In this case, all of the voltage drop, V_G, is located between the conductive plates or entirely in the oxide.

- *Depletion.* As the gate voltage increases, i.e. as V_G becomes positive, the p-type mobile charge is removed from the oxide-silicon interface. A depletion region having thickness x_d and voltage drop Ψ_S forms, where:

$$V_G = V_{ox} + \Psi_S \tag{4.1}$$

- *Inversion* As V_G further increases, also x_d increases. If the voltage applied to the gate is above a critical value, referred to as threshold voltage, V_{TH}, a layer of electrons is induced at the oxide-silicon interface, inverting the polarity of the silicon. When $V_G > 0$ but below V_{TH}, the value of the capacitance is given by the series capacitors related to the two regions, the oxide and the depleted region.

4.2 CMOS Technology Capacitors

In a typical CMOS process, several types of capacitors are theoretically available:

- Poly/n-well capacitor with low or high voltage oxide;
- PMOS capacitor with low or high voltage oxide;
- Poly1/poly2 capacitor.

For the present technological processes, the value of the specific capacitance for the n-well capacitors is around 1.5 fF/μm^2 in case of HV capacitors, 3 fF/μm^2 in case of LV capacitors, and 2 fF/μm^2 in case of interpoly capacitors. The gate capacitance of a MOS transistor equals the value of a MOS capacitor having the same oxide thickness.

The n-well capacitors are designed like p-channel transistors without source and drain diffusions (Fig. 4.3). The value of capacitance as a function of the gate voltage with grounded n-well is shown in Fig. 4.4.

Such a structure (poly/n-well) has a constant value of capacitance in both accumulation and inversion. In accumulation, a negative charge crowding is located at the bottom plate of the capacitor. The inversion zone, i.e. the positive charge crowding in the silicon, can be obtained only at low frequency, since the positive charge accumulation is a slow phenomenon. Therefore, we can deduce that the poly/n-well capacitor satisfactorily operates only if the gate voltage is positive enough.

The PMOS capacitor is instead a p-channel transistor having the drain shorted with the source, as sketched in Fig. 4.5. In this case, the inversion region is reached faster owing to the presence of p^+-type diffusions. In fact, the p^+ regions provide the positive charge for the inversion.

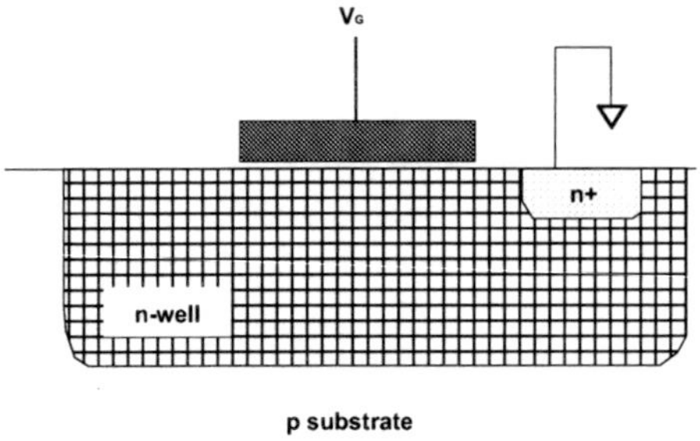

Fig. 4.3. Practical realization of a poly/n-well capacitor

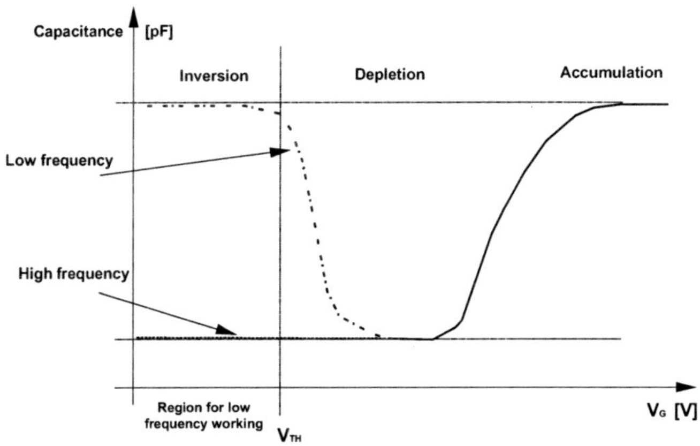

Fig. 4.4. C-V characteristic of a poly/n-well capacitor

Let's consider the practical case of a device in which both positive and negative charge pumps (see Chap. 15) are present and operate at the frequency of 30 MHz.

In the positive charge pumps, which are able to generate voltages higher than the supply voltage, the poly/n-well capacitors operate in the accumulation region since the poly gate has a higher voltage than the n-well. Hence, the majority carries (electrons) that are present at the oxide-semiconductor interface are provided by the bulk at a time rate that is negligible with respect to the switching frequency. The case of the negative pumps, which generate voltages below the ground potential, is different. In fact, in this case, the MOS system operates in the inversion region and the minority carriers (holes) can be provided only by the bulk at a rate that can be measured in seconds. Two p^+ diffusions are therefore added to the capacitor structure so as to generate minority carriers. At low frequency, the channel is in thermal equilibrium with the p^+ regions, and we can assume that the required carriers are instantaneously provided to the interface. At high frequency, the holes are not able to diffuse from the p^+ regions toward the middle of the channel with the speed required to follow the signal applied to the gate. This causes channel RC parasitic effects that become more relevant when the carrier mobility is low and the distance between the p^+ regions is great.

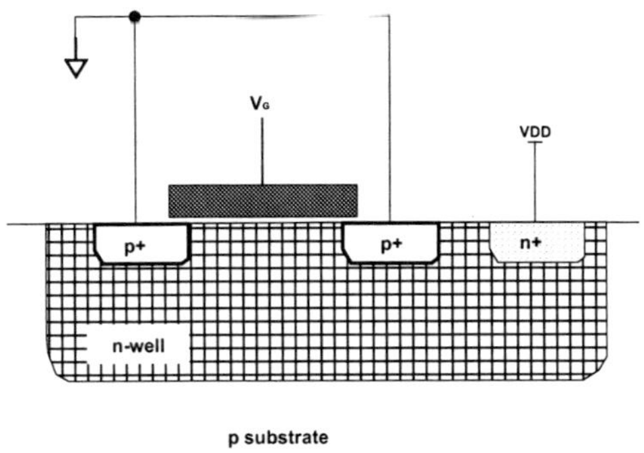

Fig. 4.5. PMOS capacitor

Let's now calculate the maximum frequency of the gate signal that can be applied to this type of capacitor in inversion. The time constant associated with the channel can be expressed as:

$$\tau_{ch} = R_{ch} \cdot C \tag{4.2}$$

where C, which is the capacitance associated with the gate oxide, can be calculated starting from the channel dimensions:

$$C = C_{ox} \cdot W \cdot L \tag{4.3}$$

R_{ch} is the channel resistance and can be obtained from the characteristic of the MOS transistor in the linear region:

$$\frac{1}{R_{ch}} = \frac{W}{L} \cdot \mu_{eff} \cdot Q_{inv} \tag{4.4}$$

Q_{inv} is the charge per unit of area in the inversion layer and μ_{eff} if the effective mobility. Based on the three foregoing equations, we can calculate the frequency associated with the channel RC:

$$f_{ch} = \frac{1}{2\pi\tau_{ch}} = \frac{1}{2\pi} \cdot \frac{\mu_{eff} Q_{inv}}{C_{ox} L^2} \tag{4.5}$$

In particular, we can notice that the frequency of the pole scales down as the square of L, which is the distance of two p^+ diffusions. This is the reason why it is very important to accurately determine the value of L of the capacitors of the negative charge pumps.

It is also possible to fabricate capacitors without using the silicon as plate, such as in the case of the poly1/poly2 capacitor. The CMOS processes that use two layers of polysilicon, i.e. the processes for non-volatile memories, offer this possibility.

Furthermore, metal layers can also be used to fabricate capacitors. The main limitation is in the value of capacitance that can be obtained. For a typical process, the thickness of the gate oxide ranges between 120 Å and 150 Å, whereas the inter-metal dielectric that forms the insulating layer between the two metal plates is some thousands of Angstroms thick. Therefore, the size of a capacitor fabricated with metal plates (i.e. metal1/metal2) is nearly 50 times larger than a poly1/poly2 capacitor having the same capacitance.

Moreover, the quality of the insulator is much different in the two cases. The gate oxide is the best possible solution offered by the present technology, since it has been studied and refined to fabricate all the transistors. Generally, the intermediate dielectric of the metal1/metal2 capacitor does not have such a high qua ity[1].

4.3 Integrated Resistors

The available resistors are all those that can be fabricated with the various layers produced by the technological process. In Table 4.1, the available layers are summarized with the value of resistance and its variation for a typical 0.35 μm CMOS process.

It is common practice to calculate the value of an integrated resistor in squares (indicated with the symbol □). In fact, the layout is a top view in which only the sides of the rectangles are dimensioned, whereas the depth is fixed and defined by the process. Therefore, we typically refer to the sheet resistance. Once the sheet resistance is known, we can obtain resistors of the required value by placing one or more squares (or fractions of squares) in series.

Table 4.1. Sheet resistence of different layers

Layer	Sheet Resistance [Ω/□]
n^+ (active area)	50 ÷ 60
n-well	600 ÷ 1200
Poly2 with silicide	5 ÷ 7
Poly2 without silicide	50 ÷ 100
Metal1	0.08
Metal2	0.04

[1] By the way, it is worth telling this story. The authors participated in the design of a device that included an A/D converter realized with metal1/metal2 capacitors. Once the devices were delivered, many of them were returned by the customers since they did not work properly and were considered as "failures". After further tests, many of them resulted to be good either immediately or after a night in the oven. The problem was due to the material used to planarize the dielectric layer between the metal plates. Such material absorbed humidity, modifying the value of capacitance due to the ions present in the water. The subsequent heating removed the charge and the devices worked fine.

It can be noted that the value of the sheet resistance is a few Ohms/□ for poly2 due to the silicide, some tens of Ohms/□ for the active area, and many hundreds of Ohms/□ for the n-well. Resistive dividers can be obtained by means of the n-well that, due to the high value of sheet resistance, reduces the current consumption.

Problem 4.1: Determine which of the resistors in Fig. 4.6 is greater.

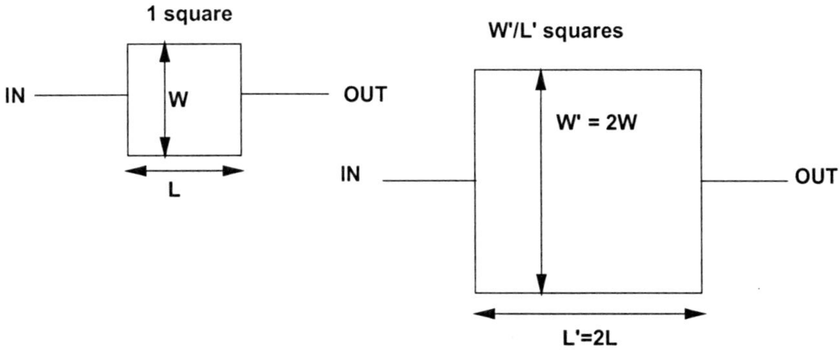

Fig. 4.6. Integrated resistors

The problem with n-well resistors is that a reverse-biased junction is used to insulate the resistor from the substrate. The voltage applied modulates the depleted region of the substrate/n-well junction changing the total value of resistance (Fig. 4.7). Therefore, the resistance is also a function of the thickness of the depleted region. If we want to realize a resistive divider, it is necessary to use values of resistance that are different from the theoretical calculations, just to compensate for such an effect.

Let's imagine an experiment with a resistive divider realized with two n-well resistors of the same size, connected as shown in Fig. 4.7.

The output voltage, V_{OUT}, would follow the characteristic V_{OUT} = VDD/2, if the value of resistance were constant with respect to VDD. Unfortunately, the depleted region in R1 increases as VDD increases; the total value of R1 increases which reduces the voltage drop on R2.

The foregoing effects are known and are accounted for during the design phase, so the value of R1 (or R2) is properly determined to obtain the required value of V_{OUT}. In practice, if V_{OUT} = VDD/2 is to be obtained, the two resistors are not designed with the same size. The depletion effect is present in all the resistors fabricated with diffused layers, even though it is more relevant in the n-well resistors, since it is a function of doping density and layer thickness. The n-well resistors are also the most utilized resistive components due to the high sheet resistance.

The problem of the depletion region imposes the use of diffused resistors having quite large width (some microns), since the depletion is present also in the lateral direction and it could completely deplete the resistance if it is too narrow.

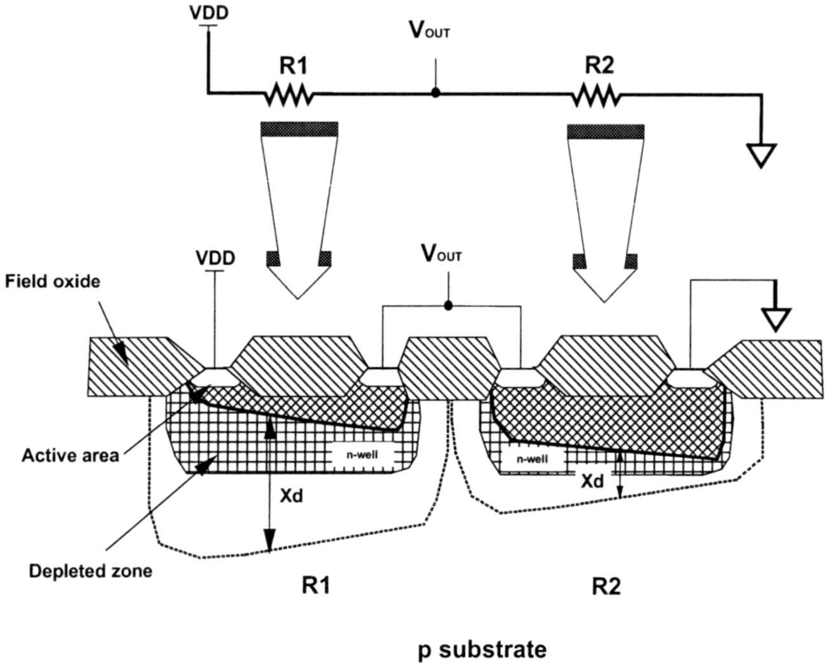

Fig. 4.7. Modulation of the n-well resistance as a function of the voltage applied

Finally, the fabrication of a resistor of a large value by means of several series resistors, cannot be achieved like in Fig. 4.8. In fact, it is difficult to determine the exact final value of resistance because of the many angles present. Moreover, the current tends to crowd near the edges, favoring possible undesired breakdown phenomena.

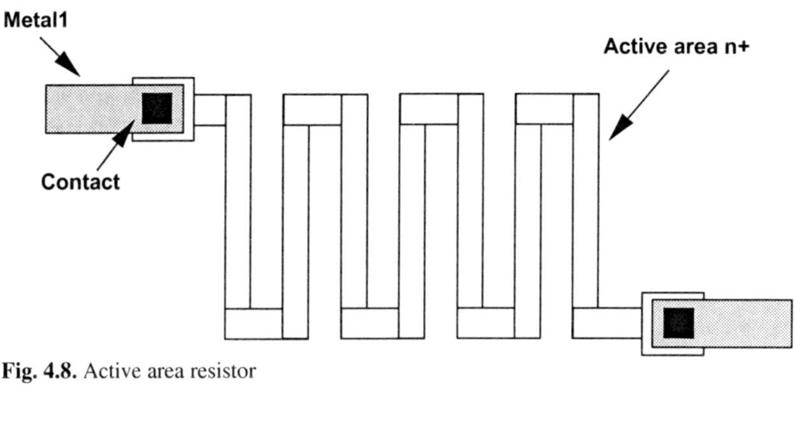

Fig. 4.8. Active area resistor

The resistors are laid out like in Fig. 4.9, in which the pieces in metal1 (it could be also metal2) have negligible value of resistance. The metal1/n$^+$ contacts are resistive: each contact may have resistance of several tens of Ohms. This requires placing more contacts in parallel so as to make the contribution of the contact resistance negligible.

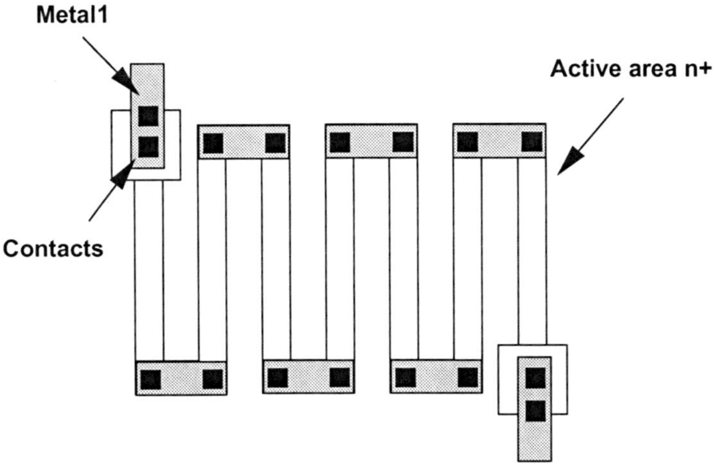

Fig. 4.9. Optimized resistor

Bibliography

R.S. Muller, T.I. Kamins, Device Electronics for Integrated Circuits. Second Edition, John Wiley & Sons, (1986).

S.M. Sze, Physics of Semiconductor Device, John Wiley & Sons, Inc, New York, (1969).

Stanley Wolf, Silicon Processing For the VLSI ERA volume 2: Process integration, LATTICE PRESS, Sunset Beach, California, (1990).

5 Fundamental Circuit Blocks

5.1 Introduction

The first step towards design consists of the study of the most used circuit blocks, from the inverter to boost concepts. A deep analysis of these circuits allows their assembly to get to the desired results. The main circuit blocks are shown always paying attention to the understanding of their functionality. The suggestion is to re-design every circuit to understand the connections between the various elements. If you have a simulator, don't try to understand the behavior of the circuit by simulating it; analyze it on paper and then verify your understanding with simulation.

5.2 NMOS and CMOS Inverters

The main structure for logic design is the inverter, which is also the fundamental block to understand the operation of both analog and logic circuits. The easiest usage of a transistor is shown in Fig. 5.1.

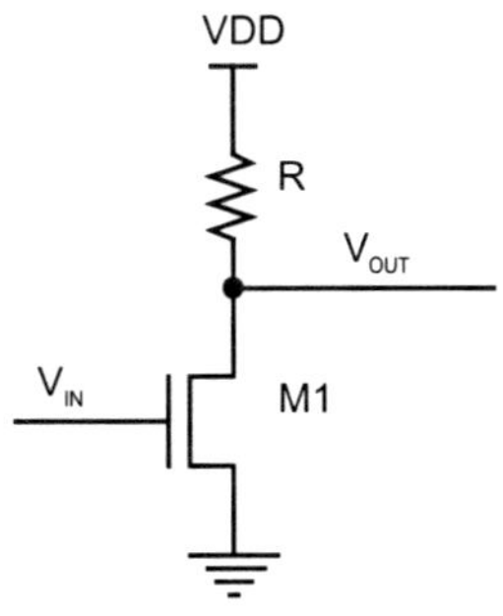

Fig. 5.1. Inverter with a resistive load

We know that by modulating the voltage of the input signal we obtain a voltage swing of the output node that can be much bigger than the one on the input. We call this phenomenon "amplification". An amplifier is a circuit that exploits an impedance variation of a device, M1 in this case, to vary the resistive ratio on the output. By modulating the channel resistance of M1 we "decide" how much current it can sink and consequently the voltage drop on the load R.

If we consider only the extremes of the range of the input signal V_{IN}, VDD and ground, we obtain a logic inverter. When V_{IN} is GND, LVS-type transistor M1 is turned off: no current flows and the voltage of the output V_{OUT} is equal to the supply voltage VDD. On the other hand when V_{IN} is VDD, M1 transistor is turned on and V_{OUT} is near to ground. We cannot say that V_{OUT} is GND, in this case, because it depends on the partition ratio between the load R and the equivalent resistance of M1.

Output characteristic of M1 transistor are shown in Fig. 5.2. Let's now consider the load line: the value of the short-circuit current (V_{OUT} = 0 V) is equal to I_{DD} = VDD/R while the open-circuit voltage (I_{DS} = 0) is V_{OUT} = VDD. The possible working points for our circuit are given by the intersections of the characteristics of M1 with the load line. We can see for instance that the minimum value of V_{OUT} is not ground, but V_{DSm}, under the assumption that V_{GS5} is equal to VDD.

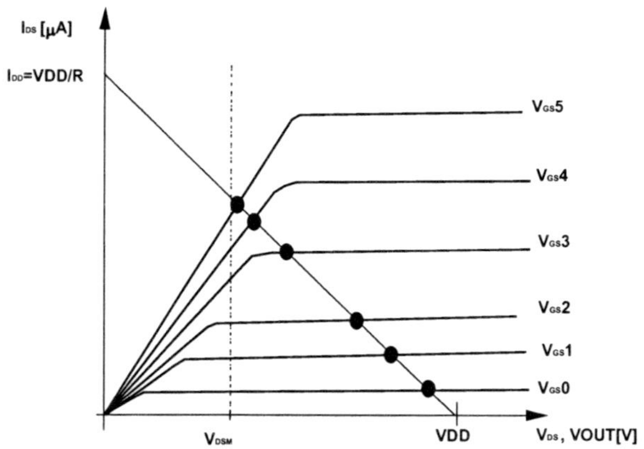

Fig. 5.2. Load line where some working points are shown ($V_{GS0} < V_{GS1} < V_{GS2} < V_{GS3} < V_{GS4} < V_{GS5}$)

Problem 5.1: An interesting exercise to carry out is the calculation of the value to assign to resistance R to be able to satisfy "reasonable" design criteria, for instance charging 1 pF in 2 ns with an associated consumption of few microAmpere. The discussion of the problem shows how inefficient is the use of a resistor to implement an integrated inverter, while it is normal to do it on a board.

The oldest integrated technology, NMOS[1], has solved this issue by replacing the passive load with an active one. A depletion NMOS transistor, i.e. a MOS whose channel is always present even if V_{GS} is zero (thanks to the negative threshold voltage) can be used as active load. Figure 5.3 shows the scheme of a NMOS

[1] This technology allows the usage of n-channel MOS transistors only.

inverter where the M2 transistor is diode-connected and has replaced resistor R of Fig. 5.1.

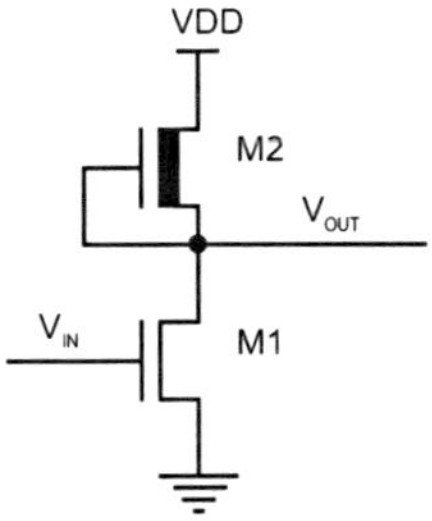

Fig. 5.3. NMOS Inverter

The first remark to do is the comparison between the load characteristics of R and M2 (Fig. 5.4). Characteristic of M2 is not linear: for high V_{GS} values the current sunk by M1 is high and M2 has $V_{GS} = 0$ V (above $V_{T,dep}$), well turned on and able to provide the required current. The important consequence is that the V_{GS} of M2 is always zero, while its V_{DS} is VDD – V_{OUT}.

If the threshold of M2 is $V_{T,DEP} = -3.5$ V, then:

µ $V_{DS} > V_{GS} - V_{T,DEP}$ is the saturation condition
µ $V_{DS} < V_{GS} - V_{T,DEP}$ is the linear zone condition.

Assuming a supply voltage of 5 V and being $V_{GS} = 0$ and V_{DS} = VDD - V_{OUT} saturation zone extends between $V_{OUT} = 0$ V to $V_{OUT} = 1.5$ V.

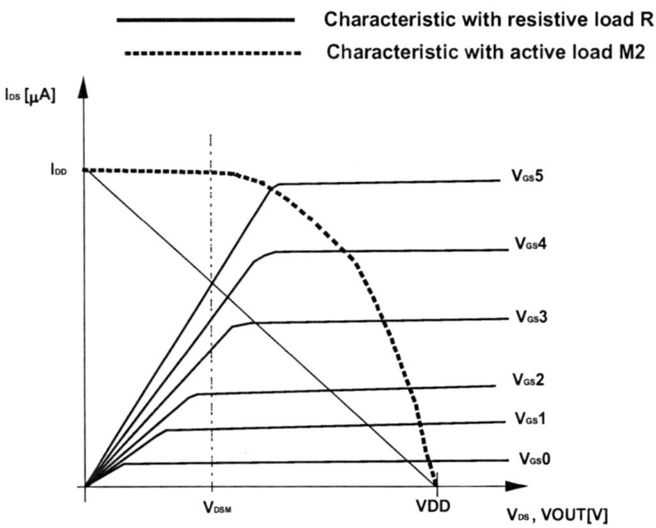

Fig. 5.4. Comparison between the characteristic of active and passive load

The advantage with respect to the behavior of the passive resistive load R is given by the fact that in the saturation region M2 can be compared to an ideal current generator, while in linear zone it behaves like a resistor. The overall effect is that it is able to provide a higher current with much smaller dimensions.

The commutation of a capacitive load is very important and it is always present in MOS devices. In Fig. 5.5 we study the commutation of the inverter composed of M1 and M2 with a capacitive load C connected to the output. Let the initial state be $V_{IN} = 0$ V and $V_{OUT} = VDD$. The capacitor C is completely charged and it must be discharged through M1. To achieve a fast discharge of C, M1 should be big; to be able to charge C, by turning off M1, in a short time, we must have a big current source, i.e. M2 must be big.

On the other hand, if M2 is too big we are no longer able to discharge C, because if V_{IN} is VDD and we want V_{OUT} voltage near to ground, M2 must be able to provide less current that the one that M1 can sink. Otherwise the two transistors get to a "ratio" condition and the output node is set to a voltage value proportional to the partition of their conductances.

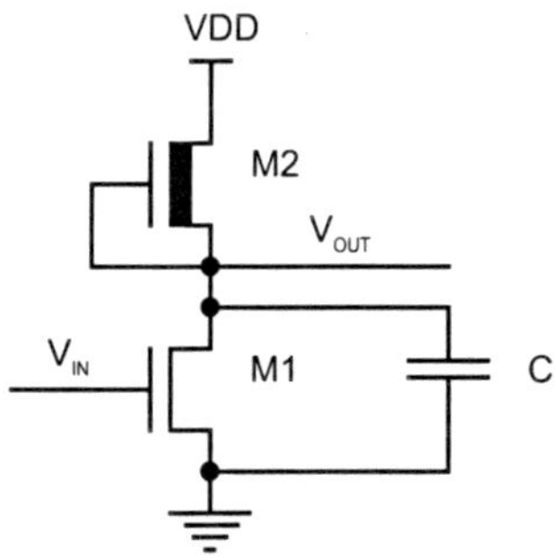

Fig. 5.5. NMOS inverter with output capacitive load

Figure 5.6 shows the equivalent circuit for the inverter of Fig. 5.5. Recalling that the equivalent resistance R_{eq} seen by looking into the source of M2 is equal to

$$R_{eq} = \frac{1}{g_{mb2} + g_{ds2}} \cong \frac{1}{g_{mb2}} \tag{5.1}$$

we can obtain the voltage gain of the amplifier stage

$$\frac{V_{OUT}}{V_{IN}} = -\frac{g_{m1}}{g_{mb2}} \propto -\sqrt{\left(\frac{W}{L}\right)_{M1} \cdot \left(\frac{L}{W}\right)_{M2}} \tag{5.2}$$

Let's recall that the dimensions of M1 and M2 are the only parameters that the designer can set: for instance, for a 2 µm process, size of M1 can be 10 µm/2 µm while size of M2 can be 4 µm/8 µm. Transistor M2 is resistive; the mean current provided by a transistor of this kind is 200÷300 µA/□, therefore M2 can provide 100÷150 µA. Our inverter has not been designed to drive big capacitances.

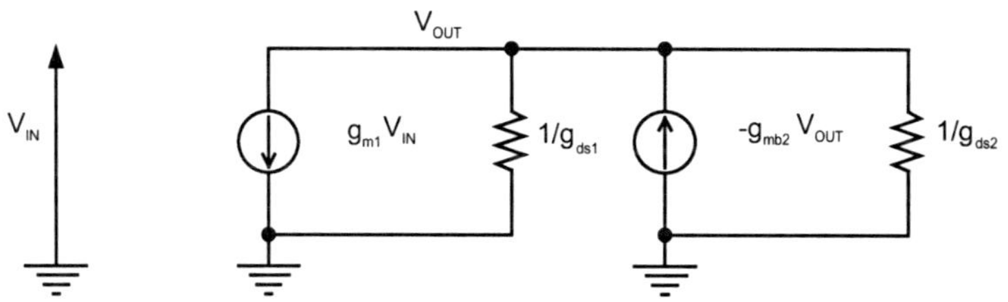

Fig. 5.6. Simplified equivalent circuit for the NMOS inverter

Maybe it is better to specify what we mean by big capacitances. The capacitances seen by the integrated circuit towards the external are comprised between 30 pF and 100 pF. Internally, biggest capacitances are given by ground and supply nodes, and they can be as high as some tens of nanoFarad. The capacitive value for a gate is in the order of some tens of femtoFarad. An internal node can have a maximum capacitance of some picoFarad. Our inverter can therefore charge in 1 ns:

$$C = I\frac{\Delta t}{\Delta V} = \frac{150 \cdot 10^{-6} A \cdot 10^{-9} s}{5V} = 30\, fF \tag{5.3}$$

that is a gate of some tens of square microns (very small indeed!).

The next step is to move to the CMOS inverter.

The main issue with the NMOS inverter is the current consumption in DC: when V_{IN} is equal to VDD, both transistors are tuned on and a current is flowing from power supply to ground. If "complementary" p-channel devices are used, the current consumption in DC is no longer present. Figure 5.7 shows the schematic of the CMOS inverter.

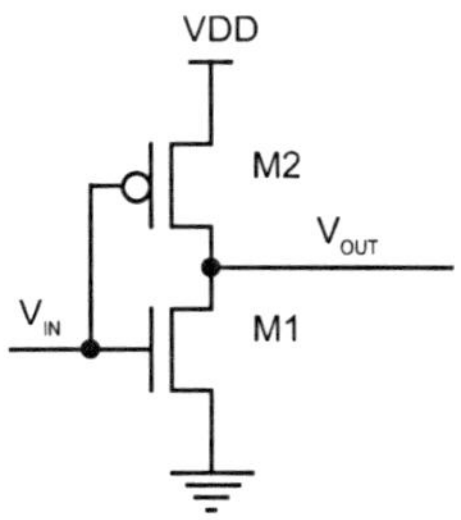

Fig. 5.7. CMOS inverter

The two gates are connected to the same input signal V_{IN}, drain junction of M1 and source junction of M2 are short-circuited to provide the output. When V_{IN} is at ground, M2 is completely turned on, while M1 is turned off and V_{OUT} is at VDD. On the other hand, if V_{IN} is at VDD, M2 is turned off, M1 is turned on and V_{OUT} is at ground. Contrary to the case of the NMOS inverter, V_{OUT} always reaches power supplies. Current consumption occurs only during commutation when, owing to the fact that V_{IN} moves in a finite time, both transistors are turned on. This current is also known as *crow-bar* current.

Problem 5.2: The issue of power consumption is particularly critical for those inverters that implement the output buffers. Design a CMOS inverter with the minimum possible crow-bar current.

In order to discuss the CMOS inverter, it is necessary to recall that the mobility of the holes is approximately 1/3 of the one of the electrons.

The expression for the current is:

$$I_{DS} = \mu C_{ox} \frac{W}{L} f(V_{DS}, V_{GS}, V_T) \tag{5.4}$$

Assuming the same kind of oxide for both the transistors (hypothesis that is always true) a p-channel and a n-channel have the same current when:

$$\left(\frac{W}{L}\right)_{PMOS} = \frac{\mu_n}{\mu_p}\left(\frac{W}{L}\right)_{NMOS} \tag{5.5}$$

A "balanced" inverter in a 0.8 μm process requires, for instance,

$$\left(\frac{W}{L}\right)_{M1} = \frac{10}{0.8}; \left(\frac{W}{L}\right)_{M2} = \frac{30}{0.8} \tag{5.6}$$

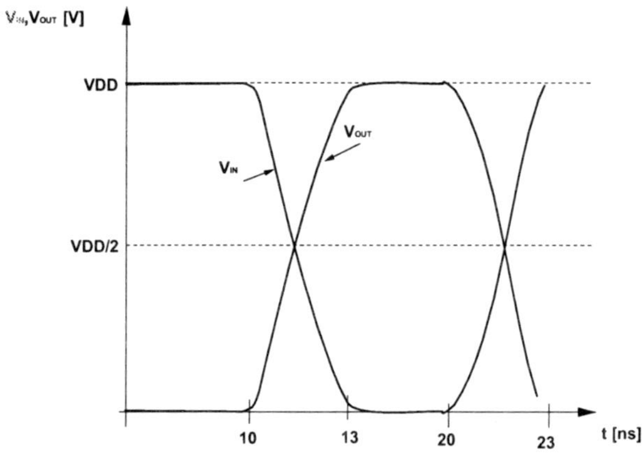

Fig. 5.8. Commutation of the balanced CMOS inverter

The result with a capacitive load is shown in Fig. 5.8, where the intersection of the input and output voltages is VDD/2. If we want to vary the crossing point of the characteristics, we need to act on sizes. If we design M1 >> M2 it is much easier to pull V_{OUT} to ground since M2 cannot provide much current and M1 is anyway able to sink it all.

In this case the crossing moves towards the lower end and it becomes lower than VDD/2. If we design M2 >> M1, M1 is no longer able to sink all the current that M2 provides, and therefore the V_{OUT} node remains high for a longer time; the crossing point moves towards values higher than VDD/2. Figure 5.9 illustrates this effect.

Problem 5.3: A popular discussion among designer is whether a NMOS inverter has a faster switching time than a CMOS: try to put an end to this argue!

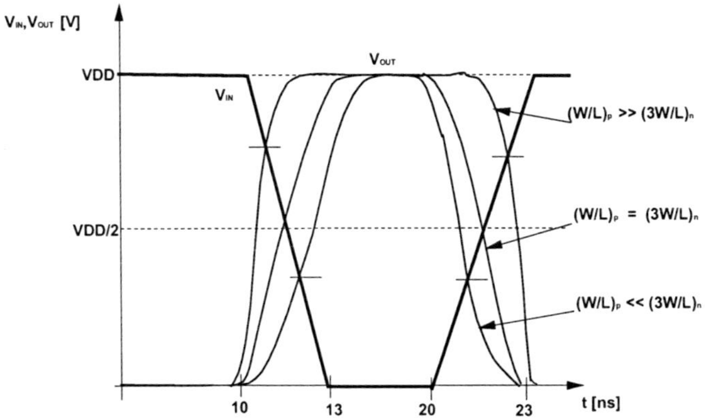

Fig. 5.9. CMOS inverter with different triggering points

5.3 The Cascode

The advantage of *cascode* configuration is the reduction of the effect given by Miller amplification of the coupling capacitance between the input and the output signal through, for instance, the gate/drain capacitance. Figure 5.10 shows the circuit where the parasitic capacitance between gate and drain is depicted[2].

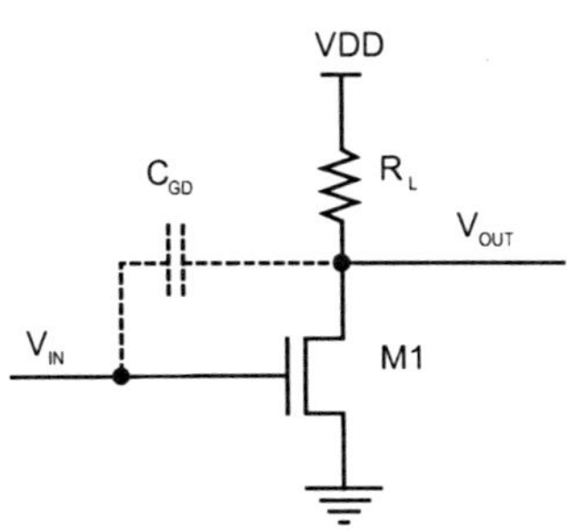

Fig. 5.10. Parasitic capacitance C_{GD} is amplified by the Miller effect

[2] Remember that between two nodes a parasitic capacitance is always present.

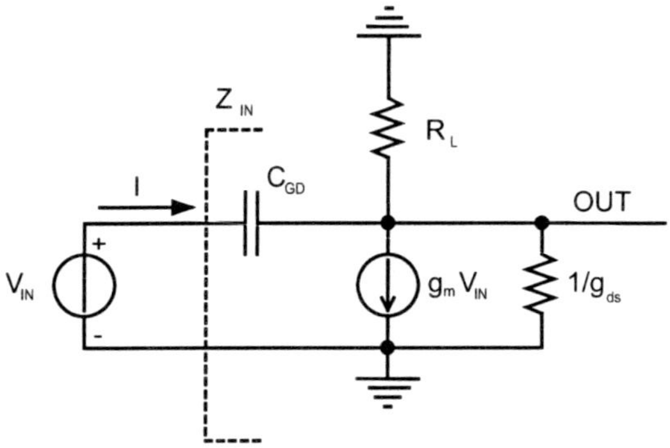

Fig. 5.11. Equivalent circuit for the inverter in Fig. 5.10

By substituting the transistor with its equivalent circuit, we get to Fig. 5.11, for which we can write that:

$$I = sC_{GD}(V_{IN} - V_{OUT}) \tag{5.7}$$

Disregarding the feedback contribution given by the current in C_{GD}, we obtain, for the output voltage:

$$V_{OUT} \cong -g_m R_L V_{IN} \tag{5.8}$$

and therefore:

$$(V_{IN} + g_m R_L V_{IN})sC_{GD} = I \tag{5.9}$$

$$V_{IN}(1 + g_m R_L)sC_{GD} = I \tag{5.10}$$

$$Z_{IN} = \frac{V_{IN}}{I} = \frac{1}{(1 + g_m R_L)sC_{GD}} \tag{5.11}$$

It is as if we had a capacitance equal to $C_{gd}(1+g_m R_L)$. If we assume that the gain of the inverter is in the order of several tens, it is easy to figure out how, thanks to Miller amplification, the small gate/drain coupling capacitance can be not so easy to drive for the previous stage. Coupling capacitance between input and output causes a decrease in the input impedance of the circuit, degrading its frequency response; to mitigate this effect, the configuration shown in Fig. 5.12 can be used.

By interposing M2 transistor, biased with a constant gate voltage, the input capacitance is decoupled from the output node. Let's evaluate the input impedance using the equivalent circuit shown in Fig. 5.13:

$$I = sC_{GD}(V_{IN} - V_X) \tag{5.12}$$

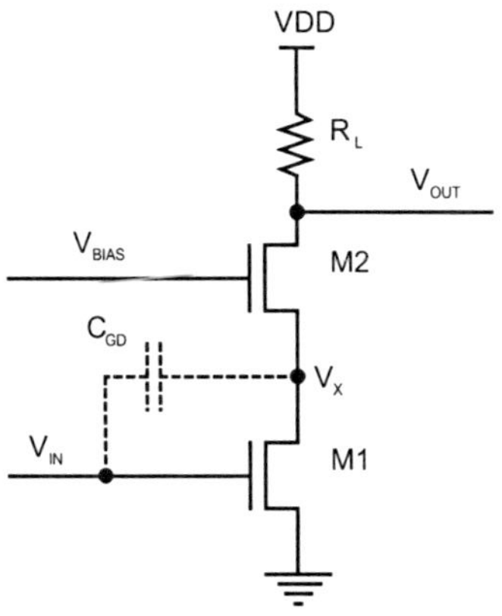

Fig. 5.12. Inverter with cascode stage

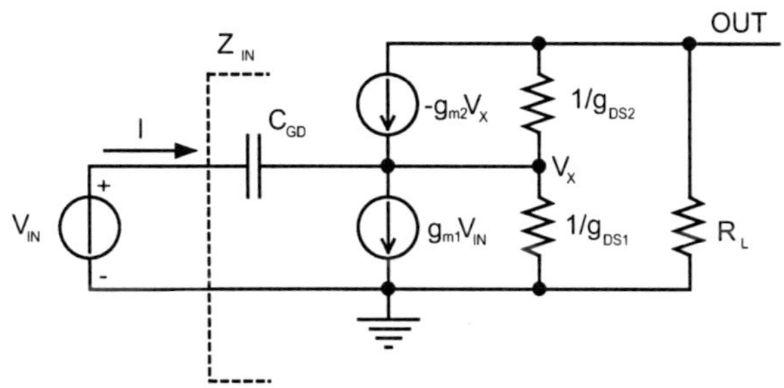

Fig. 5.13. Equivalent circuit for the inverter with cascode stage of Fig. 5.12

Disregarding the output resistances of the transistor, we can write that

$$g_{m1}V_{IN} = -g_{m2}V_X \tag{5.13}$$

$$V_{OUT} = g_{m2}V_X R_L \tag{5.14}$$

By combining the last two equations together we get

$$Z_{IN} = \frac{V_{IN}}{I} = \frac{1}{\left(1 + \frac{g_{m1}}{g_{m2}}\right) sC_{GD}} \tag{5.15}$$

In case $g_{m1} << g_{m2}$ we have $Z_{IN} \sim 1/(sC_{GD})$. Even if M1 and M2 had similar sizes, Miller amplification factor is a small number, and feedback effect has been eliminated. We can also derive the expression for the voltage gains A_{OUT} and A_X

$$A_{OUT} = \frac{V_{OUT}}{V_{IN}} = -g_{m1}R_L \tag{5.16}$$

$$A_X = \frac{V_X}{V_{IN}} = -\frac{g_{m1}}{g_{m2}} \tag{5.17}$$

A_X gain must be low to reduce Miller effect, while A_{OUT} must be high because it is the real gain of the stage.

Figure 5.14 shows a NMOS cascoded configuration where the load resistor has been replaced by an active load (M3). Same considerations as above apply, considering the output resistance of M3 instead of R_L.

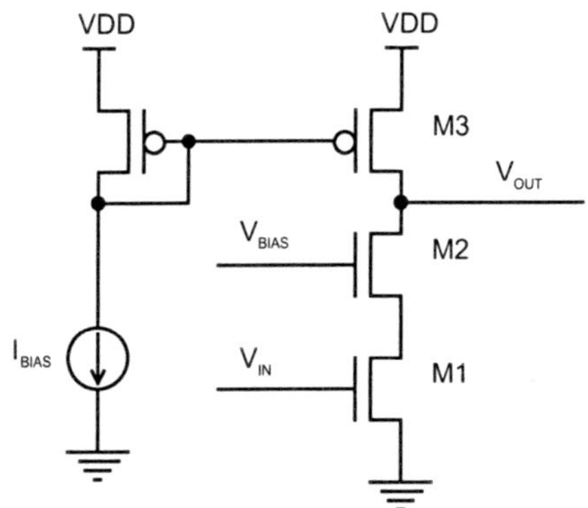

Fig. 5.14. Cascoded stage with active load

5.4 Differential Stage

In this section the fundamental concepts of one of the most important and widely used circuital configurations, the differential stage, are presented. Figure 5.15 shows the principle scheme of a NMOS differential amplifier. The two loads, M3 and M4, are considered to be equal, as well as M1 and M2; I_0 is an ideal current generator. This circuit provides a difference of the output voltages ($V_{o1} - V_{o2}$) that is greater than the difference of the input voltages ($V_{i1} - V_{i2}$), thus implementing an amplifier.

It is better to recall now what we mean when we talk of differential and common mode gain, the two main parameters that are used to evaluate how good is a differential amplifier.

It is clear that we would like to have the output difference as a sole function of the input difference, and therefore, independent of the value of the VDD, or of the possible "fluctuations" on the ground, or of the size of the load, or also of the I_0 current. Instead, the gain depends on the working point of the circuit; for instance, if supply voltage varies so much that M1 and M2 are outside saturation zone, we cannot expect the gain of the differential stage to remain unchanged. Even if supply had a little variation, thus keeping both M1 and M2 in saturation zone, because of the well-known modulation effect of I_{DS} current as a function of V_{DS}, the working point of the input transistor changes, and so does the gain.

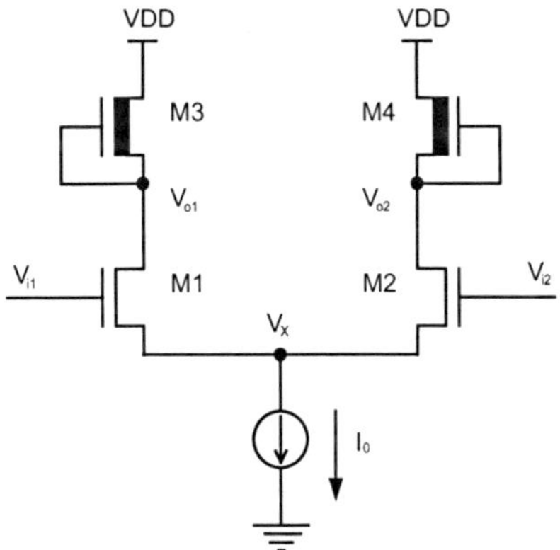

Fig. 5.15. Principle scheme for the NMOS differential stage

Let's now assume that V_{i1} and V_{i2} have a variation of $V_{id}/2$, equal as absolute value but with opposite sign. Then

$$i_{d1} = -i_{d2} \tag{5.18}$$

Since the circuit is completely symmetrical, voltage V_X cannot vary and the differential stage can be re-drawn as shown in Fig. 5.16, where the current generator I_0 is replaced by a short circuit to ground. Now we can calculate the differential gain A_{dm}

$$A_{dm} = \frac{V_{od}}{V_{id}} = -\frac{g_{m1}}{g_{mb3}} \tag{5.19}$$

Therefore, to have a high differential gain, M1 must be bigger than M3. It is worth noting that the current generator I_0 has no influence on the value of the differential gain.

Let's now analyze the common mode, as in Fig. 5.17, where two possible working points for the differential amplifier of Fig. 5.15 are shown. Biasing 1 and 2 are the values in DC of the inputs V_{i1} and V_{i2}; we would like to have the same output voltage in both cases having equal voltage differences on the input.

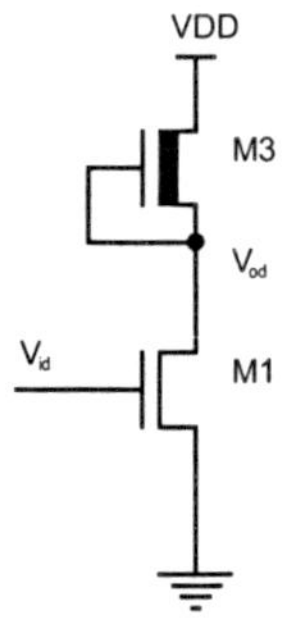

Fig. 5.16. Simplified configuration used to calculate A_{dm}

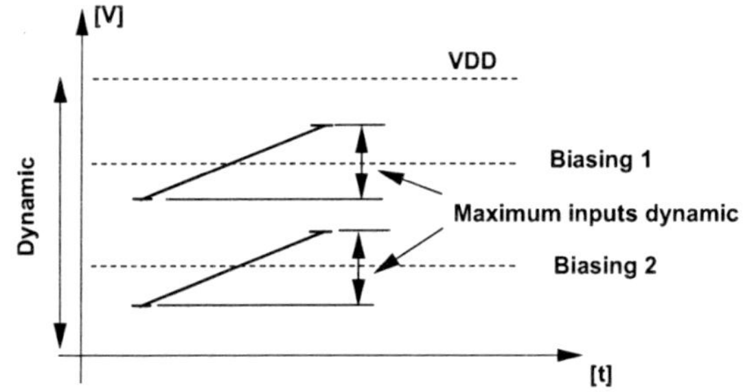

Fig. 5.17. Different working points for the differential stage

The contribution of the biasing is defined as "common mode" because it equally affects both sides of the differential stage. Let's assume to increase the input voltages V_{i1} and V_{i2} of the same quantity V_{ic} (Fig. 5.18): V_{o1} and V_{o2} would tend to decrease, but thanks to the current generator I_0, V_X voltage increases preventing the decrease of both V_{o1} and V_{o2}.

If the current generator were not present, we would not have any kind of feedback, and the output nodes would change their voltage even if the inputs were short-circuited! Therefore if I_0 is an ideal generator, the common mode gain will be zero because no variation occurs on the outputs when the inputs vary.

$$A_{cm} = \frac{\Delta V_o}{\Delta V_i} = 0 \tag{5.20}$$

The equivalent circuit for the calculation of the common mode gain A_{cm} is shown in Fig. 5.19, where R_G is the impedance of I_0 in the real case. If we assume the perfect symmetry of the circuit also in this case, it is possible to analyze just one half of the circuit, provided that we divide the resistance R_G into two parallel resistors whose value is twice the original one.

$$V_X = g_{m1}\left(V_{ic} - V_X\right) \cdot 2 \cdot R_G \tag{5.21}$$

$$A_{cm} = \frac{V_{oc}}{V_{ic}} - \frac{g_{m1} / g_{mb3}}{1 + g_{m1} 2R_G} \tag{5.22}$$

As we have previously said, the common mode gain decreases as resistance R_G increases.

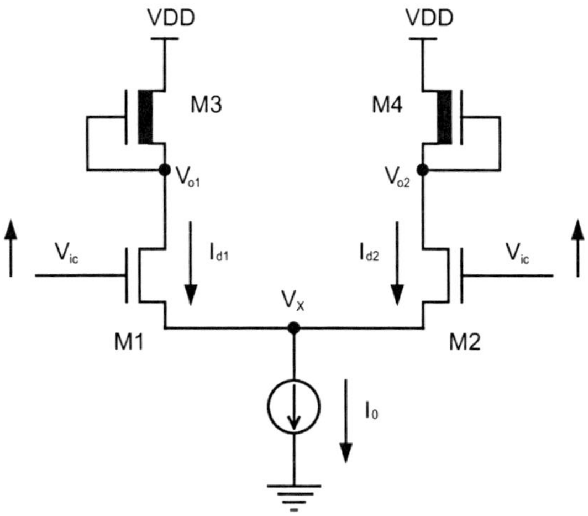

Fig. 5.18. Common mode

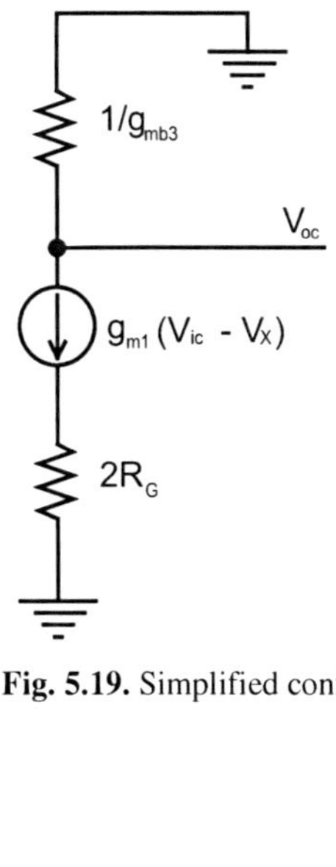

Fig. 5.19. Simplified configuration for the calculation of the common mode gain A_{cm}

An important figure of merit is the Common Mode Rejection Ratio (CMRR), ratio between the differential mode gain and the common mode one. A good differential amplifier must therefore have a high CMRR.

In the following, we will also see the use of differentials as comparators.

Problem 5.4: Analyze the differences between PMOS-input and NMOS-input differential stages.

5.5 The Source Follower[3]

A voltage buffer is a circuit that is able to transfer the voltage signal on its input directly to the output independently of the applied load. One of the most widely used schemes is shown in Fig. 5.20: it is a feedback circuit based on the concept of virtual ground. In this case V_{OUT} voltage remains equal to V_{IN} by exploiting the feedback between nodes V_{O2} and V_{OUT}. When the circuit is powered up, V_{OUT} is at ground, therefore only M1 is turned on and V_{O2} is at VDD. M3 is turned on and it starts flowing current on the load R_L and to cause V_{OUT} to increase, which turns on M2, making V_{O2} decrease, which limits the current of M3 reducing V_{OUT}. This "loop" goes on until the differential stage has reached its point of equilibrium, i.e. the currents flowing in the two resistors R are equal to $I_0/2$.

Generally the gain stages (for instance the inverter with active load) have output impedance that is too high to drive low resistive or large capacitive loads. In these cases it is necessary to add an output stage that is able to transfer the signal with low impedance. The simplest structure to achieve this result is the source follower that, unlike the structure shown in Fig. 5.20, is also a level shifter. The circuital implementation is shown in Fig. 5.21 where the output is taken on the source terminal onto which the load R_L is applied.

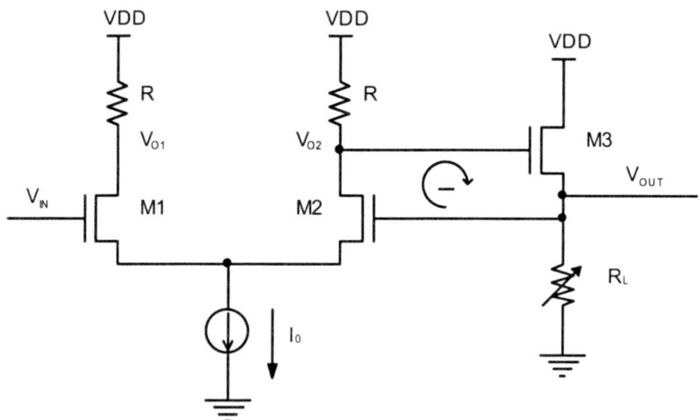

Fig. 5.20. The concept of virtual ground: V_{OUT} "follows" the value of V_{IN} thanks to the negative feedback that is present in the loop shown.

[3] When dealing with bipolars, it is called emitter follower.

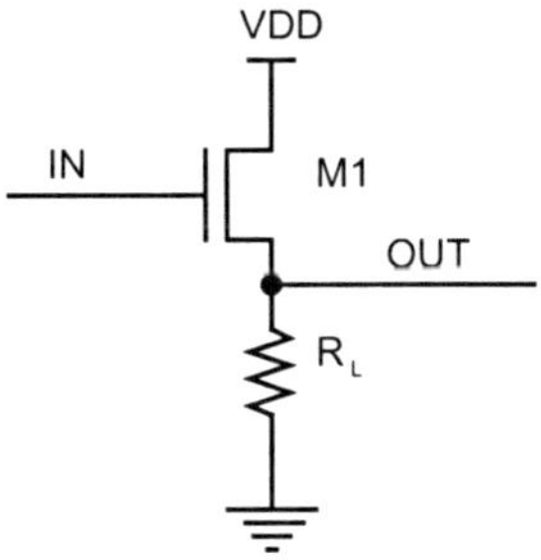

Fig. 5.21. Source follower

Figure 5.22 is the equivalent circuit for Fig. 5.21, where the voltage gain, neglecting the parasitic capacitances, is equal to

$$\frac{V_{OUT}}{V_{IN}} = \frac{g_{m1}}{g_{m1} + g_{mb1} + g_{ds1} + (1/R_L)} \tag{5.23}$$

Assuming that M1 is in saturation zone, that R is big enough to neglect the term that contains it and that the body effect is negligible, we get a unit gain, i.e. the input is equal to the output.

If we replace the load resistance R_L with an active load, as shown in Fig. 5.23, and we call r_O and C_O the total resistance and the total capacitance between the output and ground, we obtain a gain equal to:

$$\frac{V_{OUT}}{V_{IN}} = \frac{sC_{gs} + g_{m1}}{s(C_{gs} + C_O) + g_{m1} + g_{mb1} + g_{ds1} + (1/r_O)} \tag{5.24}$$

The gain in DC always tends to one, while in frequency we have a zero and a pole; if the zero is compensated with the pole, the bandwidth can become very large.

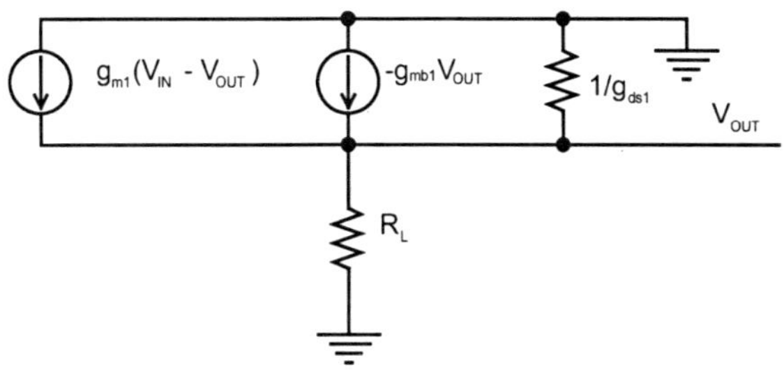

Fig. 5.22. Equivalent circuit for the source follower

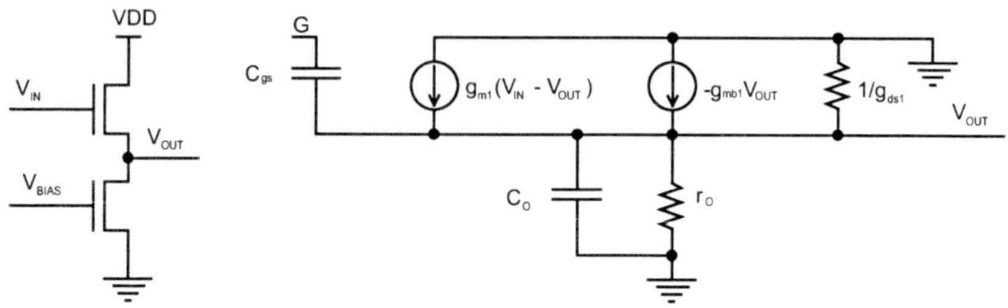

Fig. 5.23. Source follower with active load and corresponding equivalent circuit

5.6 Voltage References

The simplest implementation of a voltage reference is shown in Fig. 5.24. The partition constituted by R1 and R2 allows the generation of an output voltage that is a fraction of the supply voltage VDD.

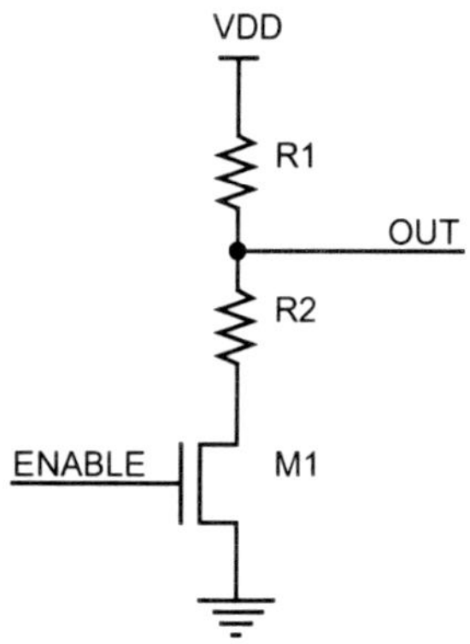

Fig. 5.24. Voltage reference based on the partition of the supply voltage

If the equivalent resistance of M1 is not taken into account, the voltage at the node OUT is equal to

$$V_{OUT} = VDD \frac{R2}{R1 + R2} \tag{5.25}$$

Transistor M1 is used to avoid consumption when the partition circuit is not active.

V_{OUT} voltage does not depend neither on the temperature nor on the process, under the assumption that the resistors are fabricated using the same layer. Dependence on supply voltage still occurs, and we will discuss this issue later. We have now to choose which of the available layers can be used to realize the resistors.

Resistivity of poly2 is too low due to the presence of silicide, so resistances that cause low power consumption cannot be fabricated with poly2.[4] The consumption of the partition circuit is in fact

$$I = \frac{VDD}{R1 + R2} \tag{5.26}$$

If VDD is equal to 5 V and the target consumption is 10 μA, the value of the resistance must be 5 MΩ. To get such a value using poly2, it is necessary to use very long stripes that occupy too much space. Similar considerations apply to active area, whose value of resistance is anyway an order of magnitude higher than the one of poly2. Therefore n-well is the remaining choice, whose resistivity is three orders of magnitude higher than the one of poly2. We have already seen how to realize a resistor in n-well and the associated depletion issue. Resistive partition circuits are used anyway thanks to their simple implementation whenever precision is not a key factor, and stabilizing, if necessary, their supply voltage.

Let's now analyze some voltage references implemented using active devices.

5.6.1 NMOS

Figure 5.25 shows a reference circuit realized using n-channel transistors; this implementation is stable with respect to variations of the supply voltage. If M1 works in saturation region (high V_{DS}) a variation of VDD moves the working point of M1, onto the characteristic $V_{GS} = 0$, causing M1 to remain in saturation region. Figure 5.26 shows the graph of both VDD and the output node V_{REF} for which we can write that

$$V_{REF} = V_{T,M2} \pm \left|V_{T,M1}\right| \cdot \sqrt{\beta_1 / \beta_2} \tag{5.27}$$

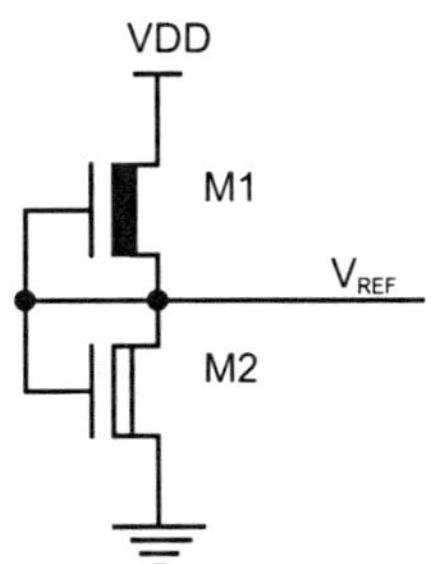

Fig. 5.25. Voltage reference stable with respect to VDD variations

[4] Recall that the overall consumption in read mode for a Flash memory device is in the order of tens of mA.

It is possible to increase the stability interval for V_{REF} by modifying the circuit as shown in Fig. 5.27. Let's observe that the stability range for the circuit depicted in Fig. 5.25, that is between 3.5 V and 5 V, is due to the fact that for low VDD the value of the V_{REF} node can only decrease, while in the circuit of Fig. 5.27 M4 transistor with its gate to VDD tends to push V_{REF} upwards for low values of VDD because less current flows across M4.

When VDD is stable, the effect of M4 becomes negligible and the circuit behaves like the one in Fig. 5.25. Transistor M3 acts as an auxiliary current generator when, being VDD low, M1 is not able to provide all the required current to M4. In this way the value of V_{REF} only depends on the ratio between M1 and M2, while the stability range is regulated by means of M3 and M4.

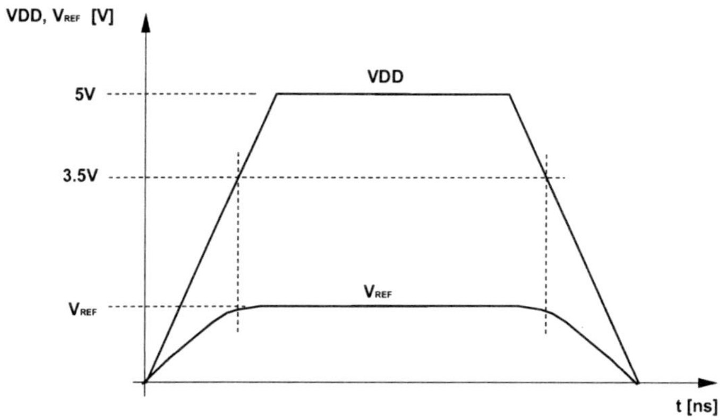

Fig. 5.26. Stability of V_{REF} voltage (Fig. 5.25) for VDD between 3.5 V and 5 V

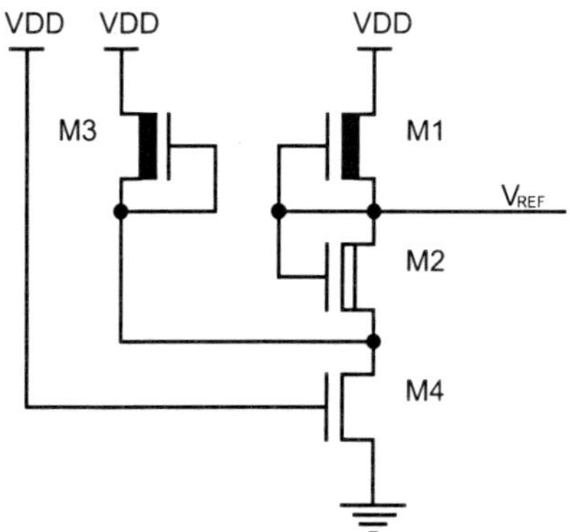

Fig. 5.27. Circuit to increase the stability interval of V_{REF} voltage (Fig. 5.25)

Another voltage reference circuit that can be realized using NMOS transistors only is shown in Fig. 5.28.

Assuming that both the transistors are working in saturation region, we can write that

$$\beta_1(-V_{REF} - V_{T,M1})^2 = \beta_2 \cdot V_{T,M2}^2 \tag{5.28}$$

and therefore

$$V_{REF} = -V_{T,M1} \pm V_{T,M2}\sqrt{\beta_2 / \beta_1} \tag{5.29}$$

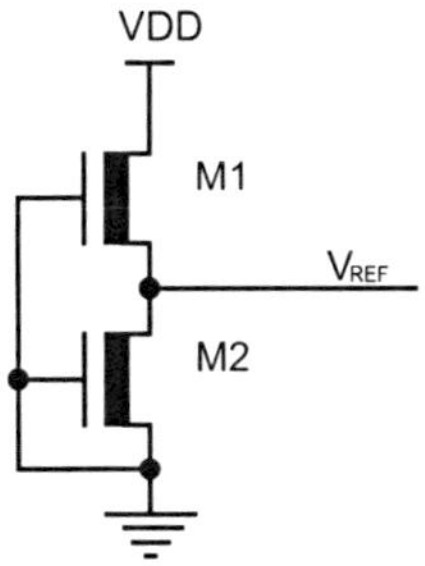

Fig. 5.28. Depletion NMOS voltage reference circuit

We must consider the smallest among the two V_{REF} values, because the maximum possible V_{REF} is equal to $-V_{T,M1}$ (the threshold voltage for a depletion NMOS is negative); for higher voltages, M1 is turned off. The threshold voltages for both M1 and M2 are explicitly shown because the body effect must be taken into account for M1.

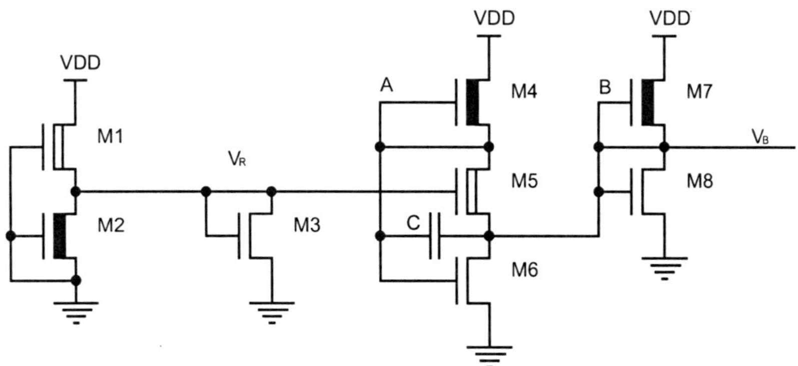

Fig. 5.29. Generator for V_B reference voltage

Another NMOS reference is shown in Fig. 5.29. Transistors M1 and M2 set the required voltage around the threshold of M1; M3 is a protection that limits V_R voltage to the V_T of M3, obviously $V_{T,M3} > V_{T,M1}$. Transistors M4, M5 e M6 constitute a feedback control stage that is able to stabilize the output node V_B. If the voltage of V_R increases, the voltage of node A tends to decrease, cutting off M6 that will bring the voltages of nodes A and B back to the fixed values, i.e. $2V_{T,M1}$ for the node B. Capacitor C acts as a filter, to absorb the current peaks that might occur when the device is switched on; finally, M7 and M8 act as auxiliary current generators for the output node.

5.6.2 CMOS

Let's now analyze some voltage references realized using CMOS transistors. The basic principles are the same previously described, with the additional advantage that p-channel transistors can be used and therefore the contribution of the body effect can be eliminated, since in every CMOS technology the substrate of the PMOS transistors can be independently biased. Figure 5.30 shows an implementation where the VDD is partitioned using two p-channel transistors: in case M1 and M2 have the same size, V_{OUT} output voltage is equal to VDD/2.

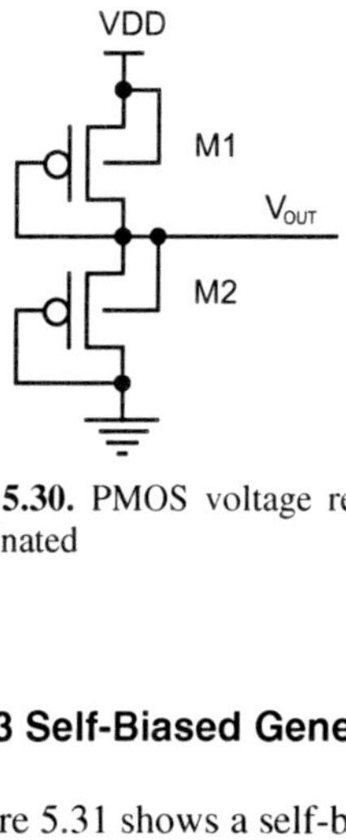

Fig. 5.30. PMOS voltage reference where the contribution of the body effect has been eliminated

5.6.3 Self-Biased Generator

Figure 5.31 shows a self-biased circuit that is more complex than those studied before; it is made of two current mirrors, one towards VDD, realized with a pair of equally-sized PMOS transistors, ant the other towards ground with a degeneration resistance. PMOS mirror forces the same current in the right and in the left branch. M3 and M4 transistors have a different size ratio where $\beta_3 > \beta_4$: since they must sink an equal drain current, their V_{GS} must be different. All the transistors must work in saturation region to ensure the desired behavior of the circuit.

Transistor M4 is for sure in saturation, since it is diode-connected and V_{GS4} is equal to

$$V_{GS4} = V_{T,N} + \sqrt{I_{REF} / \beta_4} \tag{5.30}$$

$$\beta_4 = \frac{\mu_n C_{ox} W_4}{2L_4}; K = \frac{W}{L} \tag{5.31}$$

Neglecting the body effect on M3, we can calculate the current I from the voltage drop on resistance R

$$I_{REF} = \frac{V_{GS4} - V_{GS3}}{R} = \frac{1}{R}\left(\sqrt{I_{REF} / \beta_4} - \sqrt{I_{REF} / \beta_3}\right) \tag{5.32}$$

The equation has two possible solutions

$$I_{REF} = 0; I_{REF} = \frac{2\Psi}{C_{ox}} \cdot \frac{1}{\mu_n R^2} \tag{5.33}$$

where

$$\Psi = \frac{1}{K_3} + \frac{1}{K_4} - \frac{2}{\sqrt{K_3 K_4}} \tag{5.34}$$

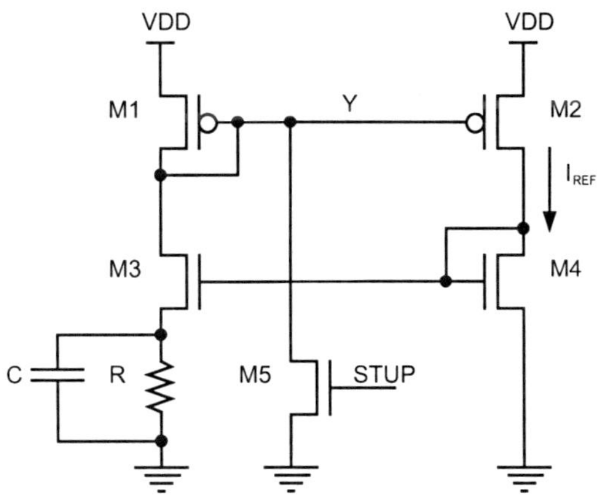

Fig. 5.31. Self-biased generator

ψ factor just depends on geometric parameters. The solution where the current is not zero (the only one to take into account) is inversely proportional to two terms that depend on temperature. Since mobility increases when temperature decreases, and the resistance behaves the opposite, the two variations can be partially compensated to reduce the variation range of the current I_{REF} with temperature. Generally, the best compensation is achieved by realizing the resistance in active area; using this layer is not always possible because of the associated large area occupation, above all when consumption of few microamperes are required. There is no dependence of the I_{REF} current with the supply voltage VDD.

Transistor M5 provides the required start-up to the circuit, keeping the p-channel turned on for some tens of nanoseconds. In this way the current forced in M1 lets a current flow in M2 that biases M4; this one turns on M3 and at this moment the STUP signal is turned off.

M3 and M4 transistors can also be of different type (for instance M3 natural an M4 LVS); in this case it is not used the overdrive difference between M3 and M4, but their V_T difference is exploited, which is applied at the terminals of resistance R.

The switching on of this circuit is quite slow as is for all the feedback circuits (in the order of 100 ns in this case) because the settling of V_Y to its equilibrium value causes a series of variations in the currents of all the transistors. To decrease the switching on transient, it is possible to "help" the nodes to get to their equilibrium value; capacitance C is added, which represents a short-circuit towards ground at power on, thus eliminating the contribution of the resistance, which in turn is the element with the most significant inertia in the power-on chain.

Problem 5.5: Add the elements that allow switching off the circuit in such a way that it does not provide current contributions during stand-by phase.

5.6.4 Band-Gap Reference

The last voltage reference that we are going to discuss is the *band-gap reference*, whose main feature is to produce an output voltage that is stable with respect to both VDD and temperature variations. This circuit is widely used in the latest generation of non-volatile memory devices, to control the voltages during program and erase operations. First generation EPROM memories, ancestors of the Flash, allowed for instance a variation of the drain voltage around its typical value of some hundreds of millivolt. Nowadays the size of the cells asks for an increased precision for the voltages, above all for drain voltage during program and gate voltage during verify operations. To carry out this requirement, the band-gap reference circuit is used. This circuit exploits the possibility of compensating temperature variation of the V_{BE} of a bipolar transistor with a voltage drop proportional to the thermal potential V_t

$$V_t = \frac{kT}{q} \tag{5.35}$$

where k is Boltzmann's constant, q is the charge of the electron and T is the absolute temperature.

V_{BE} of a bipolar transistor decreases as the temperature increases, and this behavior can be linearized around a reference temperature, thus defining a first order thermal coefficient

$$\gamma_{V_{BE}} = \left.\frac{\partial V_{BE}}{\partial T}\right|_{T=T_0} \tag{5.36}$$

$$V_{BE} \cong V_{BE0} + \gamma_{V_{BE}} \cdot (T - T_0) \tag{5.37}$$

On the thermal potential, the following relation can be easily derived

$$\gamma_{V_t} = \frac{\partial V_t}{\partial T} = \frac{k}{q} = 0.085mV/K \tag{5.38}$$

Therefore the thermal potential is proportional to temperature, while the thermal coefficient of V_{BE} depends on the reference temperature that has been chosen to calculate it.

Coefficient γ of V_{BE} lies between – 1 mV/K and - 2 mV/K; therefore γ of V_{BE} and of V_t have opposite signs. Now a V_{BG} voltage can be defined as follows

$$V_{BG} = V_{BE} + G \cdot V_t \tag{5.39}$$

For which the corresponding thermal coefficient can be eliminated by choosing a value of G such that

$$G = \frac{\left|\gamma_{V_{BE}}\right|}{\gamma_{V_t}} \tag{5.40}$$

Such a compensation technique is extremely simplified and it is known as first order because only the first order thermal coefficient of V_{BE} is compensated. There are more precise techniques that can take into account non-linear terms too. Furthermore a first order compensation is less precise as the target temperature range increases.

A widely used band-gap reference circuit is shown in Fig. 5.32.

Assuming an ideal situation (infinite gain, no voltage offset, etc.), its input voltages V^+ and V^- are equal and therefore the voltage at the terminals of R3 resistor can be written as

$$V_{R3} = V_{BE0} - V_{BE1} = V_t \cdot \ln \frac{I_{C0} I_{S1}}{I_{C1} I_{S0}} \tag{5.41}$$

$$I_C = I_S \exp\left(\frac{V_{BE}}{V_t}\right) \tag{5.42}$$

I_{C0} and I_{C1} are the collector currents of the bipolar transistors Q0 and Q1, while I_{S0} and I_{S1} are the corresponding currents of reverse-biasing of the base-emitter junction that depend on physical and geometrical parameters (it is reasonable to assume that these parameters are the same for both bipolar transistors) according to the following equation

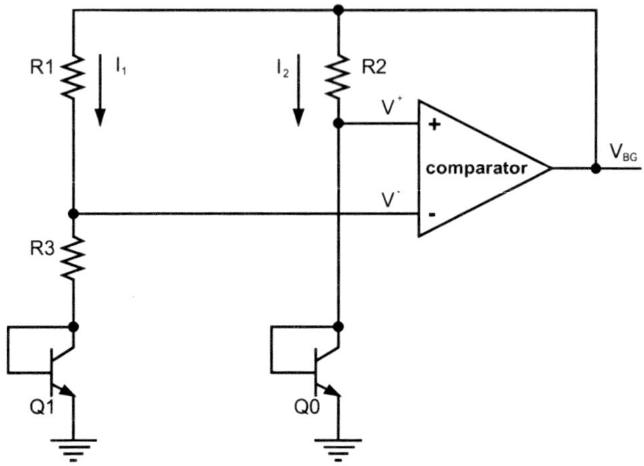

Fig. 5.32. Schematic of the band-gap reference circuit

$$I_{S0} = \frac{qD_n n_i^2}{W_B N_A} A_0 ; I_{S1} = \frac{qD_n n_i^2}{W_B N_A} A_1 \tag{5.43}$$

where

- A_0 and A_1 are the emitter areas for the transistors Q0 and Q1;
- n_i is the carrier intrinsic concentration in the semiconductor;
- W_B is the base width;
- D_n is the diffusion constant for the electrons;
- N_A is the base doping concentration.

Always assuming ideal operation, it can be noted that the voltage drop on R1 and R2 is the same and therefore we can write that

$$I_1 \cdot R1 = I_2 \cdot R2 \tag{5.44}$$

$$I_1 = I_{C1}\left(1 + \frac{1}{\beta}\right); I_2 = I_{C0}\left(1 + \frac{1}{\beta}\right) \tag{5.45}$$

where is the current gain of the bipolar transistors.

Therefore by substituting Eqs. (5.43), (5.44) and (5.45) in Eq. (5.41), the latter can be re-written as a function of resistors R1 and R2 and of the area of the bipolars

$$V_{R3} = V_t \ln \frac{R1 \cdot A_1}{R2 \cdot A_0} \tag{5.46}$$

Now, keeping in mind Eq. (5.44), it is possible to determine reference voltage V_{BG}

$$V_{BG} = V_{BE0} + R2 \cdot I_2 = V_{BE0} + R1 \cdot \frac{V_{R3}}{R3} \tag{5.47}$$

By substituting Eq. (5.46) in Eq. (5.47) we get

$$V_{BG} = V_{BE0} + V_t \frac{R1}{R3} \ln \frac{R1 \cdot A_1}{R2 \cdot A_0} \tag{5.48}$$

from which coefficient G can be derived

$$G = \frac{R1}{R3} \ln \frac{R1 \cdot A_1}{R2 \cdot A_0} \tag{5.49}$$

In order to achieve a proper dimensioning of the circuit, the first order thermal coefficient of V_{BE} must be given and then the value of G must be calculated using Eq. (5.40). Furthermore, design parameters R1, R2, R3, A_0 and A_1 must be chosen accordingly to get the desired value of G. There is not a unique possible set of these parameters, and the choice also depends on target performances like consumption, turn-on time of the circuit etc.

It is worth pointing out that coefficient G is in first approximation independent of the temperature, because we only have resistive ratios (on top of the area ratio). In fact it is possible to model the relationship between temperature and resistance using the following equation

$$R_i(T) = R_i(T_0) \cdot [1 + \gamma(T) \cdot (T - T_0)] \tag{5.50}$$

Therefore considering two resistances with the same thermal coefficient , the term inside the parentheses can be simplified and a constant ratio is achieved. The remaining issue is the non-linear behavior of the resistance with applied voltage variation; in these cases, simulations are needed to determine the right value for the resistances.

A last consideration can be done on the need for a start-up circuit. In fact, condition $V_{BG} = V^+ = V^- = 0$ is a possible, but unstable, working point where the circuit could be stuck, or anyway from which it could recover very slowly. Therefore an adequate start-up has a twofold aim: avoid that the circuit remains turned off and allow for a faster settling of the reference voltage.

Problem 5.6: Design the start-up circuit. Suggestion: put the operational out of balance adequately for a short period of time starting from a proper enable signal for the band-gap reference.

An example graph of the relationship of V_{BG} voltage with temperature is shown in Fig. 5.33: it is the result of a simulation of a band-gap circuit designed in 0.15 μm technology. It is usually possible to have one or more non-volatile registers inside the device that allow changing factor G acting on the resistive partitioning of Fig. 5.32 without requiring a photomask change. The different curves in

Fig. 5.33 refer to eight different values for the compensation factor G, assuming that three configuration registers are available.

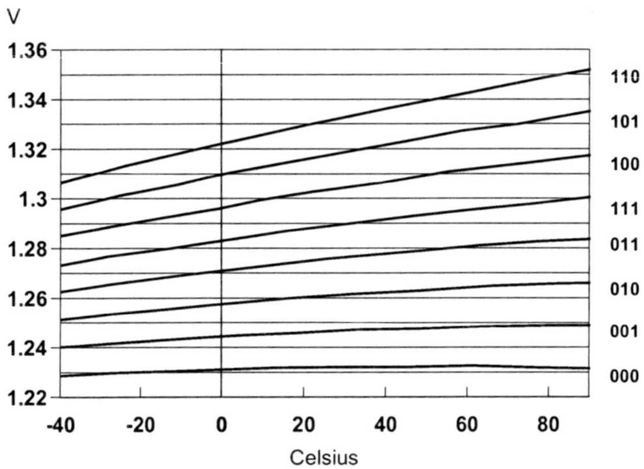

Fig. 5.33. Graphical representation of the band-gap voltage versus temperature and G compensation factor variation

The resulting value for V_{BG} is around 1.25 V. As we will see in the following, voltage regulation is achieved by multiplying the reference voltage by a proper coefficient. For instance, in order to generate 6 V, V_{BG} is multiplied by a factor 4.8. This operation has associated issues: variations of the reference voltage are amplified by the regulator and they can cause significant variations around the typical value of the output voltage. Even if V_{BG} voltage varies of, say, ten millivolt on the whole voltage and temperature range, the final result is that the variation of the regulated voltage is 4.8 times the band-gap one, i.e. some tens of millivolt. That's why it is mandatory to have a very precise and stable V_{BG}.

5.7 Current Mirrors

A widely used circuital configuration is the one known as current mirror. In the following, we will often need to generate a current equal to the reference one, or to a fraction of it, and the current mirror will let us achieve this result. Figure 5.34 shows a current mirror realized with NMOS transistors.

If we assume that M1 works in saturation zone, M1 current I_{REF}, only depends on its V_{GS}. Therefore we can write the following equation:

$$I_{REF} = \beta_1 (V_{GS} - V_{T,N})^2 \tag{5.51}$$

Assuming that M2 is working in saturation zone too, we have

$$I = \beta_2 (V_{GS} - V_{T,N})^2 \tag{5.52}$$

Since M1 and M2 are both LVS and they have the same V_{GS}, then the ratio of their currents is equal to the size ratio:

$$\frac{I_{REF}}{I} = \frac{\beta_1}{\beta_2} \tag{5.53}$$

In case the size of M1 and M2 is the same, then current I is equal to reference current I_{REF}.

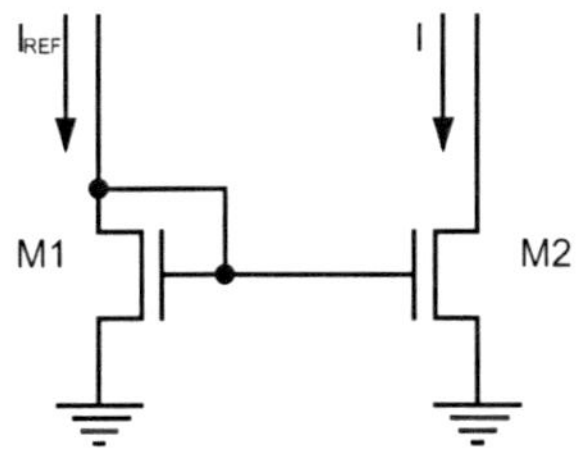

Fig. 5.34. NMOS current mirror

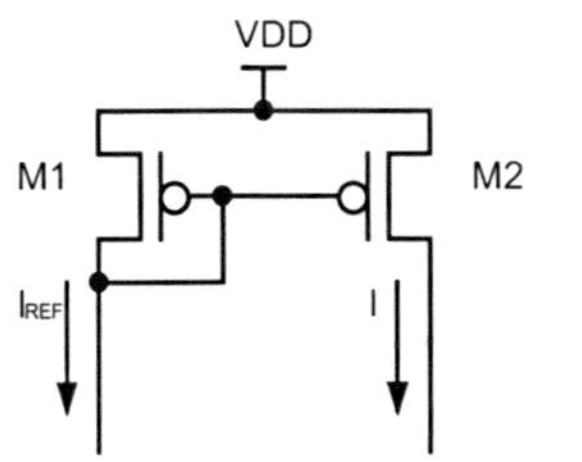

Fig. 5.35. PMOS current mirror

In a similar way it is possible to design a current mirror towards supply voltage using two PMOS transistors (Fig. 5.35). Same considerations apply.

In order to evaluate the quality of a mirror as a current generator, it is necessary to analyze its output impedance. In case of NMOS, the output impedance z_O, based on the small signal model, is equal to

$$z_O = r_{ds2} \propto \frac{L_2}{I_{DS2}} \tag{5.54}$$

If a big output impedance is required, then transistors with a long channel, and therefore with little available current, must be used. A possible solution to increment the output resistance is shown in Fig. 5.36.

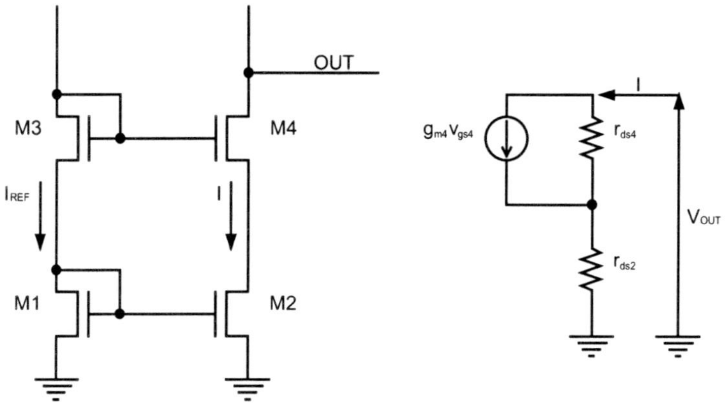

Fig. 5.36. Cascoded current mirror

Output impedance in this case is given by

$$z_O = \frac{V_{OUT}}{I} \tag{5.55}$$

$$z_O = r_{ds2} + r_{ds4}\left(1 + g_{m4} r_{ds2}\right) \tag{5.56}$$

The drawback of this circuit when compared with the one in Fig. 5.34 is a narrower swing for the output voltage. In fact, minimum V_{OUT} is equal to:

$$V_{OUT,\min} = V_{GS1} + V_{GS3} - V_{GS4} + V_{sat2} \tag{5.57}$$

Recalling that the minimum voltage required between drain and source for saturation can be written as

$$V_{sat} = V_{GS} - V_T \tag{5.58}$$

we get

$$V_{OUT,\min} \cong V_{T,N} + 2V_{sat} \tag{5.59}$$

In other words, output voltage can go below gate voltage of M4 of a maximum of a threshold voltage (body effect included). It means that if the gate of M3 is at about 2 V, V_{OUT} cannot go below 1 V (or less, because of body effect).

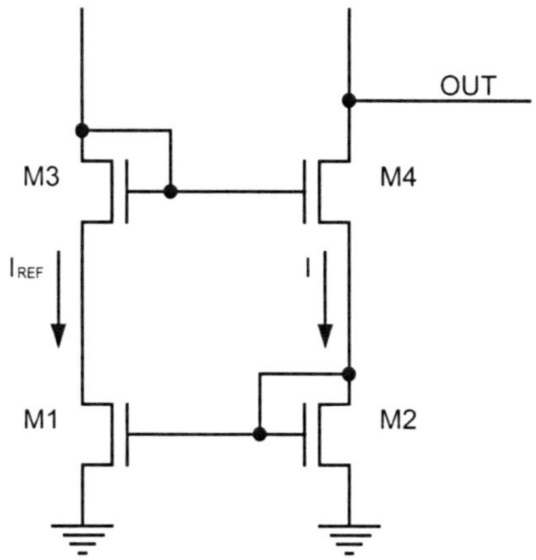

Fig. 5.37. Wilson current mirror

Problem 5.7: Discuss configuration of Fig. 5.37 comparing it with previous configurations. Design mirrors of Figs. 5.36 and 5.37 using complementary PMOS mirrors and discuss the difference with respect to the solution shown in Fig. 5.35.

5.8 NMOS and CMOS Schmitt Trigger

When we have analyzed the inverter, we have seen that it is possible to vary its triggering point modifying the dimensional ratio of the transistors. The Schmitt trigger is an inverter with two triggering points. In other words, this circuit has a high-to-low transition of the output when the input goes above a certain voltage V_{TH}^+ while the opposite transition is inhibited until the input does not go below a voltage V_{TH}^-. In this way we have an inverter that is able to stabilize the switching against rapid triggering by noise as it passes by triggering point.

Figure 5.38 represents the graph of a possible "asymmetric" transition, where V_{IN} is the input signal and V_{OUT} is the output one. The triggering points shown in the figure, i.e. the intersection of the two curves, have different voltage values; the difference between the triggering voltages is called hysteresis.

Let's analyze first the CMOS circuit that is realized to obtain the characteristic shown in Fig. 5.38, and then the NMOS one.

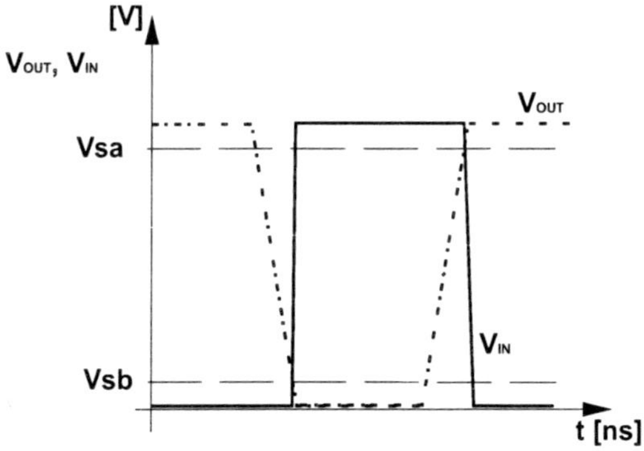

Fig. 5.38. Inverter with asymmetric input-output characteristic

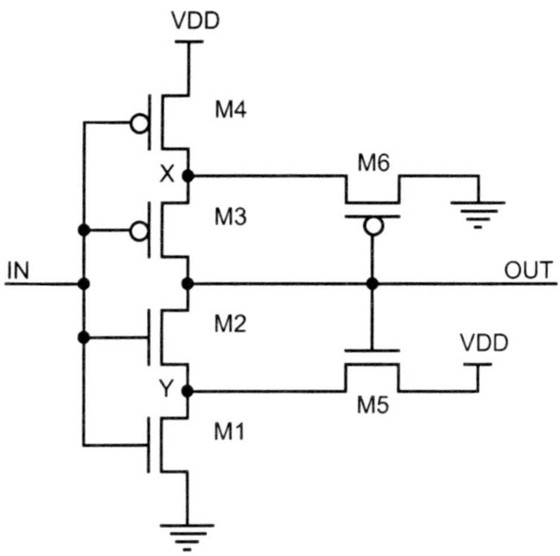

Fig. 5.39. CMOS Schmitt Trigger

CMOS version of the Schmitt trigger is shown in Fig. 5.39: hysteresis is achieved by means of the two complementary MOS, M6 and M5. When V_{IN} is at ground, M3 and M4 are turned on and the output voltage is VDD. Transistor M5 is turned on and it forces V_Y to VDD – $V_{T,M5}$ (body effect included). Assuming a supply voltage of 5 V, V_Y = 3.5 V. For sake of simplicity of the analysis, all and only the transistors involved in this phase are depicted in Fig. 5.40.

When the input voltage goes above the threshold voltage of M1 (1 V), M1 starts to conduct while M2 is still turned off. Voltage V_Y starts to decrease because of the partition between M1 and M5. As V_{IN} keeps increasing, transistor M1 continues to make V_Y decrease until when M2 turns on. At this point the circuit starts to force V_{OUT} to decrease towards ground. Triggering point of the Schmitt trigger is about 3.5 V for this transition.

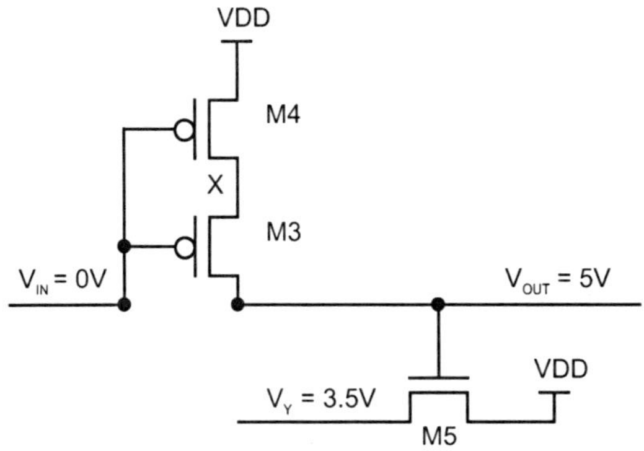

Fig. 5.40. CMOS Schmitt trigger: simplified circuit for the analysis of the biasing in case of input voltage equal to ground

The Schmitt trigger, as well as the inverter seen in Sect. 5.2, belongs to the family of the ratioed logics. Static characteristic, and therefore the value of the hysteresis, can be analytically derived by solving the equation that shows the equivalence of the currents of driver and load transistors getting V_{OUT} as a function of V_{IN}.

Assuming that voltage V_Y is able to bias M1 in its saturation region, we can calculate the triggering point for the high-low transition of the output.

$$V_{IN} = V_Y + V_{T,M4} \tag{5.60}$$

$$\frac{\beta_1}{2}\left(V_{IN} - V_{T,M1}\right)^2 = \frac{\beta_5}{2}\left(VDD - V_Y - V_{T,M6}\right)^2 \tag{5.61}$$

$$V_{IN} = \frac{VDD + \sqrt{\beta_1 / \beta_5} \cdot V_{T,M1}}{1 + \sqrt{\beta_1 / \beta_5}} \tag{5.62}$$

Let's now consider the case of input voltage equal to VDD (Fig. 5.41). Transistors M1 and M2 are turned on and the output is low. Node X is floating and it is forced by M6 to reach its own threshold voltage (1.5 V). When V_{IN} goes one threshold voltage $V_{T,P}$ below VDD, M4 turns on and starts to compete with M6. Owing to the progressive turning on of M4, the voltage of node X starts to increase until a value that lets M3 to turn on causing the transition of the output node towards VDD. In this case the triggering point of the circuit is about 1.5 V.

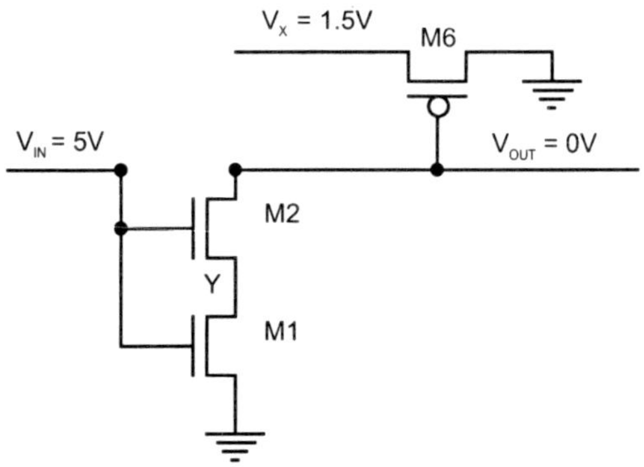

Fig. 5.41. CMOS Schmitt Trigger: simplified circuit for the analysis of the biasing in case of input voltage equal to VDD

The scheme of the Schmitt trigger realized using n-channel transistors only is shown in Fig. 5.42.

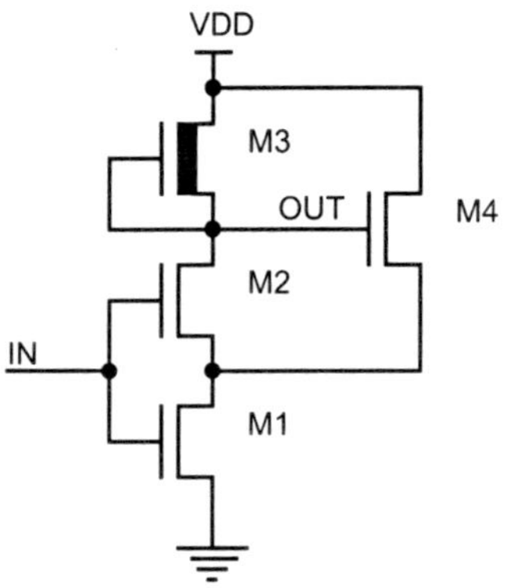

Fig. 5.42. NMOS Schmitt Trigger

Analyzing at Fig. 5.42, it is clear that in case V_{OUT} is equal to zero, transistor M4 is turned off; therefore the current of M4 is present only during one of the two transitions. In particular, when V_{IN} increases, M1 and M2 conduce and, as soon as V_{IN} goes above $V_{T,N}$, M4 is turned off. During the high-low transition of the IN signal (Fig. 5.43), M4 turns on after M2 turns off and only when the voltage at the mode OUT is higher than $V_{T,N}$ (body effect included).

Problem 5.8: Calculate the value of the hysteresis for the circuits of Figs. 5.39 and 5.42.

Another circuital scheme that can be used to get a hysteresis is shown in Fig. 5.44. It is based on two inverters that are connected in positive feedback loop. INV1 is the driver of the circuit under investigation. Let's assume that signal IN is low: M1 and M4 are turned on while M2 and M3 are turned off. If M4 were not present, low-high transition of the input voltage would immediately drive the output to low; being M4 present and turned on, it follows that V_{IN} must increase more than in the case of the driving of a normal inverter, in order to be able to push V_{OUT} low enough (so that it can turn off M4). The drawback of this circuit is the consumption of the driver.

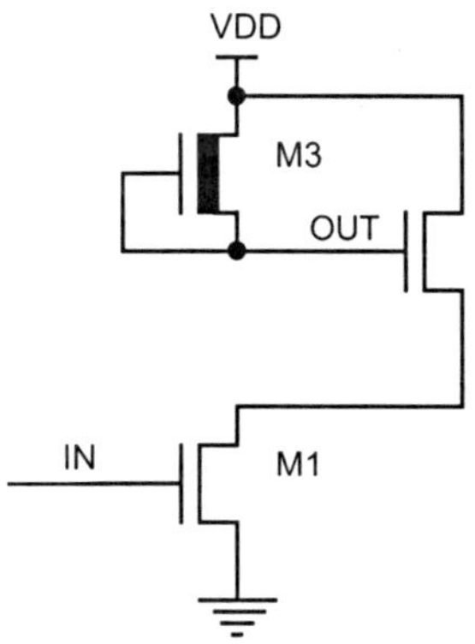

Fig. 5.43. NMOS: Schmitt trigger: simplified circuit for the analysis of the biasing in case of high-low transition of the input signal

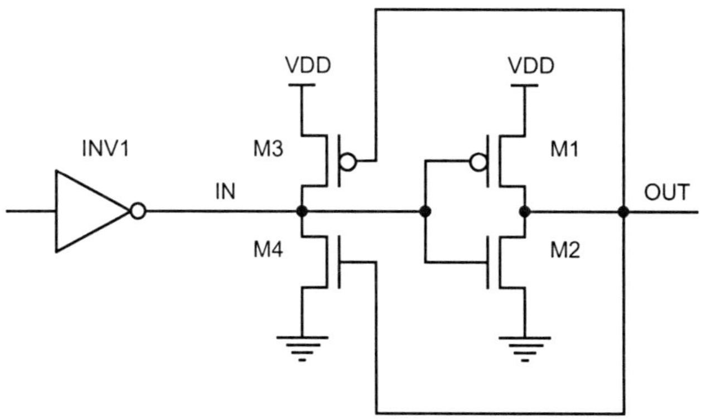

Fig. 5.44. Using a latch as an inverter with different triggering points

5.9 Voltage Level Shifter Latch

Inside a Non-Volatile memory device different voltages are present, which are used for program, erase, read, verify etc. It is therefore required to shift various signals from the VDD/GND to the VPC/GND range, where VPC > VDD; the circuit used for this task is shown in Fig. 5.45a. VPC supply voltage can be either VDD (for example 5 V) or 12 V, during the different operating modes, while the inverter that drives M2 is always supplied at VDD. The voltage swing of the IN signal is in the VDD/GND range. When M2 is turned on, the voltage of the output node V_{OUT} is ground, whereas it is VPC when M2 is turned off and M1 is on thanks to the positive reaction due to M3 and M4 transistors. To decrease the voltages applied to the transistors, thus increasing reliability of the whole structure, the circuit depicted in Fig. 5.45a is modified as shown in Fig. 5.45b. Aim of M5 and M6 transistors is to reduce V_{DG} on M1 and M2 transistors.

Positive reaction can be used to design negative level shifter as well, as shown in Fig. 5.46. HVNEG value can be either GND or negative. M7 and M8 are NMOS realized in the triple well to be able to transfer negative voltages. Power supply for the inverter and for transistors M9 and M10 can be reduced to a value lower than VDD to limit the voltage difference applied to both the junctions and the oxides.

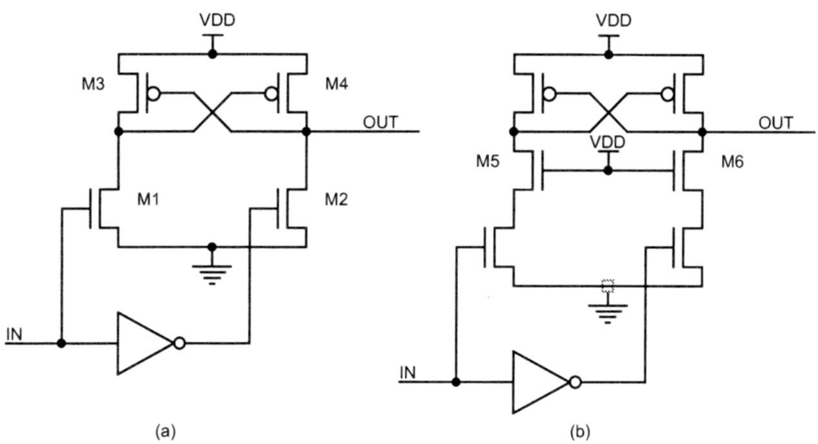

Fig. 5.45. Voltage Level Shifter Latch

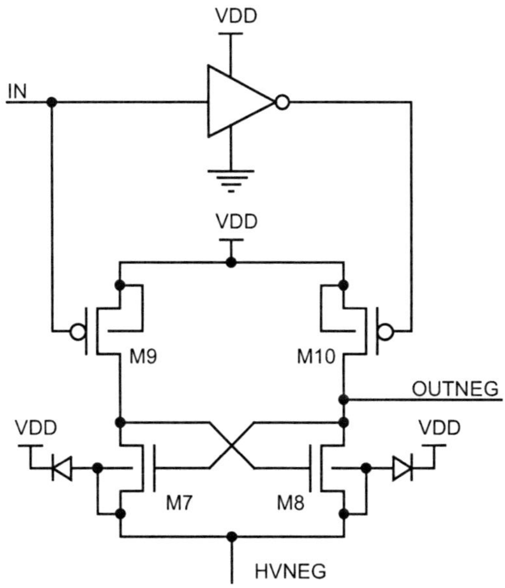

Fig. 5.46. "Cascoded" solutions for the level shifter latch

5.10 Power On Reset Circuits

A problem shared by nearly all digital devices is their correct initialization (*Reset*) as soon as supply voltage is applied, i.e. during *Power On.* A proper circuit to generate the so-called *Power On Reset* (POR) is required, whose rationale is the comparison between the ramp of the supply voltage and the value of a node inside the circuit used as a reference. To achieve correct behavior, it is mandatory that the voltage on this node reaches a stable value before the supply voltage.

In an NMOS device, having an active signal during the ramp of the supply voltage is not an issue, since power consumption during inactive phase of the POR circuit itself can be disregarded. An example is shown in Fig. 5.47: the circuit is composed of a positive reaction block, prototype of the set/reset latch, and by two other blocks, comprised of M1 and M2 (which we already know and which acts as a voltage reference) and of the series of M3, M4 and M5, whose sizes are chosen in such a way that the node "6" is equal to the supply voltage minus a fixed offset. The circuit is really simple; more complex versions can be designed in order to improve switching speed, but the rationale is always the same. Figure 5.48 shows the behavior of the main nodes; depending on the result of the comparison of the input nodes, the latch changes its state or not.

State-of-the-art devices are very demanding as far as initialization of the circuitry is concerned. The POR circuit must be very fast, must work at low voltage, i.e. the threshold is set around 1.5 V, because such memories operate at a minimum voltage of 1.8 V (and even less than that, in a near future) and therefore, at this VDD value, the circuitry must already be initialized. A certain amount of hysteresis is required as well, because the security margin is greatly reduced by the decrease of both the supply voltage and the POR triggering value.

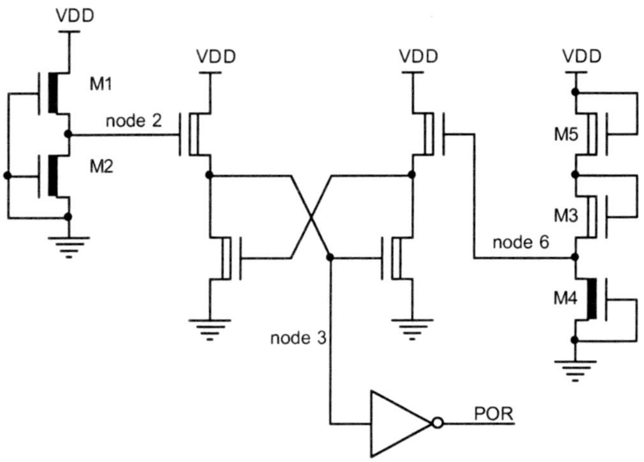

Fig. 5.47. NMOS-based POR circuit

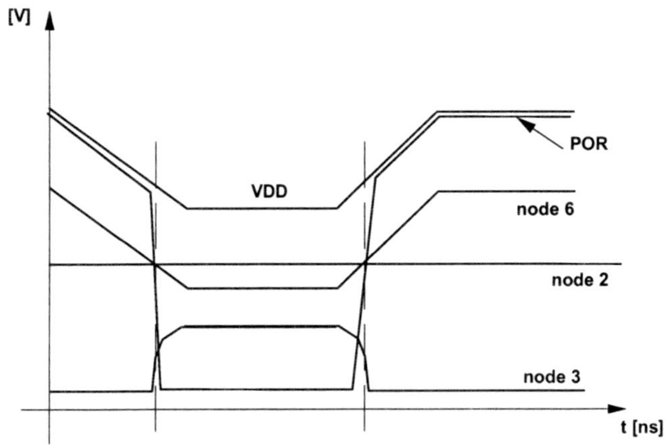

Fig. 5.48. Behavior of the nodes of the POR circuit of Fig. 5.47

Finally power consumption requirements account for a few microamperes, because the specification for power consumption in stand-by has become a top priority whenever the memory is used in a portable device.

Let's see a CMOS circuit that satisfies all these requirements.

In this case a comparator whose output, called INTPOR, directly influences the POR signal by making the comparison between the two signals, the reference and the supply follower. The scheme shown in Fig. 5.49 represents the main parts of the system. The task of the block called VDDDIV is to produce an output voltage with the same linear pattern of the VDD, but whose angular coefficient is scaled by a factor *m*; this parameter is indeed a partition ratio, whose stability must be guaranteed by the circuital implementation.

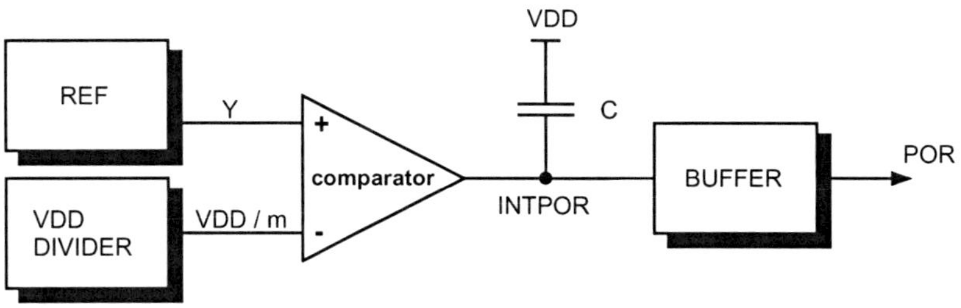

Fig. 5.49. Block scheme of the usual Power On Reset circuit

The function of the capacitor C is to allow the coupling of the INTPOR signal to the supply voltage during the ramp-up transient. The aim of the BUFFER is to de-couple the dynamic behavior of the output signal, whose range is equal to the entire VDD value, from the dynamic behavior of INTPOR, which can undergo a smaller variation; this improves the speed of the circuit. The main issue of a POR circuit realized as described above is consumption; in fact the scheme requires all the circuits to remain in the active state, to be able to re-generate the POR signal in case of a voltage drop, and this requires a non-zero current. Such a behavior is not acceptable in low-power devices, such as the single-supply memories, where a stand-by current virtually equal to zero is specified. To overcome this problem, the following requirements must be met:

1. Specific enable inputs must be introduced, so that several parts of the circuit can be switched off as soon as the POR signal has been generated;
2. The system must be able to autonomously exit the OFF state in case of unexpected supply voltage drop.

The main modifications required to achieve a zero-consumption condition are shown in Fig. 5.50. The schematics that follow depict the circuit implementation of the single blocks discussed before.

Aim of the scheme depicted in Fig. 5.51 is to provide a voltage level, tapped at the Y node, which is stable enough with respect to the possible variations in temperature, process parameters and supply voltage.

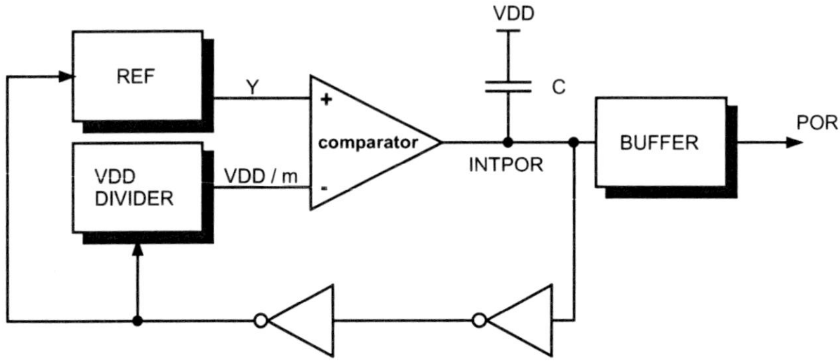

Fig. 5.50. Modified scheme of the POR to eliminate power consumption

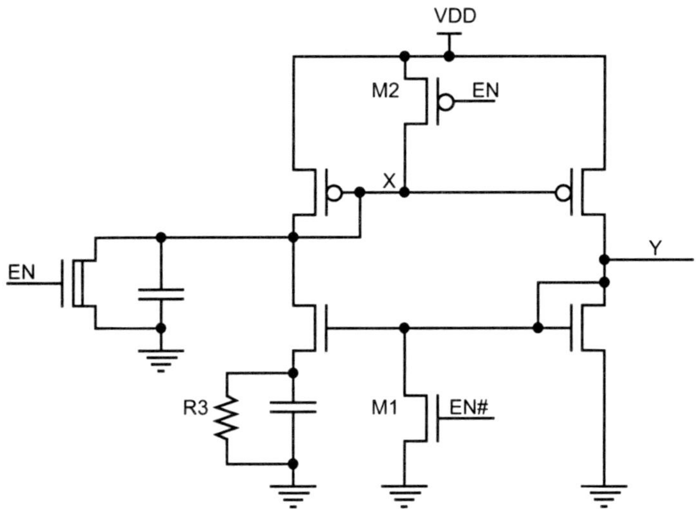

Fig. 5.51. Reference voltage generator

The circuit that partitions the VDD is realized using a resistive partition (Fig. 5.52); the two inverters are used to short-circuit the two resistors R1 and R2 to ground when the Y signal is low.

In Fig. 5.53 the circuit that compares the partition of the VDD and the reference value at the Y node is shown: the implementation is based on a differential stage whose output is suitably amplified by the two inverters I2 and I3 that can be brought to a zero-consumption stage by M3 as soon as the steady state has been reached.

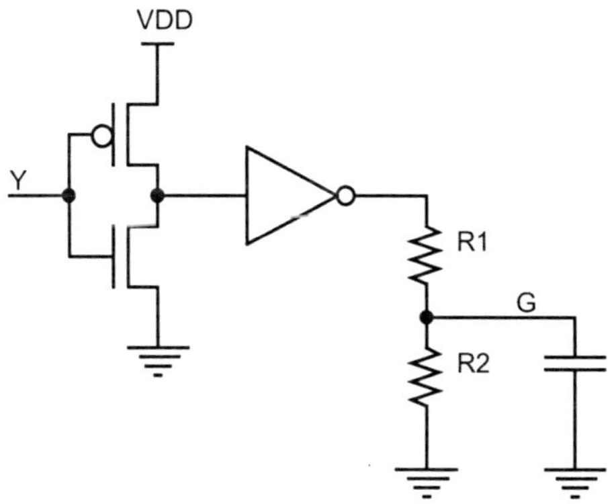

Fig. 5.52. VDD voltage divider

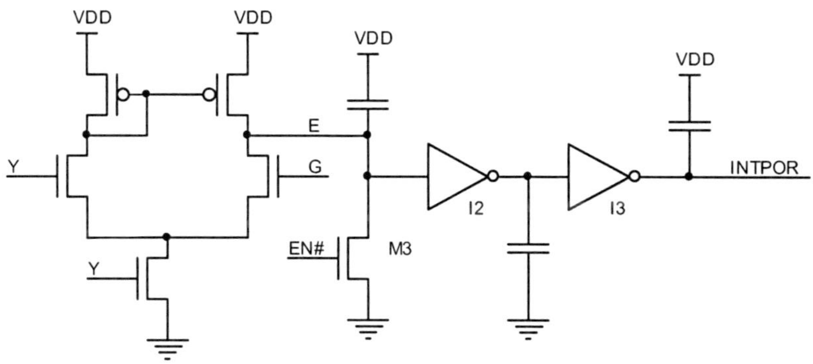

Fig. 5.53. The comparator

5.11 Analog Switch

The title indicates that the following text discusses the usage of the MOS transistor as a pass transistor (Fig. 5.54), and also the issues related to the coupling between gate, source and drain, which becomes evident in high frequency.

These subjects are dealt with in many books, so we would like to discuss the usage of the transistors as a switch in DC conditions by means of an example.

Dual voltage Non-Volatile memories (Flash and EPROM) have two pins for two different power supplies: one is the usual VDD, the other, known as VPP, must be present when the device is either programmed or erased. In any other condition it can be assigned any value between zero and a maximum as stated in the specification. The problem is that the two supplies must converge inside the device to a common node, as the supply for some circuits (Fig. 5.55).

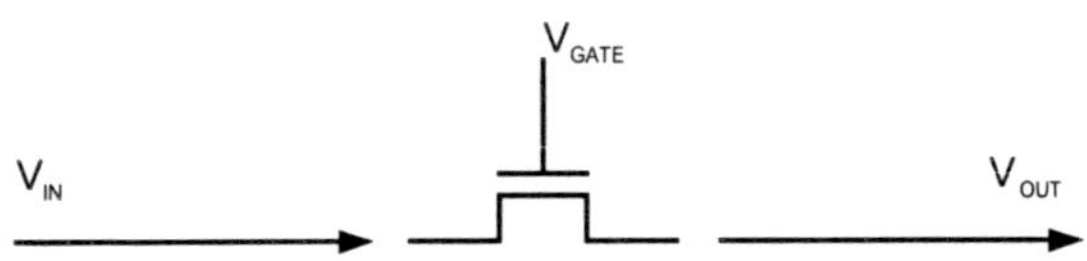

Fig. 5.54. MOS transistor used as a switch or pass transistor

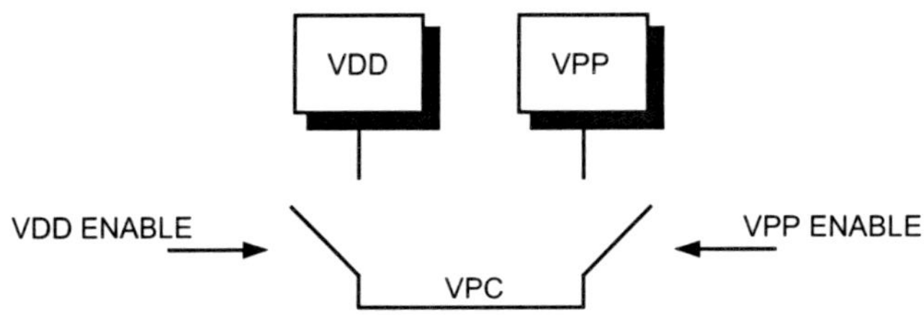

Fig. 5.55. The ideal VDD, VPP switch

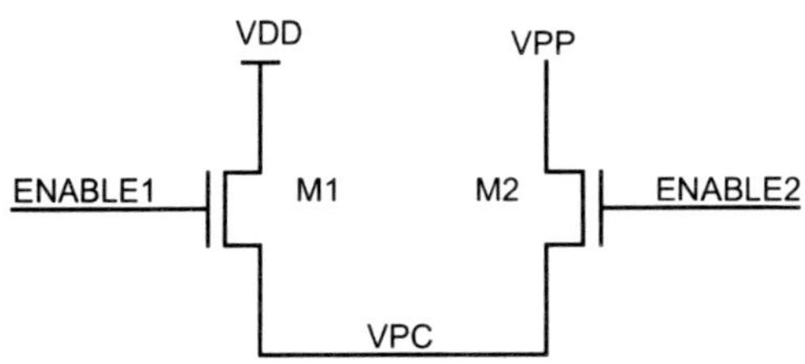

Fig. 5.56. Analog switch realized using NMOS transistors

This task would be easy if relay switches were integrated in silicon, which is not the case: therefore MOSFET must be used. In case of an NMOS process, n-channel transistors are the only choice (Fig. 5.56). In case M1 is on and M2 is off, VPC node cannot reach a value higher than power supply minus the threshold of the transistor (body effect included). In this way it does not work properly, therefore a circuit to boost the gates of M1 and M2 is needed; the section on the bootstrap describes how to achieve both VDD and VPP levels on the VPC node. In the case where p-channel transistors are available, a configuration as shown in Fig. 5.57 can be used. The question is: which is the right connection for the substrates of M1 and M2?

Remember that the PMOS transistor is inside a n-well and that both the source and drain junctions are of type p^+; to avoid forward biasing of the diode towards the substrate, the potential of the latter must always be the highest voltage present.

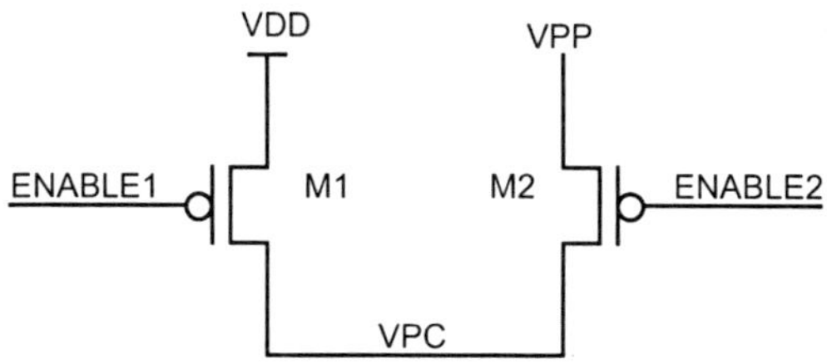

Fig. 5.57. Analog switch realized using PMOS transistors

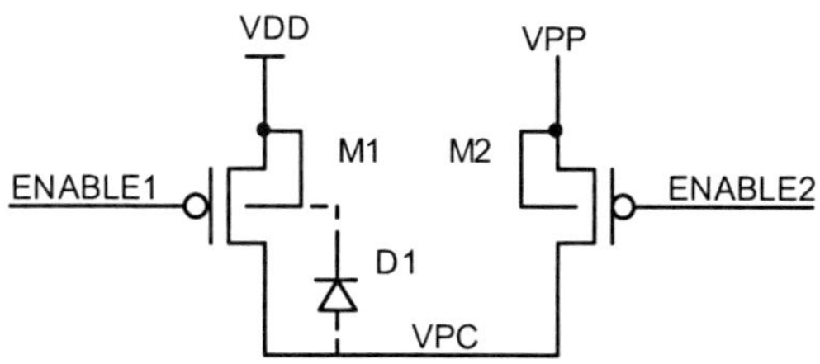

Fig. 5.58. The problem of forward biasing of the junctions in a PMOS switch

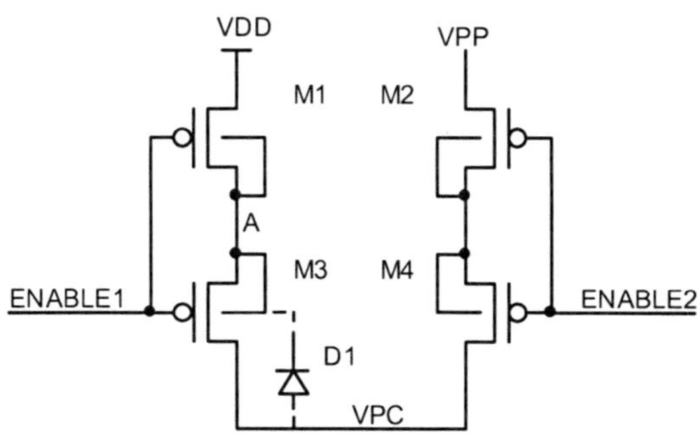

Fig. 5.59. A first attempt to isolate the n-wells of the PMOS

Suppose the substrates is connected as shown in Fig. 5.58: if VPC is equal to VDD, everything works fine; if VPC is equal to VPP, D1 diode is forward biased and nothing works. If the two n-wells are connected towards VPC node, the opposite situation occurs: in case of VPC equal to VDD, the junction of M2 towards VPP is forward biased.

Let's see if a proper control of the n-wells can solve the issue, modifying the structure as shown in Fig. 5.59. When both M1 and M3 are off while both M2 and M4 are on, ENABLE1 is at VPP and ENABLE2 is at ground, then the situation is under control for M2 and M4, while the n-wells of M1 and M3 (node A) are floating. The potential of node A, to which a parasitic capacitance is certainly associated, is initially at VDD (i.e. when ENABLE signals are in the opposite state). Parasitic diode D1 forces V_A to VPP minus the threshold voltage of the diode. The risk is related to charge injection into the substrate that could trigger latch-up effects. As soon as VPC is no longer needed at VPP, M2 and M4 are turned off while M1 and M3 are turned on. When M1 is on, node A (floating) is connected to VDD and the remaining charge stored in the parasitic capacitance is discharged, but the transient condition is difficult to control, and also the two voltages are externally controlled by the user. For example, voltage on VPC might be higher than VDD, even for a short time, when the read circuitry is on and the modify circuitry is off, thus stressing the matrix cells. The configuration shown in Fig. 5.59 is used without problems in the most recent Flash memories, where the VPP is not provided by the user, but rather it is internally generated by charge pumps. In this way the transient is controlled by the internal logic, and the user cannot inadvertently cause any improper biasing conditions.

Problem 5.9: What happens if the n-wells of M1, M2, M3 and M4 are connected to the opposite side?

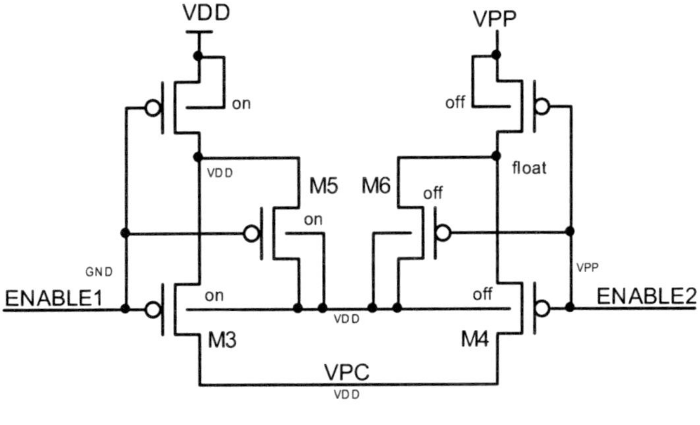

Fig. 5.60. PMOS switch when VPC = VDD. The states of the various transistors and the voltage of the n-wells are shown

The configuration shown on Fig. 5.60 solves the issue by controlling the substrate of the M3 and M4 p-channels, which present forward biased diodes and which could therefore charge the nodes of the switch to undesired potentials.

In Figs. 5.60 and 5.61 the state of every transistor and the voltages of the various nodes are shown for the case VPC = VDD and VPC = VPP respectively.

Thanks to the introduction of M5 and M6, no floating n-well is present. This solution avoids the issues of control on the transistors without decreasing the care posed in the layout of the solution, since floating nodes might inadvertently trigger a latch-up.

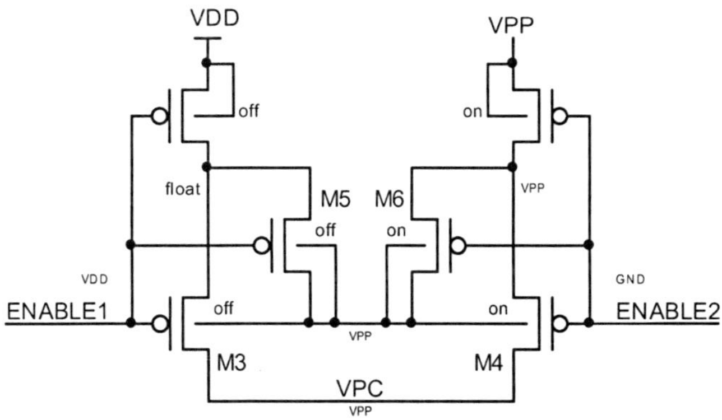

Fig. 5.61. PMOS switch when VPC = VPP. The states of the various transistors and the voltage of the n-wells are shown

The swing of the ENABLE signals is in the range between ground and VPP. Level shifter circuits similar to the one described in Sect. 5.9 can generate these signals.

5.12 Bootstrap

The word bootstrap literally means "to pull yourself up by your boots", that is to be able to rise from the ground by grasping the straps of the boots! In order to better understand this new concept, which is widely used in NMOS technology, let's recall how a capacitor is charged when a NMOS buffer is used (Fig. 5.62). A NMOS buffer cannot be realized using a depletion transistor as pull-up, because it would not be possible to turn it off.

Problem 5.10. Better elaborate the previous statement.

If V_{PD} is GND (M2 turned off) and V_{PU} is at VDD, then M1 is turned on. At the beginning, C_L capacitor is discharged and the potential of the OUT node is at ground. Initial V_{GS} of M1 is VDD as well as its V_{DS}.

The main issue is that the charge of the output capacitor obviously increases the voltage of the OUT node, thus progressively decreasing both V_{GS} and V_{DS} of transistor M1. The final outcome is that V_{OUT} does not reach the value of the power supply and that the charge of the capacitor does not occur at constant current, since the current decreases as time elapses.

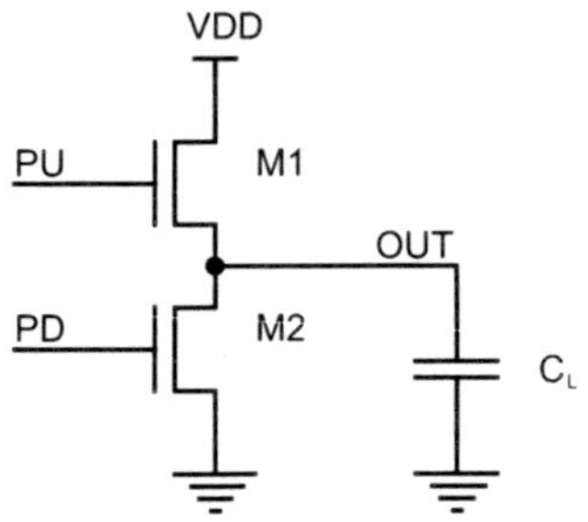

Fig. 5.62. NMOS output stage

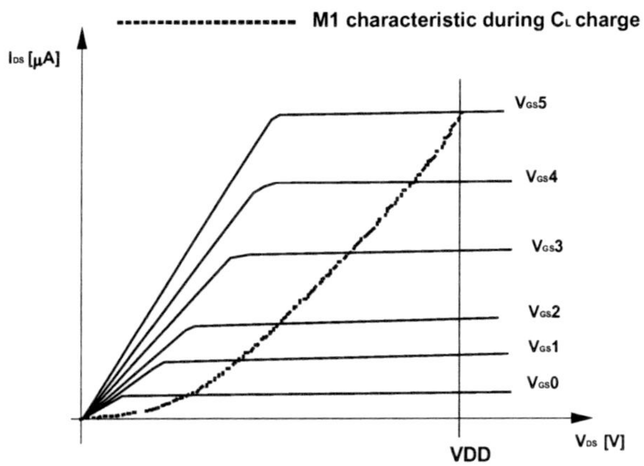

Fig. 5.63. C_L charge characteristic

We have already seen in Fig. 3.7 (see Fig. 5.63 here reported for simplicity) the characteristics of the transistor M1 and the load curve for the charge of the capacitances. This kind of buffer does not allow having a large output swing and at the same time it limits the speed of the charge of the capacitance because current decreases with time: here comes the concept of bootstrap.

Figure 5.64 shows only the pull-up of Fig. 5.62, and capacitor C_B has been added between the gate and the source of M1.

Let's assume to have "precharged" node Y to the value of the supply voltage and then to have disconnected it from a conductive path, so that C_B cannot be discharged. Now let's see what happens when M1 starts to charge the output capacitance C_L. At the beginning we will have, as is with the previous case, $V_{GS} = V_{DS} =$ VDD, being C_L discharged. M1 provides current, V_{OUT} increases, C_B maintains the voltage difference at its nodes (VDD) and therefore the voltage of the floating node Y tends to increase. The V_{GS} of M1 does not decrease while the output capacitor is charging, remaining, at least as a first approximation, constant. In reality, the characteristic of M1 is the dashed line in Fig. 5.65, where it is clear that we are not moving on a constant characteristic of V_{GS} because of the parasitic capacitor of node Y.

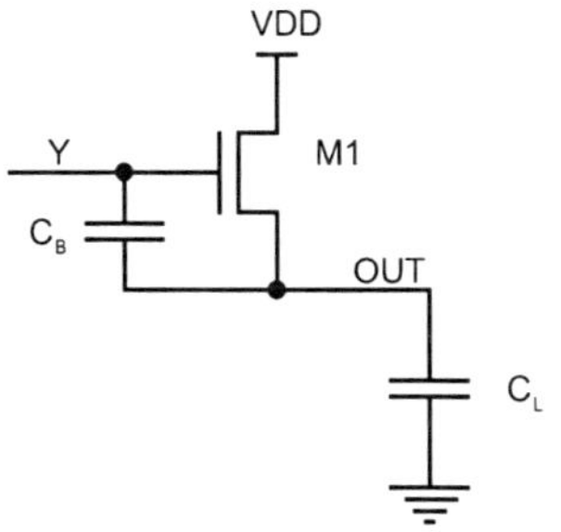

Fig. 5.64. NMOS output stage with bootstrap capacitor

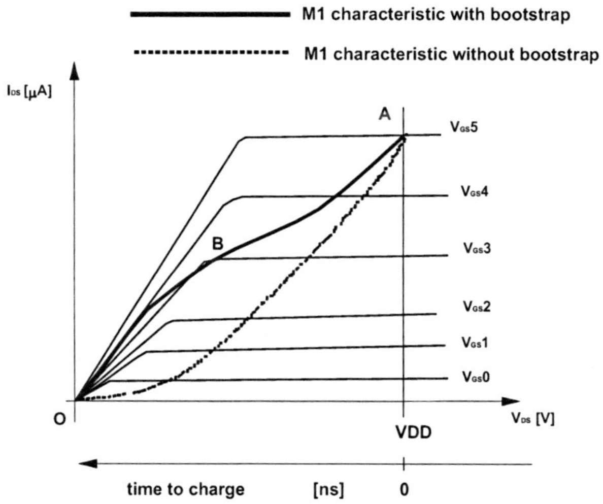

Fig. 5.65. Characteristic of M1 during the charge transient of C_L

The characteristic shown in Fig. 5.65 is broken into two distinct sections: AB, which represents the part where M1 is in saturation and B0, where M1 is in linear zone. On top of that, there is also the effect of the progressive decrease of V_{DS} voltage during the charge; moreover, the threshold of M1 varies due to body effect, therefore its current further decreases as time goes by. The proper dimensioning of the bootstrap capacitance C_B is vital to the correct operation of the circuit; before discussing the criteria for the dimensioning, let's see an example of complete bootstrap circuit. Figure 5.66 shows a driver of a row decoder for a NMOS memory. We recommend again redrawing the circuit on a sheet: redrawing helps understanding connections and it is the first step in understanding the circuit itself.

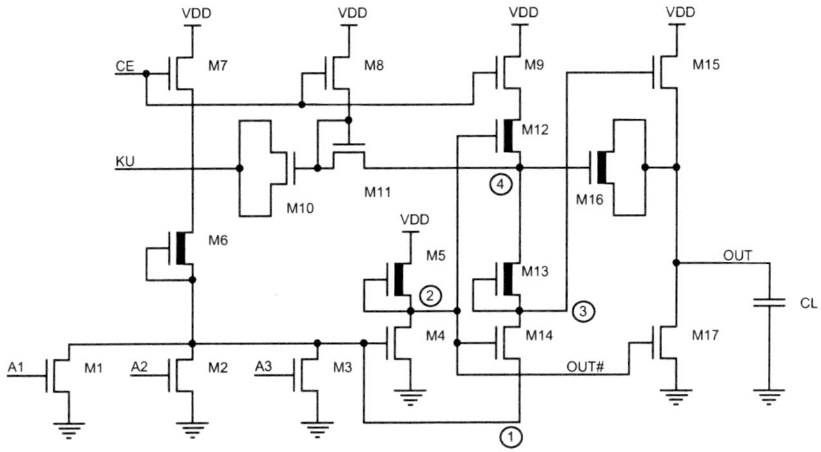

Fig. 5.66. An example of NMOS row decoding circuit where bootstrap is implemented

Aim of the circuit is to set the output signal OUT to high when A1, A2 and A3 are low. First of all, let's observe that the task of the transistors M7, M8 and M9 is just to enable circuit operation through the CE# signal, while M6 is the load of the NOR composed by M1, M2 and M3.

Let's assume that A1 be high and the circuit enabled (CE# low). M14 is turned on, the voltage of the node 1 is low and the one of the node 2 is at VDD because M4 is turned off. Therefore V_3 is at ground while V_4 is high but not at VDD because of the partition between M12 and M13. M15 is turned off and M17 is turned on, therefore V_{OUT} is at ground. If the condition A1 = A2 = A3 = 0 occurs, V_3 increases, V_2 decreases and M14 is turned off. Figure 5.67 shows the final stage of the circuit for a better understanding of the biasing conditions.

Node 3 increases in voltage as node 4, because M13 behaves as a resistor. Transistor M15 has the gate at a positive voltage and it starts providing current, charging C_L. Transistor M16 acts as bootstrap capacitance, transferring the voltage difference acquired by C_L to node 4 and, automatically, to node 3. Node 3 is floating as long as following relation holds

$$V_4 > V_2 - V_{T,M12} \tag{5.63}$$

The task of M9 is to deliver to node 4 the charge required during the idle state of the circuit, when V_{OUT} is at ground.

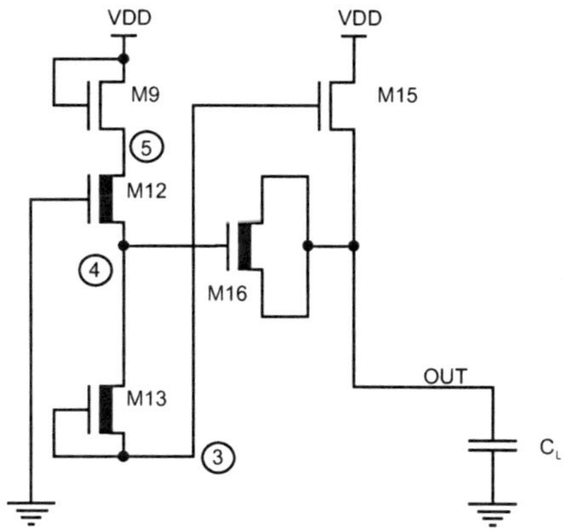

Fig. 5.67. The final stage of the circuit of Fig. 5.66 where the biasing during bootstrap phase is highlighted

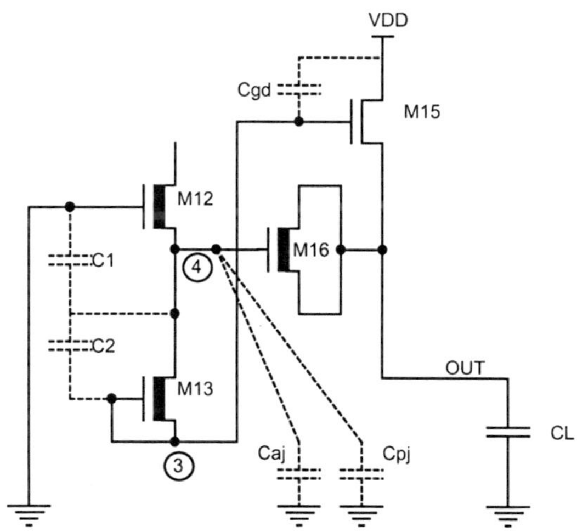

Fig. 5.68. The final stage of the circuit shown in Fig. 5.66 where parasitic capacitances are highlighted

Figure 5.68 shows the parasitic capacitances at node 4: C1 and C2 are the gate/source capacitances of M12 and gate/drain capacitances of M13 respectively, while capacitances Caj e Cpj are area and perimeter capacitances of node 4, given by the diffusions of M12 and M13. The dimensioning of the bootstrap capacitance, i.e. of transistor M16 that is capacitor-connected here, is a function of the parasitic capacitances.

Problem 5.11: Compare the different types of capacitors that can be used and compare their characteristics.

Let's call C_P the capacitance equivalent to the network of the parasitic ones of node 4 and C_B the boost capacitance (M16). Since node 4 is isolated, the initial charge Q_i that is present on the capacitors is preserved after bootstrap occurs. Assuming that node 4 has been precharged to a value equal to VDD/n, we can write:

$$Q_i = (C_P + C_B)\frac{VDD}{n} \tag{5.64}$$

$$Q_f = C_P V_f + C_B (V_f - VDD) \tag{5.65}$$

$$Q_i = Q_f \tag{5.66}$$

$$V_f = \frac{VDD}{n} + \frac{C_B}{C_P + C_B} \cdot VDD \tag{5.67}$$

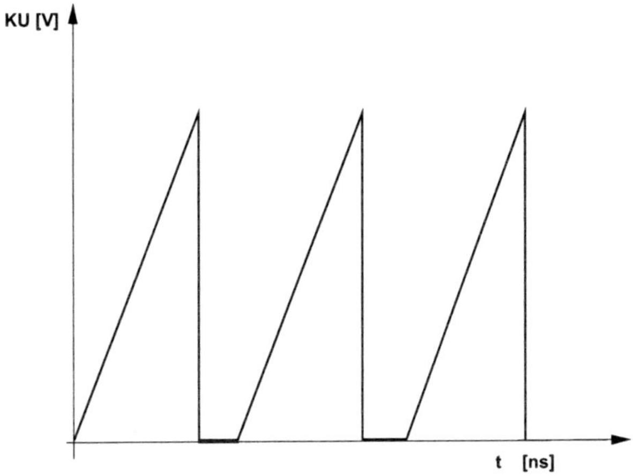

Fig. 5.69. KU signal

We made the assumption that the output node OUT can get to the power supply at the end of the bootstrap phase; if we want the voltage difference on the capacitor be integrally preserved at the end of the bootstrap, we must have $C_P = 0$.

Let's assume that we want to preserve, after the bootstrap, the 90% of the voltage difference on M16. Starting with node 4 precharged at 3 V, we can write that

$$V_f = 5V + 3V \cdot 0.9 = 7.7V \tag{5.68}$$

Using this value of V_f in Eq. (5.67), it results that C_B must be approximately 16 times larger than C_P.

The only transistors that we have not mentioned yet in Fig. 5.66 are M10 and M11. Several junctions are connected to node 4, which can be discharged by their leakage currents; moreover, in case of EPROM devices, the external light passing through the window can help discharging the floating nodes.

The task of signal KU and of transistors M10 and M11 is to "supply" node 4 with charge. In order to achieve a good efficiency, the oscillator that generated KU has a long rise time and a very short fall time (Fig. 5.69). M10 is used as a capacitor, while M11 is diode-connected, to prevent node 4 from discharging during the falling phase of KU.

Let's now examine the most common types of bootstrap.

5.12.1 PUSH-PULL Bootstrap

Precharge reaches node 3 during the time interval between the rise of node 1 and the rise of the OUT node; the delay introduced by the inverter placed between the input and the output is exploited to bias node 2. The drawback of this kind of circuit is that in case of a glitch on the input signal, occurring while the OUT node is rising, the bootstrap capacitor could be partially discharged, thus reducing the charge efficiency on C_L. Following schematic can be used to prevent such situation from happening.

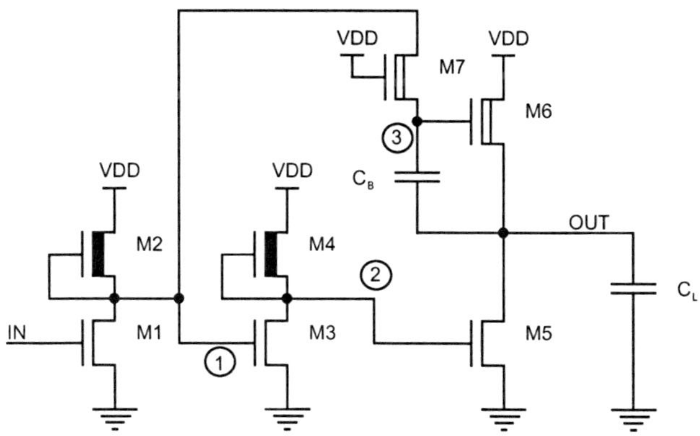

Fig. 5.70. PUSH-PULL Bootstrap

5.12.2 PUSH-PULL Bootstrap with Anti-Glitch

Precharge of node 6 takes place in the time interval between the rise of node 1 and the fall of node 4. Therefore as soon as V_{OUT} starts rising, the precharge pass transistor M1 is for sure OFF, thus preventing an incidental discharge of C_B in case of glitch on the input signal.

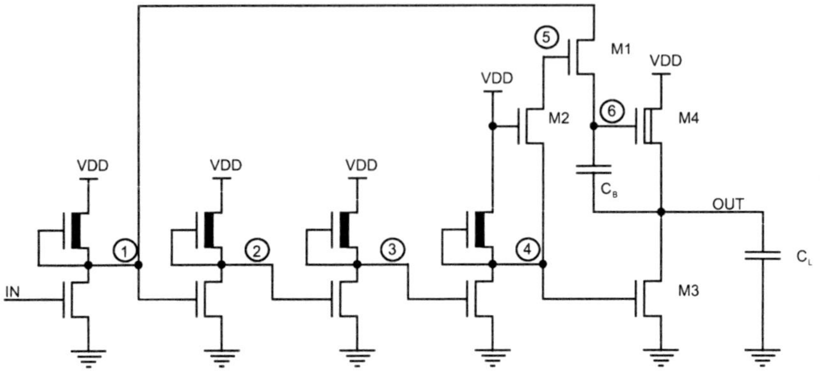

Fig. 5.71. PUSH-PULL Bootstrap with anti-glitch

5.12.3 PUSH-PULL Bootstrap for a Large Load

In case the output load is large, it is better to separate the output stage into two parts to speed up the rise of the bootstrapped node without the need of waiting the output node, thus allowing a higher charge current (Fig. 5.72).

The initial scheme of Fig. 5.66 is not immune to discharge of C_B in case of glitch on the input and, with respect to the versions of Fig. 5.70 and 5.71, it exploits the precharge for all the time when the voltage of node 1 is low, instead of a very short time equivalent to the delay of one or two stages of inverters.

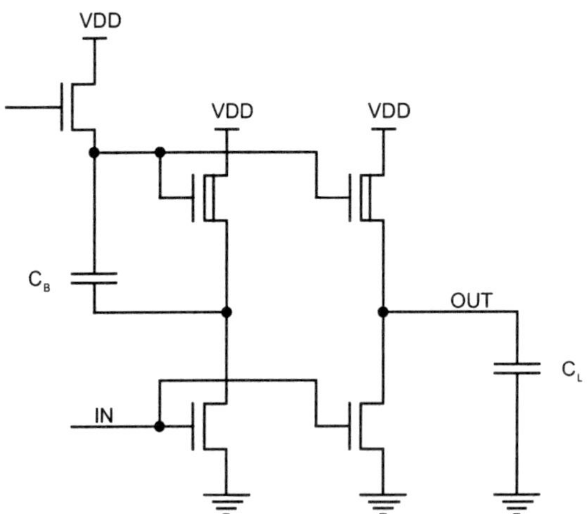

Fig. 5.72. PUSH-PULL Bootstrap configured for a large load

5.13 Oscillators

In electronic systems design, it is often necessary to generate periodic waveforms, triangular, square-shaped, sinusoidal etc. The circuits used to generate such waveforms are called oscillators; some practical implementations are shown in this chapter.

Ring oscillator is one of the well-known schemes. It is composed of a feedback chain of an odd number of inverters, as shown in Fig. 5.73. A simple analysis of the schematic shows that this circuit has a single equilibrium point, corresponding to the situation where all the inverters are biased on the trigger threshold. Under these conditions, every single inverter is biased in a region of high voltage gain, thus determining a high loop gain. As a consequence, the equilibrium point is instable and, even just because of noise, the circuit in Fig. 5.73 starts oscillating producing a periodic waveform.

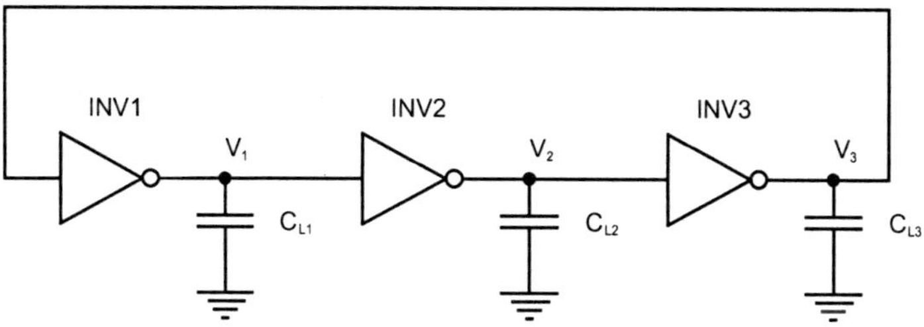

Fig. 5.73. Ring oscillator

Let's now analyze the behavior of the circuit gate by gate. When the output voltage V_1 of the first inverter toggles from the logic state low to high, the output of the second inverter goes low. The time difference between the moments when V_2 is VDD/2 and when V_1 had the same value is the propagation delay τ_2 of INV2. When the output of the second inverter goes low, the output of INV3 goes high with a propagation delay τ_3. Summing up, each inverter of the chain triggers the commutation of the following one, and the last inverter acts on the first one thus guaranteeing the continuity of the oscillation. The oscillation period T can be calculated using the propagation delays of the single inverters. Assuming the load capacitances is equal and all the inverters are identical, therefore having the same propagation delay τ_D, we can write that

$$T = \tau_1 + \tau_2 + \tau_3 + \tau_1 + \tau_2 + \tau_3 = 6\tau_D \tag{5.69}$$

Given a generic number p of inverters that compose the ring oscillators, oscillation frequency is therefore equal to

$$f = \frac{1}{T} = \frac{1}{2p\tau_D} \tag{5.70}$$

Owing to this feature, ring oscillators are often used to benchmark the speed of a given technology. In fact, it is enough to integrate an oscillator of this type and accurately measure the oscillation frequency; dividing by the number of inverters in the chain, it is possible to determine the mean propagation delay of a given technological process. Because of the high speed of the single gates in present technologies, it is necessary to implement many stages in the chain to get frequency that are low enough to be measured; otherwise, it is possible to decrease the frequency dimensioning the load capacitors C_L differently.

Let's now analyze a NMOS oscillator, used to re-integrate the charge lost by the bootstrapped node because of the leakage: this circuit is responsible for the generation of the KU signal shown in Fig. 5.69. The main feature of this oscillator is that the output voltage must have a slow rise time and a very fast fall time.

Figure 5.74 shows the schematic of the circuit and related block diagram.

It is immediately evident that the last two stages are supplied by a VPP voltage, greater than VDD, because the output signal will be then applied to other circuits that are supplied by VPP as well (during program operation). The block diagram shows a high voltage detector, a Schmitt trigger and some inverters; of course, in order to allow the oscillation, an odd number of inversions must be present. Last inverter acts as a buffer to provide the required output current.

High voltage detector compares VDD, applied to the gate of M1, with the gate voltage of M2: comparison is done in current, carefully selecting the size of the two transistors.

Problem 5.12: Find out the right size for both M1 and M2.

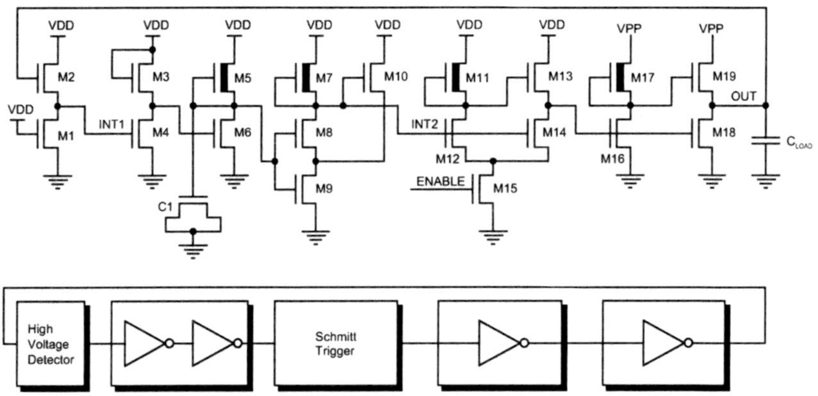

Fig. 5.74. An example of NMOS oscillator. The shape of the output signal is a voltage ramp, featuring a slow rise time and a fast fall time.

In order to generate the ramp on the output node, C_{LOAD} is charged using a constant current. When M19 is turned on, it is virtually diode-connected; working in saturation region, it is equivalent to a current generator. Generation of the output ramp is then achieved by keeping M19 turned on for the chosen time, paying attention that the output voltage does not reach a value high enough to turn off M19, thus modifying the waveform of the output signal.

V_{OUT} voltage is fed back to the input, where the comparison takes place: when the output voltage has reached the desired value, node INT1 turns on M4, that allows the charge of C1 through M5. At this point the trigger toggles, node INT2 goes low and the fast discharge of the output node through M18 is triggered: M18 is dimensioned in such a way that it can sink the discharge current completely.

Once the voltage of the output node has dropped down to ground, M1 is turned off. Before a new charge of C_{LOAD} can take place again, re-charge of C1 and the toggle of the trigger must occur. The latter has the advantage that it filters out undesired toggling induced by the noise on power supplies, since it introduces a hysteresis. Finally, M15 is an enable transistor that, if turned off, keeps the output node low. Figure 5.69 shows the qualitative behavior of the output node.

Problem 5.13: Why the enable transistor M15 is not inserted in the last stage, but it is in the previous one? Every time a choice is possible, options are never equivalent: one is always the best for our aims.

The other NMOS oscillator in Fig. 5.75 produces an output waveform that is not exactly squared, but whose edges are steep. Unlike the previous one, output load capacitance is smaller and therefore it is possible to exploit bootstrap techniques to obtain an oscillator over the full power supply swing. The study of this circuit is left to the reader. Enjoy!

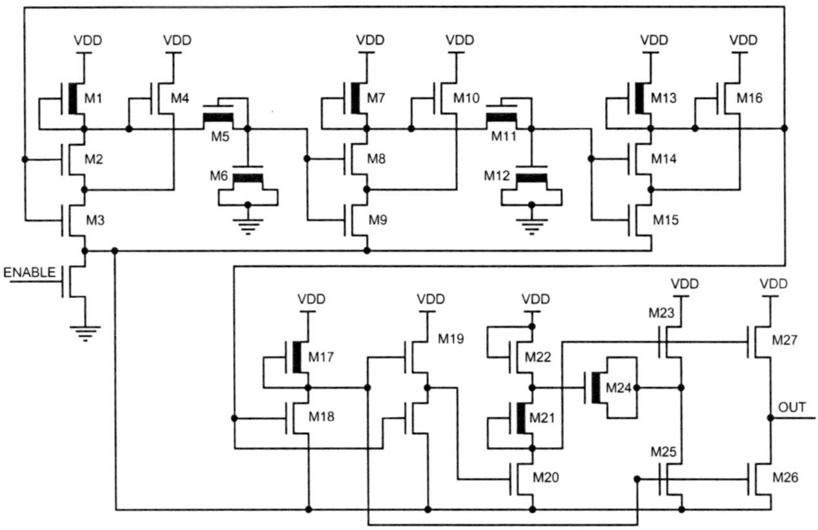

Fig. 5.75. NMOS oscillator with squared-waveform output

Let's now analyze a CMOS oscillator. Unlike the previous ones, it is realized using the voltage of the band-gap reference. This circuit is used to generate a square waveform, which allows for a variation of the frequency on the full voltage and temperature intervals of about 10%.

Figure 5.76 shows the scheme of the oscillator, based on the feedback between T1 and T2 inputs and Q and Q# outputs. If Q is high and Q# is low, transistor M2 is turned off, while its counterpart M6 is on. Capacitor C2 charges up to the supply voltage, while C1 discharges through M3 and M4 at a constant current, until it gets to the trigger threshold of INV1. At this point, the outputs of the latch composed by NAND2 and NAND3 toggle their value and the discharge of capacitor C2 starts. The time that it takes to discharge C1 and C2 determines the clock period; by changing the size of the capacitors is therefore possible to change the duty-cycle of the clock.

Let's assume C1 and C2 be identical; if a constant current discharge takes place, we can write that the clock period is equal to

$$T_{clock} = C\frac{\Delta V}{I_{disch}} \tag{5.71}$$

Parameters I_{disch} and ΔV are respectively the discharge current and the voltage difference required to toggle the output of inverters INV1 and INV3. In order to implement an oscillator that is stable both in temperature and voltage, both I_{disch} and ΔV must be kept constant.

Let's start from the voltage difference. V_{REF} voltage is set to VDD – $2|V_{T,P}|$: in this way, when the upper plate of the capacitor reaches VDD – $|V_{T,P}|$, transistor M3 (M7) turns off, voltage on node T1 (T2) abruptly falls and the inverter toggles. Indeed, we have realized a circuit that is able to detect a voltage different equal to the threshold of a p-channel. About I_{disch}, as a first approximation we can use voltage V_{BG} to drive the discharge transistors directly.

Problem 5.14: Design the circuit that drives both M3 and M7.

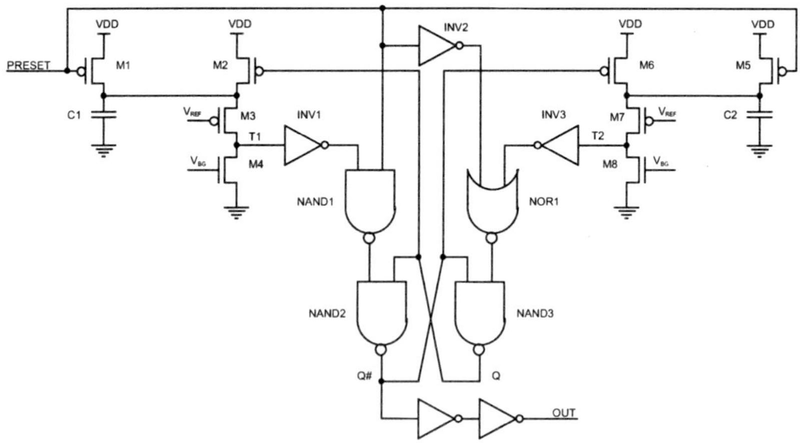

Fig. 5.76. CMOS oscillator

5.14 Circuits to Detect Third Level Signals

Communication between the internal and the external of the device takes place through the pads that are driven by structures composed of inverters, that are therefore able to accept logic values equal to either ground or VDD (CMOS interface).

How can we do, for instance, to use the addresses without triggering a new read, when any logic combination we choose maps to an address in the memory space? It is possible to design a buffer that is capable of detecting three voltage levels: ground, i.e. CMOS logic "0", power supply, i.e. CMOS logic "1" and then a voltage level higher than power supply, that we call third level.

For a device operating with a VDD equal to 5 V, the value of the third level is equal to VPP, while for 3 V-operating devices, third levels allows for lower values, around 10 V. Third levels are used by the manufacturer to place the device in those test modes that are precluded to the user. Anyway the customer has some user modes, described in the specification (like byte identifier read, temporary sector un-protection etc.) that allow for the use of third levels. The voltage value depends n both the device and the technological process used.

When a third level is detected, i.e. an input whose voltage value is much higher than power supply, a path different from the normal input buffer is activated.

The main item to take into account when designing a third level circuit is that it must not cause power consumption, or leakage towards ground, otherwise it can invalidate the normal behavior of the input buffer into which it is inserted, and whose pad it shares. A second item is related to robustness with respect to ElectroStatic Discharge, that we will cover later, according to which it is not recommended to connect drains of n-channel transistors directly to power supplies.

A possible approach for a NMOS technology is shown in Fig. 5.77, where we can recognize the same input stages of the oscillator shown in Fig. 5.74, used for the same purpose. The advantage of this kind of third level buffer is that the input is on the gate, so that we can neglect all the issues related to both ESD protection and latch-up, in case of CMOS. Unfortunately, since the circuit is based on the ratio between two currents, power consumption is not zero.

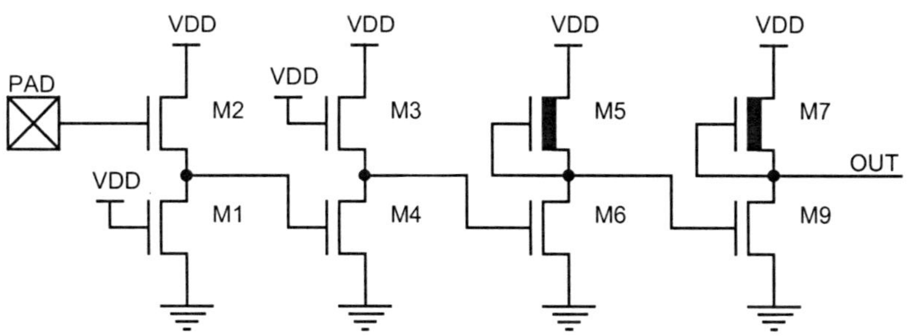

Fig. 5.77. NMOS third level buffer

A typical buffer for a CMOS device is shown in Fig. 5.78: the chain of p-channel diodes provides the triggering threshold.

The voltage of PAD must be at least equal to the supply voltage plus $|V_{T,P}|$ so that M1 can be turned on. But at this point the diode M2 drops a threshold and therefore the supply of the inverter does not allow M3 to win over M4. On the contrary: to have M3 winning over M4, apart from the dimensions that must be chosen for this aim, i.e. M4 must be resistive and M3 conductive, the voltage on the source of M3 must be at least equal to VDD. Thus the voltage on the pad must be at least equal to VDD + $2|V_{T,P}|$ plus the contribution of the given body effect plus a quantity that depends on the size ratio between M3 and M4. Assuming that VDD is equal to 5 V, the triggering threshold is around 8 V. If power supply increases, a greater voltage must be applied on PAD in order to turn the circuit on.

Finally, Fig. 5.79 shows the scheme of the third level buffer used for a CMOS device, composed of a chain of p-channel diodes that is directly connected to the pad and that is clamped to ground by a n-channel diode. Voltage on the gate of the first inverter is usually too low to cause the overall output to go high.

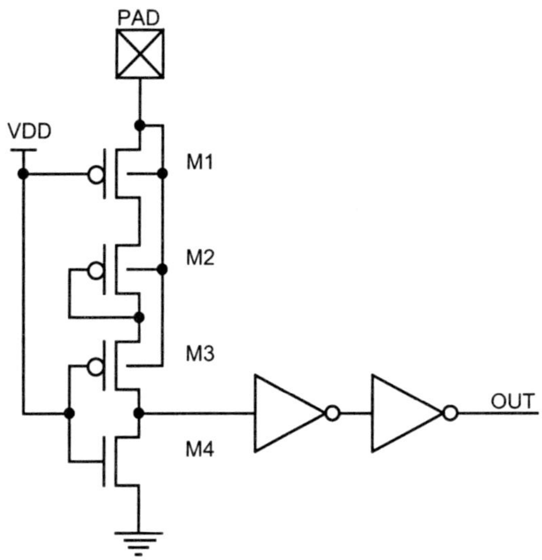

Fig. 5.78. CMOS third level buffer whose triggering threshold depends on the supply voltage

When the voltage value on PAD is greater than five p-channel $|V_{T,p}|$ plus one n-channel threshold voltage, node F can go to ground turning on the third level. In this case the value of the voltage required on the pad to turn on the third level is independent from VDD power supply.

The presence of two n-channel transistors, M6 LVS and N7 natural, as pull-down of the chain, is necessary to guarantee that the following inverter is turned off when the third level is not applied. In fact the n-channel of the inverter INV1 is of the same kind of M6 and, therefore, without N7 the input of INV1 would be exactly at the triggering threshold of the inverter itself, and power consumption might take place.

This case would be almost equal to a floating node condition, i.e. un-driven, at the mercy of spurious couplings.

Problem 5.15: Analyze the last two circuits illustrated above and decide whether it is better that the value of the triggering threshold for a third level buffer be a function of VDD power supply or not.
Problem 5.16: Analyze the nature of a floating node, define it and evaluate which damages it could cause.

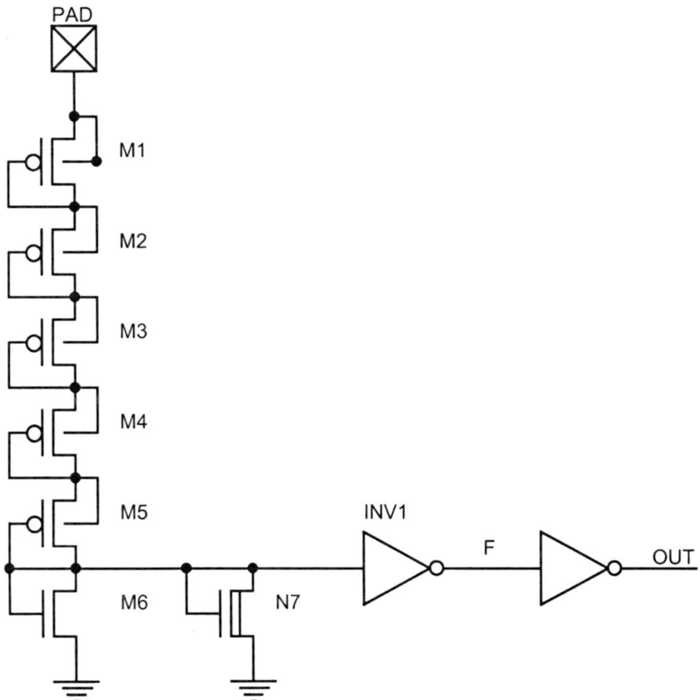

Fig. 5.79. CMOS third level buffer independent from supply voltage

5.15 VDD Low Detector

Sensibility of the flash cells calls for a careful control of the voltages applied to the matrix, in order to ensure that improper voltages be applied at the wrong times, thus damaging the cells or changing their threshold voltage. For this reason it is important to control the value of the power supply during modify operations.

If, for instance, we want to program, then program path to the matrix must be enabled and the pumped voltages are generated starting from power supply; during the operation, supply voltage must remain at the right level, otherwise program operation might not perform correctly.

For this reason, inside the device there is a circuit that controls the value of power supply and, in case it falls below a given value, a signal that stops the charge pumps and inhibits the most dangerous paths towards the matrix is activated.

The circuit can be realized using a resistive partition, whose task is to follow VDD, and a comparator whose non-inverting input is connected to the output signal of the band-gap. The output of the comparator is then sent to a Schmitt trigger to filter out spurious commutations.

Problem 5.17: Realize the circuit that has been just suggested.

Bibliography

B.K. Ahuja, "An improved frequency compensation technique for CMOS operational amplifiers", IEEE J. Solid-State Circuits, vol. SC-18, pp. 629–633, (Dec. 1993).

L.A. Akerst, "An analytical expression for th ethreshold voltage of a small geometry MOSFET", Solid State Electronics, vol. 24, pp. 621-627, (1981).

M. Annaratone, Digital CMOS Circuit Design. Kluwer Academic Publishers.

J.C. Bertails, "Low frequency noise considerations for MOS amplifiers design", IEEE Journal of Solid State Circuits, Vol. SC-14, No. 4, pp. 773-776, (August 1979).

J.Y. Chen "CMOS The emerging VLSI technology", IEEE Circuits and Device Magazine, pp. 16-31, (March 1986).

E.J. Dickes, D.E. Carlton, " A Graphical analysis of the Schmitt trigger circuit", IEEE Journal of Solid State Circuits, vol. Sc-17, No. 6, pp. 1194-1197, (December 1982).

Edit by P.R. Gray, D.A. Hodges, R. W. Brodersen, "Analog MOS integrated circuits", IEEE Press, (1980).

P.R. Gray, R.G. Meyer, "MOS Operational Amplifier Design- A Tutorial Overview" , IEEE Journal of Solid State Circuits, Vol. SC-17, No.6, pp. 969-982, (December 1982).

R. Gregorian, G.C. Temes, Analog Mos integrated circuits for signal processing, J. Wiley & Sons, (1986).

T.P. Haraszti, CMOS Memory Circuits. Boston, MA: Kluwer Academic Publishers, ch. 5, (2000).

L.G. Heller and W.R. Griffin, "Cascode voltage switch logic: a differential CMOS logic family", in 1984 IEEE Int. Solid-State Circuits Conf. Dig. Tech. Pap., pp. 16-17, (Feb.1984).

A. Kuma, "CMOS Circuit optimization", Solid State Electronics, vol. 26, No. 1, pp. 47-58, (1983).

C.M. Lee, Ellen W. Szeto, "Zipper CMOS", IEEE Custom Integrated Circuits Conference, pp. 236-239, (1986).

T. Mano, J. Yamada, J. Inoue, and S. Nakajima, "Circuit techniques for a VLSI memory," IEEE J. Solid-State Circuits, vol. 18, no. 5, pp. 463-469, (Oct. 1983).

W.L. Martino et al, "An on-chip back-bias generator for MOS dynamic memory", IEEE Journal of Solid State Circuits, Vol. Sc-15, No. 5, pp. 820-826, (October 1980).

K. Natori, "Sensitivity of dynamic MOS Flip-Flop sense amplifiers", IEEE Transaction on Electron Devices, Vol. ED-33, No. 4, pp.482-488, (April 1986).

D. Senderowicz, J.H. Huggins, "A low noise NMOS operational amplifier", IEEE Journal of Solid State Circuits, Vol. SC-17, No. 6, pp. 999-1008, (December 1982).

H. Taub, D. Shilling, Digital Integrated Electronics, McGraw-Hill, 1997.

U. Tietze, Ch. Schenk, "Advanced electronic circuit", Springer-Verlag Berlin Heidelberg New York, (1978)

Y.P. Tsividis, "Technique for increasing the gain-bandwidth product of N-M.O.S. and P-M.O.S. integrated inverters", Electronics Letters, vol. 13, No. 14, 7th, pp.421-422, (July 1977).

Y.P. Tsividis, "Design consideration in single-channel MOS Analog Integrated Circuits- A Tutorial", IEEE Journal of Solid State Circuits, Vol. SC-13, No.3, pp. 383-391, (June 1978).

E. Vittoz et al., A Collection of CSEM Papers, Electronic Design Books, Penton Publishing, (1995).

6 Layout

The aim of this short chapter is to present the main concepts that must be understood in order to read a layout, i.e. the translation of an electrical schematic into polygons. Learning to read a layout is a long and tedious process, which might take years to become proficient, because of the many cleaver methods used in developing the layout itself. Undoubtedly, it can be said that the success of a device lies in the combined skill of both the designer and the layout engineer.

6.1 Custom Layout

Designing and manufacturing integrated circuits is a very complex activity; despite the powerful machines used and the thousands of lines of software written to make such machines work, making of a chip is nevertheless a craftsman work.

Designers, layout engineers and technologists do not rely on automatic, error-free procedures; on the contrary, they become more and more similar to Renaissance Masters with their recipes and secrets orally handed down to their favorite pupils. The art of custom layout, i.e. hand-made, transistor after transistor, is probably the most astonishing activity for both the novice who observes a layout for the first time, discovering geometric arabesques and the cunning designer who cannot help being pleased observing the work to which he has contributed.

Layout artists or draftsmen translate transistors, resistors, capacitors, cells and everything on the schematics into polygons that, at the end, will constitute the photomasks to realize the device. Making of the layout requires a deep knowledge of the rules of the technological process; every layer maintains distance rules towards the other layers that are complied with and checked, by the draftsmen themselves, by means of a program known as DRC (Drawing Rules Check). At the end, another checking tool, known as LVS (Layout Versus Schematic), verifies the matching of the overall schematic with the layout.

Let's see some simple circuit and the corresponding layout.

6.2 A Three-Inputs NAND

Let's start and analyze the three-inputs NAND shown in Fig. 6.1. As we have already seen in the chapter on the process, the first layer that is diffused is the n-well[1]; then there is the definition of the areas where a thin oxide is present, as is the case with the polysilicon gate or where no oxide is present because a contact with a junction must be done. The contact is realized making a hole in the oxide in order to reach the junction. The wires to propagate the signal are[2] metal1 and metal2, connected through the vias.

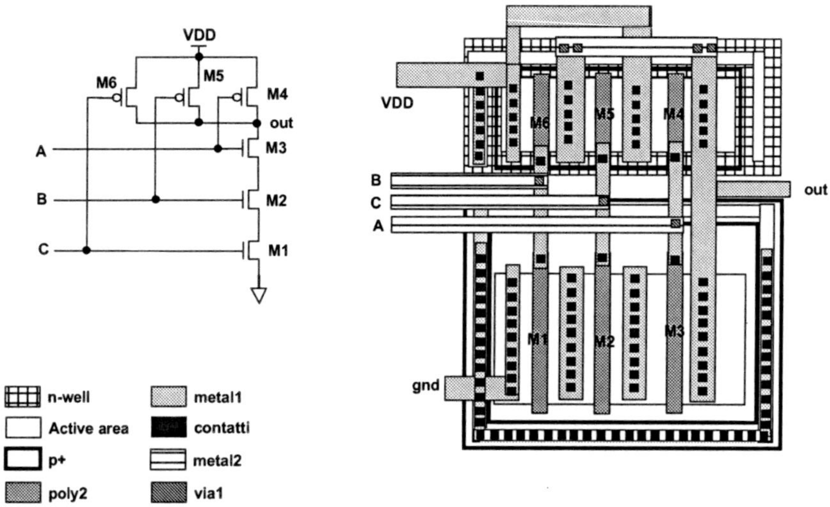

Fig. 6.1. Electric scheme and layout for a three-inputs NAND

Problem 6.1: Try and redo all the layouts shown in the present chapter using a single layer of metal.

Analysis shows the gates of the p-channels connected to those of the n-channels through a metal1 wire. It is not possible to connect the gates, for instance M1 and M6, in poly2, because the n-channel transistors are surrounded by a ring of active area that will be implanted p^+, in order to have a protection ring against the latch-up, a parasitic effect that we will describe afterwards. Poly2 cannot cross any active area; otherwise it would constitute a MOS transistor.

In the case of p-channel transistors, the ring for biasing and protection is interrupted on the side facing the n-channels. This allows additional degrees of freedom, but on the other hand it increases the risk of latch-up. The two situations must be evaluated case-by-case depending on the different requirements.

---

[1] It is important to remind that there is not a one-to-one correspondence between a layout layer and an operation (diffusion, deposition or attack). Some layers are obtained as logic combination of two or more layout layers.

[2] Throughout the layout examples, a two metal layer process is considered. Cheaper processes exists, where a single metal layer is available; on the other hand, the processes used for the most powerful microprocessors can have five or more metal layers

Among NMOS transistors, M1, M2 and M3, the active areas are contacted by metal1 polygons; the minimum distance between two poly strips is usually adopted, to decrease active area resistance. A special consideration must be done about contacts: in advanced processes, they are realized as equally sized squares[3], to overcome lithography and filling issues. When the hole that will allow metal1 to get in contact with the active area is done, the contacts are first filled with a metallic compound, and then metal1 is deposed.

Figure 6.2 shows the issues related to contact filling. We can see in Fig. 6.2a the metal1 incorrectly filling the contact, since it is thinner near the edges, so that it could break because of either mechanical or thermal stress when current flows. A more robust procedure is to fill the contact, as in Fig. 6.2b, and then have the metal1 contacting the filling material, thus ensuring a more reliable connection.

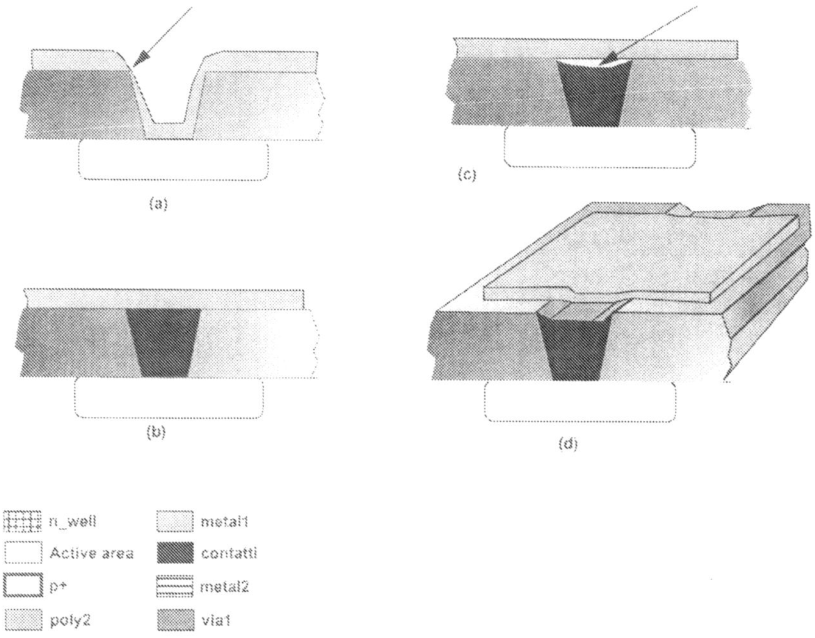

Fig. 6.2. Issues related to contact filling

The issue sketched in Fig. 6.2c is the filling of the contact that might leave a small depression in the center, thus worsening the contact with metal1. The usage of long, differently sized contacts would increase the probability of incurring a depression over the filling material and it would increase the difficulty of control during lithographic exposure (Fig. 6.2d).

[3] On silicon, they result in a round hole.

6.3 A Three-Inputs NOR

Figure 6.3 shows the layout of a three-inputs NOR. Same consideration as in the previous paragraph applies. For the sake of simplicity, the size of the transistors is not shown.

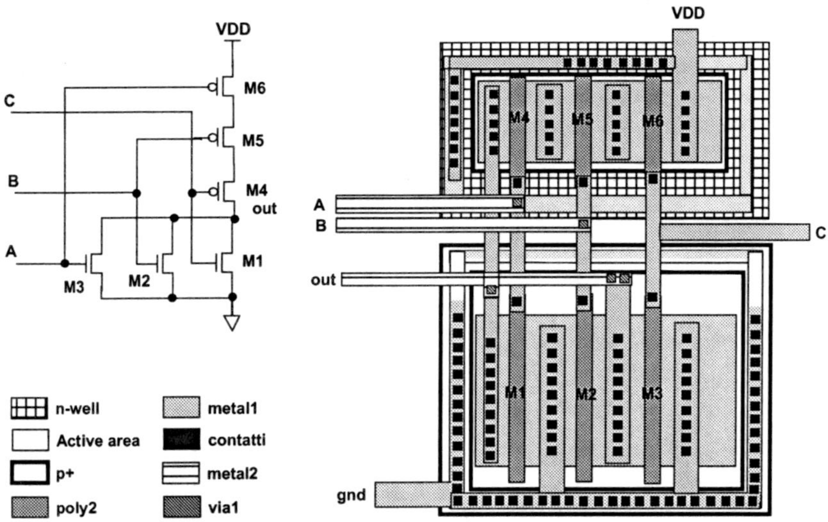

Fig. 6.3. Electric scheme and layout for a three-inputs NOR

Problem 6.2: Redraw the layout of Fig. 6.3 in such a way that every PMOS has its n-well connected to its own source, thus eliminating the corresponding body effect.
Problem 6.3: Analyze the pros and cons of the solution to Problem 6.2. Estimate the costs to eliminate body effect for each NOR and NAND.

6.4 An Interdigitized Inverter and a Capacitor

In several occasions it is required to draw transistors whose W is very large. In the case of output buffers, for instance, size might be greater than 1 mm. In these situations, it is not possible to draw a single, long polygon in poly, for both area occupation and defect-related reasons. The solution is to split the gate into different "fingers", whose maximum length is given by process rules: such pieces of poly are then connected in parallel to form the final transistor (and the resulting structure resembles a comb).

Figure 6.4 shows an inverter where both the PMOS and the NMOS have been split into three different, parallel branches.

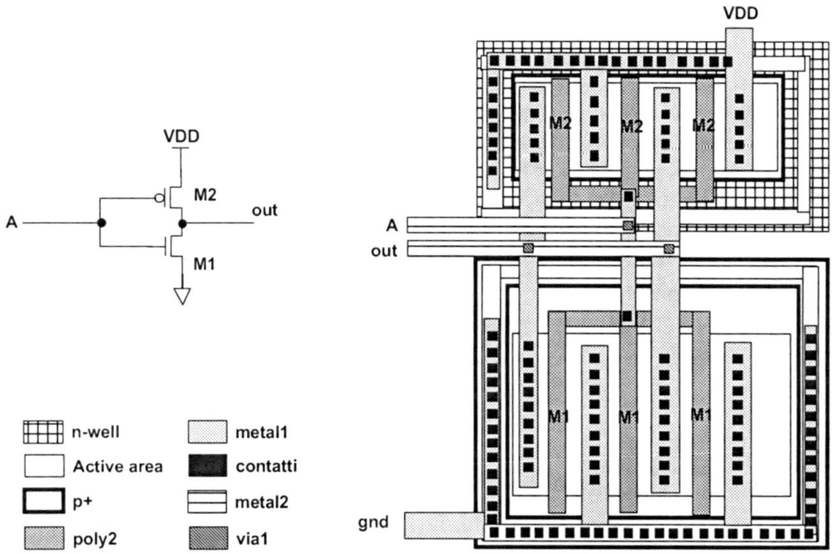

Fig. 6.4. Electric scheme and layout for an interdigitized inverter

Finally, Fig. 6.5 depicts the layout of a poly2/n-well capacitor; the upper plate is the poly2 layer, contacted through metal1 to the node B, while the lower plate A is the n-well, which is contacted through a portion of active area.

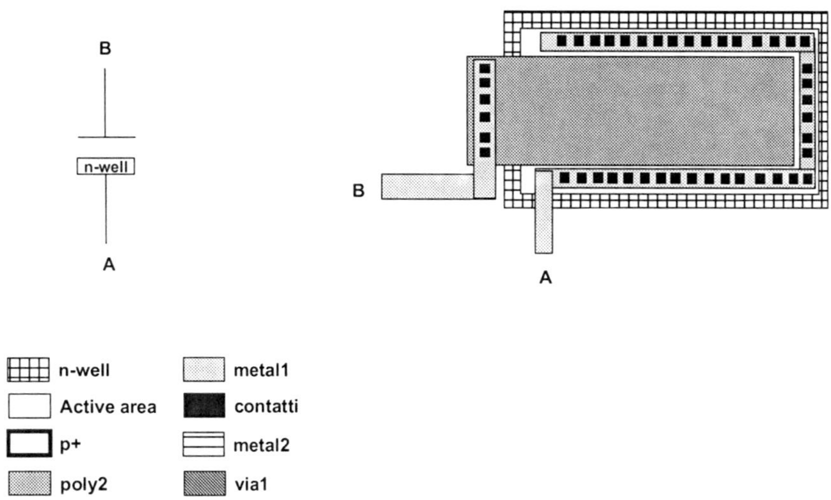

Fig. 6.5. Poly/n-well capacitor

6.5 Area and Perimeter Parasitic Capacitances

For every process, the mutual capacitances of all the different layers are calculated. Figure 6.6 shows an inverter, realized using a single "finger", and the cross-section of the NMOS transistor. Signal nodes, i.e. those connected to metal1, are affected by parasitic capacitances since they touch source and drain junctions. Such capacitances result from the sum of two contributions: junction area capacitance and perimeter capacitance; the latter is caused by the fact that at the end of every junction the field oxide starts again, and below the field oxide a p^+ junction is diffused, to isolate two active areas separated by a field oxide drawn with minimum size.

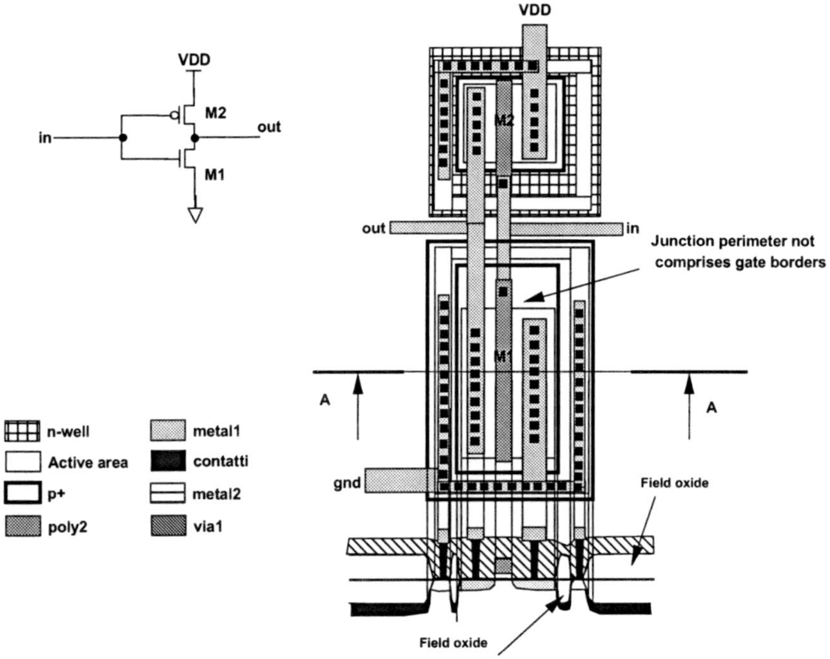

Fig. 6.6. Calculation of perimeter capacitance

Therefore the calculation of the perimeter for both a source and a drain junction is limited to the edges of the fields, and it excludes the edge between the gate and the junction itself, since p^+ junction is not present there. The perimeter to be considered is shown in Fig. 6.6. The calculation for a PMOS transistor is similar, just recalling that the capacitances of n-well with respect to p-type substrate should be taken into account also. There is a layout technique that allows eliminating the perimeter capacitance of one of the two junctions (caused, as we said, by the p^+ diffusion beneath the field). These transistors are used when breakdown value must be increased, and therefore they are called "field-less". Figure 6.7 shows one of these transistors; the gate is not a simple finger, but it surrounds source (or drain) area; in this way the field oxide at the end of the active area is not present.

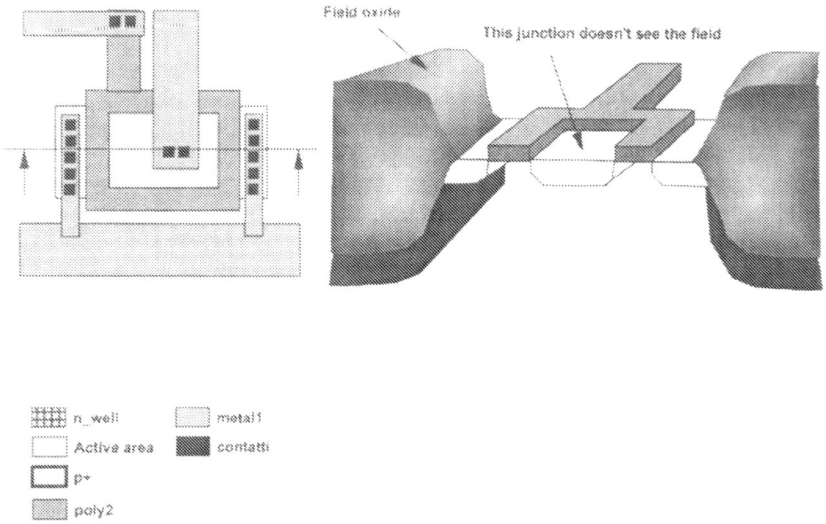

Fig. 6.7. "Field-less" transistor: the cross-section, without metal1, shows the central region of n^+ that does not face p^+ diffusions that are present below the field.

Precise calculation of parasitic capacitances is obviously very important to generate an accurate simulation of the device. Many errors can be imputed by either under or over estimated capacitances, because the resulting electrical performances of the circuit are indeed incorrect.

The big challenge is to be able to estimate the capacitance before having completed the layout: this skill makes the difference between a designer and a *DESIGNER*.

6.6 Automatic Layout

A custom layout allows a high degree of both area and performance optimization, because the draftsman has direct control on the basic components. It is possible, for instance, to pay more attention to the transmission line of certain signals depending on how critical they are with respect to both timing and noise immunity. However this approach has a high cost in terms of both time and human resources required for its implementation. In fact, the layout of a chip usually goes in parallel with the design of the schematics; therefore, a heavy modification of a schematic can bring to a complete remake of the layout, and an associated waste of time is evident.

Increased complexity of internal algorithms and the growing number of functionality supported in today's Flash memories has caused a dramatic growth of the associated logic circuitry. To compensate this effect, a standard cell based approach is more often used.

A standard cell is a circuit that execute a given logic function: three-inputs NAND, D-type Flip Flops, Multiplexers, NOR, etc. These cells constitute a standard library and they are characterized by a layout that follows a similar structure: same height, aligned power supplies, similar position for inputs and outputs.

Layout is done in such a way that all the cells have, in one direction, say Y, the same height, so that the complete circuit can be realized as an ordered sequence of cells placed onto rows to constitute a matrix, as sketched in Fig. 6.8. Of course, X size is directly proportional to the size and the number of cells on the row. Among adjacent rows of standard cells, an interconnection channel is left, to be able to connect the various logic cells as appropriate.

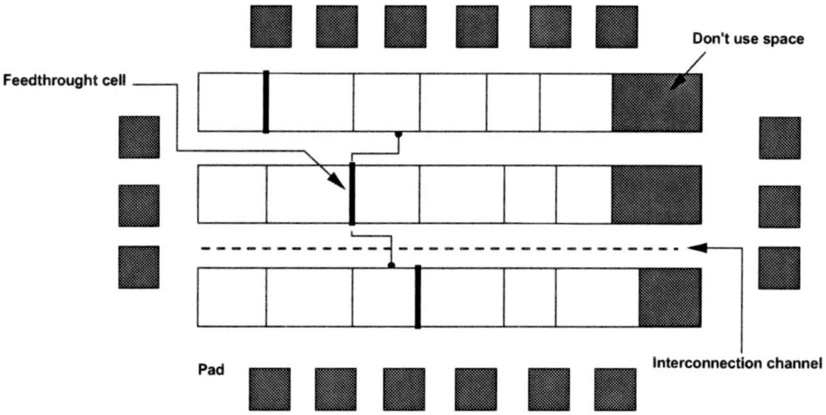

Fig. 6.8. Standard cell layout floorplan

Thanks to the availability of the standard cells library, digital design can use several software tools. The designer describes the logic network by means of a Hardware Description Language (HDL); then applying the "synthesis" operation with the proper timing and area constraints, the schematic is automatically generated. Of course, synthesis is technology dependent, since it maps the desired functionality onto the cells belonging to a library designed for the specific technological process used to design the device.

At this point, a Place&Route software starts from the synthesized schematic and picks up the predefined layout cells corresponding to the standard cells found in the schematic itself (these layout cells are in a standard library too, specific for the process as well), properly connecting them together.

"Placement" defines the physical position of every element of the circuit on the surface of the chip.

"Routing" is the following phase and defines the paths of the wires that connect electrically equivalent nodes. First automatic routing algorithm were used to resolve the trace layout in PCBs; later on, they have been adapted to solve the much more complex issues that can be found in integrated circuits. A very common concept in routing algorithms is to consider the layout as a labyrinth: finding a path between two pins on the same net is just the same as finding a way out of a maze. Anyway this is not enough, since routing must fulfill at least two main constraints: technological, because process rule must be followed (minimum distances, widths etc.); design constraints, because some signals are more critical than others from a timing point of view, or because the connection area should be as small as possible, to minimize area and thus the cost.

The number of connective layers depends of course on the type of process: in case of a two metal layer process, for instance, metal1 and poly might be used for connections inside the cells, while metal2 is reserved for cell-to-cell interconnections.

Parasitic capacitances due to interconnections are then extracted from the automatic layout, so that the designer can simulate the network again to a higher degree of accuracy, since the real loads have been *back-annotated*. In case of problems, either the schematic or the logic flow is modified and the synthesis operation is run again, so that a new layout is generated and so on, until all the defined parameters converge on the required specifications.

The most evident advantage of standard cell approach is design speed, which allows the designer to take care of the whole system instead of the single component, thus reducing time to market. Once the necessary logic functions are defined, proper CAD tools are used to automatically generate the layout using only the components that are part of the standard library.

The investigation of this topic alone would require, for sure, a dedicated book.

Bibliography

E. Charbon, R. Gharpurey, P. Miliozzi, R.G. Meyer, A. Sangiovanni-Vincentelli, SUBSTRATE NOISE – Analysis and Optimization for IC design. Kluwer Academic Publishers, (2001).

J.M. Cohn, et al., "Analog Device-Level Layout Automation", Norwell, MA: Kluwer, (1994).

R.L.M. Dang and N. Shigyo, "Coupling capacitance for two-dimensional wires", IEEE Electron Deviced Letters, Vol. EDL-2, No. 8, pp. 196-197, (August 1981).

M. I. Elmasry, "Capacitance calculations in mOSFET VLSI", IEEE Electron Deviced Letters, Vol. EDL-3, No. 1, pp. 6-7, (January 1982).

F. Maloberti, "Analog Design for CMOS VLSI System", Kluwer Academic Publishers, Boston, (2001).

W. Maly, Atlas of I.C. Technology, The Benjamin Cummings Publishing Company, (1987).

K. Ming-Dou, W. Chung-Yu, W. Tain-Shun, "Area-Efficient Layout Design for CMOS Output Transistors", IEEE, Trans. On Electron Devices, vol. 44, no. 4, (April 1997).

T. Sakurai and K. Tamaru, "Single formulas for two-and three-dimensional capacitances", IEEE Transactions on Electron Devices, Vol. ED-30, No.2, pp. 183-185, (February 1983).

N.P. Van Der Meiji, J.T. Fokkema, "VLSI circuit reconstruction from work topology", North Holland INTEGRATION, the VLSI Journal vol. 2, pp. 85-119, (1984).

7 The Organization of the Memory Array

The architecture of the memory array is one of the most complex topics in the field of electronic design. The size must be minimized and the effort to optimize the area is always a high priority.

7.1 Introduction: EPROM Memories

The organization of the array, i.e. the composition of the memory cells to form the matrix, has gained more and more importance in Flash memories in comparison to EPROMs. The array of an EPROM memory is designed taking into account the access time that imposes some constraints to row and column length[1].

This implies the necessity of realizing several sub-arrays, increasing both row and column decoders. Also the impact of electrical stress is to be considered when choosing the number of rows and columns. The typical row is composed of 1,024 or 2,048 cells, and the columns can be as numerous. Fig. 7.1 ((a) to (e)) shows some cuts of memory with a possible division into sub-arrays. This is only one of the several possibilities; in fact, the length of the row depends on the process, and larger rows can be used, for example, when silicide is used.

7.2 Flash Memory Organization: The Sectors

The difference between EPROM and Flash Memories is in the electrical erase that, differently from EEPROM, is carried out not on a byte basis but for groups of bytes called sectors. The first Flash devices were electrically erased but did not have separated sectors; thus, during the erase operation, the logic value "1" was restored on the entire array.

It is important to review the evolution of the erase methodology in order to understand the reasons that have led to different types of subdivision of the array into sectors. We can state that the division of the memory array has been done in a way to facilitate the evolving erase techniques that, on the other hand, have had to adapt to the external bias voltage which is supplied by the customer.

---

[1] To simplify the addressing, the number of bits is always a power of 2.

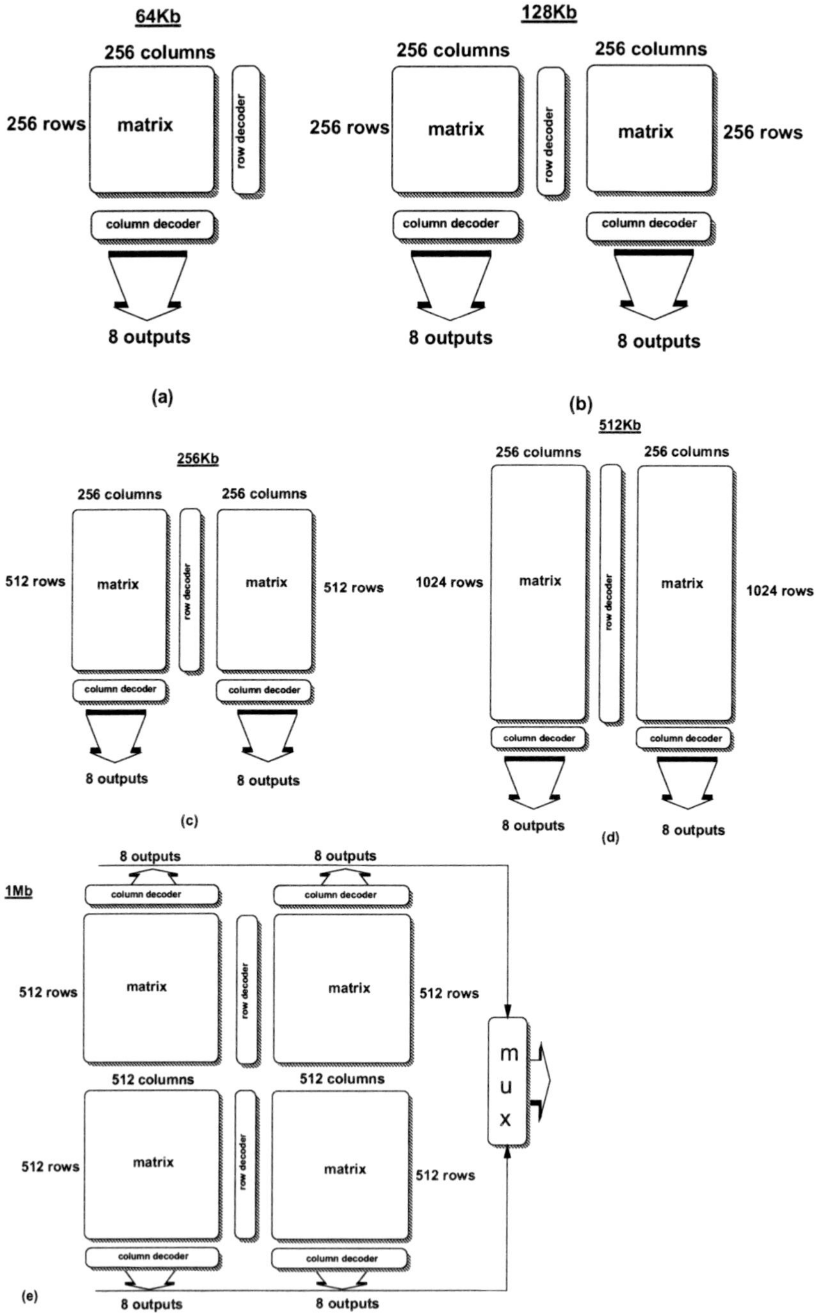

Fig. 7.1. Placement of arrays and decoding blocks for different cuts of memory

The first Flash Memories erased (and programmed) by means of an external voltage of 12 V, called VPP[2], which the customer applied to accomplish the required operation. The erase procedure biased the source to 12 V, the gate to ground, while the drain was floating. In this way, the electric field between floating gate and source required by the erase operation (several MV/cm) could be obtained.

This means that the source is the common potential during the erase phase and that the cells belonging to the same sector must have common source. The simplest implementation of the organization of the array into sectors is shown in Fig. 7.2 with reference to a NOR architecture.

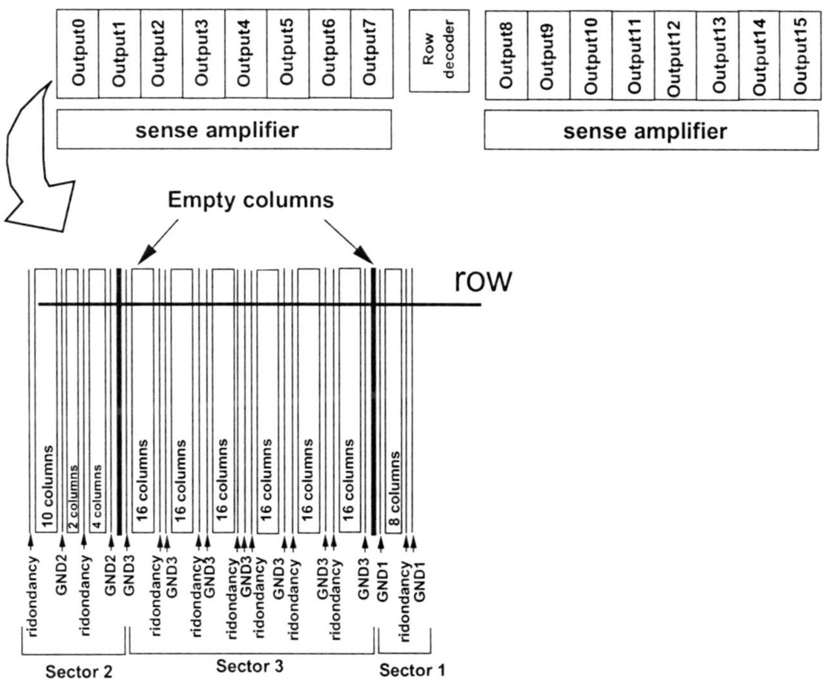

Fig. 7.2. Example of a Flash device with sectors arranged in columns. The device has 16 outputs, one of which is highlighted. Also shown is the organization of the redundancy columns distributed throughout the various sectors. The source connections, within each sector, are separated to allow selective erase. Moreover, the sectors are spaced at a distance that equals the width of a column to prevent the switch-on of the parasitic transistor that forms between two contiguous active areas.

The various sectors are located close to the corresponding output. The parasitic bipolar transistors that form between two contiguous columns require the insertion

[2] The variation allowed to the value of VPP is ± 5%, whereas it is usually ± 10% for VDD.

of a dummy column to better separate two neighboring sectors, so as to increase the distance and, thus, the voltage required to switch on the parasitic transistor.

The column decoder selects the addressed byte among the various outputs. The dummy columns provide insulation between different sectors, whereas the ground contact is realized by means of n-channel transistors that are switched on during read and program, but are switched off during the erase operation. The discharge of the nodes after an erase is a very important issue. In the case considered, the source node is charged to 12 V at the end of the erase. The parasitic capacitance associated with this node is very high, up to 1nF. If the discharge of such a large capacitor is not carried out with great care, disastrous coupling with other nodes is possible. The fast discharge of the source node could couple it with the cell gates, driving it below the ground potential, with unavoidable risks of latch-up of the row decoder. The source node is to be discharged slowly, up to a secure value, while the residual discharge can be fast.

Problem 7.1: Define the discharge speed and design the circuit to detect the source voltage in the cases of slow and fast discharge.

The placement of the sectors shown in Fig. 7.2 is not the only possible choice. A different topology is shown in Fig. 7.3. In this case, the sectors are not placed within the single outputs. With respect to the previous solution, the sense amplifier cannot be located close to the single outputs, since there is no correspondence between sectors and outputs. The output are separated by means of the column decoders and brought to the sense amplifiers.

Problem 7.2: Design the column decoders with reference to Figs. 7.2 and 7.3 (see also Chap. 9).

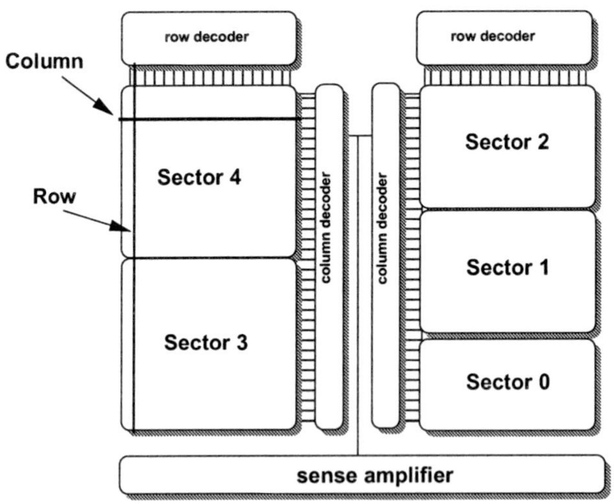

Fig. 7.3. Sectors organized by columns in which all the outputs of each sector are contiguous

As we know, the erase carried out by a positive bias of the source node is accompanied by a spurious current, the *band-to-band tunnel current*, I_{BBT}, due to the difference of potential between the n^+-type source and the p-type substrate. This current ranges between 5 and 10 nA per cell, i.e. 5÷10 mA in case of a 1 Mbit array. The device size is continuously increasing to fulfill the market demands, but the band-to-band current per cell does not decrease even as size and supply voltage scale down, since the thickness of the thin oxide does not scale down proportionally due to charge loss phenomena at room temperature. Moreover, the electric field for erasing is constant and, thus, the value of I_{BBT} remains nearly unchanged over the various technological generations.

The reduction of the supply voltage and the elimination of the VPP pin, so as to have devices operating with a single supply voltage, have required the inclusion of additional internal circuitry to obtain all the potentials necessary for the various operations. At the same time, the reduction of the device size has exacerbated the problems related to stress resulting in the need to redesign the sector organization.

Thus, the first generation of single supply Flash memories with a VDD of 5 V was created. In the case of a single supply, the voltages above VDD are produced on-chip through charge pumps. We repeat here the fact that the current-voltage characteristic of a charge pump realized with diodes and capacitors can be approximated to a line, the slope of which is the pump output resistance R_{OUT}. The supplied current decreases as the output voltage increases. The value of R_{OUT} amounts to tens of KΩ, whereas the maximum current supplied amounts to mA.

The voltage applied to the source during programming has been 12 V for devices with VPP externally supplied, and this voltage must be reduced to values in the 8 to 9 volt range to enable the charge pump to supply the required current. Furthermore, the necessity of having small sectors and an elevated number of erase-program cycles has led to the sector organization "by row" shown in Fig. 7.4. In this way, the sectors are completely insulated from each other, reducing the stress; moreover, this architecture makes it easy to obtain the potential needed for erasing by applying the negative voltage to the rows of the sector. The floating gate reaches the required potential due to the source (positive), and the gate (negative). Notice that the local column decoder (divided bit line architecture) eliminates the drain stress induced by programming the other sectors connected to the same main bit line.

Let's now examine the structure of the row decoder that was used in the devices of the first generation to drive the rows to a negative voltage during the erase.

Normally it is not possible to apply a negative voltage to an n-channel transistor without forward-biasing the n^+/p-substrate junction (see fig. 7.5). If n-channel transistors with insulated triple-well substrate are not available, it is not possible to apply negative voltages to the NMOS terminals.

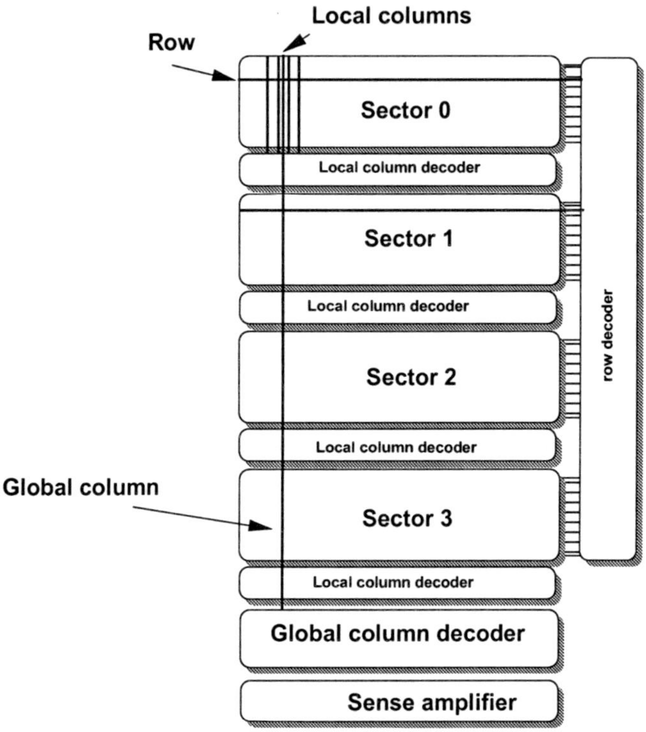

Fig. 7.4. Division by rows with the columns of each sector insulated with respect to the main bit line by means of the local decoder of the single sector

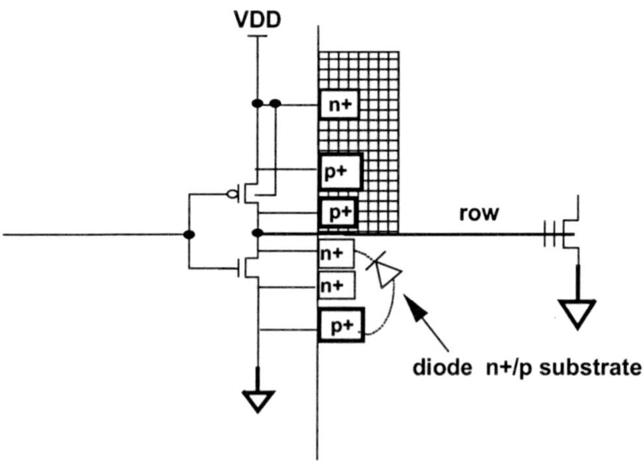

Fig. 7.5. The n^+/p-substrate diode clamps negative voltage on the row

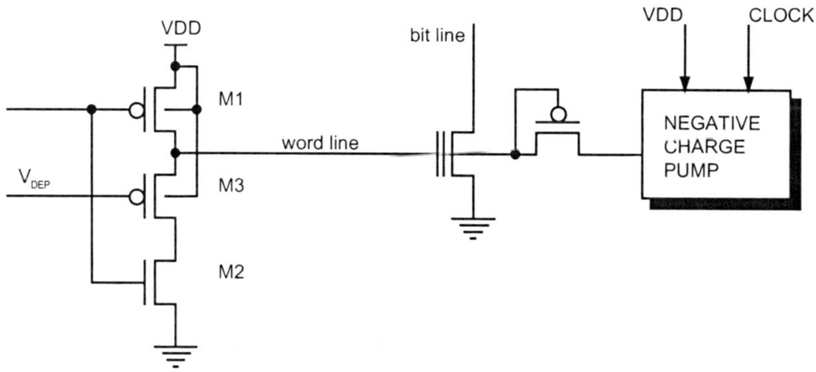

Fig. 7.6. A possible solution to apply negative voltages to the row without using NMOS transistors with insulated substrate

Modifying the final driver of the row as shown in Fig. 7.6 solves the problem.

The solution is realized by "masking" the negative voltage to the NMOS transistors. In this way all the structures that generate and control the negative voltages consist of PMOS transistors. The final driver of Fig. 7.6 has a PMOS transistor, M3, between M1 and M2, and the connection to the row is realized by means of two PMOS. The negative voltage is transferred through a p-channel diode-connected transistor located at the end of the word line.

When a negative voltage is applied to the row, M3 is switched off, preventing the negative voltage from being applied to the drain of M2. The problem is now that the rows that are not selected must be tied to ground during erase and program. In this case, M2 is "on" but M3, with its gate tied to ground, would transfer a voltage that can at most equal the magnitude of $|V_{T,p}|$. In order to drive the unselected rows to the ground potential, the voltage of M3, V_{DEP}, must be negative. The introduction of the triple-well has allowed simplifying the row decoding, eliminating the M3 transistor and the final diode.

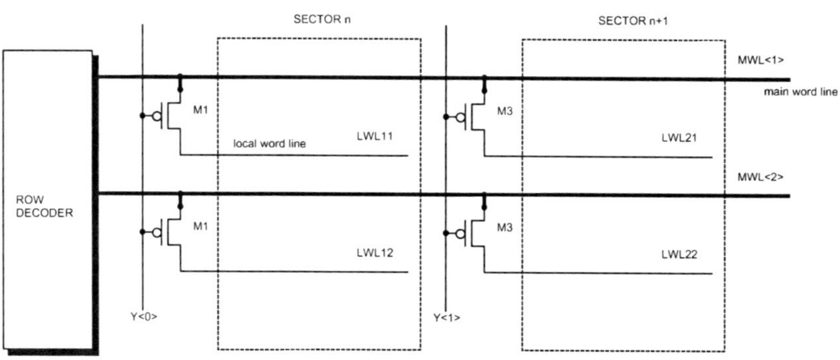

Fig. 7.7. Sector organization obtained by dividing the row with a local decoder

Beside this type of organization of the memory array, obtained by dividing the columns with a local decoder, there is also the possibility of dividing the row. In Fig. 7.7 a possible realization of this type of architecture is shown.

In this case, we have global rows, fabricated with metal2, as metal1 is already used to fabricate the columns, and a PMOS transistor that connects the global to the local row. During program (or read), the addressed global row is driven to VPP (or VDD) and the selection of the row is carried out by means of the Y<n> signals. For example, if we want to drive LWL12 to VPP (VDD), we will drive Y<0> to –1 V, and Y<1> to VPP (VDD). The drawback of this kind of approach is related to the Y<n> signals that must be driven to a negative voltage to prevent local rows of the addressed sector from being in a floating state.

7.3 An Array of Sectors

Customers' demands are heading toward smaller sectors and increased number of program-erase cycles in memories of bigger size. These customer demands can be met by one solution that consists of organizing the sectors as an array, thus creating small arrays by means of hierarchical row and column decoders. This kind of solution allows us to further develop this idea by creating a local decoding that utilizes the triple well and associates the source to the single small array.

The concept of an array, in which the single element is identified as the intersection of a row and column address, takes on an expanded meaning by considering the sector and not the cell as the basic element. Inside the sector, the array is decoded further by means of a row and column address to identify the memory cell.

This kind of organization incurs a cost in terms of area for the local decoders. It is necessary to remember that the advanced processes aim at diminishing the cell size and not the transistor size. Thus, it happens that the transistor of the final inverter of the row decoder does not fit in the cell pitch but, on the contrary, occupies the area of two or three pitches. This requires a significant enlargement of the row decoder since there must be a final inverter for each row of the array. In the case of the described sector organization, the global row decoding is greatly reduced, since we have a global row driver every 4, 8, or 16 sectors, whereas the selection of the single row is determined by the local decoder. While the global decoders are large, the local decoders are small and in the end the solution is surprisingly efficient.

With regards to the columns, the goal is to make the size of the transistors in the decoder such that a minimum V_{DS} drop at the transistor terminals is present when the cell drains current. The voltage drop on the column selectors reduces the effective voltage at the cell terminals. The power consumption during the programming operation has always been considered the main issue. The present design and technological techniques have dramatically reduced the consumption, so that it is now comparable with the dissipation during the read, with the further advantage of diminishing the size of the transistor of the selectors by 75% with respect to the past.

The design of an array of sectors also simplifies the redundancy. In the case of very large arrays (1 Gbit), the traditional row and column redundancy was accompanied by the sector redundancy that allows substituting an entire sector at a time.

The logic size of the sectors can be varied by using dedicated non-volatile registers. From the customer's point of view, the size of the sector is defined based on the erase. The sector is composed of cells that can simultaneously be erased; if two distinct sectors are erased simultaneously, they can be regarded as a single sector.

The logic composition of the different sectors is stored in non-volatile registers written for the user or by the user, and read during the system boot, so as to be reconfigurable over time (it is sufficient to give the customer the possibility of erasing and re-programming the registers).

The last topic deals with the division of the *read-while-modify* architecture into sectors. For a Flash memory, the main limitation related to the modifications of the content of the memory is the time necessary to erase, which amounts to nearly a second. Such time is enormous even for a microprocessor that operates at only 100 MHz. Many applications require modifying the code while the memory is being read. The traditional memories do not permit this kind of contemporary operations. This limitation has been overcome through the introduction of the concept of *read-while-modify*. In this case, the memory is divided into banks (groups of sectors) on which it is possible to operate independently from each other. In order to avoid having as many independent circuits as the number of banks, many of the blocks are dedicated to specific functionalities (read or modify). For example, only two groups of sense amplifiers are present, one to read, and the other to check the modifications. If a bank is modified, its column decoder is connected to the group of sense amps dedicated to the modifying, while the other group of sense amps is used to read from the other banks. The foregoing considerations apply also to other circuits. On the contrary, the switches that connect to the supply voltage are local to each bank, so as to transfer the proper voltage to read or modify the addressed bank. The read and modify paths are separate and, as a consequence, they can be optimized for their particular functions. The read path is optimized in terms of speed, whereas the program path can be slower and includes all the test mode circuitry.

Problem 7.3: Design the architecture for a Flash memory to fulfill the requirements of read-while-modify following the foregoing description.

7.4 Other Types of Array

Device size and the program time[3] are parameters of fundamental importance for a memory. Architectures of arrays different from the standard NOR architecture have been developed to realize smaller arrays which can be programmed in a shorter time.

---

[3] The program time is defined as the time necessary to load the array with code and data. If we imagine a 32 Mbit, i.e. 4,194,304 bytes having an average program time of 10 μs each, we obtain nearly 42 s, a very long time for an assembly line.

As described in Chap. 3, programming through hot electrons, typical of the NOR cell, has the serious limitation of power consumption. This problem is, of course, more serious in the present single voltage memories that have to generate high voltages by means of charge pumps. The drawback of the program current can be overcome by programming with the same physical mechanism used to erase, i.e. FN tunneling. The same solution is adopted in the EEPROMs. In this way, the current is very small but the electric fields and, thus, the voltages are high. The programming operation by FN tunneling is much slower than programming based on hot electrons (milliseconds instead of microseconds). The problem can be solved by programming a high number of bits in parallel (even 4,096), which is possible due to the reduced current consumption.

From the standpoint of device size, the reduction of the drain contact is one of the direction along which research has been conducted. Examining the layout of a cell, it can be seen that the drain contact is one of the largest elements, and additionally the distance between this contact and the gate must be preserved. Beside the traditional lithographic and technological reduction, architectures in which arrays of cells have been realized without standard contacts have been designed, increasing the cell density with some degradation in electrical performance.

Let's now consider three arrays that operate exploiting the FN tunneling principle.

7.4.1 DINOR Arrays (Divided Bit Line NOR)

In this case (Fig. 7.8), three polysilicon layers are deposed, one for the floating gate, one for the control gate, and the other to fabricated the local column to which the drains of 64 cells are connected. These groups of cells are insulated from other groups by means of the hierarchical column decoder. The advantage of this solution is the usage of the so-called direct contacts between polysilicon and active area, which occupy less space than the standard contacts between metal and drain diffusion.

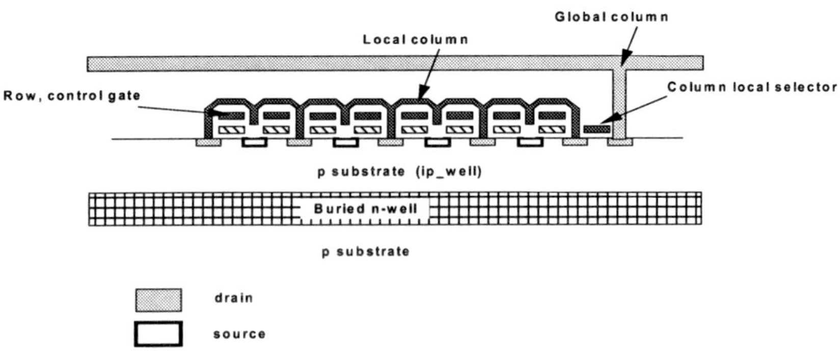

Fig. 7.8. DINOR cells architecture

In Fig. 7.9 the electrical scheme of a portion of the array is shown; the drain, source, and gate decoders are highlighted. In Fig. 7.10, the bias that is necessary to program and erase, without channel current (i.e. through tunneling), is shown. The read is carried out like in a standard Flash, by enabling the transistor of selection and driving the column potential to approximately 1 V.

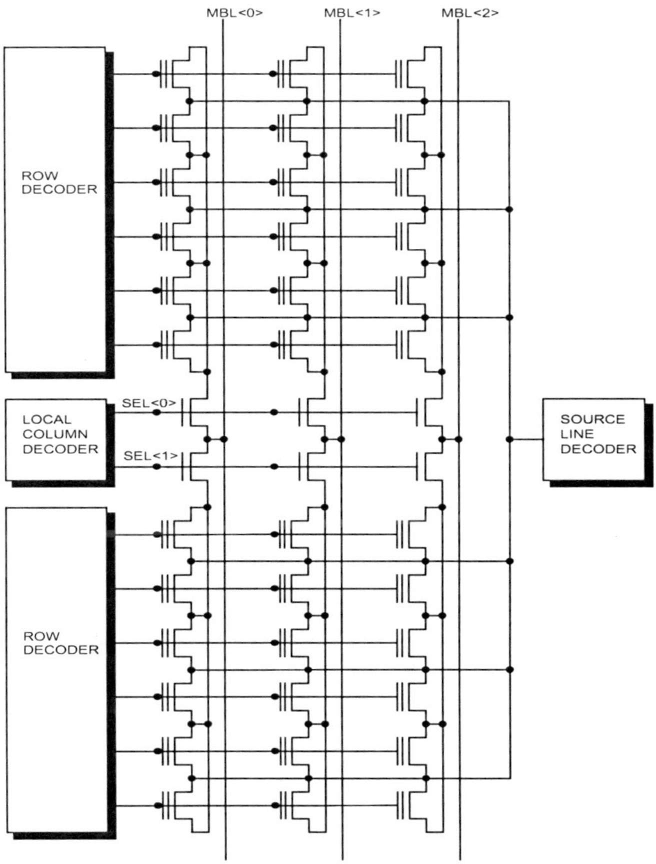

Fig. 7.9. Circuit scheme of the DINOR organization

Up to 256 cells can be programmed in parallel exploiting the tunneling effect. In this case the program operation requires the discharge of the floating gate and, therefore, corresponds to the erase of a Flash.

Problem 7.4: Why is it necessary to implement the erase like the operation that corresponds to the programming of a NOR Flash?

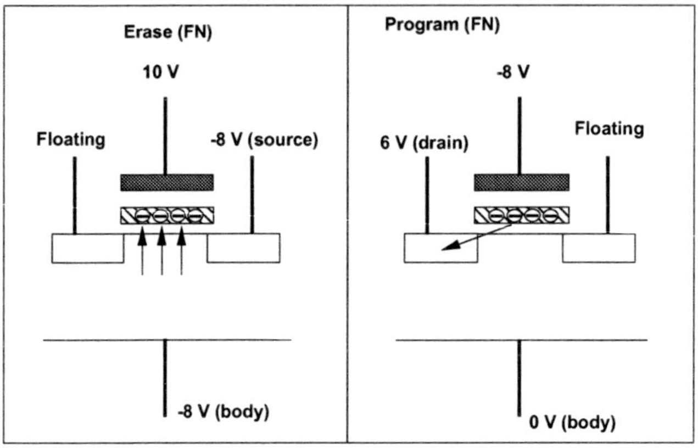

Fig. 7.10. Erase and program mode for the DINOR architecture

7.4.2 AND Arrays

In this case, the contacts are completely removed and the cells are connected through the active area. Considering the resistivity of this layer, it is easy to guess that one of the main consequences is the reduction of the available current. The cells are located in parallel between local bit lines (LBL) and local source lines (LSL); they are selected through proper transistors driven by the SEL signals (Fig. 7.11).

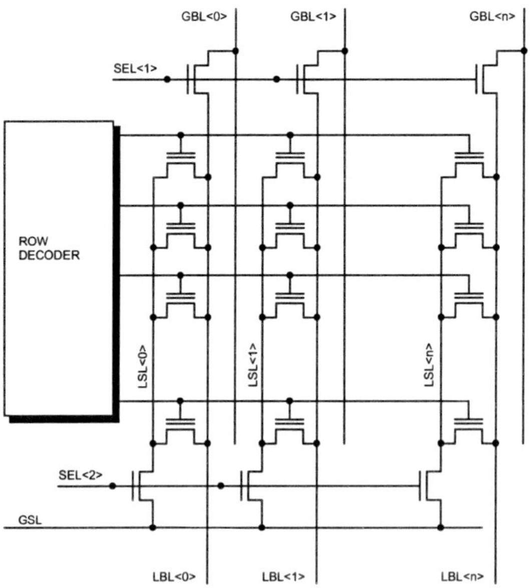

Fig. 7.11. Architecture of an AND memory

GBL is the Global Bit Line, whereas GLS stands for Global Source Line. The main advantage of this type of array consists of the reduction of the pitch of the word line. Program and erase are carried out by means of tunneling (Fig. 7.12), like the DINOR. In this case, the substrate is not biased, and the triple well structure is not needed.

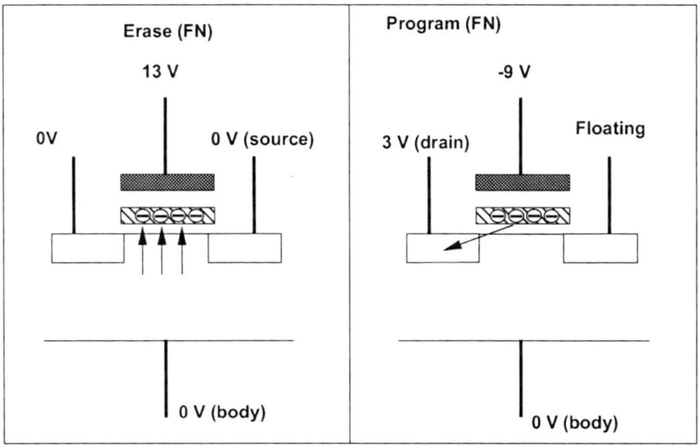

Fig. 7.12. Program and erase for the AND cell

7.4.3 NAND Architecture

Another possible solution to eliminate the contacts is the NAND architecture (Fig. 7.13), in which the cells are stacked like in the NMOS part of a NAND CMOS gate. Typically, 16 cells are connected together in series; the string obtained is separated from the rest of the array by means of two transistors of selection, one for the bit line, the other for the source. The fact that the cells are connected in series implies that all the other unselected cells must be "on" during the reading phase, differently from the NOR architecture. In other words, the unselected cells operate as pass transistors so as to transfer the read voltage to the selected cell. The erased cells have negative threshold while the programmed cells have positive threshold. The potential of the selected word line can equal the ground potential, guaranteeing that only the erased cells drain current. The sensing is usually based on the precharge and evaluation mechanism. The column maintains the potential applied of 2 V approximately only if the cell is programmed. The unselected cells must transfer the drain and source voltages to the selected cell and, thus, must be "on"; therefore, a voltage greater than the threshold of the programmed cell (V_{PR}) is applied to the word line. It is obvious that, because of reliability issues, it is necessary to accurately check the programming in order to control the value of V_{PR}.

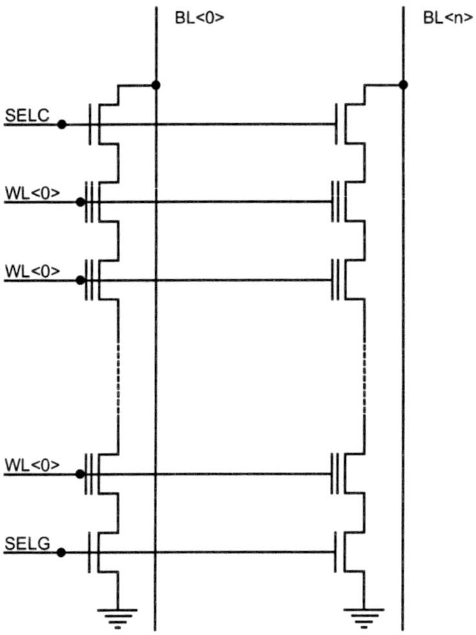

Fig. 7.13. NAND architecture

The modify operations are carried out by FN tunneling effect. The erase is completely driven by the substrate that is biased at 21 V, whereas the word lines are tied to ground. Obviously, it is necessary to employ a triple-well technology to be able to drive the array substrate. Cells with negative threshold voltage are the result of such a modify operation.

The program operation is carried out by applying approximately 20 V to the word line and grounding the channel, so as to trigger the FN tunneling from the channel to the floating gate. The channel is grounded through the bit line: the unselected gates have an intermediate potential (10 V) and the substrate is tied to ground.

Table 7.1. Biasing voltages for NAND

	READ	PROGRAM	ERASE
SELC	5 V	VDD	floating
Unselected WLs	5 V	10 V	0 V
Selected WL	0 V	20 V	0 V
SELG	5 V	0 V	floating
Selected BL "0"	2 V	0 V	floating
Selected BL "1"	2 V	VDD	floating
Unselected BLs	floating	floating	floating
Bulk	0 V	0 V	21 V

The transistor driven by SELG is always "off" during programming, while the WL signal operates on the selection transistor on the bit line. The bit line, corresponding to the cell that is to be programmed, is forced to the ground potential and SELC is driven to VDD so as to transfer the voltage to the channel of the selected cell. In order to avoid programming the cells that share the same word line, it is necessary to tie the bit line to VDD, so as to switch off the transistor of selection of the column. At this point, we have a cell insulated from the bit line with 20 V on the control gate. Due to the coupling between gate and channel, the potential of the channel itself reaches 8 V approximately, thus reducing the electric field across the tunnel oxide to a value below the FN tunneling threshold.

In Table 7.1, the voltages required for the different operations in a NAND array are summarized.

Problem 7.5: Why are the gates of the unselected cells biased at 10 V during program?

Bibliography

L. Bedarida, G. Campardo, A. Silvagni, G. Fusillo, "EEPROM memory device with simultaneous read and write sector capabilities", USA patent No 5,748,528, (May 5, 1998).

A. Bergemont, "NOR Virtual Ground (NVG)- A New Scaling concept for Very High Density FLASH EEPROM and its Implememtation in a 0.5um Process", IEEE/ IEDM, (1993).

G. Campardo, R. Micheloni "Architecture of Non Volatile Memory with Multi-bit cells" 12th Bi-annual Conference June 20-23, INFOS2001, Invited paper, (2001).

G. Campardo, A. Silvagni, L. Bedarida, G. Fusillo, "Non volatile memory device having sectors of selectable size and number", USA patent No 5,793,676, (August 11, 1998).

G. Campardo, et al., "An overview of Flash architectural developments", IEEE Proceeding of the, Vol. 91, No. 4, pp. 523-536, (April 2003).

Y.S. Hisamune, "A High Capacitive-Coupling Ratio (HiCR) Cell for a 3V-Only 64 Mbit and Future Flash Memoires", IEEE/ IEDM, (1993).

M. McConnel et al., "An Experimental 4-Mb Flash EEPROM with Sector Erase", IEEE Journal of Solid-State Circuits, Vol. 26, NO 4, pag 484-489, (April 1991).

Carver Mead, Lynn Conway, Introduction to VLSI system, Addison-Wesley Publishing Co., USA, (1980).

8 The Input Buffer

The first block that the input signal passes through to enter the device is the input buffer, a circuit featuring several functions of fundamental importance to eliminate all the possible disturbances that the external world might cause to propagate inside the chip.

8.1 A Discussion on Input and Output Levels

At first glance, input and output buffers may seem the easiest circuits to design since their only aim is to accept input logic values in the former case and to provide logic output values in the latter. On the contrary, designing both input and output buffers is one of the trickiest challenges of the entire device.

The first topic of concern is related to the voltage values that the input and output levels receive. The devices that we design use CMOS logic, but they must be able to inter-operate with other devices that may implement non-CMOS logic. For this reason, it is a requirement to have two voltage ranges that are conventionally identified as a logic "0" (or logic level low) and as a logic "1" (or logic level high) respectively.

In the case of TTL logic, input levels are considered a "1" if they are higher than 2 V (V_{IH}), while a "1" output level must be higher than 2.4 V (V_{OH}); a logic "0" is detected as an input if it is lower than 0.8 V (V_{IL}), while a valid output is lower than 0.4 V (V_{OL}). A logic gate is usually driven by another gate of the same kind. Therefore it is convenient to define a noise immunity margin (ΔN) as the difference between the input and output values for both the logic states:

$$\Delta N_H = V_{OH} - V_{IH} = 0.4\ \text{V (TTL)} \tag{8.1}$$

$$\Delta N_L = V_{IL} - V_{OL} = 0.4\ \text{V (TTL)} \tag{8.2}$$

Generally speaking, memories can handle both CMOS and TTL input levels (voltage ranges for both input and output levels are declared inside the product specification). The evaluation of both the input and output logic signals is done considering V_{IH} and V_{IL} in the case of TTL compatibility, and 50% of VDD in case of CMOS as shown in Fig. 8.1.

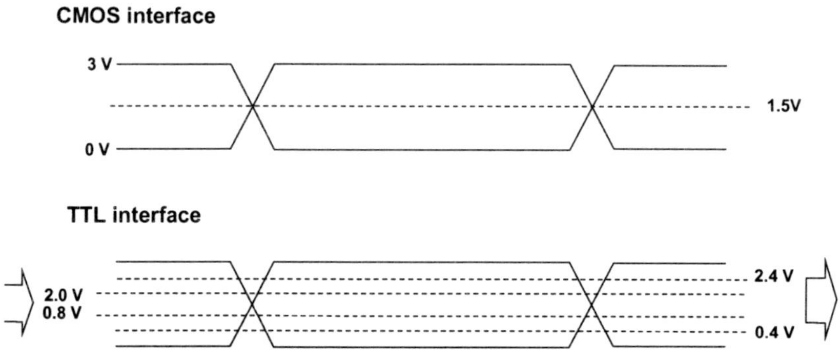

Fig. 8.1. The two available interfaces CMOS and TTL

8.2 Input Buffers

Input buffers are present on all address and control pins, such as CE#, OE#, BYTE#, etc. Non-volatile memory processes are typically fabricated in CMOS technology, and therefore there are no compatibility issues when interfacing to external CMOS logic signals. In the case of TTL inputs, a TTL to CMOS input buffer is required.

In an input buffer, the easiest circuit scheme for a TTL to CMOS conversion is one composed of a chain of inverters. These are all powered by the internal VDD where the size ratio of the first inverter is unbalanced so that its commutation threshold is lowered. The other inverters in the chain are balanced, so that both rising and falling edges of the signal provided to the downstream circuitry are symmetrical.

In this solution, the first inverter can have both transistors in conduction when receiving TTL inputs (see Fig. 8.2)[1] even if the output of the circuit is driving the right logic value. Therefore a high current consumption can occurs, especially in the case that VDD is at its maximum value.

In order to ensure a satisfactory noise margin, the input buffer is dimensioned so that, under typical VDD and temperature conditions, the voltage range between V_{IL} and V_{IH} is centered halfway in the indetermination zone between TTL input levels, i.e. 1.4 V. Let's imagine applying a V_{IH} (i.e. the minimum voltage recognized as "1") of 1.6 V.

Under these conditions, the n-channel is turned on "strongly" enough to pull the output node of the inverter to ground, regardless of the fact that the p-channel is turned on as well. As VDD increases, V_{GS} of the n-channel does not change, while V_{GS} of the p-channel increases in modulo. Therefore the value of V_{IH} increases. Similar considerations hold true for the V_{IL} level.

[1] The picture considers a CMOS process, but the concepts are the same in case of NMOS process too.

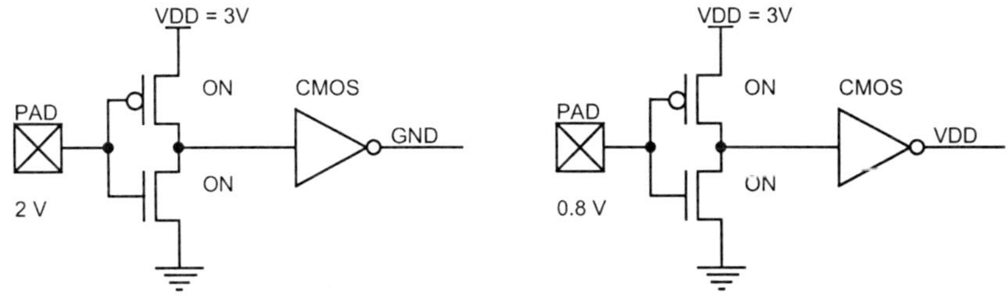

Fig. 8.2. Applying TTL inputs, the first inverter is switched on for both V_{IL} and V_{IH}. It is the size ratio that "drives" the second inverter (whose ratio is set for CMOS levels) which can provide a full output swing over VDD

To sum this up, V_{IL} and V_{IH} levels increase as VDD increases and therefore it is mandatory to control these values and keep them in the range given by the specification. It is also necessary to maintain a sufficient margin for noise that can be present on both VDD and GND. The technique that is usually adopted is to control the current of the p-channel on the input inverter (see Fig. 8.3) through the biasing of a pass PMOS. The current flowing in M8 can be controlled as a function of the VDD or, equivalently, the voltage of node B can be kept constant as VDD varies.

This trick allows minimizing both V_{IL} and V_{IH} variations with respect to VDD, but it also causes several issues that may be unacceptable.

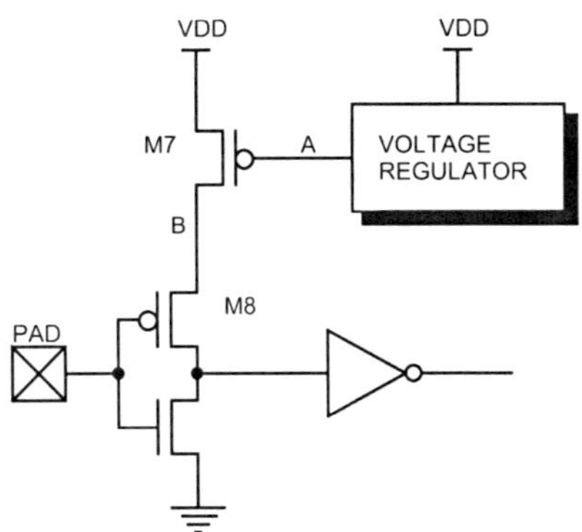

Fig. 8.3. To limit variation of V_{IL} e V_{IH} levels with respect to VDD, the current that flows in the input inverter is controlled

First of all, the problem of power consumption is worsened by the fact that, there is normally not a unique regulation block for all the input buffers but, for instance, there may be a block for every two buffers. A single regulation block can not quickly drive a large capacitive load and thus multiple regulation circuits are used to drive the set of input buffers. Furthermore the circuit that controls the current shown in Fig. 8.3 cannot be left on in stand-by mode, but on the other hand it must respond in a few nanoseconds to immediately propagate new addresses inside the device.

8.3 Examples of Input Buffers

Unlike CMOS buffers, an input buffer that is implemented in NMOS technology can be significantly different depending on the specific usage. The reason for variation is that the bootstrap can be used or not used (simplifying the design a lot), and the load that must be driven influences this choice. Figure 8.4 shows both the electrical and the principle schematic for an address buffer in an NMOS process. It is evident when there is no attention paid to the variation of V_{IL} and V_{IH}, solved only by correctly dimensioning the input inverter. This is possible because NMOS devices (that are now quite old) allowed a supply voltage range limited between 4.5 V and 5.5 V, which is far above the voltage range of the TTL.

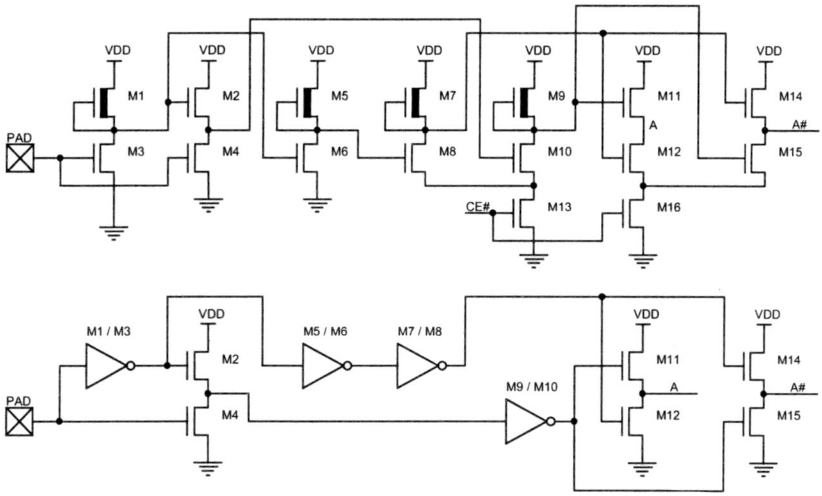

Fig. 8.4. Input Buffers for the Addresses in NMOS Technology

Detailed analysis is left to the Reader. The fact that the final drivers, which are realized using LVS transistors, do not allow A and A# outputs to reach the supply voltage is very important. In this case, we suppose that the address and its complement have a quite reduced load, since they just need to propagate from the buffers to the predecoders. These predecoders have the task of bringing the signal on lines whose capacitive load is very high. Finally, the enabling of the buffer via CE# signal allow it, during stand-by, to tie both outputs, A and A#, to zero.

Let's go back to Fig. 8.3 to discuss today's CMOS buffers. These buffers work on very wide supply ranges, for instance from 1.8 V to 3.6 V, therefore allowing almost 2 V of swing as compared to the 1 V of the previously described NMOS case.[2] Furthermore, the decrease of supply voltage causes a reduction in noise immunity and greater attention is paid to the control of V_{IL} and V_{IH} values.

If VDD tends to increase and the input voltage is the same, then the current flowing in the input inverter grows as well. One way the variation of the supply voltage can be compensated is to have a regulator circuit able to raise is to raise driving voltage of node A as VDD increases, thus partially turning off M7 PMOS and limiting the current.

Problem 8.1: The input inverter does not work exactly as a CMOS inverter because it is never turned off when TTL values are applied at the input. Is it still correct to call it an "inverter"?

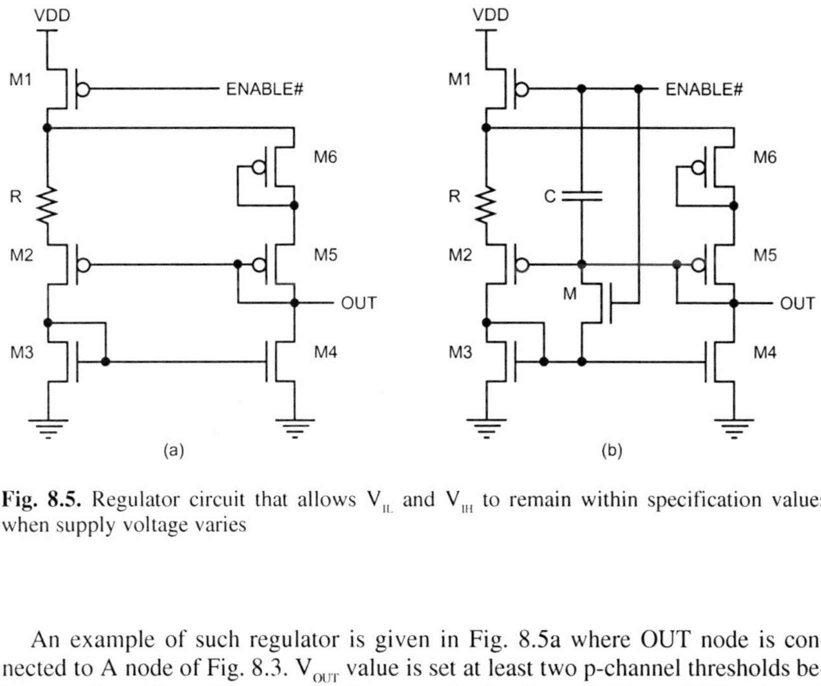

Fig. 8.5. Regulator circuit that allows V_{IL} and V_{IH} to remain within specification values when supply voltage varies

An example of such regulator is given in Fig. 8.5a where OUT node is connected to A node of Fig. 8.3. V_{OUT} value is set at least two p-channel thresholds below supply voltage, thus guaranteeing an output voltage that keeps M7 p-channel switched on. If the supply voltage grows, the current flowing through M1, M2 and M3 and, in the right branch, M4, M5 and M6 increases. The output voltage increases as well.

[2] Can you explain why a range for the supply voltage is given, instead of imposing a precise value to the customer?

Let's see what happens when temperature varies. Suppose that the current imposed by M4 is constant. A temperature increase causes not only linear decrease of the thresholds, but also exponential decrease of mobility. According to Eq. (3.9), M5 and M6 must increase their V_{GS} in modulo to bear the current imposed by M4. The overall result is a lowering of the output value of (VDD – $2|V_{T,p}|$). On the other hand, the value of R resistor increases with a temperature increase, thus making the current in the left branch decrease, M4 to drive less current, and the output node to rise. This compensates for the fall caused by M5 and M6. Figure 8.5b shows how to speed-up the power-up phase of the circuit by pre-charging all the critical nodes.

Problem 8.2: Discuss the use of both the capacitor and the transistor added in Fig. 8.5b.

8.4 Automatic Stand-By Mode

In current devices, which are more and more used in portable applications, consumption, in particular during stand-by phase, has become an outstanding parameter. This is a key decision factor for purchasing. There is also an automatic stand-by, also known as Automatic Sleep Mode (ASM), that puts the circuit in a "nearly stand-by" condition. This means that not all the circuitry is switched off and the allowed consumption is not a few microamperes, but some tens of microampere. The device automatically enters ASM mode if it is not accessed for a predefined time ranging from 200 to 300 ns. From this state, the device automatically "wakes up" as soon as at least one address or CE# vary, and the first read takes the usual access time to complete.

Also in this case, the input buffer must be really fast to turn on and able to keep consumption under an acceptable limit.

One way to solve consumption issues is to remove current control, allowing a moderate variation of V_{IL} and V_{IH} with respect to supply voltage and a reduced noise margin. Low consumption specifications are usually very tight about dissipation. ASM signal is expected to de-activate most of the circuits of the chip, which in turn reduces the internal noise on VDD and GND making the chip less noisy and making it possible to accept a lower noise margin.

This choice allows the design of a circuit that satisfies both consumption and power-on time requirements. It is only during the stand-by phase that the input buffer is turned off, and it will be almost instantaneously turned on as soon as the CE# signal is asserted again.

ASM consumption specification is satisfied turning off those transistors that usually constitute the input inverter and turning on other transistors to bring the IN signal to CMOS values even in case of TTL inputs (see Fig. 8.6). The first network shows two very conductive switches, MP1 toward VDD and MN1 toward GND, used during normal operation. The second network has two corresponding resistive switches, MP2 and MN2, which work in a low consumption state.

In other words, it is possible to limit crowbar current that flows through the first TTL inverter of the input buffer chain by properly selecting the supply path to connect to the input buffer itself. This selection of the working condition is performed by a proper ASM signal that detects the timeout of a pre-defined address latency range. When the buffer works in low consumption mode, it is still able to switch without timing penalties. It is the switching of the buffer that brings the device's working condition back to normal mode, de-asserting the ASM signal.

The schematic in Fig. 8.6 also depicts two capacitors C1 and C2 and two diode-connected natural transistors D1 and D2, which are part of the low consumption network. The role of these components is explained by the fact that a certain amount of time (about 10 ns) is required after the first switch of the input buffer in low consumption mode to let ASM to be de-asserted, thus having the chip back to normal mode. During this time, other switches of the inputs are possible, but MP2 and MN2 alone are not enough to guarantee the current required to transfer such switches to the output before ASM reaches GND. In other words, MP2 and MN2 are deliberately resistive to limit crowbar current, but this resistance prevents the buffer from switching within the required time. C1 and C2 are useful charge reserves, just in case ASM transition is slower than expected because of a badly estimated internal delay.

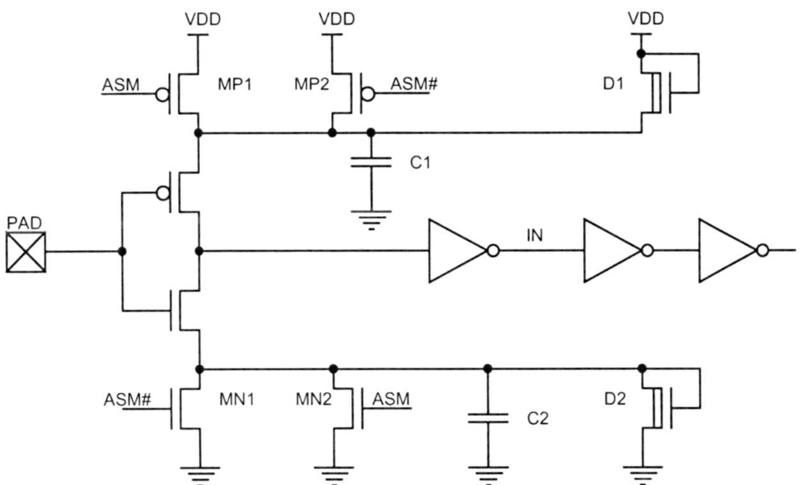

Fig. 8.6. Comprehensive electric scheme for an input buffer capable of complying with the most restrictive ASM specification

The task of D1 and D2, highly conductive and normally turned off transistors, is to prevent the discharge of C1 and the charge of C2 beyond a certain limit, even if ASM transition should never occur.

The CE# buffer deserves a special mention: it cannot have stand-by disable control, since CE# causes this mode itself; therefore this buffer must be carefully designed from a consumption point of view.

Bibliography

T. Kobayashi et al., "A Current-Controlled Latch Sense Amplifier and a Static Power-Saving Input Buffer for Low-Power Architecture", IEEE Journal of Solid State Circuits, Vol. 28, No 4, (April 1993).

H. Miyamoto et al., "Improved Address Buffers, TTL Input Current Reduction, and Hidden Refresh test Mode in a 4-Mb DRAM", IEEE Journal of Solid State Circuits, Vol. 25, No 2, (April 1990).

C. Wang et al., "A 3.3-V/5-V Low Power TTL-to-CMOS Input Buffer", IEEE Journal of Solid State Circuits, Vol. 33, No 4, (April 1998).

C. Yoo et al., "A Static Power Saving TTL-toCMOS Input Buffer", IEEE Journal of Solid State Circuits, Vol. 30, No 5, (May 1995).

9 Decoders

Row and column decoders represent one of the most crucial challenges in the design of all memories, including EPROM, Flash, or RAMs. The complexity of the decoder circuitry increases with the introduction of new functionalities, the inclusion of different operating voltages, and the continuous growth of the size of the devices. The memory devices that were once a few Kbits in the early '80s have grown to hundreds of Mbits at the end of the century, and the Gigabit (in NOR architecture) is certain to be fabricated in the near future.

9.1 Introduction

Let's consider a 4 Mbits Flash memory, i.e. 4,194,394 memory cells[1]. On the silicon die that contains the circuit, a very compact area exists where the memory cells are fabricated as close to each other as possible. If the cells could not be addressed one by one to read, write, and erase, we would not be able to use the memory. In order to reach all the cells of the array, a system of paths must be realized, so as to identify each cell inside the array itself, by means of the address.

The type of array organization we will deal with in the following paragraphs is the so-called NOR array, because of the way in which the cells are connected to each other. Just to discuss about a real case, let's suppose that a 1 Mbit array is to be decoded with unity form factor, i.e. 1,024 rows (word lines) and 1,024 columns (bit lines). Each column, fabricated with metal, has 1,024 cells connected, whereas the 1,024 rows are composed of 1,024 cells having common gate. All of the sources of the cells are connected to ground through equally spaced metal diffusions.

Figure 9.1 reports the scheme of a 1 Mbit memory, structured as eight outputs. A new concept has thus been introduced here: we do not decode a single cell but a byte, i.e. a group of 8 cells, or a word, i.e. 16 cells in parallel.

Now, in order to address 1 byte (8 cells), the corresponding row must be driven to the read voltage, and 8 columns must be enabled. The choice to realize the byte with bits that belong to the same row is preferred, since, in this case, one single row must be addressed to read the bits of the same byte, which are placed on different columns, with consistent performance in terms of time.

---

[1] The cuts of the Flash memories currently on the market range from 1 Mbit to 256 Mbits, 64 Kbits to 32 Mbits for EPROMs.

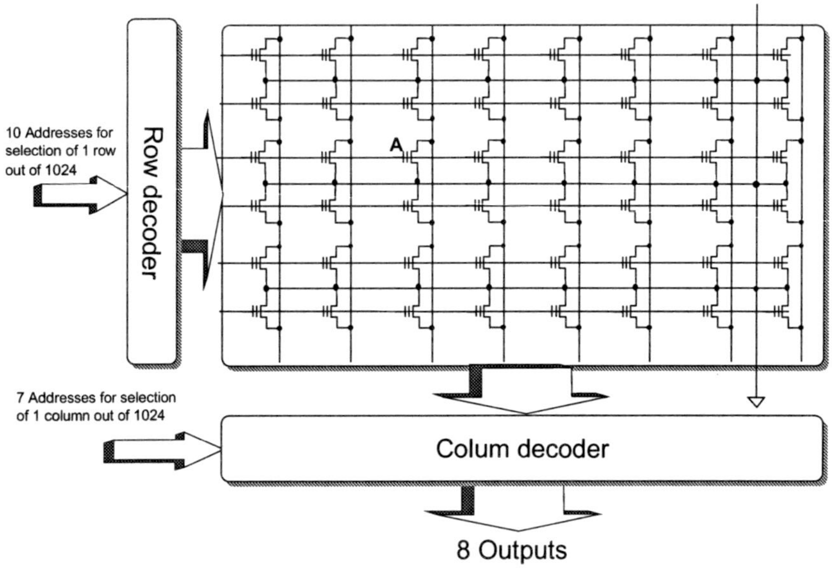

Fig. 9.1. A 1 Mbit memory structured as eight outputs

As it can be noted in Fig. 9.2, the bits that form the byte are not close to each other, but they are placed at the maximum distance allowed. In this way, potential problems of noise during the read or write, which could affect contiguous cells, are minimized. Furthermore, the resistive paths toward ground are equivalent for all the cells of the byte (or word).

The array in Fig. 9.2 is divided into eight sub-blocks that represent the eight outputs of the device. The row decoder has to address 1,024 rows and, hence, 10 bits are necessary (2^{10} = 1,024 combinations); seven bits are required for the column decoding (2^7 = 128 combinations).

Row and column decoders are generally composed of two sections placed in cascade. The first, often referred to as "pre-decoder", realizes the logic operation required, so as to address row and column, as specified by the user through the input pins. This part is usually biased at VDD. The second stage, the actual driver, is responsible for the transfer of the analog voltage, which is required for the different operations, to only the selected cells. In general, the second stage is commonly referred to as "decoder", even though the term "selector" would be more appropriate. As we will detail in the following, the difference is not always so clear, since, sometimes, also the logic selection is carried out at high voltage.

Theoretically, having n row addresses, the pre-decoding can be realized by means of 2^n logic AND gates with n inputs, through all the possible combinations of the inputs, inverted or not. Obviously, in practice such a solution is hardly feasible in terms of area occupied and complexity of the layout. Usually, a hierarchical approach is adopted.

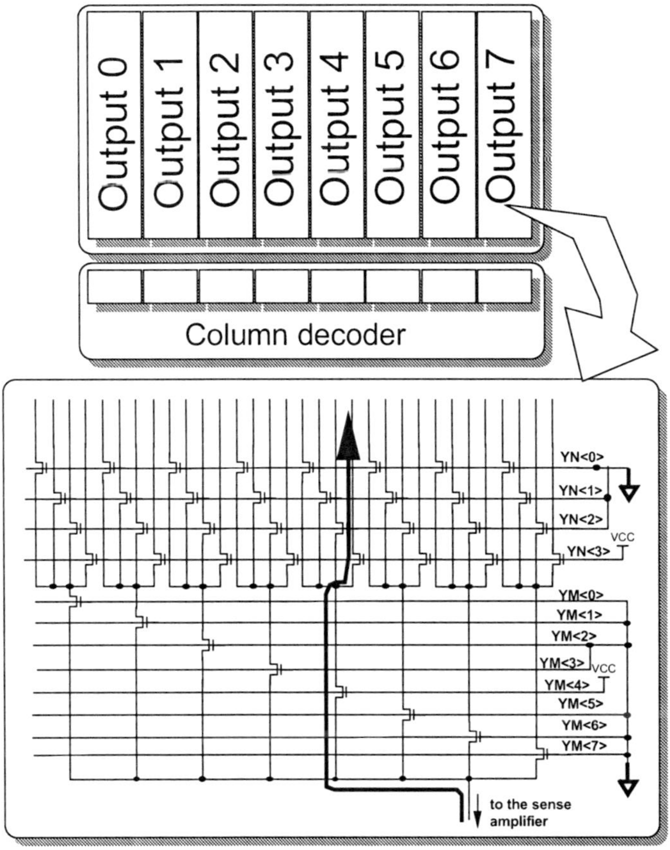

Fig. 9.2. Scheme of an output. Enabling YM<4> and YN<3> allows accessing the bit line indicated, and this happens for all the outputs. The bits that form the byte are thus placed at the distance of 127 columns.

The bits of the address are divided into subgroups and independently decoded. In Fig. 9.3, an example with 4096 rows (n = 12) is shown; the bits are divided into 4 groups named LX, LY, LZ, and P. The overall number of AND gates that are required for the addressing is still 4096, but the total number of inputs has decreased from 12 to 4 by inserting 32 AND gates with 3 inputs.

Let's now examine the impact of the hierarchical approach on the complexity of the layout (Fig. 9.4). Suppose that the 4096 rows are divided into 8 sectors of 512 rows each. The signals named LX<7:0> could be used to identify the sector. LX<3> must physically reach only sector number 3 and all the AND gates of that sector have LX<3> as input.

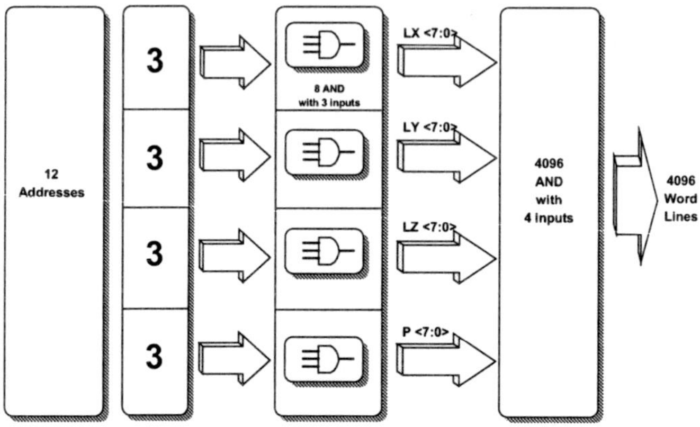

Fig. 9.3. Stage of logic pre-decoding with hierarchical structure

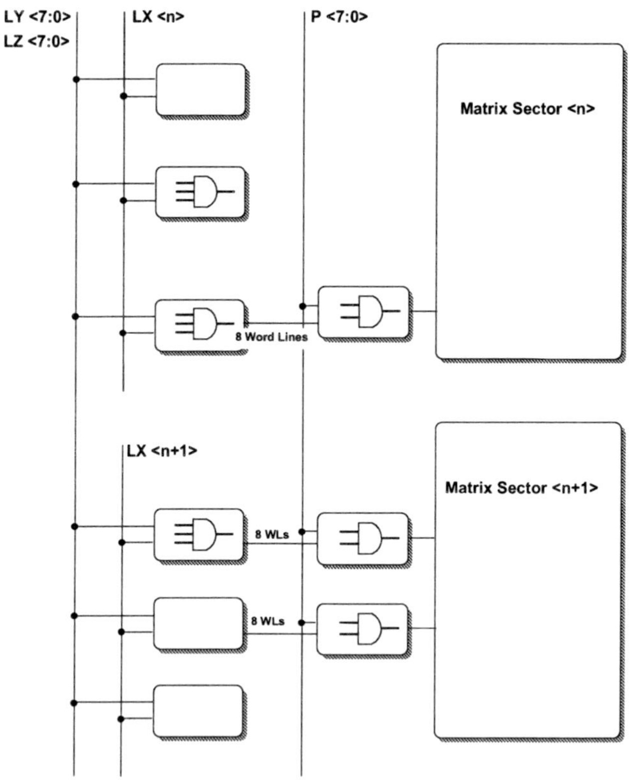

Fig. 9.4. Organization of the row decoder

Performing the AND logic operation between each LY and each LZ, 64 groups of 8 rows each can uniquely be identified. By means of the P<7:0> signals, the final selection of one row out of eight can be carried out. In other words, only the LY, LZ, and P signals cross the entire memory. Obviously, the hierarchical partition of *n* addresses is a possibility that is at the designer's disposal. For example, the fact that a single LX signal identifies an entire sector is not a casual choice, since, as we will detail in the following, the LX are directly used in all the activation circuitry of the switches that drive high voltages to the selected row during a program or an erase.

9.2 Word Line Capacitance and Resistance

A memory designer knows that the first section of the memory to design is the row decoder. Such structure is the most compact portion of the device, apart from the array. In fact, the final driver of the row decoder is an inverter that must fit into the row pitch, which is not trivial at all due to the reduced size of the memory cell (Fig. 9.5). Starting from the load that must be driven, i.e. the polysilicon word line, let's analyze the row decoder.

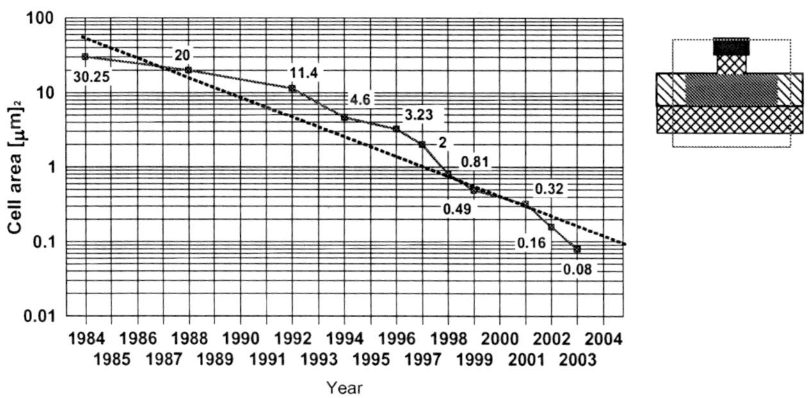

Fig. 9.5. The area of the memory cell has progressively been reduced over time as a result of the evolution of the integration technologies

The row is composed of a set of cells with common gates, as sketched in Fig. 9.6. The estimation of the load due to the row involves the evaluation of the row resistance, which is due to the polysilicon used to connect the control gates, and, on the other hand, the evaluation of the capacitance due primarily to the gate of each cell of the row.

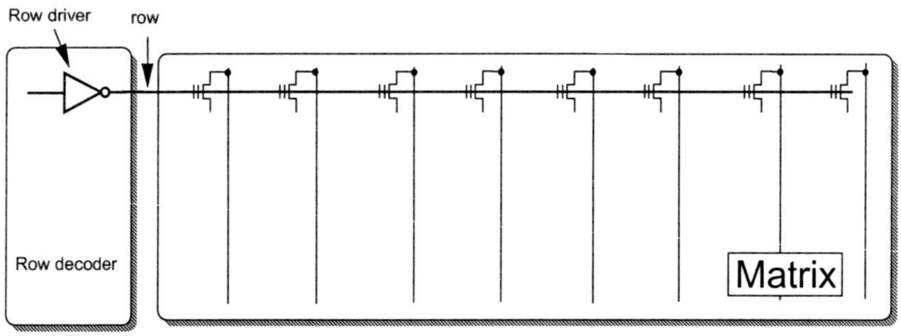

Fig. 9.6. The load of the row driver is due to a distributed RC

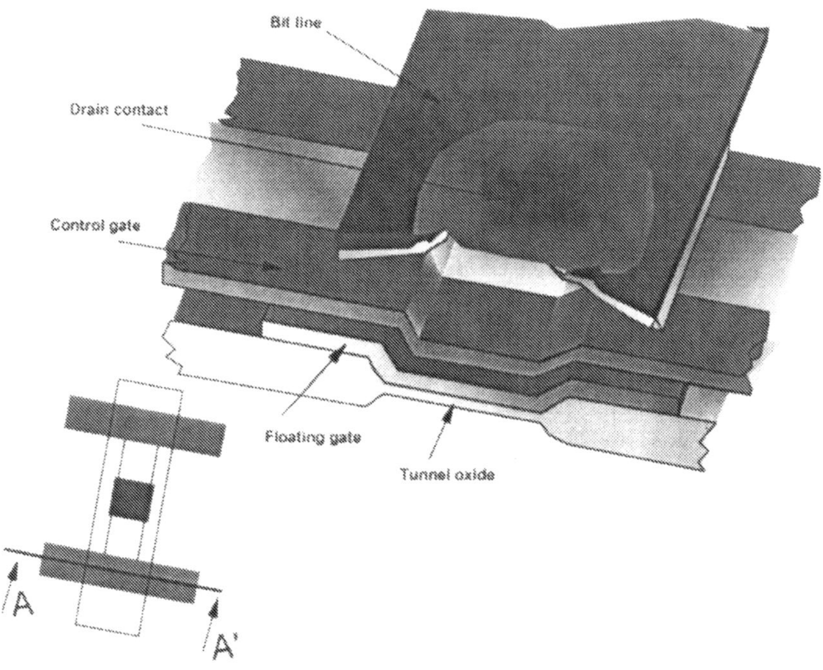

Fig. 9.7. Section of a Flash cell along the row. It is possible to distinguish the oxides, the floating and control gate, the dielectric between poly2 and metal1 of the column, which penetrates into the drain contact, and, at the bottom, the drain junction

Figures 9.7 and 9.8 show the top view of the cell layout and the 3-dimensional cross-section view. In order to calculate the capacitance of the cells, we will refer to the AA' section. We can use the hypothesis that the floating gate, control gate, and substrate form the plates of two capacitors having plane and parallel plates. The contribution of the poly2 capacitor on the substrate can be neglected, since the field oxide that separates them has a thickness that is one order of magnitude greater than the oxide used to insulate the floating gate. Figure 9.9 shows the schematic of the cell that is suitable for this purpose.

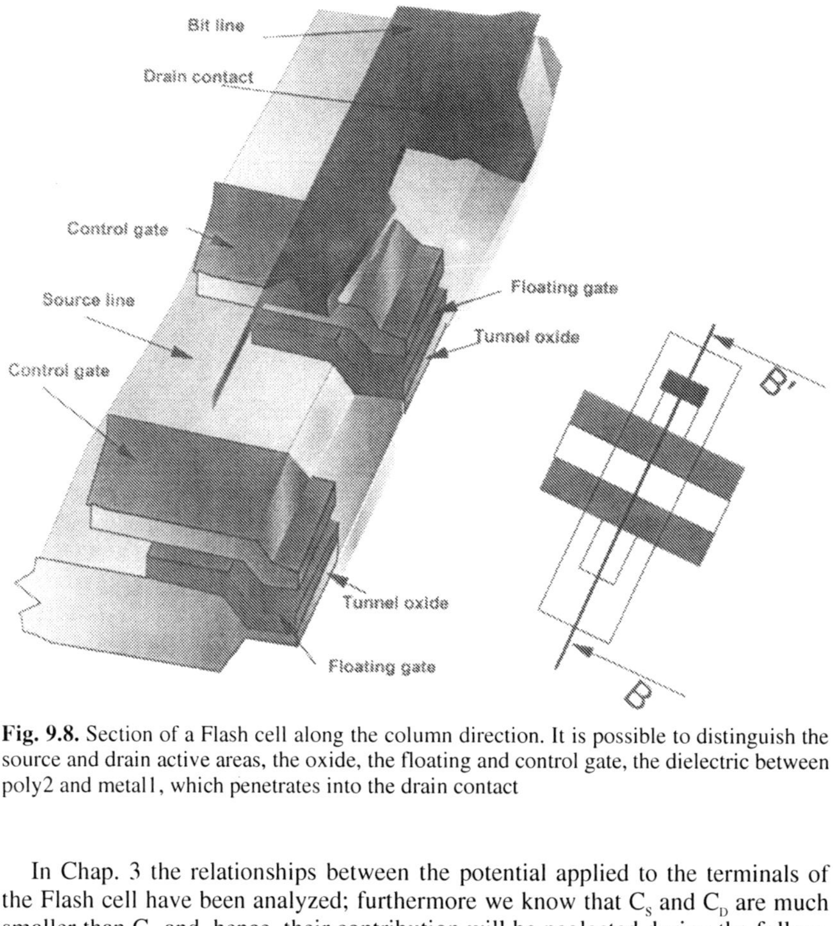

Fig. 9.8. Section of a Flash cell along the column direction. It is possible to distinguish the source and drain active areas, the oxide, the floating and control gate, the dielectric between poly2 and metal1, which penetrates into the drain contact

In Chap. 3 the relationships between the potential applied to the terminals of the Flash cell have been analyzed; furthermore we know that C_S and C_D are much smaller than C_B and, hence, their contribution will be neglected during the following analysis[2].

Let's suppose that the cell has a length and a width of 0.8 μm, and the thickness of the oxide between floating gate and substrate, also called thin or tunnel oxide, is

[2] Obviously, it depends on the process and analysis we want to carry out. For example, during the erasing phase, the capacitance between the source and floating gate plays a fundamental role and cannot be neglected.

120 Å. The capacitance of a single cell, with respect to the control gate, is given by the series of the two capacitances C_{FC} and C_B. Thus, given an interpoly oxide of 200 Å, the capacitance with respect to the control gate is:

$$C_G = \frac{C_{FC} \cdot C_B}{C_{FC} + C_B} \tag{9.1}$$

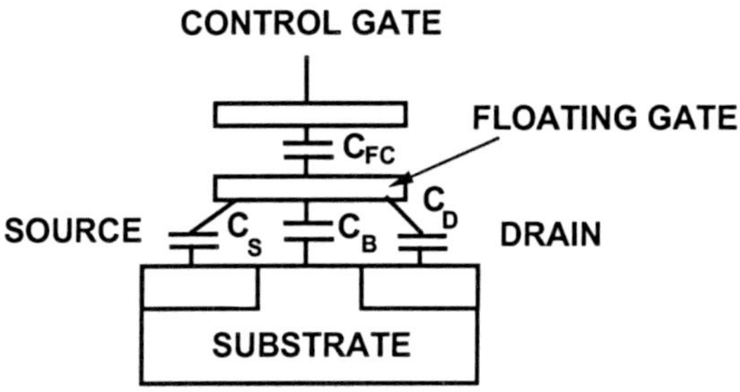

Fig. 9.9. Schematic of the capacitances of a Flash cell

If the calculation is carried out with the indicated values, the result is smaller than what is actually measured on silicon. Re-examining Fig. 9.7, it is possible to notice that poly1 extends beneath poly2 beyond the active area (i.e. over the field oxide). This extension of poly1 is called a "wing", and it is useful in order to increase the capacitive ratio and improve the write efficiency. Indicating with C_{wing} the capacitance associated with the poly extension, Eq. (9.1) can be rewritten as:

$$C_G = \frac{(C_{FC} + 2 \cdot C_{wing}) \cdot C_B}{C_{FC} + 2 \cdot C_{wing} + C_B} \tag{9.2}$$

$$C_{FC} = \frac{\varepsilon_0 \varepsilon_{ox}}{200 \cdot 10^{-10} m} \cdot 0.8\mu m \cdot 0.8\mu m = 1.1 fF \tag{9.3}$$

$$C_B = \frac{\varepsilon_0 \varepsilon_{ox}}{120 \cdot 10^{-10} m} \cdot 0.8\mu m \cdot 0.8\mu m = 1.84 fF \tag{9.4}$$

Supposing an extension of 0.5 µm, C_{wing} can be calculated:

$$C_{wing} = \frac{\varepsilon_0 \varepsilon_{Si}}{200 \cdot 10^{-10} m} \cdot 0.8\mu m \cdot 0.5\mu m = 0.69 fF \tag{9.5}$$

Using Eq. (9.2), we obtain a capacitance of 1.06 fF per cell. Hence, a row with 1,024 cells produces a capacitance due to the gate only that amounts to nearly 1.1 pF. Such a value is to be added to the contribution of the layers above poly2, i.e. metal1 and so on, according to the process and design realized. In our example, we can estimate an overall capacitance of 1.2 pF.

The calculation of the row resistance is simpler, since it depends only on the number of squares of the polysilicon row. Considering a row with 1,024 cells having pitch in the row direction of 3 μm, and polysilicon having resistance of 3 Ω/□[3], we obtain an overall word line resistance R_{wl} equal to:

$$R_{wl} = \frac{3\mu m}{0.8\mu m} \cdot 1024 \cdot 3\Omega = 11.52K\Omega \tag{9.6}$$

Then, with the calculated row capacitance and resistance, the time constant of the row is:

$$\tau_{wl} = R_{wl} \cdot C_{wl} = 1.2 \cdot 10^{-12} \cdot 11.5 \cdot 10^{3} s \cong 13.8ns \tag{9.7}$$

Thus, the time required for the row to reach 63% of the final value is about 15 ns. If VDD is 5 V, after 63% of the charge time, the voltage of the row is 3.15 V. For typical values of threshold of virgin cells, this can be considered as a "good" value at which the cells can start draining the current required, so as to have the reading circuit (sense amplifier) work.

Reasoning in terms of pure RC equivalent, like we have done so far, can lead to a 20 to 30% error. Once the parasitics associated to the word line have been estimated, it is necessary to tune the equivalent circuit that accounts for the real behavior in terms of load, delay, and so on. In the case of materials having high resistivity, a model having lumped parameters (one single capacitor and resistor) could be quite inaccurate. One of the alternative models, widely used in such cases, is sketched in Fig 9.10, even though it is suitable just for computer simulations.

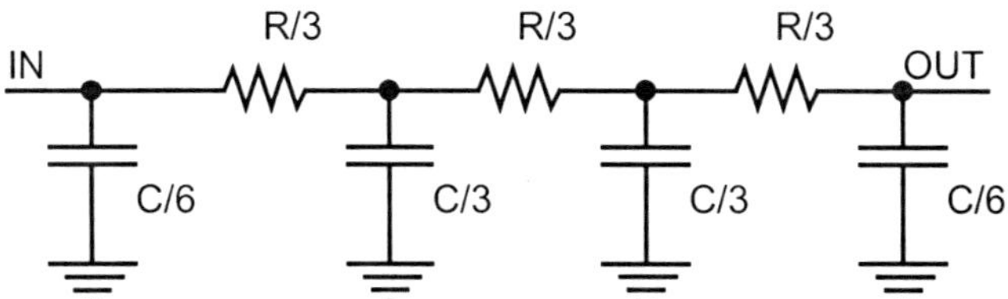

Fig. 9.10. Equivalent circuit of the word line with distributed parameters

[3] This is true under the hypothesis of Self-Aligned-Silicide (SAS); otherwise the poly2 resistance could be even 50 Ω/□. The materials used to form silicide are titanium or cobalt. The metal is deposed onto the poly2 layer so as to form silicide ($TiSi_2$ and $CoSi_2$) through thermal treatment. The word line is therefore composed of the parallel of the two layers.

9.3 Row Decoders

Let's summarize the characteristics of the row decoder:

- A good memory designer should dimension the row decoder immediately after the array;
- The row decoder is the most compact part of the device after the array. The decoding has to fit into the row pitch and it is convenient to design the entire layout using the minimum rules[4].
- Due to the high load that is to be driven, the row decoder is one of the crucial sections in order to obtain a good access time. As a consequence, the layout of the final drivers is drawn with great care.
- The complexity of the row decoder is also due to the fact that the row is to be tied to VDD or to a boosted voltage during the read, to VPP during programming, negative during erase, and, finally, to ground when it is not selected.

Let's now imagine that we want to design the row decoder and concentrate on the single row. We need to design a row driver, i.e. an inverter, supplied with a variable voltage, e.g. VDD in read and VPP in program[5]. In this way, we design the final driver and consider two rows in program as shown in Fig. 9.11. The VPC bias equals VDD in read and VPP during the program. The block named "LOGIC", controlled by means of the P<7:0> signals, selects one of the rows, whereas the first NAND selects the block to which the examined rows belong, by means of the LX, LY, LZ signals.

In the logic chain that starts from the row address, there must be a point of separation between the low voltage VDD and the high voltage VPP. Generally, this stage is the NAND gate of the row decoder. If VPC equals VPP and WL<1> is selected, the gate of M1 and M2 are driven to ground. This portion of the circuitry works fine. It is also required that WL<7> is tied to ground like the other rows. Nevertheless, if the gate of M3 and M4 are driven to VDD, the p-channel, M3, is still "on", since the difference of potential between source and gate is greater than its threshold voltage. Thus, there is a problem with M3 being still on and not allowing WL<7> to be pulled completely to ground.

Thus, it is clear that, in order to deselect the word line, it is necessary that the driving voltage of the inverter is at least equal to the difference between its supply minus the V_T of the p-channel. Hence, the block called LOGIC, which is interposed between the NAND (at low voltage) and the final inverters (at high voltage), cannot be realized by means of simple logic gates: it must contain something that is able to deselect all the rows that are not addressed. In other words, it is necessary to have a level shifter, like in Fig. 9.12.

---

[4] In all the other cases, the rule of thumb suggests not designing at the minimum distance for what concerns different layers. On the contrary, it is better to use relaxed rules to increase the device reliability, still having the possibility of making some modifications in a later phase.

[5] During the program phase the voltage applied to the row reaches 10 V.

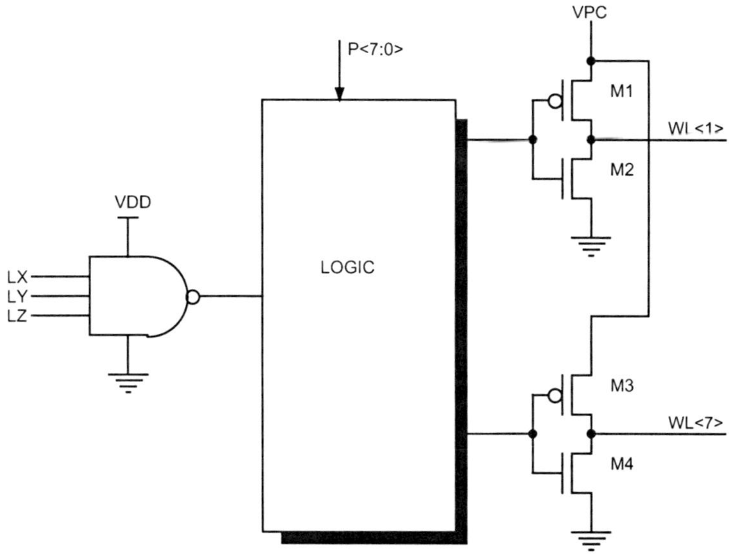

Fig. 9.11. The final drivers of two rows belonging to the same group of eight rows. The row decoder addresses one single row in read and program, all the others must be set to ground.

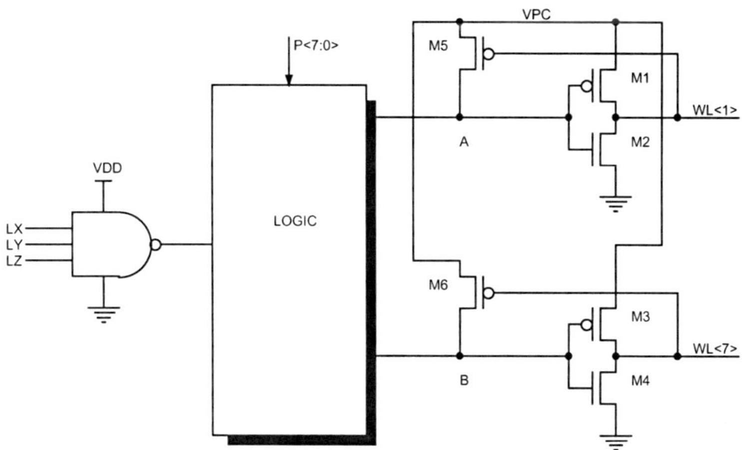

Fig. 9.12. The p-channels in a feedback configuration that realize the level shifter

The M5 and M6 transistors transmit the signal of the row to the input inverter. If WL<1> is driven to 10 V, VPC must be 10 V, the A node has to be driven to ground and B to 5 V, i.e. to VDD. In this way, M2 results in being "off", M1 "on", and WL<1> is biased by VPC. On the other hand, if V_B equals VDD, both M3 and

M4 are “on”, and WL<7> rises to an intermediate value between ground and VPC. Yet, also M6 is “on” and the voltage of the B node is driven toward VPC, because of M6. Hence, M4 conducts more and M3 less, and, as a result, the potential of WL<7> tends toward ground, making M6 more conductive and driving V_B to a higher potential. The positive feedback that has been triggered comes to an end only when V_B reaches the maximum possible value, i.e. VPC, switching M3 off and driving WL<7> to ground.

Although everything seems to be fixed, a serious problem of isolation between low and high voltage still exists. If the NAND is biased at 5 V and the p-channel in reaction drives the input of the inverter, which is also the output of the NAND, to 10 V, the drain-body junctions of all the p-channels of the NAND itself are forward-biased (Fig. 9.13).

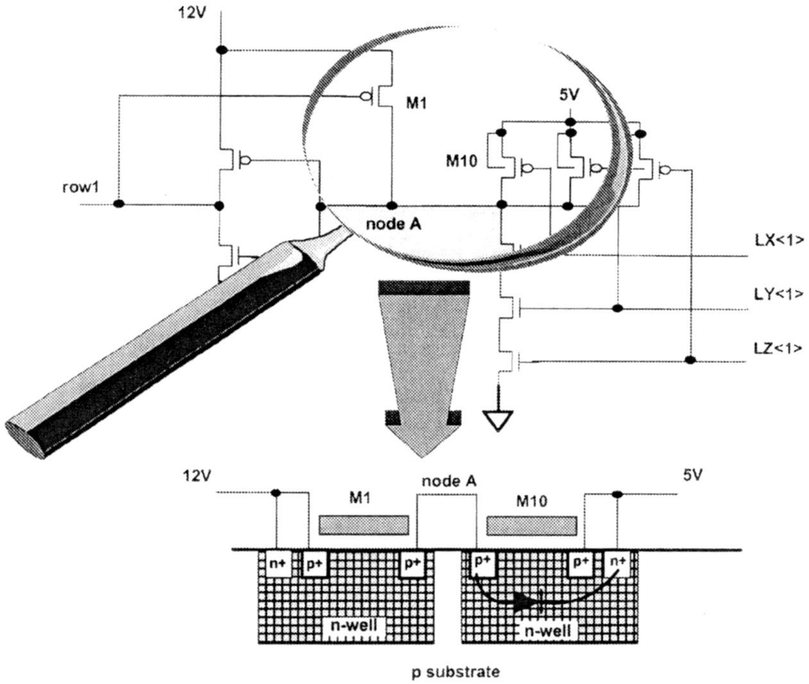

Fig. 9.13. When the voltage of the A node equals VPP, the drain-body junctions of the p-channels belonging to the decoding NAND are forward-biased

Thus, it is necessary to separate the parts biased at different voltages. The simplest way to achieve the required result is shown in Fig. 9.14. When V_C is driven to ground, it is followed also by V_A, and the output node of the inverter, i.e. the row, can rise up to VPC.

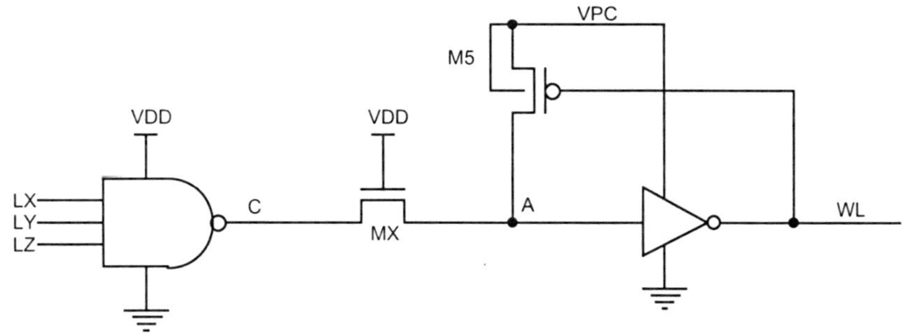

Fig. 9.14. The insertion of the n-channel pass-transistor MX, with positive threshold, fixes the problems of separation of the two different supplies, even though it introduces a penalty in terms of speed.

When the C node is driven to VDD (5 V), A does not follow it up to the maximum value because the n-channel, having its gate at VDD, pulls it up to VDD minus the V_T of the n-channel, i.e. to about 3 V, assuming that the threshold voltage of the n-MOS equals 1 V.

Problem 9.1: Why do we maintain that the C node initially goes to 3 V if the threshold of the MX n-channel is 1 V? It should go up to 4 V, shouldn't it?

In this way, the feedback due to M5 starts working up to point in which A is charged to VPC, i.e. 10 V. Now the MX pass transistor can be regarded as belonging to the opposite path, from A toward C. The A node tries to reach 10 V, but MX limits the voltage drop to $V_{GS} - V_T$, and, hence, C is still driven by the NAND gate to VDD. The M5 p-channels and those of the NAND, separated by MX, do not suffer from the forward-bias problems any further.

Problem 9.2: The insertion of the MX transistor poses some problems to the dimensioning of M5. Can you find out why?

The block named LOGIC, highlighted in Fig. 9.12, has been omitted from the analysis, so far. If the NAND selects a group of rows, eight for instance, there must be a further level of decoding so as to select the correct row in that group. A first type of decoder is shown in Fig. 9.15. The NAND gate selects a first group of 8 rows, driving the potential of A to a low value. In the second block of 8 rows V_B remains high. If WL<0> is addressed, in the first block there must be P<0> low and P<7:1> high. Hence, C goes low, allowing the inverter, INV1, to drive WL<0> high. At the same time, D is driven high by P<7>, and rows from 1 to 7 are driven low.

In the second block, and in all the other blocks of the row decoder, the first row is low because the equivalent of the A node is high and, hence, with P<0># high, also the equivalent of the C node is high, and the inverter drives WL<8> to ground. The remaining rows of the block, instead, have P<7:1> high and, as a consequence, drive the corresponding rows low.

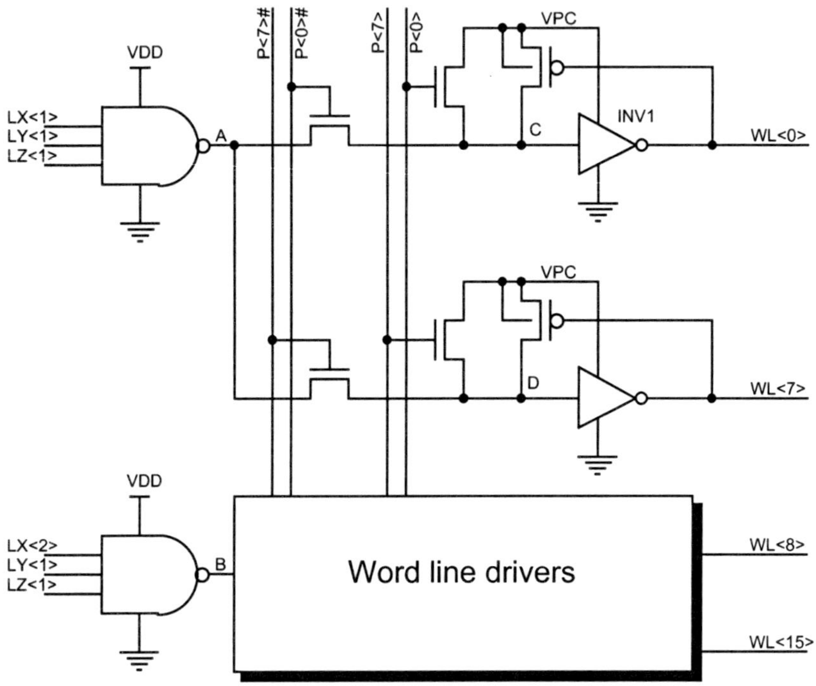

Fig. 9.15. The LX, LY, and LZ signals divide the rows of the array into blocks of eight (in this case). Signals P<7:0> select one single row of the block.

Just to summarize: the address is divided into row and column address. The address of the row is used as an input of the pre-decoder that typically generates the signal that have been referred to as P and L. By means of the P signals, the correct row is selected inside the group (Fig. 9.16).

Finally, we have been able to select one single row among all the rows of the array. Figure 9.17 shows the voltage versus time waveform of the selected row, measured at the end of the row itself. Each time the row is selected, we have to wait until the previously selected row is deselected. In Fig. 9.17, we suppose that at time t_1, the row selected with voltage V_u has a potential that is high enough to start reading the cell. This means that the word line has a potential that is high enough to consider an erased cell (or a virgin cell in case of EPROM) as "on". In the same figure, we can see that the row deselected at t_1 has voltage equal to V_d. If the cell addressed in the deselected row with gate voltage V_d is still "on", we cannot say that one cell is uniquely addressed at t_1, because also the cell that has previously been addressed is able to draw current. In practice, it is not sufficient to consider a cell as "on" to start reading, but it is also necessary that the cell addressed in the previous read cycle is "off".

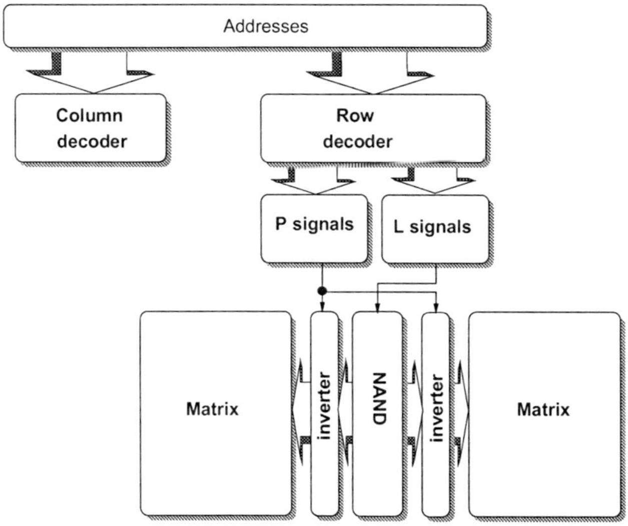

Fig. 9.16. Distribution of the P and L signals from the addresses to the array

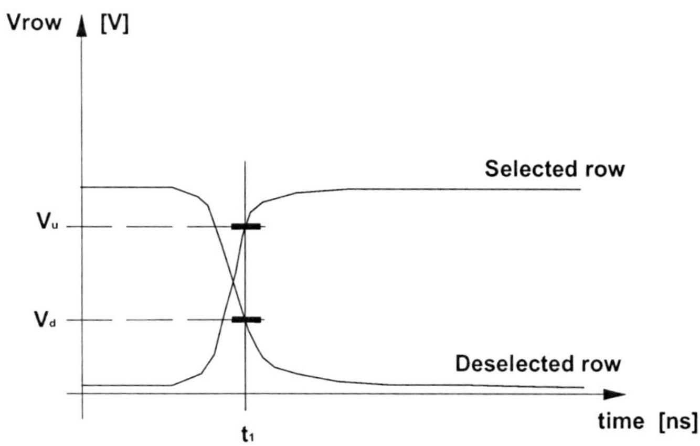

Fig. 9.17. Voltage versus time waveform of a selected and deselected row

A method to reduce the delay during the access is to unbalance the final inverter of the row, designing a larger n-channel which decreases the fall time of the row that has to be deselected.

Problem 9.3: In light of the latest consideration, discuss the dimensioning of the circuits highlighted in Fig. 9.15.
Problem 9.4: Can the charging of the row be accelerated without limits by increasing the size of the charging transistor?

Let's now examine some possible types of row decoding.

9.4 NMOS Row Decoder

Let's start analyzing an NMOS row decoder. The main problem is related to the fact that, in this technology, we do not use complementary p-channel transistors and, hence, the final driver of the row can be realized only by means of n-channel LVS transistors. In Sect. 5.12 an example of a row driver with bootstrapping is given.

Problem 9.5: Why is it not possible to use a typical NMOS inverter with a depletion load as final driver of the row?

The A1, A2, and A3 signals of Fig. 5.66 can be substituted for LX, LY, and LZ; the corresponding output signals, OUT and OUT#, are then used as input of a further decoding block controlled by the P signals, as shown in Fig. 9.18. Supposing that OUT1# and OUT2# are at the high logic level, all the LVS transistors are "on", whereas all the natural transistors are "off", under the hypothesis that all the natural transistors have positive threshold voltage. Thus, none of the P signals is enabled to be transferred to the rows. On the contrary, all the rows are driven to the VHB voltage level that, in this case, is supposed to be tied to ground. In case a block or a row is selected, the OUT signal is biased at VDD and the row having the corresponding P signal high is selected.

The problem is that, if the natural transistors have a slightly negative V_T (–100 mV), as it often happens, these transistors cannot be switched off by simply driving their gates to ground. In the case of the Flash devices of the latest generation, this drawback could be solved by using a negative voltage source so as to drive the OUT signals to a negative potential that is at least equal in magnitude to the V_T of the natural transistor. In the case of EPROMs belonging to old generations, instead, it was not strictly necessary to drive the rows to the ground potential, accepting a gate stress that was negligible for those processes[6].

The signals of the row decoder are repeated so as to have only one group of rows having the OUT signal high and the P signal, the one that identifies the row to activate, at VDD. However, other groups of rows have the OUT signal low and the corresponding P high. Since it is not possible to switch off the pass transistors through the gate signal, any possible paths between the P signal (high) and ground must be eliminated, so as to limit the power consumption. This implies the use of a voltage slightly higher than ground for the low logic level.

[6] One of the problems that arose passing from EPROM to Flash designs was the necessity of driving the row to the ground potential after reading. In the first generation EPROMs, the addressed row was driven to the ground potential during a subsequent read operation. At the end of reading, if the address did not change, only the sense amplifiers were switched off. With the technology of that time, the gate stress did not have any impact on the reliability. The same strategy applied to a Flash memory caused the failure of the device in a short time due to charge losses.

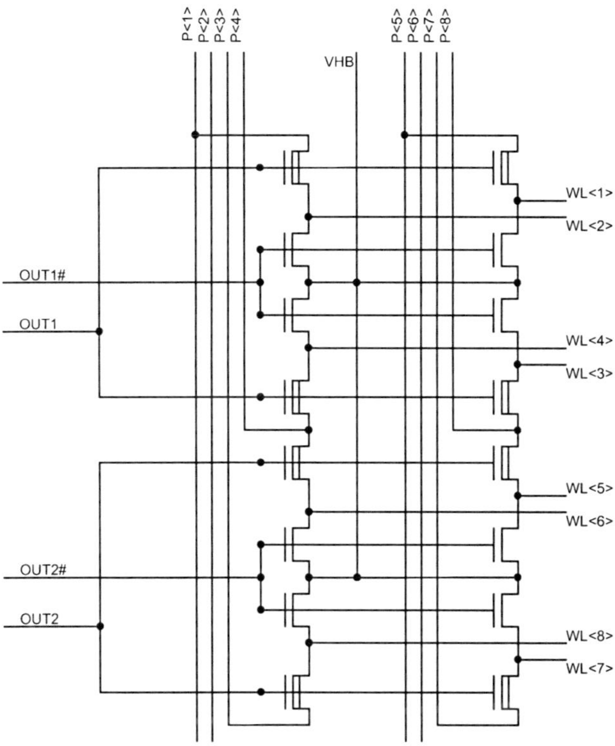

Fig. 9.18. The combination of P and OUT signals selects one single row

The OUT signals always have a voltage swing between VDD and ground, whereas the P signals can range between VDD and VHB, which is a voltage of about 200 mV, i.e. twice the magnitude of the threshold voltage of the natural transistor[7]. The minimum value of the voltage that is necessary to switch off the natural transistors is $V_{T,NAT}$; the voltage above the threshold voltage is useful as safety margin for the design. The possibility of correlating the VHB voltage to a process parameter such as the transistor threshold, allows following the process variations maintaining, at the same time, the minimum value of VHB. Figure 9.19 shows the point of insertion of the circuit in the decoding. Thus, if row WL<1> is to go high, the OUT and P<1>signals must be high, whereas P<5> must be at VHB. At this point, WL<2> is slightly positive, even though not high enough to be able to switch on the erased cells that, in case of an EPROM, have a dense distribution around the UV threshold.

[7] This circuit is detailed in Chap. 5, in the section dedicated to the voltage sources.

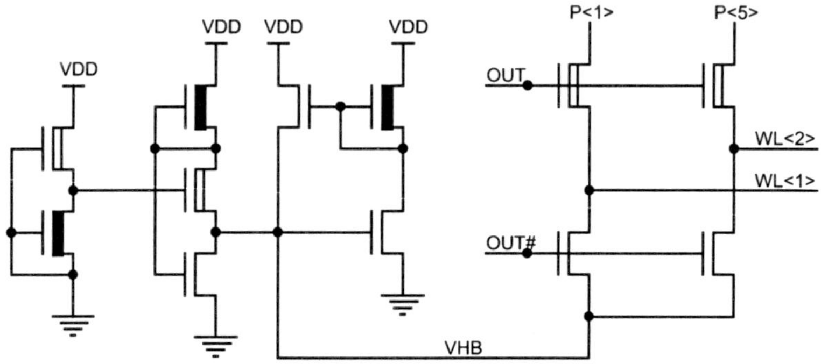

Fig. 9.19. The use of the VHB voltage source solves the problem of the switch-off of the natural transistors of the decoder

The P signals are common to many pass transistors, since they are repeated every 16 rows, and we want to be sure that, when OUT is at ground, the natural transistor acting as a pass device for the P signal is "off" (Fig. 9.20).

If VHB were at GND, M1 would be "on" even with grounded gate, since its threshold voltage is negative. This would cause power consumption due to the current flowing from VDD to VHB through the driver of WL<2>. Considering that such a condition applies to each block of rows, it is easy to figure out what the effect on the overall device power consumption would be.

When the memory is switched on, all the rows go to the VHB potential, charging the parasitic capacitances involved. During reading, the selected row is biased at VDD and the previously addressed row is driven to VHB again.

Problem 9.6: Is it possible to use a simple resistor instead of the bias circuit, accepting the consequent loss in terms of correlation between process variations, temperature, etc?
Problem 9.7: Describe how the VHB voltage source operates as the temperature varies.

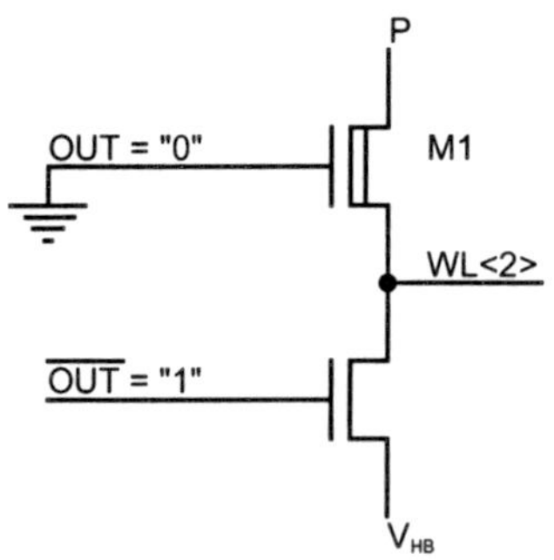

Fig. 9.20. The configuration of the OUT and OUT# signals shown, in which VHB is at GND, is not able to switch M1 off

9.5 CMOS Row Decoders

Let's now examine a CMOS decoder. The basic difference with respect to the previous case is that the final inverter that drives the row is now a CMOS inverter and, hence, we do not need any bootstrap to drive the row to VDD.

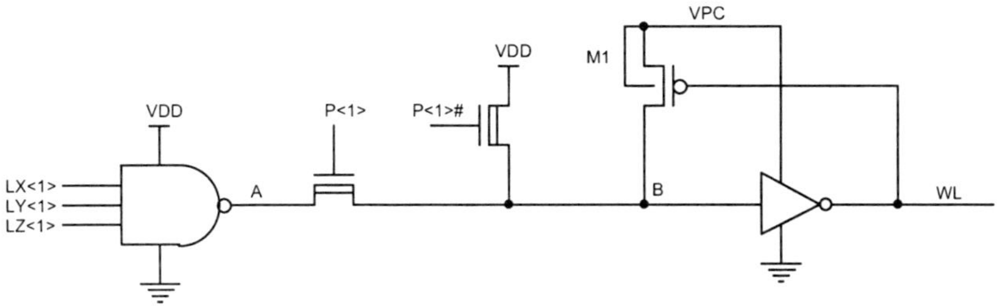

Fig. 9.21. An example of decoding in which the P signals act as pass to separate high voltage from low voltage

As an example (Fig. 9.21), we can realize a decoder in which the P signals act as pass signals and, hence, provide the decoupling function between high and low voltage, or, as an alternative, we can design a decoder in which the voltage of the row is driven by the P signals and the function of the p-channels in the feedback configuration is to correctly switch off the p-channel that is driven by the P signal (Fig. 9.22).

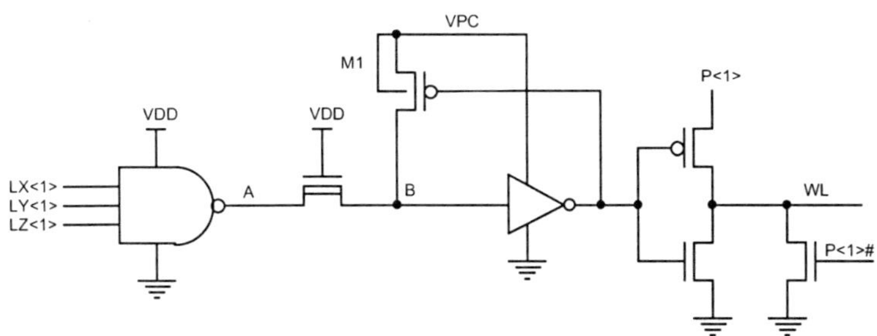

Fig. 9.22. A CMOS decoder in which the voltage of the row is supplied by the P signals

In this case, it is evident that the P and P# signals must be able to reach high voltage. We recall that the row decoder is a block that is placed very close to the array, often in the middle of two array blocks, and, for this reason, the generation of the P signals is located in an area of the device that is external to the row decoder, where it is possible to design with more relaxed constraints and, thus, with more freedom.

A different decoder is shown in Fig. 9.23. It is highly compact since the contact of the M1 and M7 p-channels to the word line has been removed, even though static consumption for the p-channel in the feedback path of the addressed row has to be accounted for. If WL<1> is high, P<1> is also high, as well as the corresponding LX, LY, LZ signals, and, therefore, there is current consumption from M1 through the NMOS branch of the NAND.

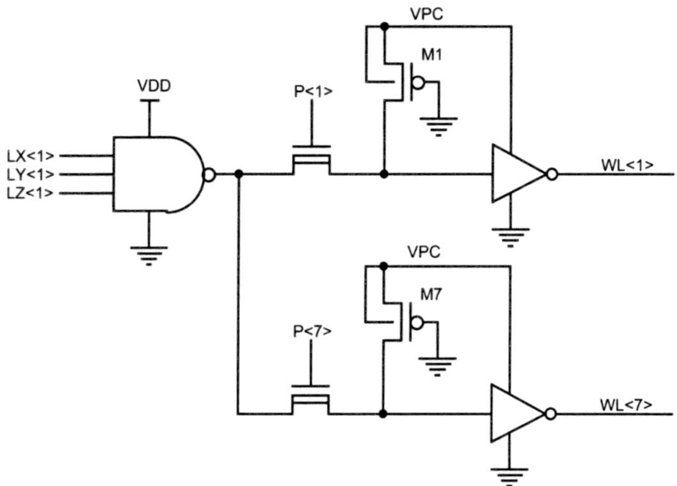

Fig. 9.23. The elimination of the connection to the word line of the p-channels of the feedback path allows compacting the decoder and using non-inverted P signals only

The main difficulty that one may come across during the design of the row decoder is due to the fact that the decoding must fit into the pitch of the array and, hence, the available space is always very limited. Moreover, in the case that the device must operate at low supply voltage, e.g. 2.5 V, it is convenient not to use pass transistors since they are less efficient as they transfer the signal voltage minus the transistor V_T. In order to solve this problem, a decoder like the one shown in Fig. 9.24 could be employed. In this case, no p-channels are present in the feedback path of the row and there are no pass transistors. Let's try to understand how this configuration works and, then, we will analyze the cost that has to be paid to achieve such result.

In Fig. 9.24, the schematic of the driver and the corresponding NAND is shown. The L signals, as already mentioned, drive the NAND that selects which of the sub-blocks must be active. When all the L signals are high, the NAND output is low and, due to the following inverter, INV1, the M1 transistor is switched on, which enables all the inverters that are driven by the P signals. Thus, if the P signal is high, the respective row is also high. The sub-blocks that are not enabled have the NAND output that is high and, as a consequence, the output of INV1 is low. Therefore, M1 is "off" and the inverters driven by P (high) have floating outputs, as well as the corresponding rows. In order to overcome this drawback, p-channels directly driven by INV1 have been introduced. The goal is to have the output of the inverters driven by the P signals in the "high" state when the NAND output is high. At this point, we have realized a decoder without the p-channels in the feedback path of the row, thus decreasing the complexity of the layout, without pass-transistors, improving the low voltage operations and, finally, only the P signals are used, with no need to provide circuitry for the generation of inverted P signals like in the previous decoders. This results in further saving in terms of area.

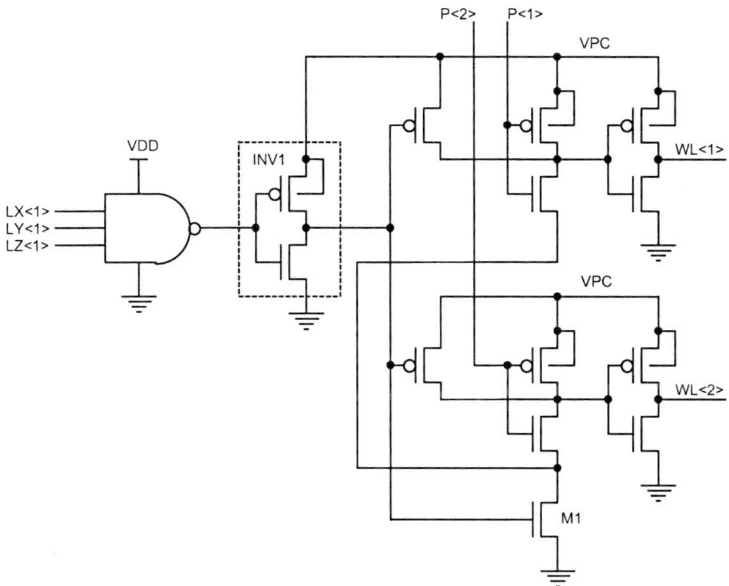

Fig. 9.24. A CMOS decoder without p-channels in the feedback path and without n-channel pass-transistors. Try to dimension the transistors.

The price to pay is that the P and L signals must be at high voltage. The separation between low and high voltage has been moved outside of the decoder, to the so-called "row pre-decoder", which generates the P and L signals from the address. We will deal with such circuit blocks in Sect 9.9.

9.6 A Dynamic CMOS Row Decoding

Until now, we have dealt with "static" decoders, which means operating in asynchronous mode: the voltage state of a row is a logic function of the P and L signals, with no signal that provides timing for the reading phase. Nevertheless, "dynamic" decoding, in which a timing signal regulates the various steps of reading is also possible. As an example, in the case of memories embedded into a microcontroller, i.e. a microprocessor with some peripherals inside, a clock signal is always active during any active phase. In such cases, it is straightforward to exploit the clock signal to obtain a dynamic row decoder. In Fig. 9.25, an example of dynamic decoder is shown.

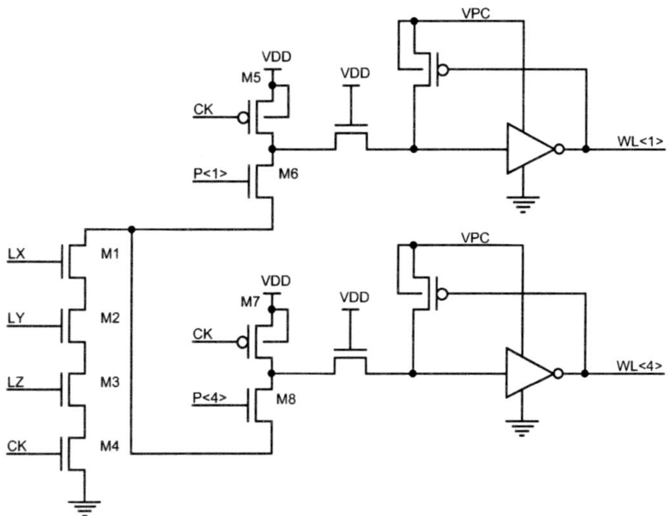

Fig. 9.25. A dynamic CMOS decoder. The NAND composed of the M<4:1> n-channels has only one p-channel. The circuit operates in two phases: precharge and evaluation

The feedback circuits that realize the translation from low to high voltage, the separation pass, and the P and L signals can be recognized. The essential difference of the dynamic decoder is that the NAND driven by signals L does not have p-channels.

Let's focus on WL<1>. The reading phase is split into two steps, the first of which is called "precharge", whereas the second is called "evaluation". During the precharge, the CK signal is low and M4 is consequently "off", whereas M5 and M7 are "on". All the rows are low and the NAND is not enabled to operate. The P and L signals are driven to the proper logic level. Finally, the evaluation phase is carried out, when the CK signal goes high, switching the p-channels off and enabling the NAND.

The main advantage of the dynamic decoding is the reduction of the number of transistors. In fact, the NAND gates driven by the L signals are realized with only one p-channel. The drawback is in the necessity of a correct timing that, in the stand-alone memories, i.e. not embedded into a microcontroller, has to be generated from asynchronous signals.

Problem 9.8: Discuss the timing of a dynamic decoding for a memory that is not connected to an external clock.

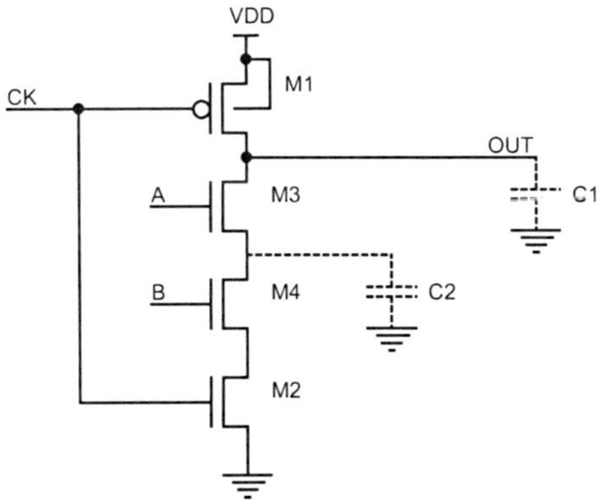

Fig. 9.26. Charge sharing problems in case of dynamic decoding

A second problem related to dynamic decoders and, more in general to precharge circuits, is shown in Fig. 9.26. When CK is low, we are in the precharge phase: M1 is "on" and M2 is "off"; hence, the output signal, OUT, is high. CK is then driven low and the output node is then floating, i.e. connected to the parasitic capacitor C1 only. M2 is "on" and the evaluation phase can be carried out. If M3 and M4 are "on", the output node is driven to ground, discharging C1; otherwise, C1 is still charged, maintaining the potential of the OUT node high. Let's now imagine that the precharge phase has been carried out and the A and B signals have respectively been driven high and low. We have a charge sharing, i.e. a redistribution of the charge stored in C1 into C2 through M3, and, if the parasitic capacitance C2 is comparable with C1, the potential of the output node will decrease. In such a condition, the following circuit might detect a voltage switch and, as a consequence, propagate incorrect information.

In conclusion we can affirm that dynamic structures need more care during the design phase in order to prevent incorrect operations.

9.7 A Semistatic CMOS Row Decoder

Together with static and dynamic decoders, also hybrid decoders, which we will refer to as "semi-static", can be designed. In Fig. 9.27 an example of this type of decoder is shown. The NAND is static but the feedback path from the row to the input of the last inverter, due to the M1 and M7 p-channels, has been eliminated, thus facilitating the layout.

The gate of M1 and M7 are driven by a signal called PULSE# that is synchronous with the ATD, which, as we will detail in Chap. 11, is the signal that detects any variations of the address. Bringing back to mind that the main problem that forces us to use the PMOS in the feedback path is due to the program phase during which the pass transistors driven by P<7:0> have the task of separating the low voltage of the NAND from the high voltage, VPC. In this situation it might be difficult to make the final inverter switch and, in any case, undesired consumption would result. If all the inputs of the final inverters are kept at VPC during the switching phase of the P and L signals, the problem of the difficult switch of the last inverter is eliminated.

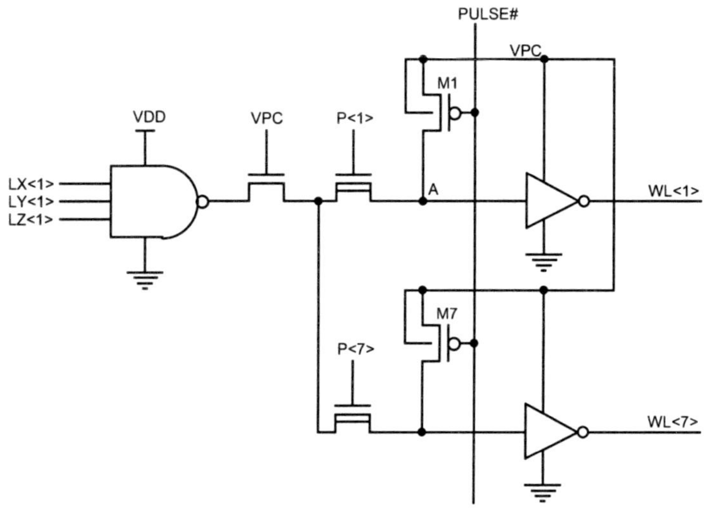

Fig. 9.27. The precharge of the A node allows the feedback of the semi-latch that drives the row. The main problem is due to the requirement of perfect synchronism of the PULSE# signal with respect to the P and L signals of the row decoder.

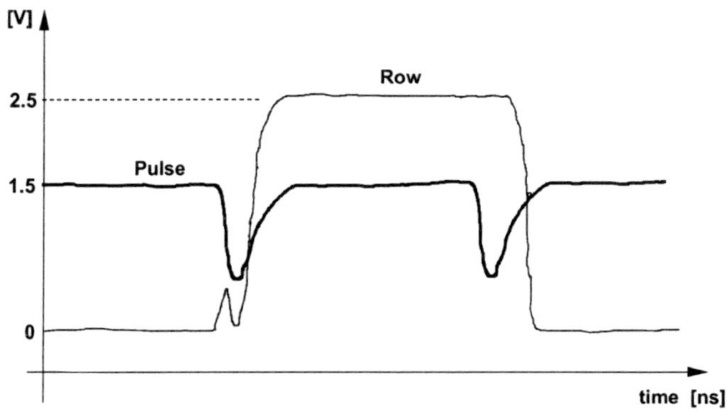

Fig. 9.28. The effect of the PULSE# signal on the rise of the row, in the case of incorrect synchronism with respect to the P and L signals

The main problem with this kind of architecture is the perfect synchronism that the PULSE# signal must have with respect to the P and L signals, which is not easy to achieve, especially if the device works with a boosted row decoder and, hence, the dynamic operations take place also during the read phase and not only during the program phase. Figure 9.28 shows the transition of a row that is biased at VDD through a variation of the address in the case of incorrect synchronization. The PULSE# signal is kept at the VDD voltage minus a V_T, so as to prevent the A node (see Fig. 9.27) from being left floating.

Problem 9.9: Explain completely the operation as stated in the foregoing statement.

The PULSE# signal goes low when the row has already begun to go high and, hence, the row is first driven to ground. Then it begins to go high, as soon as the P and L signals take control of the decoder again. It is very important to synchronize exactly the precharge signal with respect to the asynchronous switch-on of the row, so as to prevent the risk of increasing the read access time.

9.8 Row Decoders for Low Supply Voltage

The evolution of the electronic devices over the last several years has headed toward lower values of the supply voltage. The devices of the last decades required supply voltage of 15 V or even more; subsequently, 12 V were used and, finally, 5 V. Presently, the most common value of supply voltage is 3 V, even though also the 1.8 V supply voltage is being used in some cases. The advent of portable systems and the increasing process scalability means linear reduction not only of the geometric size of the device, but also of the oxide thickness, with consequent reduction of the maximum electric field and, hence, of the supply voltage that can be applied. This is provoking a very quick decrease of the supply voltage so that the possibility of designing 0.9 V devices is already being evaluated.

The importance of this topic deserves a thorough discussion. In order to give an example, we will design a decoder for a Flash memory device working up to a 1.8 V supply voltage, exploiting and further expanding some concepts introduced in the foregoing paragraphs.

Let's begin by recapping some of the ideas already discussed.

The most commonly used circuit configuration for the row decoder is shown in Fig. 9.29, in which M1 realizes the positive feedback that allows switching off the p-channel of the inverter that is not enabled, using an input signal equal to VDD, whereas the supply of the inverter itself is VPC, being VPC > VDD. There is a second effect that acts on the NAND that enables the decoder. In fact, this gate is biased at VDD, whereas its output, due to the feedback, is at VPC, which switches on the drain-bulk junction of its p-channels. In order to prevent this, the M2 NMOS transistor is inserted.

Such a solution is optimal for supply voltage of 5 V, but is hardly suitable for the low voltage operations, since the voltage drop through the terminals of M2 and M3, which are used as pass transistors in common gate configuration, is too high.

It is important to notice that, in such a scheme, it is not possible to use a CMOS transmission gate, since the high voltage would propagate back to the p-channels of the NAND through the PMOS of the CMOS transmission gate. It is not even conceivable to use the triple well for the pass transistors, which would solve the problem of the threshold variation due to the body effect for the n-channel, since the decoding density does not permit adopting such a solution that is very costly from the standpoint of the area.

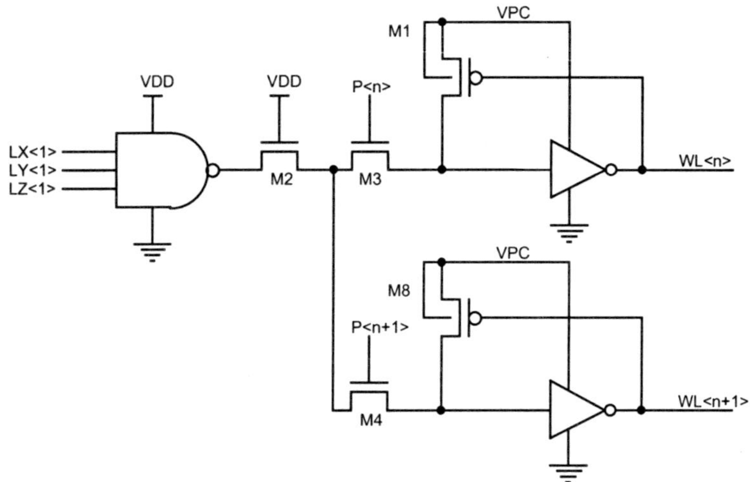

Fig. 9.29. Classic configuration of the row decoder. The last inverter is biased by VPC, which is greater than VDD during the program phase.

In order to eliminate the p-channels in the feedback configuration and the pass transistors, it is necessary to realize all the row decoding at high voltage and place the circuit that produces the high voltage outputs, but is driven at low voltage, outside the decoder. Let's focus on the decoder, first.

In Fig. 9.30 the schematic of the row decoder is shown. As can be seen, the entire circuit is biased at VPC. The LX, LY, and LZ signals select a group of rows, on both the left and the right side, switching on the M1 and M2 n-channels. At this point the P<n> signal, which is at the high voltage level, drives the corresponding row high, for both the right and the left half of the array. The overall organization is shown in Fig. 9.31: only one of the two blocks addressed by the L signals is active, whereas all the remaining have M1 and M2 "off". The division of the sectors in half sectors is useful to decrease the RC of the word line.

The final inverter does not have the NMOS connected to ground, but to a potential called VGC that is different for each sector. It is connected to ground during the read operation, and to a negative potential during erase. Such NMOS transistor must be fabricated in triple well. It is possible to decrease the size of the decoder preserving the performance by modifying the schematic of Fig. 9.30 to that of Fig. 9.32. Each decoding block operates on 16 rows on the right and 16 on the left, but the P signals are 16, not 16 for the right side and 16 for the left, since one out of two inverters driven by the P signals has been removed, and the output signal has been connected back.

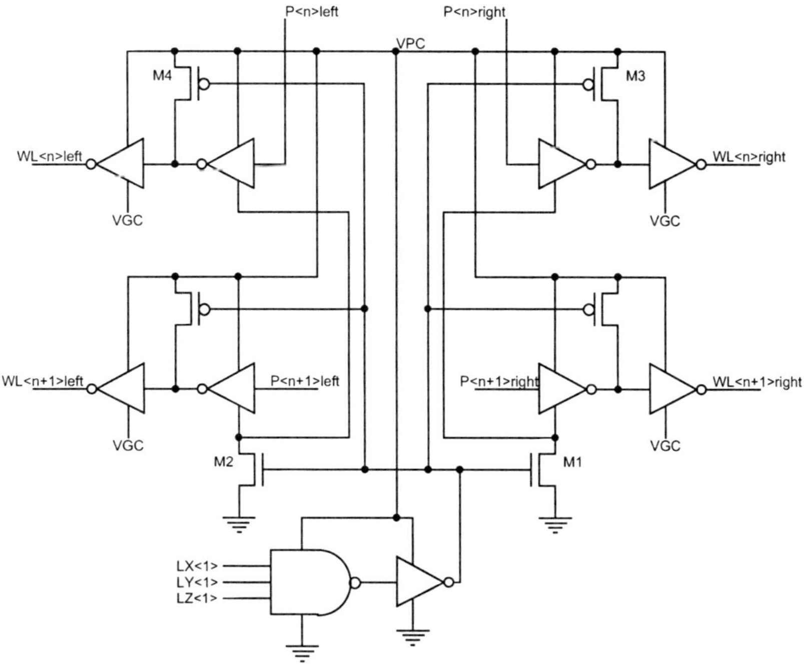

Fig. 9.30. The row decoder without PMOS in the feedback configuration on the row and without pass transistors. The NMOS transistors of the final drivers from the same sector have the source biased at VGC. Each sector has its own supply voltage, VGC, that is driven negative during erase

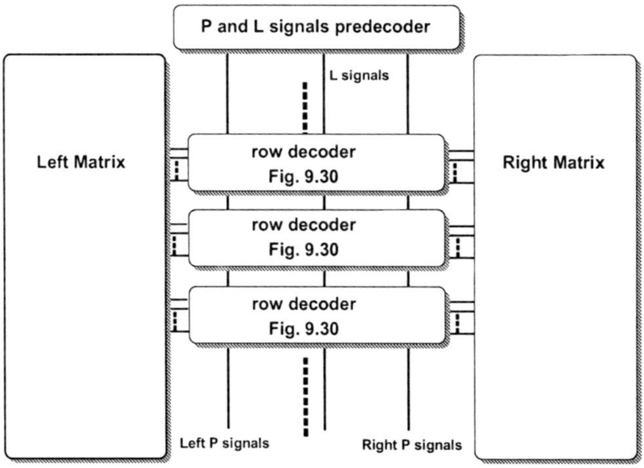

Fig. 9.31. Overall organization of the row decoder. The pre-decoder enables only one circuit like the one shown in Fig. 9.30 through the L signals and the row is then selected by means of the P signals. The read voltage biases one row in each half of the array

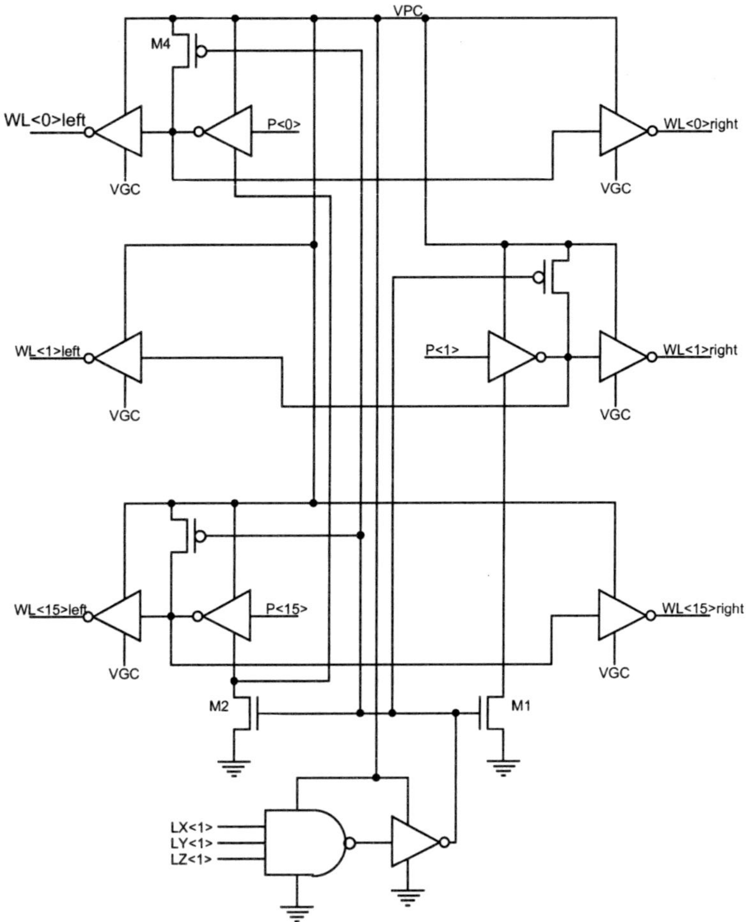

Fig. 9.32. Schematic of the row decoder modified to optimize the area

9.9 Row Pre-Decoder at High Voltage

Let's now design the row pre-decoder, i.e. the generators of the P and L signals. Figure 9.33 shows the main part of the generators of such signals.

In the schematic, two different paths exist to drive the output. During the read phase, the level shifters are not enabled because the READ signal is high and, hence, M1 is "off", the output node, O2, is low, whereas O1 is driven through the CMOS transmission gate IN is the predecoding signal (LX, LY, LZ or P) in the VDD/GND range. When the READ signal is low, the CMOS transmission gate is switched off and the level shifters are activated. O1 is then driven by level shifter 2. Level shifter 1 can be common to more level shifter 2.

This double path is necessary since the transitions of the level shifter are very slow, tens of nanoseconds, and, thus, it is not convenient to place them directly on the read path. On the contrary, it is suitable that the CMOS transmission gate is in the read path, whereas it cannot be placed in the program path, as previously explained. The usage of the two paths that drive the same output node, O1, meets both the requirements.

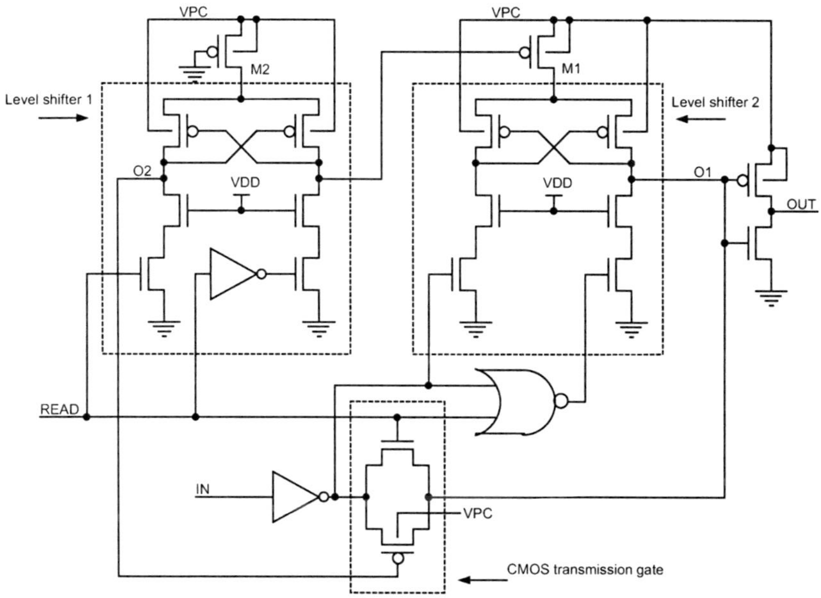

Fig. 9.33. The row pre-decoder consists of the circuit shown in the figure, with multiple copies to generate each individual L and P signals. Notice the two different paths from the address to the output

9.10 Sector Decoding

The analysis of the specifications shows that it is always the higher part of the address that selects the sector, e.g. A<18:15> in case of an 8 Mbits divided into 16 sectors and organized in 16 outputs. In our example, the sectors are organized in rows, which means that the sector address is also the row address and, hence, the column address is the lower part of the overall address, i.e. from A<0> up.

The internal signals that select the sector have been referred to as LX, and they are generated using the 4 highest bits that allow 16 different combinations to uniquely identify the sector. At first glance, the order of the bits of the address to scan rows and columns seems to be absolutely irrelevant. First of all, let's try to understand what is the difference between using the lower part of the address to select the row and the higher to select the column or vice versa. In the case where the lower part of the address drives the rows, if we refer to a binary addressing, moving from one byte to the following byte we select the subsequent row on the same column. On the contrary, in the case where the row address is represented by the higher part of the overall address, moving from one byte to the following byte we select the same row, which means that the (n+1)-th byte belongs to the same row as the n-th, but on the subsequent column. The program that the microprocessor executes which is stored in the non-volatile memory is written in serial locations, e.g. the first byte is written in the location corresponding to the 00000 hexadecimal address, the second in 00001, and so on.

If the lower part of the address corresponds to the row address, the microprocessor will scan the memory by column when sequentially executing the code. If we consider that one of the most frequent causes of failure is the single cell that shorts the column to ground, then the part of the code related to the shorted column would be a single word would be corrupted and the rest of the code is uncorrupted. If we have to store the control code of a dangerous machine, we could write several times in the memory the portion of the code that forces the operation in a safety state in case of danger. In this way, even if a part of program became corrupted, the microprocessor would be able to recover the missing code in another location. On the contrary, if the memory were scanned row after row, a failure in a column would corrupt portions of code across the entire memory, making it less likely to be usable.

The addressing of EPROM memories followed the foregoing mapping criterion, while the addressing of Flash memories allows selecting the sectors by means of the higher part of the address.

We already know that it is very important to pay attention to the cell geometry, since asymmetries could have remarkable impact on the behavior of the cells themselves. Thus, it is fundamental for the designer to know the physical location of each single bit with respect to the electrical address. In practice, if the byte is written in the 00000 hexadecimal address, it is important to know where such location is placed, e.g. if the row that is written is the first or the last of the array, with respect to the output pads. This is not very important to the customer, as what matters to him is that the bits of the sector are all physically connected together, so as to be erased simultaneously. Within the same sector, a one-to-one correspondence between cells and addresses exists, independent of the spatial location. How could we verify such correspondence? In order to carry out the test, a passivated device is taken and completely programmed. Afterwards, by means of ink spots, some signs are marked on the array; subsequently, the sample is put in an UV eraser that erases only the cells that are not protected by the ink. A read operation is then carried out and a map of the array is derived, according to the assumed correspondence between physical position of the cells and the electrical addresses. The map of the memory shows the contents of the memory with one color showing erased locations and another color for the unerased cells, e.g. a green pixel on the computer monitor indicates a “1”, a red pixel indicates a “0”. If the physical-electrical correspondence is correct, the map of the green pixel is identical to the map of the ink spots.

9.11 Memory Space for Test: The OTP Rows

The space in the Flash memory is increased by adding some One Time Programmable (OTP) rows, which can be used by both the manufacturer and the customer through a particular procedure that is generally not reported in the Data Book, but is available only to the so-called "key-customers", those customers that require an intimate knowledge of the non-volatile memory.

The use of these words produces a sort of pedigree that records the history of the device, manufacturing date, lot number, assembly site, and the most important electrical characteristics. By means of such data, it is easy to track the history of the device in case of returns by the customer[8], so as to improve test activities and to prevent marginal devices from being delivered over time. The customer could also exploit the OTP to include a password that prevents the software from being copied without authorization.

These rows are called One Time Programmable but, as a matter of fact, they can be electrically erased since the cell threshold of the UV erase is a parameter that can be modified; for example it could be increased to fix some problems in program and (or) erase. If it were not possible to erase the OTP cells, they would result all written during a low voltage read.

Let's now discuss the real OTP rows, the ones that are used in the EPROM that does not have the window for the UV erase. We know that EPROM memories are similar to Flash devices but cannot be electrically erased and have to be exposed to UV radiation to restore the neutral electric state on the insulated gate. This fact forces the manufacturer to assemble the device in a specific package, with a quartz window to allow ultra-violet light to penetrate. The package must be ceramic and not plastic so as to hold the window. This package makes the EPROM device very expensive for customers who seldom erase the EPROM. To accomplish the erasure, it is necessary to remove the memory from the board, running the risk of breaking it either mechanically or for Electro Static Discharge (ESD). In practice, it is necessary to erase the EPROM only during the development phase when many iterations of the final code must be tried in the application hardware. Once the production phase has begun, it is unnecessary to have an erasable device, or better yet it is wasteful. EPROM manufacturers provide the same device in an inexpensive plastic package without the quartz window, instead of the ceramic package. Such devices are called OTPs and, of course, they cannot be erased but can only be programmed once.

The problem is the testing of the OTPs: the EWS are not critic since the erasing is carried out by placing the entire slice in an UV eraser, but how is it possible to verify that a device still can be written after it is packaged in a plastic package without a window?

In addition, how can the speed class be determined if a significant pattern can not be written into the EPROM since it is not possible to erase it afterward? The customer must have completely blank (erased) devices, rather than devices that al-

[8] The return by the customer is measured in ppm, i.e. parts per million and today a level of 2 ppm is demanded.

ready contain data, as in the case of ROMs (Read Only Memory) that have fixed data written in during manufacturing.

The issue of yield of the EPROMs, i.e. the loss of devices having failures that could not be detected by the limited testing, is addressed by using a discount to the customer, which is a function of the failure rate. However, the problem of the determination of the speed class and the verification of the internal circuitry is solved by adding a number of OTP rows and columns that lay outside the space of the memory that customers are allowed to address. Through such additional rows and columns, which are written after the sample has been enclosed into the package, it is possible to measure the access time and determine the other characteristics of the EPROM device.

Another issue arising from the use of a plastic package is related to the passivation of the devices. Unlike the ceramic packages, plastic packages are permeable to water. The problem is that water contains many impurities and once the moisture enters the package, impurities might penetrate through the pad openings or along the perimeter where the device has been separated from the wafer or, finally, through the possible fractures of the passivation and eventually move across the array.

These ionic contaminations can form a shield with respect to the charge on the insulated gate, causing detrimental effects on the data reading. Additionally, the temperature tends to scatter them and, hence, a "mobile failure" is also possible inside the array. One solution could be the modification of the passivation layers so as to make them more resistant to the possible micro cracks produced by thermal stress. On the other hand, in this way, it becomes more difficult to erase the wafers by UV light, since the passivation layers could tend to absorb more radiation.

9.12 Hierarchical Row Decoding

The continuous technological evolution determines a progressive reduction in the cell size. This does not apply to oxide thickness; as a consequence, program and erase voltages do not scale. In this condition, high voltage row decoder transistors cannot be shrunk and the problem of drawing the row driver in the pitch of the single cell becomes more complicated. The usage of a hierarchical decoder is a possible solution.

One of the characteristics of the hierarchical approach is the organization of the array in very long rows, called Main Word Lines (MWLs), divided into several Local Word Lines (LWLs), the size of which does not cause gate stress issues, even in the case of erase with negative voltage. The LWL is made of polysilicon while the MWL is realized of metal.

The connection between Local and Main Word Line is realized through the local selector sketched in Fig. 9.34. NCH, PCH, and DISCH are the control signals of the local selector and are generated so that, if the LWL is addressed, the voltage of the MWL is applied to the LWL. On the contrary, if the LWL is not addressed, it is biased at 0 V. The VPCX signal, which acts to bias the body of the p-channel transistor, corresponds to the positive supply of the row driver. Thus, VPCX equals V_{READ} during the read operation, whereas it is pulled to the VPC voltage during program, when VPC ranges between 1 V and 9 V. During the erase phase, VPCX is 1.8 V, in order to limit the voltage drop across the oxide of the p-channel transistor of the local selector and prevent breakdown phenomena. HVNEG, besides biasing the source of the n-channel transistors controlled by the DISCH signal and the pull-down of the global row decoder, biases the p-well of the two n-channel transistors fabricated in triple-well. This signal is always grounded except during the erase operation, when it reaches the voltage value of –8 V.

Let's now analyze the functioning of such a decoder in detail.

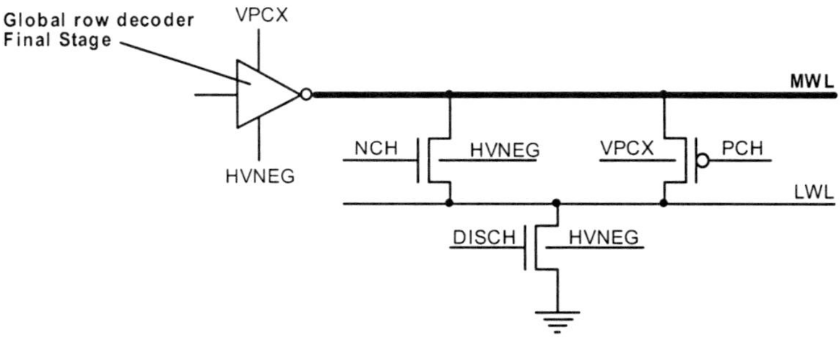

Fig. 9.34. The local row decoder circuit

9.12.1 Read & Program

In the addressed sector, with the MWL biased at VPCX, we want to pull only one LWL at VPCX. The gate of M1 is pulled to ground so that this transistor transfers the voltage of the global row to the local row. The M2 transistor, controlled by the DISCH signal, must necessarily be switched off in order not to ground the LWL. The bias condition of M0 is, in this case, irrelevant, but it is better to switch it on so as to diminish the equivalent resistance of the switch. This resistance cannot be neglected when the cell has to be programmed with a voltage of 1.5 V. In fact, it is true that M1 does not suffer from the body effect, it is connected to VPCX, but its threshold voltage is around 0.8 V (it is a high voltage transistor) and, hence, its overdrive is only 0.7 V in the case of 1.5 V programming voltage. In any case, the transistor driven by the NCH signal contributes to the charge of the LWL up to VDD – $V_{T,n}$.

In the case in which the MWL is at VPCX and the LWL is at GND (the sector that has not been addressed), M5 must be switched off and, therefore, the corresponding PCH<1> signal must be at VPCX. Also M4 must be off not to pull up the local row. Instead, M3 must be "on" so as to pull the row to ground. To this purpose, the DISCH<1> signal must be pulled at VDD. On the contrary, if the MWL and the LWL are tied to ground, the bias condition of M12 and M13 is irrelevant. In order to tie the local row to ground it is possible to act on both the M11 and M8 pair and the M9 and M10 pair by means of a signal biased at VDD.

These three cases are summarized in Fig. 9.35. The dynamic range of the PCH signals is between ground and VPCX. NCH and DISCH range between ground and VDD. Note also that NCH, PCH, and DISCH are in common to all the rows of the sector.

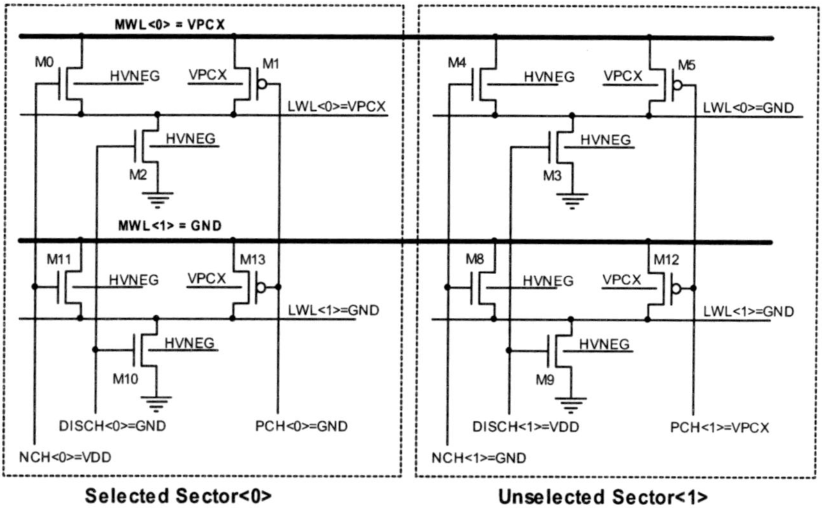

Fig. 9.35. The positive voltage VPCX is transferred to the array row: the bias of the addressed sector is highlighted along with the bias of the non-addressed sector

9.12.2 Erase

During erase all the gates of a sector are simultaneously driven to negative voltages.

In the sector to erase the bias condition of M1 and M13 is irrelevant, since the voltage that is applied to the local row is negative. The gates of M0 and M11 are at VDD, pulling the LWL to HVNEG. M2 and M10 are switched off to prevent charge from flowing from ground to the negative voltage source; DISCH<0> equals HVNEG. In the non-addressed sector M4 and M8 are switched off, whereas M3 and M9 are active so as to prevent local rows from being floating, by forcing them to ground. In Fig. 9.36 the erase configuration is summarized.

The assembly of these circuits allows designing a hierarchical row decoder in which each MWL is connected to several LWLs (e.g. four). Each LWL is selected through the corresponding command signals, NCH<(n-1):0>, PCH<(n-1):0>, and DISCH<(n-1):0> (Fig. 9.37). The symbol <(n-1):0> generally indicates an n-bit bus. In this case, this symbol is used, similarly to what usually happens in the schematics, to indicate n signals that can assume independent values with respect to each other, even though they have the same function.

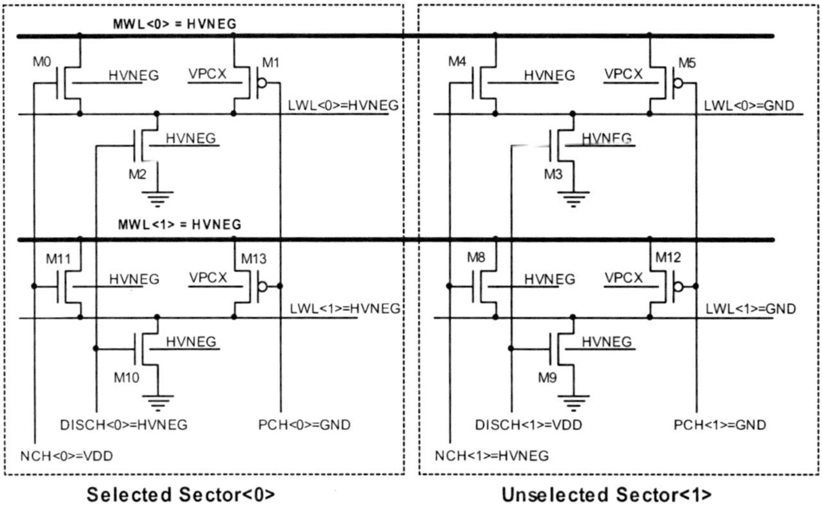

Fig. 9.36. The negative voltage HVNEG is transferred to the array row: the bias of the sector to erase is highlighted along with the bias of the unselected sector

The use of a hierarchical row decoder (4 LWLs for each MWL) implies several advantages. In fact, it is possible to eliminate the necessity of using a local metal connection in the row pitch, thus reducing the value of the parasitic coupling capacitance between word line metals that run contiguously, and, consequently, improving the settling time of the MWL. In this way, the probability of short-circuit between metal stripes is reduced, which increases the production yield of these circuits.

Another advantage is the fact that it is not necessary to place the MWL drivers in the word line pitch, since a space corresponding to four word lines is now available. This allows studying and designing very efficient drivers with reduced current consumption. The fact that a larger space is available to fabricate the peripheral circuitry of the array, such as the MWL drivers, has a fundamental importance in technological processes that constantly reduce the cell size but not the size of the transistors that are external to the array.

At this point, we can analyze how the hierarchical row decoding impacts on the charging time of the word line.

The realization of a MWL fabricated of metal makes the parasitic resistance practically negligible, but this does not hold for the parasitic capacitance. We can sketch the global row, the hierarchical decoder, and the local decoder as in Fig. 9.38. The MWL capacitance is determined by three contributions:

1. bus capacitance due to the coupling between metal2 (MWL) and the underlying layers;
2. capacitive coupling due to the overlap of the main bit line, realized of metal3, over the MWL;
3. junction capacitance of the transistors of the local selector, which connect the Main to the selected Local Word Lines.

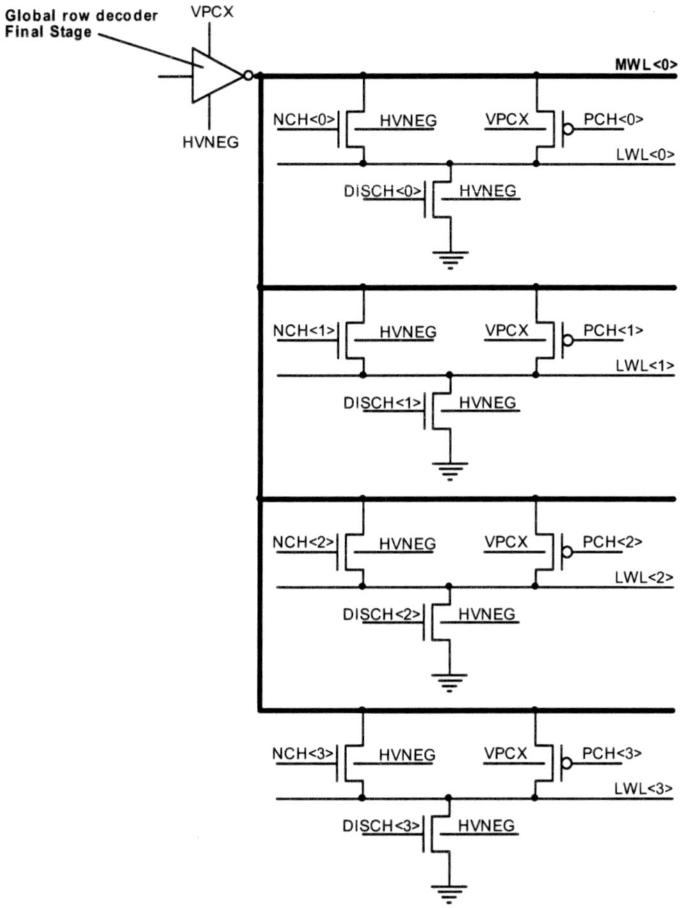

Fig. 9.37. Four-level hierarchical row decoder

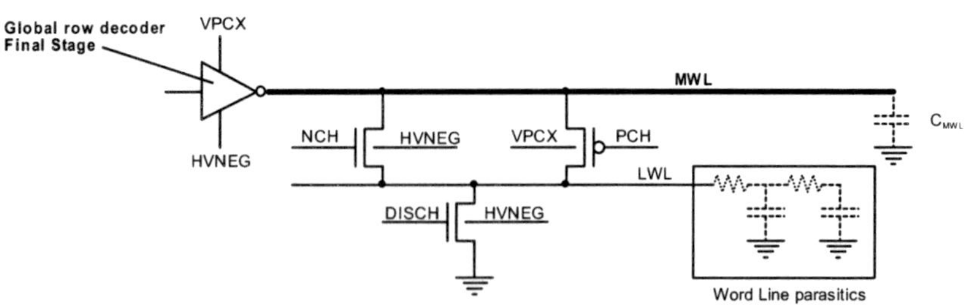

Fig. 9.38. Global row (MWL) and local row (LWL) with the respective parasitic load

One possible solution that allows diminishing the charging time is shown in Fig. 9.39. The M4 transistor allows reducing the time to bias the local row, by exploiting the charge of the two sides of the LWL.

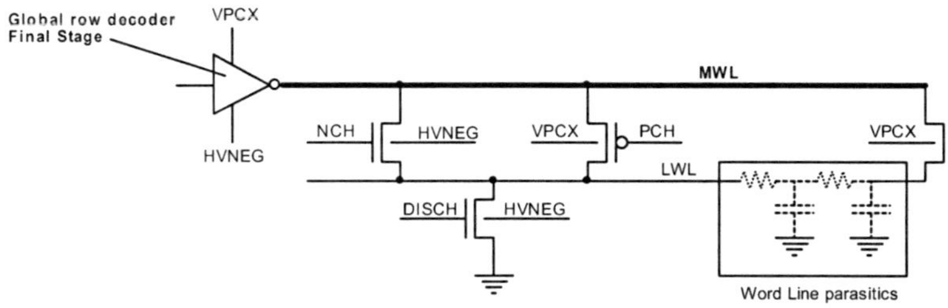

Fig. 9.39. Solution to reduce the charging transient time of the local row

9.13 Low Switching Consumption Row Decoder

In the case of multilevel memories (see Chap. 13), the read voltage, V_{READ}, can be high, up to 5 V or 6 V. The management of such a voltage is complicated in devices with a single supply voltage of 3 V or less. The internal voltages are, in these cases, generated by means of charge pumps that can be represented by means of the equivalent Thevenin circuit shown in Fig. 9.40.

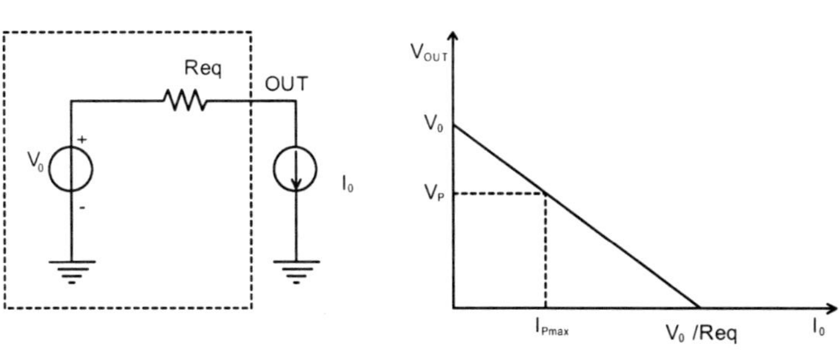

Fig. 9.40. Equivalent circuit and output characteristic of a charge pump

Analyzing the output characteristic of the charge pump, it is possible to note that the current sunk cannot be greater than I_{Pmax} if an output voltage not lower than V_P is required. Usually, pumps fabricated in CMOS technology can source current lower than 1 mA. On the word lines, a voltage of 6 V is required during reading, with a maximum ripple of 50 mV. For this reason, a voltage regulator, placed at the pump output, is necessary. The available current must therefore be divided between regulator and row drivers that, typically, sink current from the VPCX node during the row selection and de-selection phase. In fact, the circuit structures that we examined in the foregoing sections do not have any DC current consumption. The existence of feedback loops causes high dynamic current consumption that increases as the difference between VDD and VPCX increases, due to the crow-bar current.

One possible solution to this problem is indicated in Fig. 9.41. HVNEG equals the ground potential during program and read, whereas it equals –8 V during erase. The functioning of the circuit is explained hereafter. In order to pull up the output (selected row), it is necessary that both the inputs, IN1 and IN2, are forced to a high voltage value. At this point, the NAND gate switches M4 off and M62 on. The M61 transistor starts charging the output only after the time due to the delay of INV3 and M1. In order to limit the current through M3, the size of M1 is increased so as to switch off M3 faster, thus reducing the current through M61 and M62. The crow-bar current through M1 is limited by reducing the overdrive of M52.

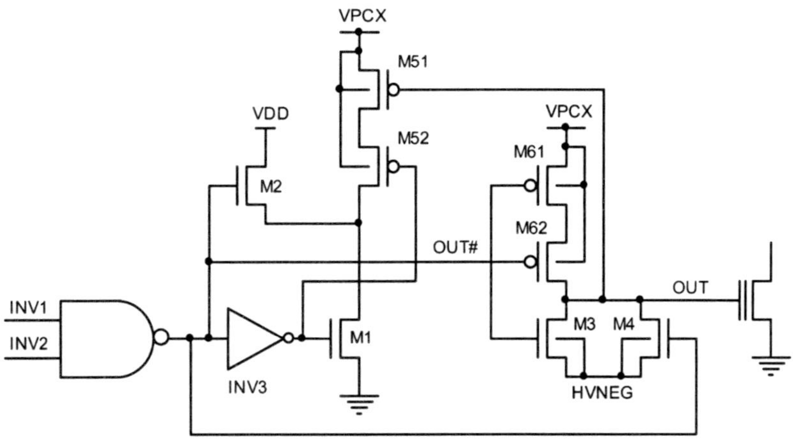

Fig. 9.41. Row driver with reduced current consumption during the toggling

During the opposite transition (unselected row), the problem is due to the slow rising of the OUT# node, since M51 and M52 have small size (they belong to the feedback path). This fact might cause large current consumption through M61, M62, and M3. In order to limit this phenomenon, the overdrive of M62, which is driven directly by the output of the NAND gate, is reduced and the M2 transistor is added to speed up the charging of OUT#. When M2 is switched on, the $V_{OUT\#}$ voltage reaches VDD – $V_{T,M2}$. During the erase phase, the driver must be set to the condition of unselected row. The OUT# node is pulled up to VPCX and M3 is on, thus transferring the HVNEG to the output node.

In Fig. 9.42, a possible modification of the driver is presented, to increase the driving capability. In this case, a driver composed of M10, M11, and M12, is added outside the feedback path, and the two series transistors are NMOS-type. The OUT node corresponding to the solutions presented above may be directly the polysilicon word line or the Main Word Line in the case of hierarchical organization.

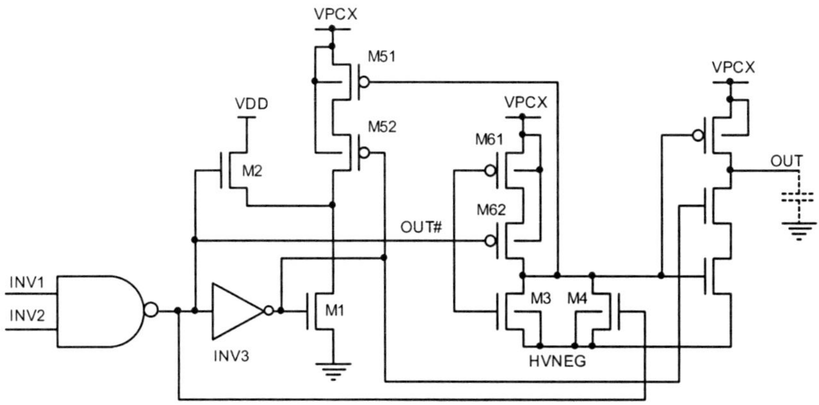

Fig. 9.42. Driver of Fig. 9.41 modified to increase the driving capability

9.14 Column Decoders

In the previous chapters we studied the column decoders, since they do not have a complex structure, but are constituted of simple pass transistors that enable the path from the sense amplifier or the program load circuitry to the selected cell.

Also in this case, it is possible to use a hierarchical structure composed of Main Bit Line (MBL), and Local Bit Line (LBL) as indicated in Fig. 9.43. The MBL is fabricated with a metallization (e.g. metal3), from which four metal stripes of metal1 (LBL) start for each sector, which can be selected by means of the four corresponding NMOS generally indicated as YO. Only one of such YO transistors is on; all the other YO transistors of the unselected sectors are off. The selection of the MBL is carried out by the transistors called YM and YN. The drains of the cells are connected directly to the LBL. It is possible to transfer the voltage of the node called COLOUT to the cell drain through the path selected by the YM, YN, and YO transistors, which are on. To detect the status of the cell, the current is sunk from the COLOUT node. The V_{COLOUT} voltage equals 1 V during read, while it is driven to nearly 5 V during program. When the cells are not selected, and during erase, the COLOUT node has high impedance.

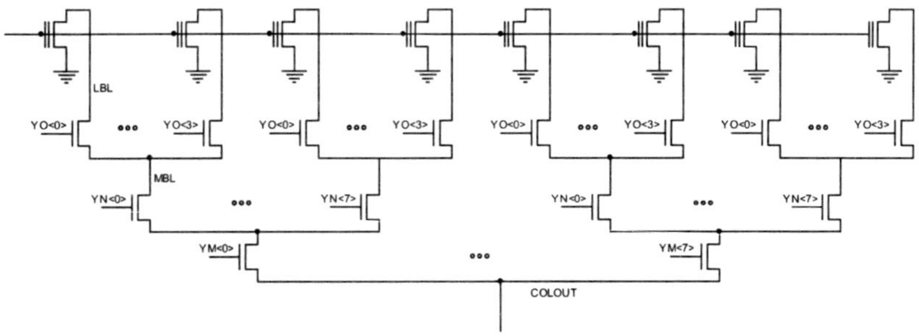

Fig. 9.43. Hierarchical column decoder

When the array is designed, the column decoders often represent a problem because the YO transistors, one for each column, must be placed in the cell pitch. Because of the small size of the cell, this is not always possible and, hence, specific layout solutions must be found to minimize the area occupation. In Fig. 9.44, the typical solution adopted is shown: the YO decoder is divided in two parts, placed above and below the sector so as to design the YO transistor within the pitch of two cells.

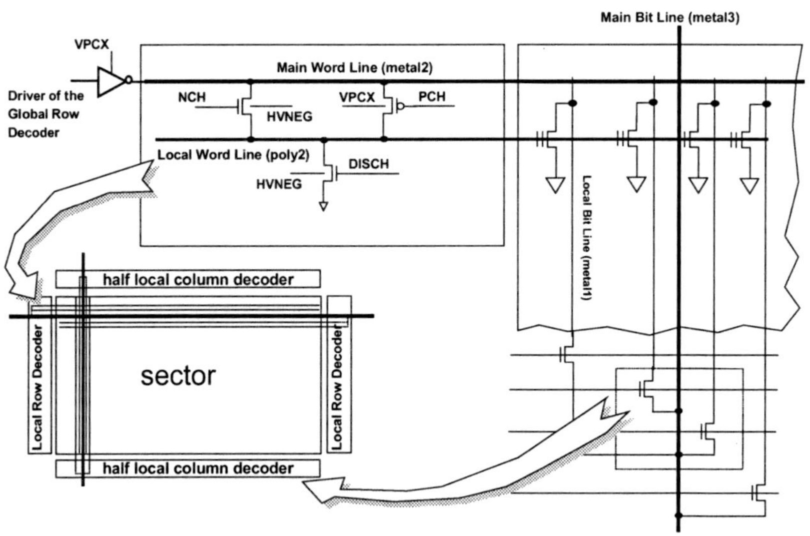

Fig. 9.44. Row and column hierarchical sector decoders

Furthermore, the problem of the area is emphasized by the fact that the transistors of the column decoders must be high voltage type, often with non-minimum length, since, during program, voltages greater than the supply voltage value must be transferred. Finally, it is not possible to increase the distance between gate and contact or use areas with different doping density, since they would occupy more space.

Beside area occupation issues, also electrical issues must be accounted for. The control of the programming voltage can be achieved directly only on the COLOUT node, whereas the other nodes are not directly accessible, and, thus, controllable by means of a feedback path, due to their high connectivity.

The current consumption of the cell may be higher than 100 μA during program. This may cause a voltage drop between the COLOUT node and the LBL that may amount to several hundreds of milliVolts. Since, as we have seen, it is not possible to increase the size of the transistors, one can act on the voltage by which the gate is biased, to diminish their equivalent resistance. To this purpose, usually a dedicated voltage called VPCY exists, which equals VDD during read and is around 8 V during program. The final inverter, which drives the YM, YN, and YO signals, cannot be a standard inverter. The problem of controlling some inverters biased at VPCY by means of signals belonging to a VDD/GND logic poses again. All the solutions we have already examined with reference to the row decoder are still valid.

Bibliography

H. Arakawa, "Address Decoder Circuit for Non-Volatile Memory", United States Patents 5,039,882, Aug 13, (1991).

S. Atsumi, A. Umezawa, M. Kuriyama, H. Banba, N. Ohtsuka, N. Tomita, Y. Iyama, T. Miyaba, R. Sudoh, E. Kamiya, M. Tanimoto, Y. Hiura, Y. Araki, E. Sagakami, N. Arai, and S. Mori, "A 3.3V-only 16Mb flash memory with row-decoding scheme", 1996 IEEE Int. Solid-State Circuits Conf. Dig. Tech. Papers, pp. 42–43, (Feb. 1996).

G. Campardo, R. Micheloni, S. Commodaro, "Line decoder for memory devices", USA Patent No. 6,018,255, (January 25, 2000).

G. Campardo, R. Micheloni, "Row decoder for a flash-EEPROM memory device with the possibility of selective erasing of a sub-group of rows of a sector", USA Patent No. 6,122,200, (September 19, 2000).

G. Campardo et al., "Circuit structure for providing a hierarchical decoding in semiconductor memory devices", USA patent No. 6,515,911, (February 4, 2003).

J. C. Chen et al., A 2.7V Only 8Mx16 NOR Flash Memory, Symposium on VLSI Circuits, Digest of TEchnical Papers, IEEE, (1996).

Kobayashi el al., "A 3.3V-Only 16 MB DINOR Flash Memory", FLASH MEMORY SESSION, ISSCC95.

R. Micheloni, et al., "Row decoder circuit for an electronic memory device, particularly for low voltage applications", USA patent No. 6,069,837, (May 30, 2000).

R. Micheloni et al., "Line decoder for a low supply voltage memory device", USA patent No. 6,111,809, (August 29, 2000).

R. Micheloni et al., "Nonvolatile memory device, in particular a flash-EEPROM", USA patent No. 6,351,413, (February 26, 2002).

Y. Miyawaki et al., "A New Erasing and Row decoding scheme for low supply voltage operation 16-Mb/64-Mb Flash Memories", IEEE Journal of Solid State Circuits, Vol. 27, N 4, (April 1992).

A. Umezawa et al., "A 5V-Only operation 0.6um Flash EEPROM with Row decoder scheme in triple-well structure", IEEE Journal of Solid State Circuits, Vol. 27, N 11, (November 1992).

10 Boost

10.1 Introduction

In order to ensure both a proper operation of the matrix during read operations and good reliability during cycling, it is mandatory to set some constraints on the distribution of the threshold voltages of the flash cells. In particular, the distribution of the erased cells must lie between about 0.5 V and 2.5 V. The lower limit is dictated by the need to insure that no sub-threshold leakage current is present on the bit line. The upper limit is set by the intrinsic width of the distribution.

The read operation of the Flash cells is performed by biasing the row of the cell to VDD and by "reading" its current. If the cell is written, then its threshold voltage is about 5 V and it does not sink current: on the contrary if it is erased, then its threshold voltage has to be low enough to allow a current to flow. The issues related to read operations are clear when low VDD devices are taken into account.

In the case of the power supply equal to 2.5 V, all the cells, both written and erased, do not sink current and therefore the sense amplifier is not able to distinguish the state of the cell. This issue can be overcome by using a row boost technique, which provides the gate of the addressed cell with a voltage that is higher than the power supply.

10.2 Boost Techniques

In Fig. 10.1 an ideal boosting circuitry is shown together with the required timing of all the signals involved. BN is the node whose voltage is to be boosted.

At the beginning, the auxiliary boost capacitor (C_{BOOST}) and the parasitic one (C_{LOAD}) are precharged to the supply voltage via the p-channel MP. When the boost of node BN is required, transistor MP is turned off and, then, signal B is driven low. In this way the lower plate of the capacitor C_{BOOST} is biased at a voltage equal to VDD. Since node BN is isolated, we can calculate its potential by imposing charge preservation. The charge Q_i that is present when precharge occurs can be written as

$$Q_i = (C_{BOOST} + C_{LOAD}) \cdot VDD \tag{10.1}$$

Final charge, i.e. when boost has occurred, is equal to

$$Q_f = C_{BOOST} \cdot (V_{BN} - VDD) + C_{LOAD} \cdot V_{BN} \tag{10.2}$$

By imposing $Q_i = Q_f$, it follows that the value of the potential on node BN after boost operations equal to

$$V_{BN} = VDD + \frac{C_{BOOST}}{C_{BOOST} + C_{LOAD}} \cdot VDD \qquad (10.3)$$

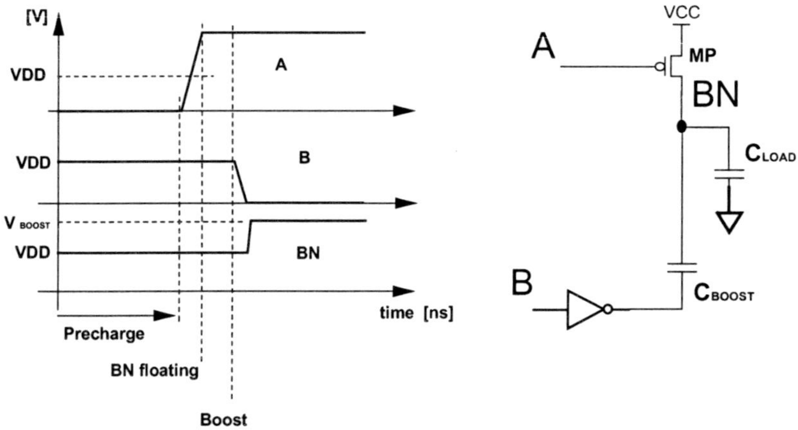

Fig. 10.1. Boost approach. Ideal scheme and main signals timing

There are two ways to implement the boost. The former is the continuous boost, and Fig. 10.2 shows the ideal scheme. Thanks to a switch, the voltage on the row can be equal either to VPP (program) or to VDD (read). The boost circuit is used in read and it is activated each time a rising (or falling) edge of a given clock signal CK occurs. The boosted node is the supply of the row decoders, whose parasitic load is represented by C_{LOAD}. The selected row is therefore biased at a voltage value that is higher than the power supply, as is required.

The advantage of this approach is that the boost capacitor is not required to be very large, since the over-voltage is produced by a series of small increments. For the same reason the time required by the initial charge of C_{LOAD} is very long (microseconds) and therefore it is not acceptable in case the recovery time from stand-by must be less than approximately one hundred nanoseconds.

In order to reduce the the recovery time from stand-by, it is possible to use an auxiliary boost, smaller than the main one, that keep capacitor C_{LOAD} charged during the stand-by phase; of course this additional consumption should not cause the total power consumption to exceed the limit indicated in the device specification.

The latter type of boost is the pulsed boost (also known as one-shot boost), whose ideal scheme is shown in Fig. 10.3. This solution calls for a large boost capacitor: by means of a single boost operation, C_{LOAD} must be charged to the proper value. In this way the contribution of boost to both consumption in stand-by and access time becomes negligible. Since the size of C_{LOAD} is in the order of several hundreds of picoFarad in the case of large-sized memories, it is possible to calculate the required value for C_{BOOST}. In this case, the issue is related to the large area occupied by the boost capacitor and by the driving circuitry.

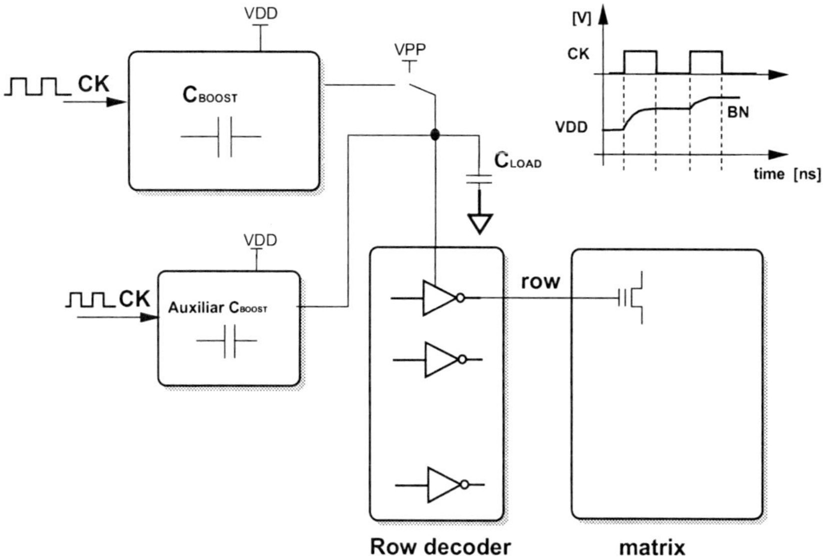

Fig. 10.2. Block scheme of the continuous boost. An external clock (CK) provides the charge to the capacitor C_{LOAD}, boosting the supply voltage of the row decoders, and therefore the row itself, during read operations

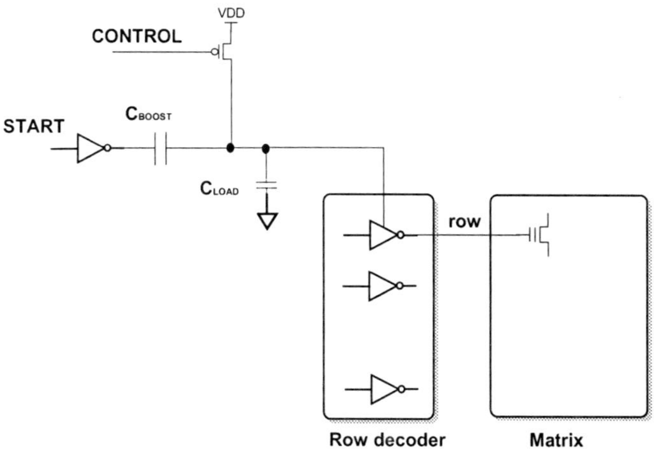

Fig. 10.3. Ideal scheme for the pulsed boost. The supply node for the row is initially forced to VDD. When a read operation is required, the signal CONTROL isolates node BN and the signal START provides a pulse that allows the voltage of the word line to be pushed above power supply

Let's now analyze the way to generate the CONTROL signal shown in Fig. 10.3, which is used to turn off the precharge transistor MP. In Fig. 10.4 a scheme for the boosted voltage generator is shown, which is able to generate both the turn off signal for the transistor MP and the boosted voltage, starting from a unique input signal.

The circuit is activated by the high to low transition of signal IN. When IN is high, the output of the chain of three inverters is at ground, while the other plate of the capacitor C_{BOOST} is at VDD, thus precharging the capacitor to a voltage difference equal to the power supply. When IN goes low, it turns off the NMOS transistor M4 and it turns on the PMOS M5, then it flips the chain of three inverters causing the signal START to go high. The capacitor tends to preserve the voltage difference at which it has been charged before, andd therefore it "pushes" the voltage of the output node above the power supply voltage.

The output signal OUT is transferred, through M5 that is turned on, to the gate of transistor MP, keeping it turned off. At this point node OUT is floating, which is necessary for the boost operation to work properly.

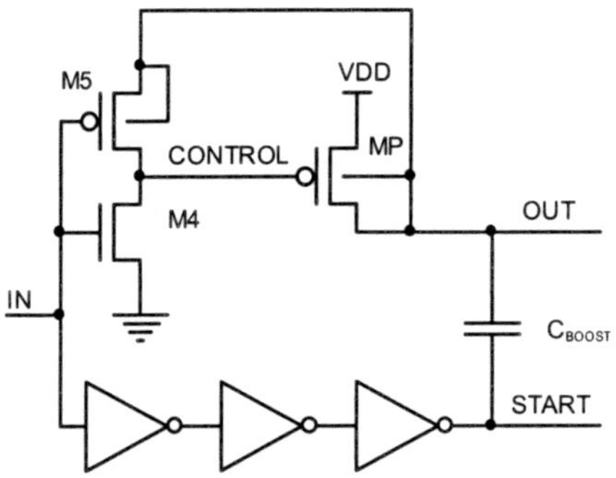

Fig. 10.4. Boosted voltage generator

In the next section we will analyze a solution that preserves the advantages of the two solutions presented in this section: an acceptable size of the boost capacitor that does not cause a degradation for neither power consumption nor access time. There is a disadvantage however that limits the achievable value of the boosted voltage.

10.3 One-Shot Local Boost

We have shown that the boosted node, BN, is indeed the supply of the row decoder so that the boosted voltage can be directly transferred to the selected word line. The supply of the row decoder is usually indicated as VPCX or VPC to distinguish it from the supply of the column decoder, also known as VPCY.

We are now going to discuss the design of a boost that drives the VPCX node for a portion of the row decoders.

Let's assume that the sectors of the matrix are organized by row and that they are totally isolated from each other by means of a hierarchical column decoding scheme (Fig. 7.4). In this case, addressing a row means selecting a particular sector. The task of the circuit called LOCALBOOST is to apply the boosted voltage to the supply of the selected sector only. The overall distributed boost architecture is shown in Fig. 10.5.

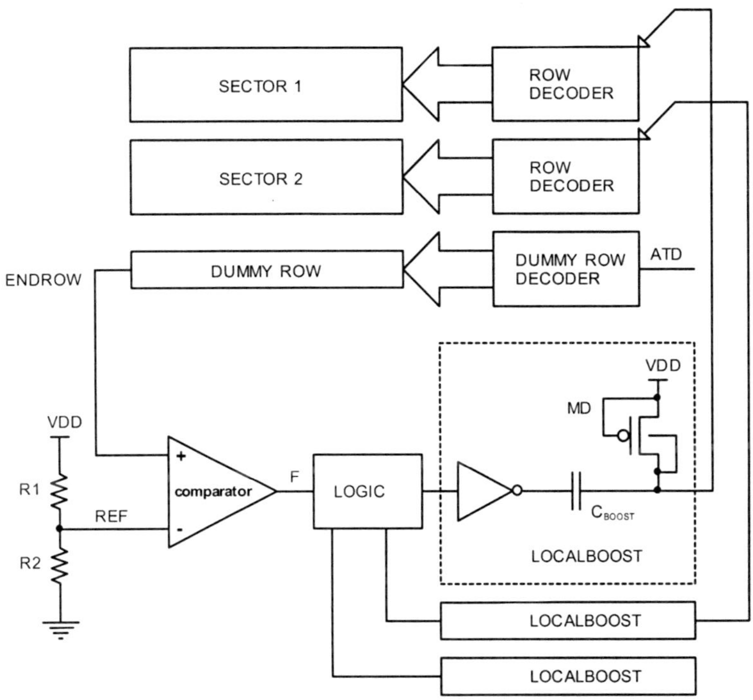

Fig. 10.5. Distributed boost architecture

We have seen that the value of the boosted voltage is a function of the pre-charge potential of C_{BOOST} and C_{LOAD}. It is therefore necessary to introduce a circuit that detects the condition when the selected row has reached a voltage near to VDD.

Problem 10.1: If N capacitors are available, which is the maximum voltage value that can be obtained using the one-shot boost technique?

In the case of boosting of the row, the timing sequence of the events is made complex by the fact that these events are not synchronous, since different delays are associated with the various signal paths involved. Therefore, an equivalent circuit is used to generate delays that are the same as those in the actual row circuitry; the "end" of this dummy path, ENDROW, is used to emulate the voltage that is present on the true word line (row). Therefore there are two decoders: the row decoder, driven by signals P and L (see Chap. 9) and the special one of the dummy row alone, both of which are activated in parallel on equal loads. Of course, in order to reproduce the load of the matrix, the dummy word line must be as long as the matrix word lines.

Concerning access time, there is a tradeoff between the actual boosted voltage and the speed of the boosting operation. In Chap. 9 we have seen that a matrix row can be modeled as a distributed *RC*; after a time constant, using a lumped parameter model, the voltage has already reached the 63% of its maximum value, while it takes 3 time constants to get to values above 95%. The resistive partition R1-R2 in Fig. 10.5 aims at selecting the fraction of VDD that triggers the boost on the word line.

Selection of the dummy row is done by the ATD signal that, as we will see in Chap. 11, detects a change in the addresses and, therefore, the need for a new read operation. At this time, it is useful to point out that the LOCALBOOST circuit (both capacitor and driver) is repeated for each sector, while the circuitry required to decide when the boost should occur can be unique for the entire device. The compact size of C_{BOOST} allows for an easy placement in the layout of the device; for instance, it can be realized with a long strip of poly2 placed between the matrixes of the different sectors.

In Fig. 10.6 the behavior of the main signals involved in a distributed boost is shown, together with the addressed word line.

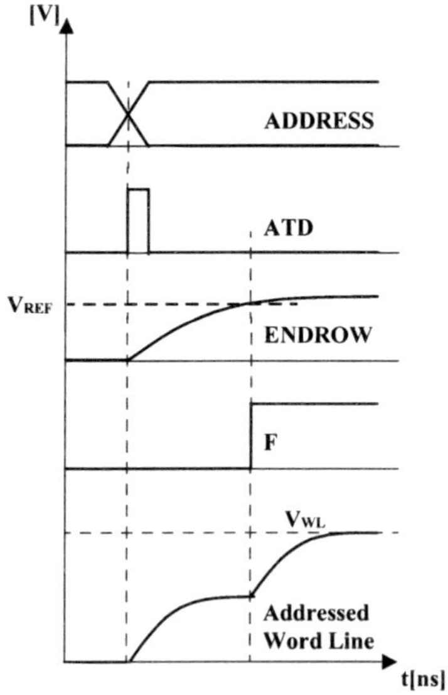

Fig. 10.6. Timing of the main signals involved in the distributed boost architecture

Let's now see in detail how the LOCALBOOST circuit is connected to the row decoder. Signal CONTROL in Fig. 10.7 turns off the p-channel transistor MP that provides the supply to the final inverters of the row decoder. CONTROL must turn off MP just before the boost occurs so as not to dissipate the charge that had been stored in C_{BOOST} through the VDD power supply.

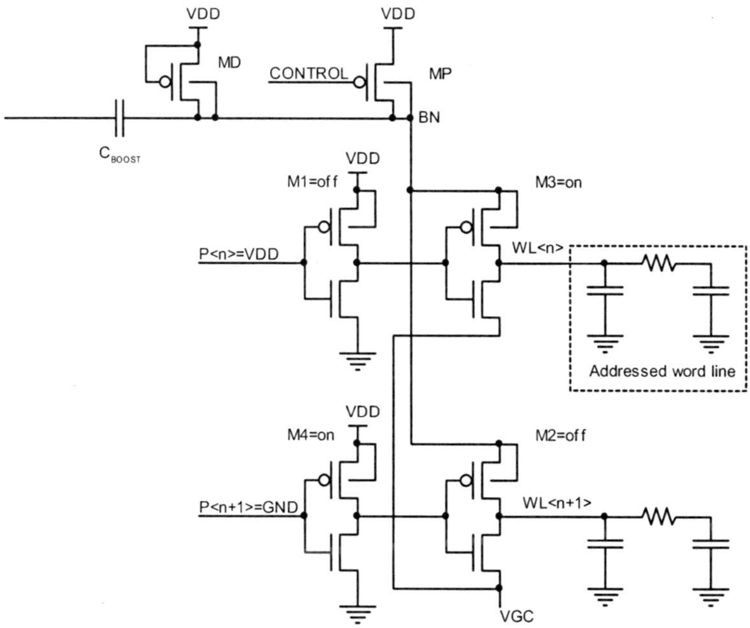

Fig. 10.7. Clamping of the voltage of the boosted node to V_{CLAMP1} is necessary to prevent simultaneous multiple row selection

If we want to have WL<n> high and WL<n+1> at ground, we must guarantee that M3 is turned on and M2 is turned off. Transistor M2 is driven by an inverter whose power supply is VDD, therefore its gate will be at VDD; so M2 is turned off until its source, i.e. the boosted node, does not exceed a threshold voltage above VDD. Referring to Eq. (10.3), we know that the value of the boost voltage is a function of both the capacitive partition and the actual value of power supply; if, for instance, the boost pulse occurs when the power supply is equal to 3.8 V, the voltage of the node can go above 7 V. It is therefore necessary to introduce a component that cuts off the over-voltage on node BN when it reaches the value, in magnitude, of one $V_{T,p}$.

Clamping of the voltage on node BN is achieved by means of the transistor MD, which is diode-connected; this transistor must be of the same type as those used in the row decoder, so that they share the same threshold voltage. Then we can define a voltage V_{CLAMP1} equal to

$$V_{CLAMP1} = VDD + \left|V_{T,p}\right| \tag{10.4}$$

Clamping diodes are placed near the capacitor C_{BOOST} and far from the node to boost so that the charge in excess is really absorbed by the diodes themselves, preventing the boosted node from exceeding the voltage V_{CLAMP1}. It is mandatory that the boosted node does not exceed the threshold of the p-channel, otherwise the final inverters of the decoding of the same sector would be turned on instead of being off: in fact, the boost is applied only to the final inverter, the previous one being supplied at VDD.

10.4 Double-Boost Row Decoder

Clamped boost technique described above can be applied even more extensively. Figure 10.8 shows a row decoder where two separate boost voltages are applied: one to the word line driver and one to the previous inverter, driven by the predecoding signals indicated as P<n>. Furthermore, the presence of two boost capacitors can be noted, C_{BOOST0} e C_{BOOST1}, together with the corresponding control signals. For the MINIBOOST nodes, the considerations explained in the previous section are still valid: its voltage is clamped at V_{CLAMP1}. In order to understand the operation of this particular decoding scheme, we have to assume that the potential of the node indicated as BULK cannot exceed VDD by more than twice the threshold voltage of the p-channel transistor. In the following discussion, we will refer to this additional clamping voltage as V_{CLAMP2}.

$$V_{CLAMP2} = VDD + 2 \cdot \left|V_{T,p}\right| \tag{10.5}$$

The row decoder shown in Fig. 10.8 is designed to drive the two final inverters with different boosted voltages. The next to the last inverter is connected to a voltage V_{CLAMP1}, whereas the actual rowl driver biases the selected word line at a voltage V_{CLAMP2}. In this way, each inverter of the row decoder is able to keep the p-channel of the following inverter completely turned off. Even in this case, in order to clamp the above-mentioned boosted voltages, p-channel transistors that are identical to those present in the decoder are used, in such a way that they have the same threshold voltage $V_{T,p}$. We have also seen that it is necessary to keep the unselected row to ground in order to avoid charge loss from the boost capacitor and prevent gate stress of the memory cells. Therefore it would be safer to provide the next to the last inverter with a supply voltage V_{CLAMP1} + ? (? being a positive value), thus ensuring a margin with respect to the turn on voltage of the final inverter. As we have seen in Chap. 3, the threshold voltage of a transistor can be increased by exploiting the body effect. Decoding of Fig. 10.8 is therefore modified as shown in Fig. 10.9: the PMOS that are boosted by the MINIBOOST signal have their n-well connected to the signal called BULK[1], in such a way that they can be boosted to a higher value because their threshold is increased by the contribution of the body effect.

[1] The name of the signal should be self-explanatory.

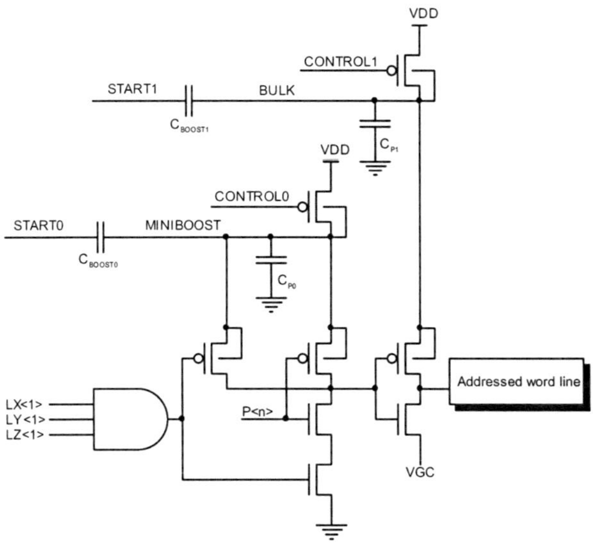

Fig. 10.8. Row decoder where the power supplies of the two last inverters placed before the word line are boosted with different voltages

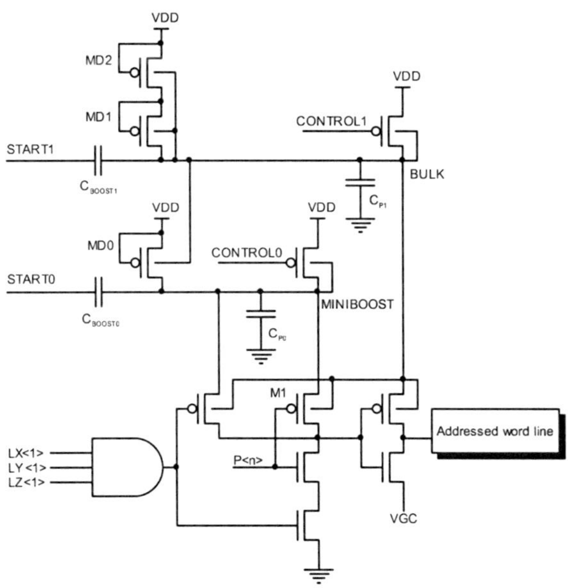

Fig. 10.9. In order to increase the $|V_T|$ of the second last p-channel (M1), its body is connected to the node called BULK that reaches a higher voltage than MINIBOOST

At this point the PMOS MD0, which is diode-connected in the MINIBOOST circuit of Fig. 10.9, must have its own n-well connected to the BULK signal so that its threshold voltage is modified in the same way as that of the PMOS in the decoder and it can clamp the voltage to a higher value. Figure 10.10 is a simulation that shows the different voltages reached by the MINIBOOST node before and after the connection of the n-well of transistor M1 of Fig. 10.9 to the boosted node called BULK.

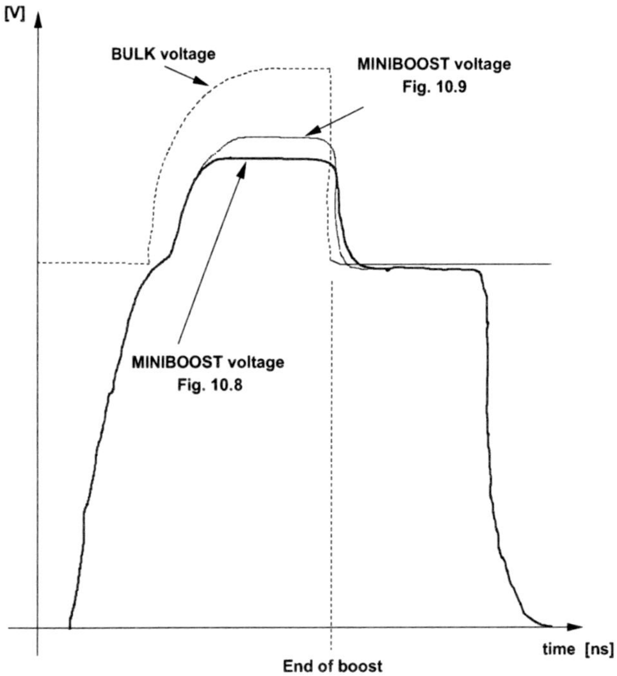

Fig. 10.10. Graphical representation of the voltage on the node MINIBOOST, using or not the n-well driving technique of the transistor M1 of Fig. 10.9

The schematic shown in Fig. 10.11 depicts the generator circuit for the signals MINIBOOST and BULK. Capacitors C_{BOOST0} and C_{BOOST1} are the one shown in Fig. 10.8. The circuit is composed of two structures similar to those shown in Fig. 10.4. We can also identify the diode-connected transistors used to clamp the boosted voltages. Task of MD0 is to clamp the voltage on node MINIBOOST to the value V_{CLAMP1}, while both MD1 and MD2 are used for the clamping of V_{BULK} at V_{CLAMP2}.

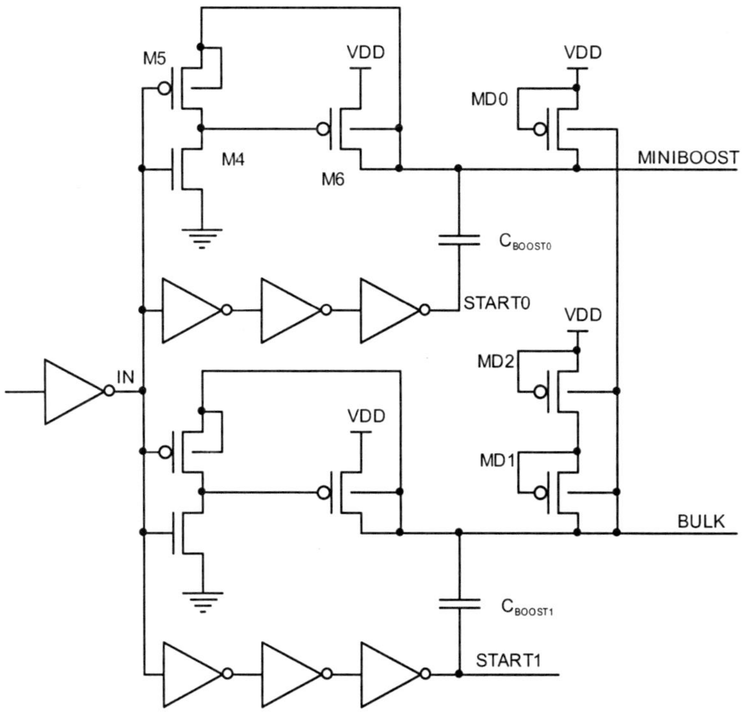

Fig. 10.11. Circuit for the generation of boosted voltages BULK and MINIBOOST

10.5 The Issue of the Recharge of C_{BOOST}

If we use a pulsed boost to limit the power consumption during stand-by, two additional problems arise:

1. The start of the boost operations must be controlled and synchronous with the rise of the row, which corresponds to the node to be boosted. This problem is solved, at least conceptually, by implementing a dummy path that imitates the real read path and provides the means to synchronize the boost operation..
2. The second issue, very important as well, is related to the recovery of the initial conditions which must be restored at the end of each read operation. Figure 10.12 shows a synthetic flow diagram of the operations performed during a read. To work properly, C_{BOOST} must be precharged to VDD before the boost pulse start. It can be seen that it is necessary to have a signal that indicates that the read operation is over, which is the ENDREAD signal that will be described in the next chapter.

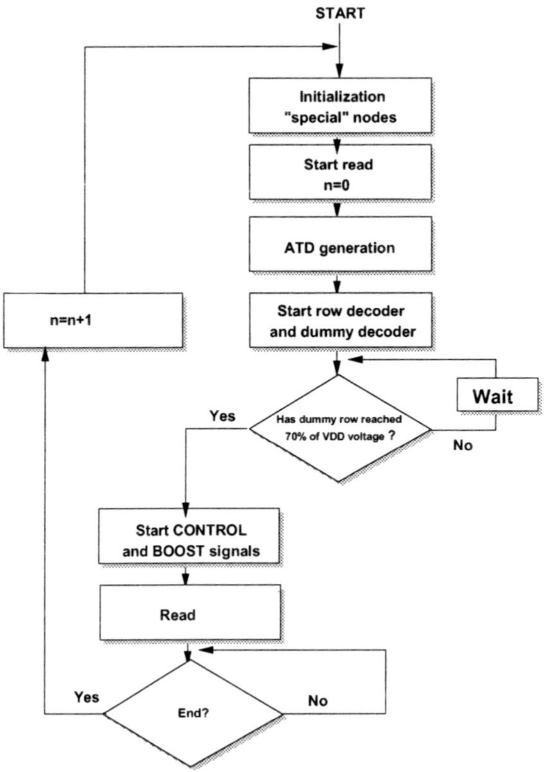

Fig. 10.12. Flow diagram of the operations required to end a read and get ready for the next one

Going back to Fig. 10.9, let's just consider the part indicated in Fig. 10.13. In order to operate properly, capacitor C_{BOOST1} must be precharged at VDD before the boost pulse is applied; therefore signal START1 must be at ground and V_{BULK} must be equal to VDD. Afterwards the boost occurs: START1 goes to VDD and, since the signal CONTROL0 turns off MP1, node BULK is boosted. The boost voltage is transferred to the selected row and the sense amplifier can read the state of the memory cell. At the end of the read operation, a signal will cause the circuitry to restore the initial conditions in preparation for the subsequent read cycle.

The fact is that C_{BOOST} must be[2] in the order of some tens of picoFarad, and there are two ways to change the voltage of such a large capacitance:

- slow discharge of node BULK towards the power supply VDD. This solution prevents the node from undershooting, but it takes a very long time (too long);
- fast discharge of node START1, causing a bounce on node BULK and therefore a waste of time to recharge it.

[2] The exact value of C_{BOOST} will be calculated as a function of the parasitic load C_{P1} of the row decoding.

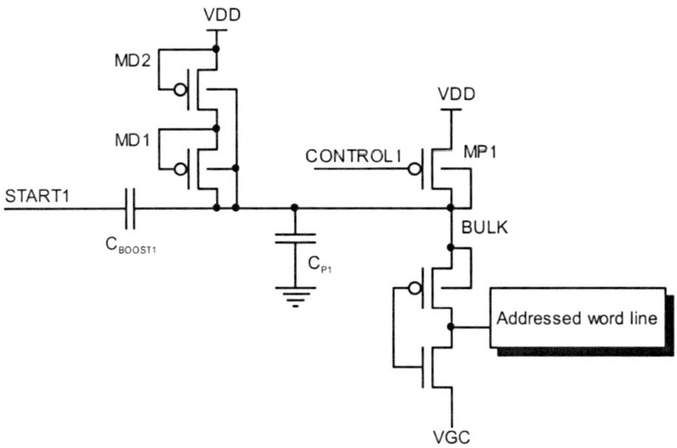

Fig. 10.13. Portion of Fig. 10.9 used to describe the issue of the recharge of the boost capacitor

The former method is not acceptable because the time that it takes to achieve a slow discharge that does not cause bounces is in the order of 50ns! (access time of present memories hardly goes above 100 ns). Therefore we must go for the fast discharge and devise some trick not to significantly increase the access time.

We know that node BULK is boosted to the voltage V_{CLAMP2}, i.e., under typical conditions, about 1.7 V above VDD; node START1 starts at ground and then increases until it reaches the power supply voltage. This means that, when the boost has taken place, the voltage difference between the plates of the capacitor is equal to $2|V_{T,p}|$: the charge in excess is in fact discharged through the diode-connected transistors MD1 and MD2.

When the nodes are then brought back to their initial condition, the voltage step of node START1 is equal to VDD. The capacitor tends to maintain the voltage difference between its terminals unaltered and, therefore, V_{BULK} wants to decrease by an amount equal to VDD.

Voltage on node BULK, once the lower plate of the capacitor C_{BOOST1} is discharged, will be equal to:

$$V_{BULK} = VDD + 2 \cdot \left|V_{T,p}\right| - VDD = 2 \cdot \left|V_{T,p}\right| \tag{10.6}$$

Ultimately, node BULK does not go back to VDD and therefore it must be recharged before being ready for a new boost. Unfortunately the recharge time of this node, because of the large attached boost capacitance, is in the order of 30-40 ns. This time cannot be reduced by increasing the size of the p-channel transistor used for the recharge, because this transistor works with a small V_{DS} and therefore is able to provide only a small current , resulting in a slow recharge.

In case a second read should start when the voltage on node BULK has not yet been restored to VDD, we would get a smaller final boost voltage and therefore at the end of the second read phase node BULK would go even lower... and so on. After some reads under these conditions, the device would stop working properly[3].

Problem 10.2: Is the issue related to the insufficient recharge of the boost capacitor also present in case of continuous boost?

A solution that might come to mind is to use a NMOS transistor whose gate is boosted to recharge node BULK, but this solution is too complex and not efficient. The solution that we want to explore is to double boost circuits.

10.6 Double-Path Boost Circuitry

The circuit that contains the duplicate boost circuitry is shown schematically in Fig. 10.14. Starting from the ATD signal, a BSTSTART signal is generated by the block called LOGIC. The BSTSTART signal is transferred to C_{BOOST1} and C_{BOOST2} alternately.In this way, a read operation is performed by using one of the two boost capacitors, while the other capacitor has sufficient time to recharge, so that it is ready when the next read occurs. The mechanism described above works only if it is possible to activate the signal EN1 and EN2 alternately, so that either one or the other capacitor can be accessed. D-type flip-flops are used which are edge-triggered on the rising edge of the clock signal, CP.

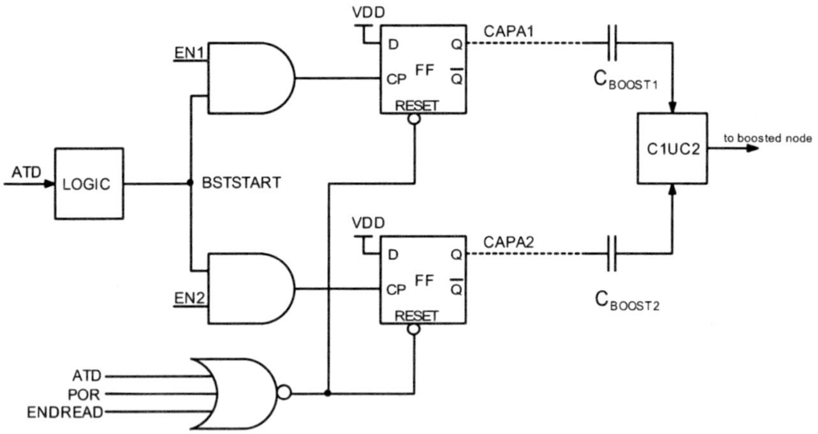

Fig. 10.14. Duplication of the boost circuitry is necessary in order to solve the issue of the recharge of C_{BOOST}

[3] When these circuits are studied, it is important not to limit the analysis to a single read: several consecutive read operations must be verified, to insure the correct restore of the nodes at the end of each phase.

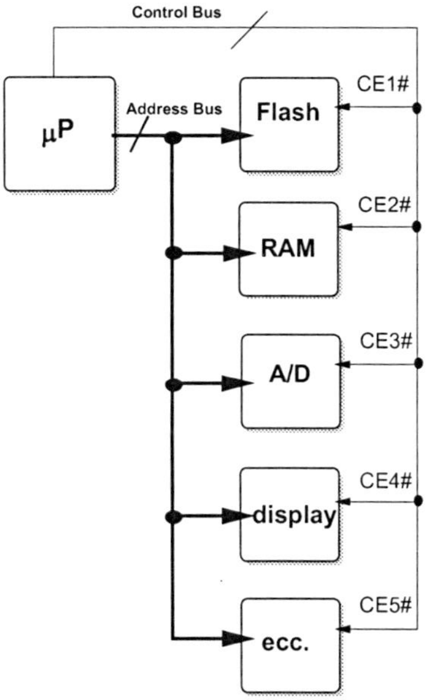

Fig. 10.15. A system where the Flash memory shares the address bus with other devices, all driven by the same microprocessor

As shown in Fig. 10.14, there are three signals that can reset the flip-flop, thus causing the boost capacitors to recharge. The signal POR (Power On Reset) is generated during the rise of the power supply voltage and its task is to guarantee that all the logic circuits are correctly initialized (see Chap. 5). ENDREAD is the signal that indicated that the sense amplifier has completed a read operation, i.e. the moment when it is possible to start restoring initial conditions.

The presence of the ATD signal in the reset of the two flip-flops is needed so that the boost chain can be ready for every address change. Let's assume that the memory shares the address bus with other peripherals as shown in Fig. 10.15. The microprocessor is using the A/D converter sending and receiving signals on the bus. The only chip enable that is asserted, of course, is CE3# (which is low). Now the microprocessor needs to access the Flash memory. No specification provides a timing protocol for selection/de-selection of the peripherals. The microprocessor disables the A/D and enables the Flash memory at the same time by means of both CE1# and CE3#. Throughout the time interval between the activation of CE1# and the sending of the addresses on the bus (both operations performed by the microprocessor), the bus has the previous address value. The Flash memory, now active, starts working: ATD is generated, boost takes place etc. Unfortunately the read is incorrect. If the microprocessor sends the correct addresses before the read is done, a new ATD takes place without the recharge of the boost capacitors. By including the ATD signal in the reset condition for the flip-flops, the restore of the proper initial conditions is guaranteed whenever a new ATD occurs. When, on the other hand, the address value is the "true" one, boost will start and the reset is caused by the ENDREAD signal after the "real" read has taken place.

Let's now design the circuit that generates EN1 and EN2 complementary signals. When the device is turned on, EN1 is enabled. The first ATD passes through the AND driven by EN1 and the CAPA1 signal is activated. At this point, without losing the value of CAPA1, we can bring EN1 low and activate EN2, preparing C_{BOOST2} for the next ATD. Ultimately, signal EN1 and EN2 must toggle by means of a signal that must be active when the boost occurs, or shortly after the store of the CAPA1 and CAPA2 signals by the flip-flops.

Let's assume that EN1 path is enabled, and let's consider the following cases:

1. boost occurs and afterwards ENDREAD signal is activated;
2. boost occurs and a new ATD pulse arrives before the end of the current read;
3. a new ATD arrives before boost occurs.

In the first two cases it is necessary to change the path to give time to the capacitors on the EN1 path to recharge; in the third case we don't need to change path, since boost did not take place and, therefore, it is not necessary to recharge the capacitor. In summary, it is necessary to use the alternate path only if either the read has completed correctly or the read has been interrupted after the boost occurred. The signal that confirms the start of the boost can be, for instance, the START1 signal shown in Fig. 10.11, which is driven high when the boost occurs.

We can conclude that the path must be changed if the following expression holds true:

$$(\text{ENDREAD})\ \textbf{OR}\ [(\text{ATD})\ \textbf{AND}\ (\text{START1})] \tag{10.7}$$

The implementation for the circuit is shown in Fig. 10.16. There are two START1 signals (A and B), one for each boost circuitry. Let's recall that both the ATD and ENDREAD signals are considered as pulses while START holds its value (it is the lower plate of the C_{BOOST} capacitor and it remains at VDD throughout the boost phase). The combination of the ATD signal with the START signal into the NAND gate is required because the condition for changing the path on the START signal is boost already occurred; therefore START signal is static at a logic 1 level, and the ATD is used as a pulse.

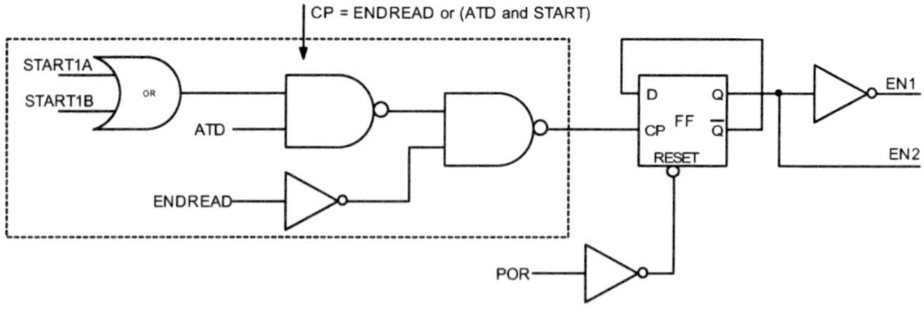

Fig. 10.16. Generation of signals EN1 and EN2 of Fig. 10.14 which allow to switch the boost path

Before moving on, it is appropriate to summarize the concepts discussed above. We have described the issues related to the execution of consecutive read operations, because of the need of recharging the large capacitive nodes are used during boost phases. This is a structural problem and it is present whenever the pulsed boost is used to read. In order to overcome this issue we have introduced a double boost circuit so that we can recharge one capacitor for the following read operation while we use the other one for the current read. This complex system calls for the duplication of every boost capacitors and their associated drivers. The analysis of the possible iteractions between the signals has shown that the enable strategy for the two paths must be carefully designed; finally, we have understood that the boost path must be toggled if and only if either the read has completed or it has been aborted once the boost occurred.

10.7 Boosted Voltages Switch

Once we have solved the electrical issues, it is necessary to take care of the area occupation due to the doubling of both the boost capacitors and related circuitry. In Fig. 10.14 we have seen that it is necessary to devise a circuit, called C1UC2 in this case, that is able to select and transfer one of the two boosted voltages on to the supply of the row decoder. Ultimately, we must design a switch that selects between two voltages and the simplest way to do it is to use pass transistors as shown in Fig. 10.17. Recall that the nodes to be boosted in the row decoders are indeed both BULK and MINIBOOST. Let's start designing the circuitry required to handle the pass transistors of the signal BULK, indicated in the following as MS1 and MS2. Similar considerations apply for MINIBOOST as well.

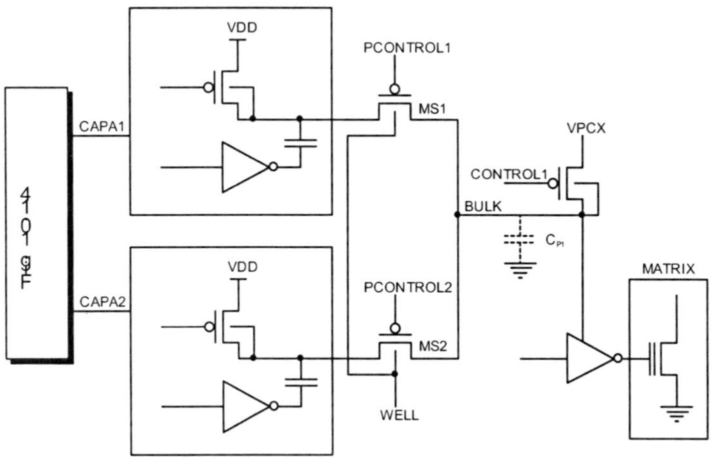

Fig. 10.17. Ideal scheme of the switch that allows the transfer of one of the two boosted voltages to the row decoder

The task that the pass transistors have to carry out, i.e. the transfer of the charge stored inside the boost capacitors to the row decoder, must be efficient, i.e. without losing charge through undesired paths. Let's assume that boost occurred on node BULK1. In order to transfer the charge of C_{BOOST1} onto BULK, PCONTROL1 signal must go low, while MS2 must be turned off; to prevent the charge from redistributing on C_{BOOST2} as well (causing a loss of efficiency of the boost), MS2 must be completely turned off before MS1 turns on. It is important to note, at this point, that it is not only a matter of timing, but also of drive voltage: in order to turn MS2 off, it is not enough to bias its gate at VDD, but it is also necessary to have a boosted voltage. In other words, block C1UC2 is an analog multiplexer, but it also calls for boosted command voltages.

Therefore we must design a boost circuit, used to drive the gate of the pass transistors of the switch, which starts synchronously with the boost of the nodes of the row decoder. This circuit must also be able to provide an output voltage equal to VPCX. In fact during programming (which usually takes some microseconds to complete), boost capacitors are not biased at VPCX. As we will see later, in this case there are specific charge pumps that provide the required voltages and are able to deliver the proper amount of current. VPCX, which is about 10 V during programming, is directly transferred to the supply node of the row decoder, while both V_{BULK1} and V_{BULK2} must remain at VDD: in this way a voltage equal to VPCX is required to turn off the pass transistors. Boost capacitors, if connected to VPCX, would be indeed a useless capacitive load for the charge pumps, slowing down their transients; furthermore, their oxide would be stressed in vain.

Let's summarize in Table 10.1 the values that the gate voltages of MP1 and MP2 must have in the different working conditions:

Table 10.1. Values of the voltages of the main nodes of Fig. 10.17 during read and program

	BULK1	PCONTROL1	BULK2	PCONTROL2	VPCX	BULK
Read-BULK1	V_{CLAMP2}	GND	VDD	V_{CLAMP2}	VDD	V_{CLAMP2}
Read-BULK2	VDD	V_{CLAMP2}	V_{CLAMP2}	GND	VDD	V_{CLAMP2}
Program	VDD	VPCX	VDD	VPCX	>VDD	VPCX

A fundamental node is missing in the table: the n-well of the pass transistors, called WELL in the scheme of Fig. 10.17. It is important, as a general rule, that the n-well of a PMOS always be the terminal with the highest voltage, in order to avoid forward biasing of either the source or drain junction towards the body. Node WELL must therefore be boosted at the same voltage as the word line in read, and be brought to VPCX during program. Also in this case, a boosted voltage generator as the one shown in Fig. 10.4 can be used.

The continuous quest for space optimization leads us towards the definition of the overall architecture. A fundamental requirement is not to compromise boost efficiency when layout is realized: special care must be taken in regards to parasitic capacitances. Referring to Fig. 10.17, parasitic capacitance denoted with C_p, i.e. the one that must be boosted, is due to the p-channel transistors (body, junctions, etc.) of the row decoder and to the metallizations used to transfer the boosted signal. It is important that capacitance C_{P1} is not further increased by the contribution, for instance, of the capacitance of the metal that goes from the pass transistor MS1 (MS2) to C_{P1}. Layout must therefore be carefully controlled to ensure that the connections are as short as possible.

It is interesting to describe the circuit that can be used to bias the node called CONTROL at three different voltage values: ground, VPCX and a boosted voltage. This circuit is shown in Fig. 10.18 and it is essential to generate the signals described in Table 10.1. When the ENABLE signal is forced low, the output of the NOR port in Fig. 10.18 only depends on the value of the IN signal; at the same time the p-channel M1 acts as a pass transistor between the gate of M2 and the CONTROL signal.

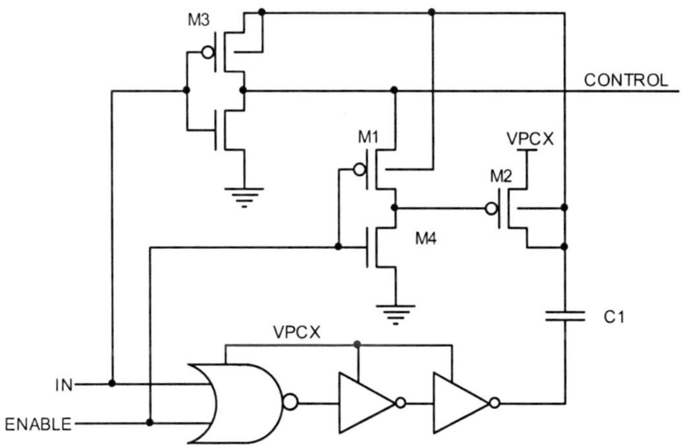

Fig. 10.18. CONTROL signal can be biased at ground, VPCX or a boosted voltage

In this condition, the CONTROL signal is at ground if IN is high, or it can be boosted if IN is low. In fact, the gate of M2 is connected, through M1 and M3, only to the upper plate of the capacitor C1 that is therefore not connected to the power supply. In this way, the transition toward the low state of the input signal causes the boost on the source of M3, and therefore of the CONTROL signal to occur.

When the ENABLE signal is high, IN signal is no longer able to influence the output of the NOR port. The gate of M2 is forced to ground by M4 while M1 is turned off. In this way, the source of M3 is at VPCX and the CONTROL signal becomes the complement of the IN signal, in the GND-VPCX logic. Table 10.2 summarizes what we have just explained. We have realized an inverter with three output levels!

Table 10.2. Values of the CONTROL signal of Fig. 10.18 for different combinations of both the IN and ENABLE input signals

ENABLE	IN	CONTROL
GND	VPCX	GND
GND	GND	Boosted voltage(>VPCX)
VPCX	VPCX	GND
VPCX	GND	VPCX

When the design of the analog switch, that can be used to transfer boosted voltages, has been carried out, we can devise a way to implement the architecture shown in Fig. 10.19 where, for the sake of simplicity, only the pass transistors for the BULK signals are shown. It can be seen that, in this way, it is possible to share the drivers for both BULK and MINIBOOST signals among different sectors. Focusing on the single sector, we have seen that the issue of the recharge of C_{BOOST} calls for the doubling of the boost circuitry. Thanks to the sharing of the boost circuits, the total number of the circuits shown in Fig. 10.11 is equal to the number of the sectors inside the device. Of course, the architecture described in this chapter can be used also in the case where the single sector is divided, for instance to reduce the time constant associated with the word line, into two or more sub-sectors.

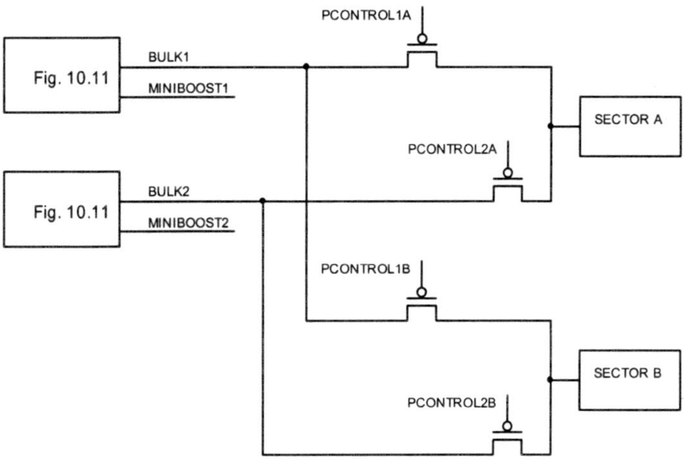

Fig. 10.19. Circuit to share the boosted voltages BULK and MINIBOOST among different sectors

10.8 Leakage Recovery Circuits

When we deal with boosted nodes, some precautions should be taken not only for the pull-up phase but also during the subsequent holding phase. The isolation of the boosted nodes is a fundamental requirement for success operation. The leakage current, i.e. the current lost in any junctions, opposes the boost operation. In our case, there is leakage due to the reverse-biased current of the n^+ junction flowing to the p-substrate or of the p^+ junction flowing into the n-well, and its order of magnitude can be estimated around fractions of pA/μm^2.

Specifically, the nodes that we want to protect are the terminals of the n-wells of the PMOS pass transistors which should be closely tied to the boosted voltage to avoid the latch-up phenomenon.

Unfortunately, the leakage current cannot be eliminated. Instead, we can utilize a leakage recovery circuit, i.e. a circuit that replaces the charge lost from leakage current on the most critical nodes. In order to do that, we need a free running oscillator, which continuously oscillates once the device has been switched on and the device is no longer in stand-by. The value of the oscillation frequency and the stability with respect to both temperature and supply voltage are not very important. To this end, it is possible to use a ring oscillator of the same kind as the one described in Chap. 5.

The circuit used to replenish the lost charge is shown in Fig. 10.20, where it is possible to identify the structure of the boosted voltage generator of Fig. 10.4. The charge transfer takes place during the falling edges of the oscillator output voltage. The boosted voltage generated on node OUT is transferred to node WELL of Fig. 10.17 by means of a natural transistor; N1 is diode-connected in order to prevent charge from being taken away from node WELL when the upper plate of the boost capacitor is brought back to VDD.

As the leakage current can amount to some microAmpere, our oscillator should be realized to provide 10 μA.

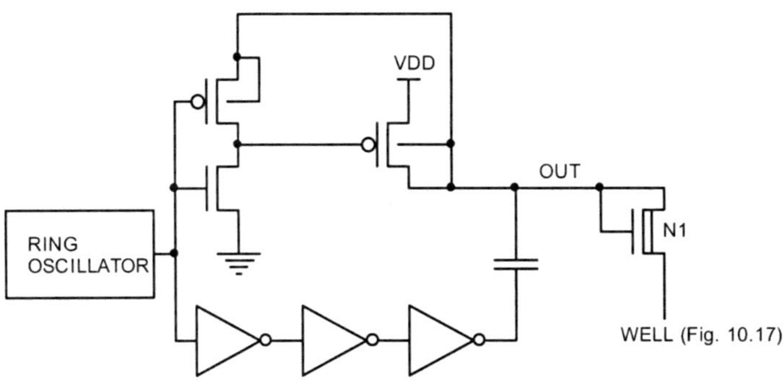

Fig. 10.20. Circuit to refurbish the charge lost by the boosted nodes because of leakage currents

Bibliography

A. Bellaouar et al., "Bootstrapped Full-Swing BiCMOS/BiNMOS Logic Circuits for 1.2–3.3 V Supply Voltage Regime", IEEE Journal of Solid-State Circuits, vol. 30, no. 6, (June 1995).

G. Campardo, R. Micheloni, S. Commodaro, "Low supply voltage nonvolatile memory device with voltage boosting", USA patent No. 5,903,498, (May 11, 1999).

O. Khouri, R. Micheloni, I. Motta, and G. Torelli, "Word-line read voltage regulator with capacitive boosting for multimegabit multilevel Flash memories", in Proc. European Conf. Circuit Theory and Design 1999, vol. I, pp. 145–148, (Aug.-Sept. 1999).

H. Morimura and N. Shibata, "A Step-Down Boosted-Wordline Scheme for 1-V Battery-Operated Fast SRAM's", IEEE Journal of Solid-State Circuits, vol. SC-33, No. 8, (August 1998).

N. Otsuka and M.A. Horowitz, "Circuit Techniques for 1.5-V Power Supply Flash Memory", IEEE Journal of Solid-State Circuits, vol. 32, no. 8, (August 1997).

T. Tanzawa and S. Atsumi, "Optimization of word-line booster circuits for low voltage flash memories", IEEE J. Solid-State Circuits, vol. SC-33, pp. 410-416, (March 1998).

T. Tanzawa and S. Atsumi, "Optimization of word-line booster circuits for low-voltage flash memories", IEEE J. Solid-State Circuits, vol. SC-34, pp. 1091–1098, (Aug. 1999).

11 Synchronization Circuits

The synchronization circuits are fundamental to the correct behavior of the device. They are a bit like the traffic lights that regulate the flow of the signals that propagate across the chip on roads that sometimes are dedicated, sometimes are shared among groups of signals. It is obvious that the system works properly if the different signals do not come into conflict with each other.

The main problem is that the memory is an asynchronous device; hence, it is necessary to include circuits that synchronize the internal operations, especially in the case of dynamic circuits.

11.1 ATD

It is important to generate a signal that triggers all the circuitry and synchronizes all the blocks of the read path.

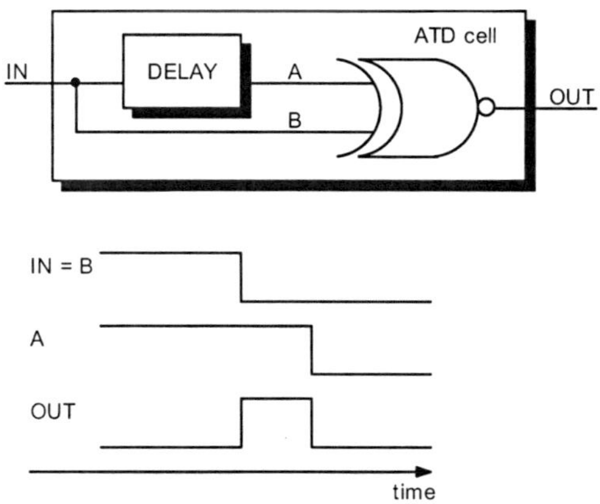

Fig. 11.1. Block diagram of the circuit that detects the variation of the IN signal. The resulting waveforms are reported

In reality, our memories, which are usually not connected to any external clock, are asynchronous with respect to the external world during the read phase. Internally, however, it is possible to generate a pseudo-clock that is able to provide one single pulse synchronously with the basic event of the read phase, that is either the address or the chip enable (CE#) transition. It is necessary to realize the circuitry to sense the change of the address, CE#, BYTE# or WORD# signals, and eventually trigger the reading. First of all, let's concentrate only on the address transition. In Fig. 11.1, the principle of generation of a pulse corresponding to a transition of an input signal is shown. If we realize a circuit like this (called ATD cell) for each address, and then combine all the outputs with a distributed NOR (Fig. 11.2), we obtain a signal each time the device is accessed for reading.

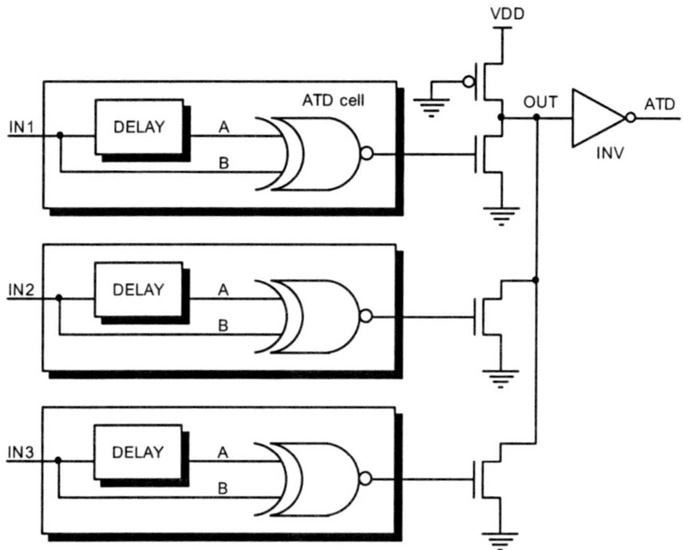

Fig. 11.2. The composition of several control blocks allows detecting the transition of the input signals. The various outputs are combined by means of a distributed NOR

The final output node provides a pulse any time at least one of the input signals changes. The duration of the OUT signal generated can be enlarged as necessary by means of the addition of buffers and monostable circuits. The final signal is called ATD, acronym of Address Transition Detector. It is the signal that triggers the reading and the only timing reference we can rely on for the entire circuitry.

One of the most used schemes for the ATD cell is shown in Fig. 11.3. The AX signal and its opposite are the inputs. When AX toggles from the low to the high logic state, C2 is rapidly charged by M2, which has a high aspect ratio, while C1 is slowly discharged by the pull-down of the INV1 inverter, due to the low W/L ratio of the n-channel. Therefore, the output of INV3 goes high as soon as the address changes and remains high until the potential of the top plate of C1 pulls the NAND output high.

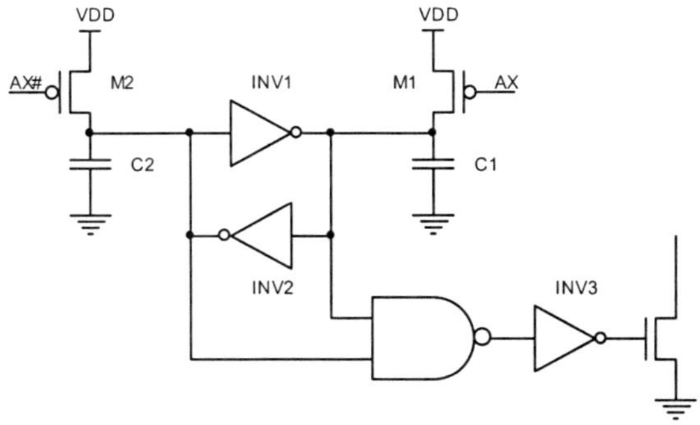

Fig. 11.3. Schematic of the ATD cell

11.2 Multiple ATD Management

The ATD has a fundamental importance in current devices that must continuously optimize the performances, for example in terms of access time. A synchronization signal allows adopting a number of different techniques to precharge and equalize the sense amplifiers, as detailed in Chap. 12. The ATD is generated as a consequence of the change in the address. It is the signal that triggers the read operation. Now the question is: how is the address applied?

To provide the correct answer, the problem known as address skew will now be investigated. Let's consider a microprocessor that is connected to a Flash through the bus and simultaneously transmits all the signals that represent the address. The routing of the connecting lines is not identical for all the signals. Possible geometrical dissimilarity determine different capacitive, resistive, and inductive loads. As a result, the signals start together but arrive at the pins of the Flash memory in different instants. Two situations are then possible:

1. the ATD pulse is extended because the OUT node in Fig. 11.2 is still low;
2. a sequence of ATD pulses is generated.

The first problem can be solved by using the falling edge of the ATD as trigger for reading. The impact on the access time is quite evident, but the risk of starting the read operation when the address is not yet stable is prevented. Let's examine the second possibility, multiple ATD pulses. If the addresses were simultaneously applied at the device pins, or changed after a time longer than the access time (see Fig. 11.4), there would not be any problem. Unfortunately, there are no specifications to force the customer to adopt such a precaution. Customers are allowed to change the address randomly, and the valid address of which the device must present the output data is therefore the last one (see Fig. 11.5). The frequency of variation of the addresses could even generate a single ATD pulse, for example of one second. This would be purely by chance, but the reading would start on the only falling edge of the ATD.

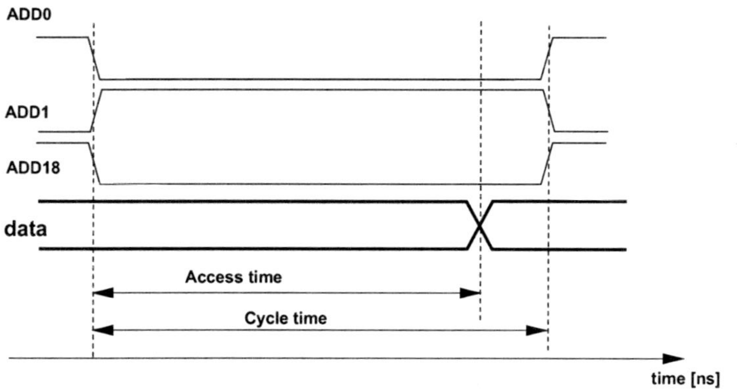

Fig. 11.4. Cycle and access time

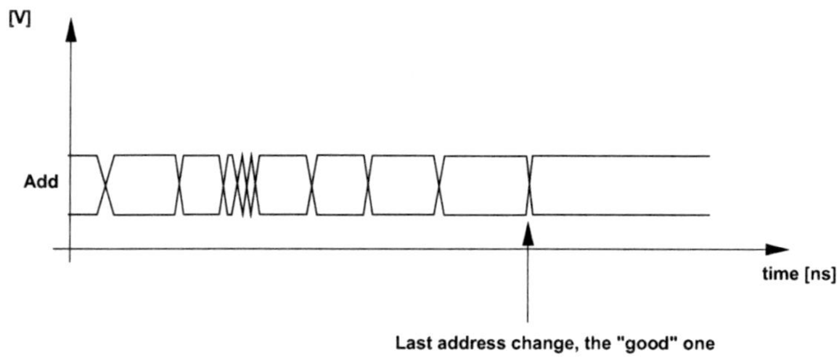

Fig. 11.5. The address could change at any rate before settling at the valid value

How could we know when the last valid address is applied? It is not possible to derive this information, and, hence, the device must be ready on any address transition since it could be the valid one. A completely static design, having no precharge circuitry, can operate without any problems in this situation. Major problems arise, however, when the read circuitry relies on precharged nodes and boosted circuits. Let's analyze the most complex case, a device that reads with a one-shot boost and also utilizes signals that must be preset to certain voltages. How can we realize the circuitry in order not to miss any address transition and correctly start all the internal operations?

Problem 11.1: It would be important if the reader could answer without reading the solution.

The solution can be similar to the one related to the problem of the recharging of the boost capacitors. We can use two parallel timing chains alternately and, finally, merge the outputs in a single signal that will be used throughout the circuitry. The basic schematic is shown in Fig. 11.6.

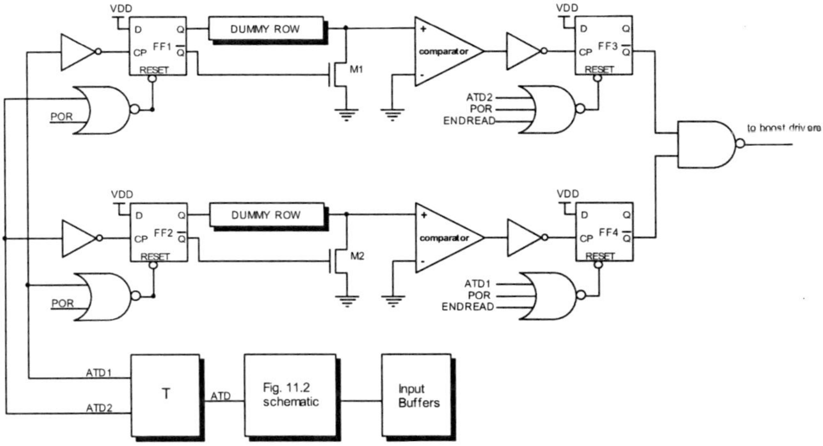

Fig. 11.6. Schematic of the double timing path aiming at not missing any address transition

The circuit in Fig. 11.2, located at the end of the input buffer, generates an ATD pulse having a short time duration. The T block fulfills the task of "diverting" the input signal toward either the ATD1 or ATD2 output, alternatively. In order to properly work, the T block has to operate in a predictive way. During the rise of the supply voltage, i.e. during the device switch-on, the circuitry of the T block enables the path toward ATD1. On the falling edge of the first ATD, the T block disables the path of ATD1. Subsequently, the path toward ATD2 will be enabled and ready for a new ATD pulse. In practice, the T block has always a path ready for the subsequent ATD.

11.3 Let's Connect the ATD to the Boost Circuitry

One practical difficulty is not in the generation of the ATD or other synchronization signals, but their correctly usage. For this reason we will try to explain how such signals can be plugged into the read path.

The block diagram of a portion of the read path is shown Fig. 11.7 which summarizes the example of Chap. 10 which utilizes the one-shot boost[1]. The ATD1 and ATD2 signals are generated in the T block (Fig. 11.6). They are latched in FF1 and FF2 and applied to circuitry that models the real read path to obtain the same propagation delay.

[1] For the moment, the ENDREAD signal will be regarded as a positive pulse generated at the end of the read phase.

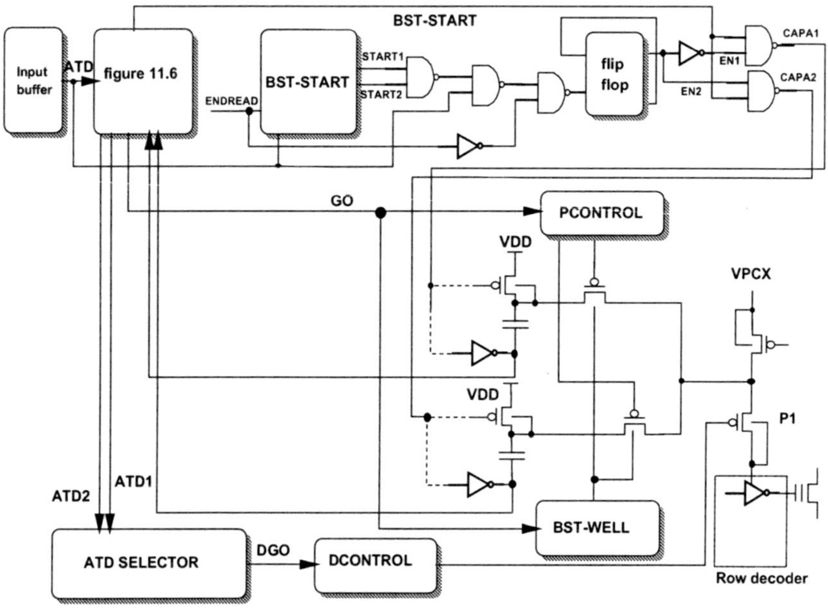

Fig. 11.7. Schematic of the boost circuitry and its connection to the ATD

The four flip-flops in the figure are a sort of "decompression chamber" within which the ATD pulses can propagate at any frequency and do not vanish because of the initial and final flip-flops. The two paths generate the BST-START signal that is used to trigger the boosting. Meanwhile, the signals of the sector decoder, activated by the GO signal, have already switched on the PMOS pass transistor on which the boost operates. We recall that also the well of the p-channels is boosted by means of the same signals so as to prevent them from being forward-biased. It is necessary to switch off transistor P1 of the row decoder so that the supply voltage of the final inverter is floating. This operation must be carried out in advance; hence, a parallel dummy path must be designed with a dummy load equivalent to half a row of the array. The purpose is to switch off P1 before the boost event is triggered.

Problem 11.2: Draw the waveforms of the various nodes of Fig. 11.6 between one address transition and the subsequent transition.

The M1 and M2 transistors of Fig. 11.6 are worth mentioning. In order not to miss any ATD pulse, the two dummy paths must not have any slow node. The dummy row introduces an RC parasitic so that rising and falling delays are equal. In order to guarantee that the discharging of the row is faster than its charging, the M1 and M2 transistors are included. Therefore, the row is now charged through the dummy decoder located at the side of the row, whereas it is discharged through both the dummy decoder and the M1 transistor, reducing the fall time.

Moreover, the Power On Reset (POR) signal is used to set the output value of all the flip flops, even though it would not be necessary, at least theoretically, since at least one of the two paths is active and therefore is always able to operate.

11.4 Equalization of the Sense Amplifier: SAEQ

Two main operations are performed using the ATD signal: the boost is triggered and the nodes of the sense amplifier are equalized. The ATD pulse is suitably enlarged to obtain the SAEQ signal that is applied to the sense amplifier, as we will detail in the chapter dedicated to the read circuitry. The purpose of the equalization is the preparation of the critical nodes of the converter, so as to minimize both the time for the read transition and length of any indeterminate state, thus preventing oscillations.

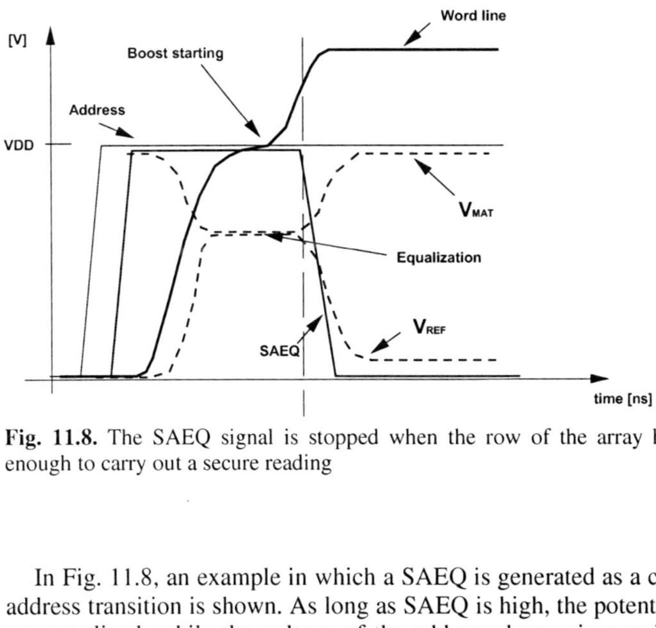

Fig. 11.8. The SAEQ signal is stopped when the row of the array has been pulled up enough to carry out a secure reading

In Fig. 11.8, an example in which a SAEQ is generated as a consequence of an address transition is shown. As long as SAEQ is high, the potentials V_{REF} and V_{MAT} are equalized, while the voltage of the addressed row rises as it is boosted. The reading, i.e. the separation of the potentials, occurs when SAEQ goes low. When SAEQ is high, the control of the potentials V_{REF} and V_{MAT} is guaranteed by the equalization circuitry, whereas when SAEQ is low, such a control is carried out by both array and reference cells. In other words, the equalization allows minimizing the voltage swing of the MAT and REF nodes. Moreover, sensing nodes are always reset to the same potential independently of what happened during the previous reading, guaranteeing the same transient at any memory access.

The problem that is typically encountered is how to generate and control the SAEQ equalization signal. The following aspects must be considered to determine the duration of the SAEQ pulse:

1. In order to maximize the effectiveness, it is necessary that the equalization continues until the addressed cell draws the correct current; in other words, the comparator must be unbalanced immediately in the correct direction;
2. The equalization must not last more than necessary to avoid penalties in terms of time;
3. The nodes of the converter must have sufficient time to reach the required voltage.

From the foregoing considerations it can be determined that the lower limit of SAEQ is given by the time necessary to bias the cell correctly, in terms of both gate and drain voltage. The drain bias is usually faster than the row bias since the voltage is applied to a metal stripe that does not suffer from the typical RC parasitic effects of the polysilicon word line.

Let's now examine some of the solutions that have been used to generate the SAEQ. Herein, we will discuss the falling edge of SAEQ, since the rising edge is determined by the ATD. The simplest solution is the realization of a delay chain that, starting from the ATD signal, determines the duration of the SAEQ. Once the time constant of the word line has been calculated, it is possible to evaluate the performance of the sense amplifier through computer simulations and establish the optimal value of SAEQ. Such a solution has a serious drawback: it is not possible to automatically follow the variations of the technological process for the fabrication of the memory. The time to charge the word line is directly related to the resistivity of poly2. As the deposition of polysilicon having resistivity of a few Ohms/square is one of the most critical process steps, it follows that the deviation of such a parameter is not negligible. The solution with the delay chains is not flexible from this point of view, since hardware intervention, requiring the modification of one or more masks, are necessary to obtain proper variations. Moreover, the duration of the delay is tuned according to the minimum temperature and maximum voltage condition to have enough margin in any operating conditions. In this way, at the opposite corner of the operating region (high supply voltage and low temperature) the time duration is always longer than necessary, with a considerable waste of time.

A more flexible solution is the usage of a dummy row, perfectly matched to the rows in the array but not belonging to the normal addressable space. Each time a new ATD is generated, the dummy decoder and row are activated in parallel to the path in the array. The end of the dummy row is connected to a comparator that detects when a given percentage of VDD is reached. The voltage is measured at the end of the dummy row to emulate the behavior of the array cell located farther from the row decoder. At this point the SAEQ signal can be generated by means of a set-reset flip flop; the ATD operates on the set, the comparator output determines the reset and, hence, the end of the SAEQ pulse.

Let's now examine how to modify this structure in the case of devices with boosted read voltages.

11.4.1 Word Line Overvoltage: One Shot Boost

Let's start with a device with pulse-boosted word line, as presented in Chap. 10. In this case, the RC parasitic of the row is not the only important element to determine when the cell starts sinking the correct current: it is also fundamental to account for the correlation with the boost of the word line. Recalling the example of Chap. 10, we must select the node called BULK as reference, which is boosted to $2|V_{T,p}|$ above VDD.

Let's design a circuit that detects the boost condition on BULK and, as a consequence, pulls SAEQ to ground. The signal obtained is active (high) only for the bare minimum time. The SAEQ rising edge is synchronous with the ATD. The achievement of the full boost value on the BULK node is detected by means of the derivator circuit shown in Fig. 11.9.

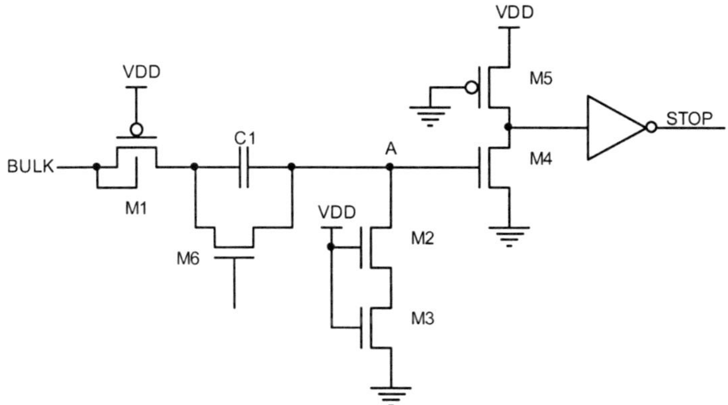

Fig. 11.9. The derivator circuit used to detect when the potential of the BULK node equals VDD plus the $|V_{T,p}|$ of M1. STOP is used to end the SAEQ pulse

The BULK potential is normally VDD and, hence, M1 is off, the A node is pulled to ground by the M2 and M3 transistors, and STOP is low. When BULK is pulled above the value of VDD by a threshold voltage, M1 starts conducing in spite of M2 and M3 that tend to pull A to ground, provided that the ratio between M1 and the series of M2 and M3 is properly sized. M4 is on and the STOP node is pulled to the high logic state. After the input step has finished, the A node is discharged with a time constant that is given by C1 multiplied by the series resistance of the channels of M2 and M3, thus restoring the original conditions. During the discharge transient of the BULK node, C1, which is necessary to decouple M2 and M3 from the boosted node in DC, causes a boost of A below ground that might forward bias some junctions, with the consequent risk of latch-up.

In order to reduce the undershoot of V_A, it is possible to add a transistor (M6) to short circuit the capacitor terminals, thus redistributing the charge of C1 and diminishing the difference of potential during the discharge.

11.4.2 Word Line Overvoltage: Charge Pump

In the case of a single supply voltage, all the voltages above VDD must be generated inside the device by means of charge pumps. The main problem with such circuits is the limited capability of sourcing current that barely amounts to a few milliamps. If very stable voltages are needed, it is necessary to insert a voltage regulator between the pump and the downstream circuitry. Since the power consumption of the regulator and the row decoder during the toggling is very high, there is no place for other circuits. With such considerations it is clear that the previous simple solution for the timing of SAEQ, i.e. the row and the dummy comparator biased with the same read voltage, cannot be applied.

The foregoing problem has a fundamental importance in case of multilevel devices in which reading is carried out at voltages higher than 5 V, the supply voltage being 3 V or less. In fact, the word line potential must be determined very precisely due to the small separation between contiguous distributions of cell threshold voltages. The solution to the problem is based on the consideration that the transient of the word line does not depend upon the absolute value to reach but merely upon the associated time constant. In Sect. 9.2 we stated that a C_{wl} and a R_{wl} can be associated to the word line under the hypothesis of lumped model. The voltage of the word line, V_{WL}, can be calculate as follows:

$$\frac{V_{WL}}{V_{READ}} = 1 - e^{-t\left(R_{wl}C_{wl}\right)^{-1}} \tag{11.1}$$

In other words, the time that is necessary to reach a given percentage of the read voltage does not depend upon the voltage itself. Therefore, if one wants to detect the instant at which two time constants have elapsed, it is possible to bias the dummy circuitry at VDD, as results in Eq. (11.1), instead of V_{READ}.

Unfortunately, this does not suffice for multilevel reading, the precision of which demands to know the exact instant when the word line fully reaches V_{READ} (e.g. 6 V). This means that the dummy comparator, biased at VDD, should detect the instant when the dummy word line reaches VDD. In this case, it is not possible to use a standard comparator that would operate out of its dynamic range. The comparator is hence modified as shown in Fig. 11.10.

The ENTIMER signal activates the timing circuit when it is in the high logic state. The SABIASN signal is analog (it could be the voltage generated by the band-gap reference) and is used to bias the current source of the comparator. The M3 transistor is used to bias the comparator output to the high logic state when the circuit is off to avoid power consumption. During the circuit operations, the potential of the FOLLOWER node equals VDD and biases M1 that has its gate connected to the signal coming from the dummy word line. M1 operates in follower configuration and transfers the signal from the gate to the source, shifted by a threshold voltage. The M2 transistor biases M1 allowing it to operate as a source follower.

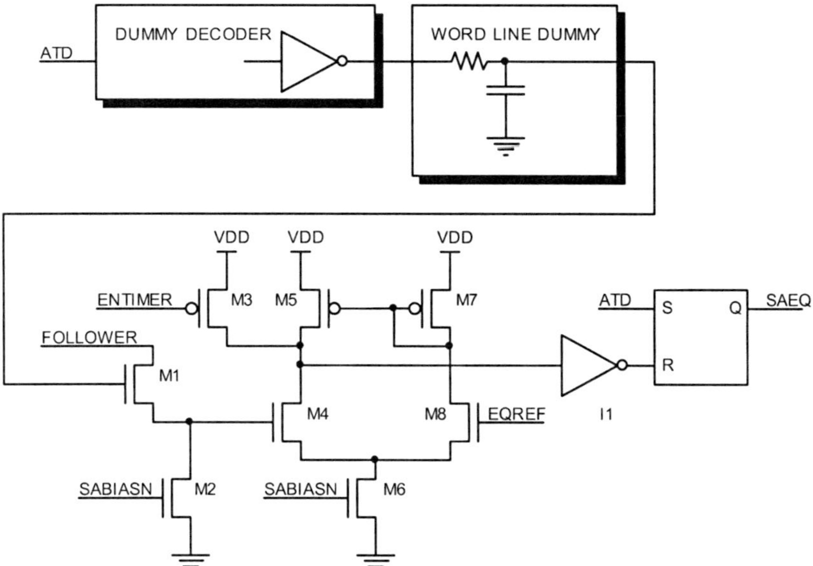

Fig. 11.10. Timing path for multilevel memories

The EQREF signal is generated in a similar way by means of M15 and M24 (Fig. 11.11). Also M15 operates as a source follower, thus transferring the reference signal shifted by a threshold voltage to the comparator. V_{REF} is obtained by partitioning VDD and can be tuned to the required value (e.g. 90 ÷ 95% of VDD). Since we use a differential pair to detect when the dummy word line reaches the steady state value, it is necessary to realize the circuitry as symmetrical as possible to minimize any offset. In particular, M2 and M24 must be identical as well as M1 and M15. The I1 inverter is used to convert the comparator output to logic levels.

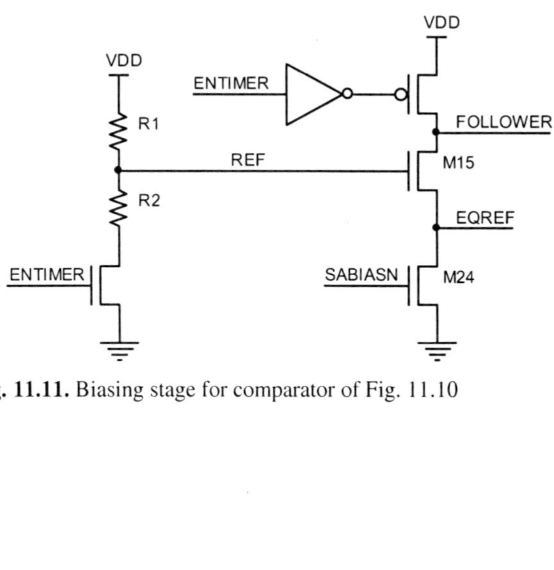

Fig. 11.11. Biasing stage for comparator of Fig. 11.10

11.5 The ENDREAD Signal

From the input pads to the equalizers of the sense amplifiers, the path of the signals is comprised essentially by logic gates and RC delays. The determination of the memory cell content by the sense amplifier is, instead, an analog operation, the success of which is strongly dependent on the cell, VDD, temperature, initial bias of the internal nodes, disturbs, and so on. On the other hand, it is important to have a signal, which will be referred to as ENDREAD, to determine when the sense amplifier has completed the discrimination phase between "1" or "0". In fact, as we will see in Chap. 19, to prevent noise induced by the switching of the output pads from propagating through the internal nets, it is better to latch the data before driving the output pads. Moreover, if an internal end read signal is available, it is possible to restore the initial precharge condition of all nodes before the subsequent read operation.

The chapter dedicated to the sense amplifier will show how its behavior depends on a large number of factors. How can we predict the settling time in order not to waste time? The technique typically used in these cases is based on the use of dummy elements. In other words, an external path to the array is designed so as to reproduce the worst-case behavior during reading.

The first problem is the definition of the dummy read element. It is not possible to know in advance whether a "1" or a "0" will be read. Therefore, it is necessary to use two dummy paths, one with an erased cell, and the other with a programmed cell. Therefore, the two dummy paths will be read and the result will be combined, so as to obtain a signal that becomes active only when both the read operations finish. This is the ENDREAD signal.

How can we determine when the two read operations have completed? In this case we know the result in advance, since the value of the threshold voltage of the dummy cells is predefined during the testing phase. Before reading, the comparators of the two dummy paths are forced to the complementary logic state with respect to the final value, as shown in Fig. 11.12.

The two dummy sense amplifiers have the same circuit structure as the real sense amplifiers, apart from the transistors that are required for the initial bias of the output node. The ATD signal triggers the dummy read path and the circuit blocks are as similar to the real ones as possible, including the layout, so as to reproduce the same parasitic load. The output of the sense amplifier (SAOUT) that reads the erased cell is initially high, and goes low after the equalization phase. The output of the sense amplifier that reads the programmed cell, which is initially low, is pulled to the high state (Fig. 11.13).

Finally, the outputs of both the sense amplifiers are routed to a logic block where they are combined, generating a pulse that is latched in a flip-flop and then transferred to a mono-stable latch that produces a new pulse having longer time duration. In this way, a suitable time duration is guaranteed to ENDREAD, since the signal must route across the chip and must not be filtered due to parasitic capacitance.

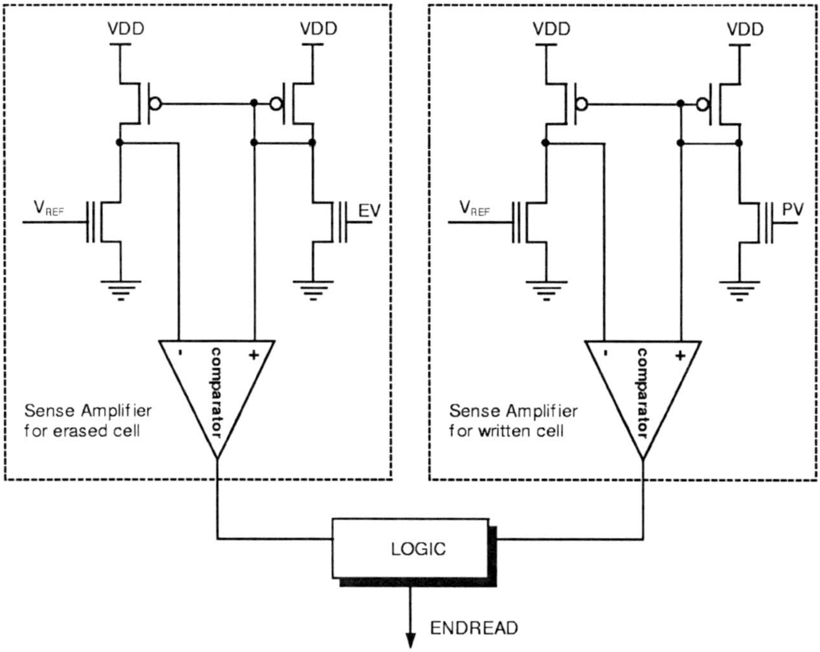

Fig. 11.12. Two dummy sense amplifiers generate the ENDREAD signal

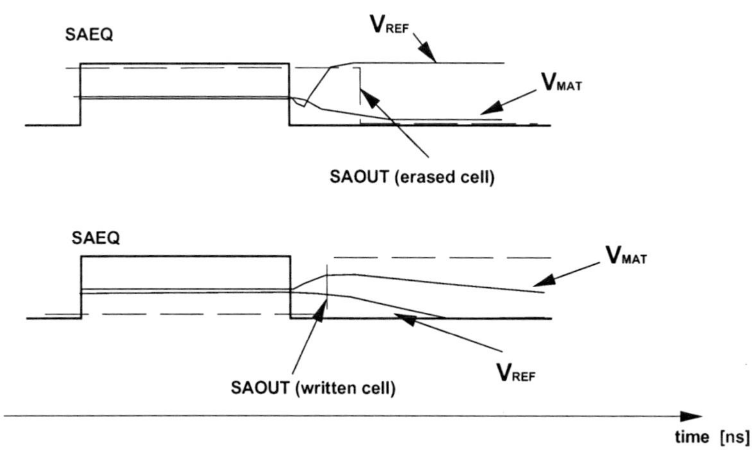

Fig. 11.13. Each time the array is accessed, the dummy sense amplifiers read cells whose V_T is already known

11.6 The Cells Used by the Dummy Sense Amplifiers

The cells used in the dummy sense amplifiers to generate the ENDREAD signal have the fundamental importance to track the actual read operation. The ENDREAD signal causes the sampling of the output of the read sense amplifier. If this operation is not executed at the right moment, the read operation can be incorrect. The ENDREAD signal controls the timing chains that have already been discussed, and concludes the boost phase, resetting the nodes to the proper voltage value before the subsequent read. Finally, ENDREAD enables the output buffer to present the data on the output pins of the memory. In order to correlate ENDREAD with any generic cell of the array, such a signal must be associated with the commutation of the slowest programmed or erased cells, so as to represent the worst case transition. ENDREAD must always arrive together with the signal of the comparators of the array or later, to insure correct reading.

We can use the cells that are already present in the reference array (often referred to as small array)[2], which are used in erase verify and program verify mode. The erase and program phases are followed by a read phase, called verify phase, to check whether the operation has correctly been carried out. This read operation is executed with more strict sensing to insure a certain margin, and this is accomplished by means of a specific reference as we will see in Chap. 12.

Therefore, the cell used as a reference in the erase verify phase, called EV, represents the cell having the worst erased V_T, i.e. the one that is able to sink the minimum current during the read phase. All the erased cells have threshold voltage lower than the EV cell. Similarly, the cell used for the program verify (PV) phase, represents the worst programmed cell, i.e. the one with the lower V_T. All the programmed cells have higher threshold voltage than PV. In this case, EV and PV have V_T of 2.5 V and 5 V respectively. The cells that are read by the dummy sense amplifiers are the verify cells. This is the best choice to account for the variation of the read time of the cells of the array. The cells of the dummy sense amplifiers must be read like the cells of the array to reproduce the worst case conditions during the read operation. The last detail that must be clarified is related to the load of the drain of the dummy cells. In order for the dummy path to be as close to the real one as possible, a column of the array must be connected to the dummy cells during the read phase.

11.7 ATD – ENDREAD Overlap

What happens if a new ATD arrives during the ENDREAD pulse (Fig. 11.14)?

In our case, the first ATD triggers all the event sequence that ends up with the ENDREAD generation. Suppose that the address changes before the time necessary to satisfy the access time specification has been achieved, and a new ATD is

[2] The small array is external to the actual array and contains reference cells that are programmed or erased to the required threshold voltage value during the testing phase. We will go into detail about this in Chap. 12.

generated and superimposed on the ENDREAD. It is mandatory that no ATD transition is missed because it could correspond to the real address being applied. This is the reason why the conflict must be resolved in favor of the ATD. In the structure that controls the timing of the double path to activate the boost signal as shown in Fig. 11.6, the ENDREAD signal is not applied to the RESET of the FF1 and FF2 flip-flops. In the case the ENDREAD pulse includes a complete ATD pulse (ATD1 or ATD2), we would miss the ATD and, as a consequence, the possibility of reading. The problem persists for the FF3 and FF4 flip-flops in the case where ENDREAD is superimposed on ATD1 or ATD2, and we can only rely on the delay introduced by the dummy chains and the comparators.

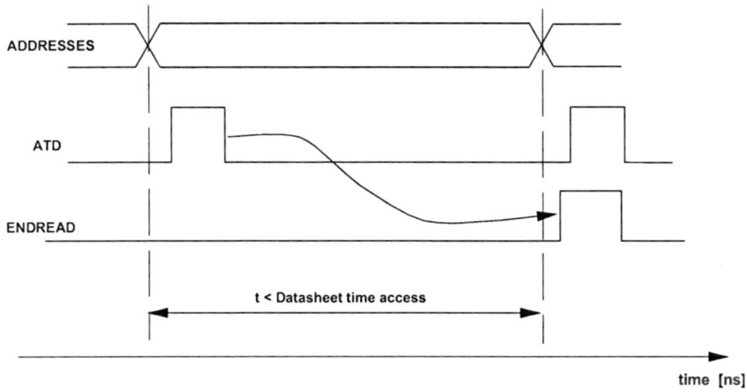

Fig. 11.14. What happens if a new ATD signal arrives when ENDREAD is active? Such a condition occurs if the address changes before the access time has elapsed.

In the condition of maximum supply voltage and minimum temperature, the propagation might be so fast that the ATD1 (or ATD2) could arrive at the input of FF3 (or FF4) when the flip-flop is still in the reset state. To overcome such a problem, the ENDREAD signal, which operates on FF3 and FF4, is conditioned so that a new ATD can pull it down immediately, and the read can restart in any situation.

11.8 Sequential Reads

The time necessary to retrieve code or data from the memory is one of the basic parameters that system designers take into account when they choose a specific memory device. With the structure of the memory we have examined in the previous text, which requires a read cycle triggered by the ATD signal for each access, increasing system performances means reducing the access time of the memory. This operation has an important impact on the design of the entire device, including the array organization, decoders, sense amplifiers, and so on, with direct effect on size, power consumption, and device cost. The Asynchronous Page Mode and

Synchronous Burst Mode are two modes that provide a tremendous increase in terms of system performances without requiring severe reduction of the access time to the memory. Such techniques are based upon a few considerations about the system that will be briefly detailed later.

Studies and analyses on the correlation of the access to the memory by a large number of different processors that execute a large variety of programs have pinpointed that accesses to contiguous locations are very likely. Intuitively, this is due to the fact that programs are translated into sequences of instructions, stored in neighboring memory locations. This also holds true for both elementary and composed data structures. Furthermore, it must be considered that a large part of the execution time is spent on a limited number of code cycles or functions (about 90% of the time spent on 10% of the code). The foregoing observations, together with the fact that the cycle time of modern processors has been reduced to a few nanoseconds whereas the access time to an external memory is around some tens of nanoseconds, give origin to system architectures composed of several levels of memory. Typical structures combine a Flash memory of large size, slow and not expensive, with a faster memory. The entire code is copied from the Flash to the internal fast memory, before being executed. In other configurations, one or more levels of very fast cache memories are also present. The inclusion of a cache memory strongly affects the way the processor accesses the memory and the way in which the external memory is used. A large number of configurations of caches have been implemented in processors as well as in systems using processors. Typically, a cache memory is organized as a table that contains a given number of entries. Each entry contains a DATA field and a TAG field. The DATA field is made up of N contiguous words that are aligned with respect to the memory address (page). The TAG field contains the information about the address to which the data are associated, along with several flags related to the status of validity of the data stored. Without delving into implementation details, it is important to describe the paged organization of the cache memory. When the data that are required by the processor are not present in the cache, the Memory Management Unit (MMU) generates a signal of "cache miss", providing the selection of one of the pages of the cache that must be substituted for a page coming from the external memory. This means that each access to the external memory is actually carried out by means of a sequence of accesses. Different types of processors adopt different techniques to update the content of the cache. The simplest procedure requires updating the page starting from the first location (aligned mode), and only at the end of the transfer is the processor allowed to continue. The most sophisticated versions demand the immediate load of the word that caused the "miss", freeing the processor from the hold state, and, afterwards, completing the page load (not aligned and wrapped mode). These reasons compel the designer to seek novel Flash architectures that provide higher throughput when sequential transfers of data or pages are required. The Asynchronous Page Mode and the Synchronous Burst Mode are two possible solutions to such a requirement. The Asynchronous Page Mode has a minor impact on the interface toward the system, the Synchronous Burst Mode, which is more complex, provides the best performances. We recall that the Asynchronous Access Mode is still the simplest read mode, and is usually the one enabled after the system boot. Therefore, the Asynchronous Page

Mode and the Synchronous Burst Mode are generally implemented in addition to the functionality of the Asynchronous Access Mode.

Starting from the architecture of an asynchronous memory, we will discuss how to transform it and implement both of these more advanced modes.

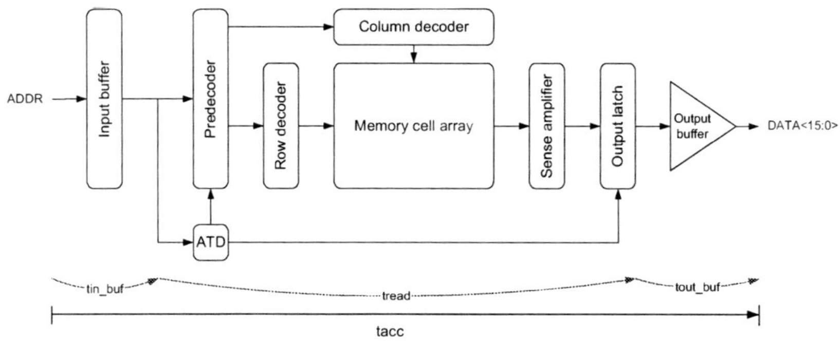

Fig. 11.15. Representation of the blocks of the memory involved in the asynchronous read mode

In the representation of Fig. 11.15, the basic blocks involved in the read phase of a typical Flash Memory (16-bit word) are shown. The input buffer, address pre-decoder, ATD block, row and column decoder, cell array, 16-sense amplifier block, output latches, and output buffers form the read path. The ATD block detects the address transitions and triggers the sequence of events. First of all, the predecoder and decoders select the memory cells, then the sense amplifiers are switched on and equalized to read, finally the output latches sample the data, concluding the read operation. The sense amplifiers are then switched off and the array is set in the stand-by condition, while the output buffers are activated to transfer the data to the output pads.

The read access time is determined by the sum of the settling time of the input buffer, the time to read the array, which is controlled by the ATD, and the settling time of the output buffers in the maximum load condition.

11.8.1 Asynchronous Page Mode

In the Asynchronous Page Mode a parallel data structure is used (see Fig. 11.16) on the read path. The memory has a paged organization in which each page contains a set of words. Each time the user requests a word, two possibilities may exist. The first possibility is that the word belongs to a different page from the previous read; the entire page is read with the standard asynchronous read flow and stored in a "page latch". In the second case, the requested word belongs to the previously addressed page; the page latch already contains the requested word that does not need to be read from the array but just selected in the page latch and driven to the output. In Fig. 11.16, the structure of a memory with Asynchronous

Page Mode Read and 4-word pages is shown. The original basic structure of Fig. 11.15 has been modified as follows. The address bits (ADDR) have been divided in two groups: the most significant bits select the page to read and are monitored by means of the ATD, so as to detect if a different page is addressed and consequently trigger the sequence of the ordinary asynchronous read. The less significant bits are used to select the word inside the page. The predecoder, and the row and column decoders select the cells that belong to the page of memory, while the four blocks (SA0, SA1, SA2, SA3) of sense amplifiers (16 for each block) allow the parallel reading of the four words that form the page. A "page latch" stores the result of the read operation and a "word selector" allows selecting the word to present as output on the device pads, according to the word address, i.e. the less significant bits of the entire address.

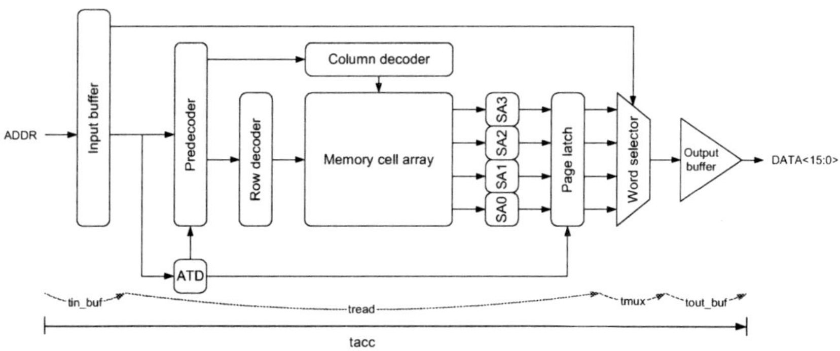

Fig. 11.16. Representation of the blocks of the memory involved in the Asynchronous Page Mode

The access time (see Fig. 11.17) to read a word belonging to a different page from the previous read is quite similar to the ordinary asynchronous read access time, since only a multiplexer is present in addition to the standard read path. The access time for a word belonging to the same page can be calculated as the sum of the settling time of the input buffer, plus the transition time of the output selector and output buffer.

The Asynchronous Page Mode does not require any specific device configuration and is totally compatible with the standard asynchronous access mode.

The increment in terms of speed obtained by means of the Asynchronous Page Mode with respect to the standard asynchronous access is evident. Two remarks should be made. First, the memory core is active only during the page switch and idle during the transfer of the single words. Secondly, the Asynchronous Page Mode allows accessing the data in the page buffer in any order whatsoever.

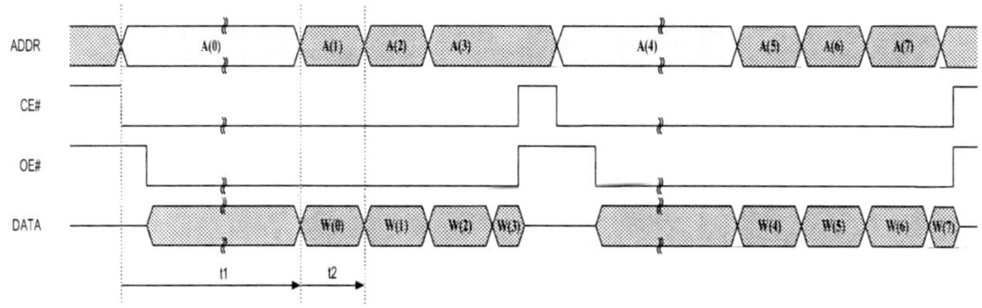

Fig. 11.17. Access to the memory in Asynchronous Page Mode. Access time in case of page switch (t1). Access time without page switch

11.8.2 The Synchronous Burst Mode

The Synchronous Burst Mode permits a further increase in the data transfer rate with respect to the Asynchronous Page Mode. In this read mode, synchronization between memory and controller is introduced, like a pipeline architecture, and the additional constraint of burst transfer is imposed. The synchronization is generally obtained through a clock signal (the rising edge of CLK in the following examples) and an activation signal (Address Valid – ADV#) that must be provided by the system. In this context, burst means that a sequence of data is presented as output of the memory with an order that is a function only of the starting address. This method eliminates the application of external addresses for all the subsequent data accesses.

In burst mode, the starting address (ADDR) is latched on the first active edge of the clock when ADV# is active. In these examples the starting address is A(1). After a time equal or greater than the access time and corresponding to L clock cycles (latency time), the W(1) data corresponding to A(1) is output and can be sampled by the controller on the L clock edge. The subsequent data are sequentially sampled on the successive edges. There are different ways by which data can be output, and they are reported in Figs. 11.18, 11.19, and 11.20:

- Synchronous Continuous-Word Burst Read Mode (Fig. 11.18). In this mode, the burst duration is not pre-defined. After the first data corresponding to the address is output, the memory continuously outputs data corresponding to the subsequent addresses, until the entire content of the memory is read.
- Synchronous N-Word Wrapped Burst Read Mode (Fig. 11.19). In this mode, the burst length is N words and the memory can be regarded as organized in N-word pages. The starting address specifies the page and the word inside the N-word page that one wants to obtain as first output. If the first required data are the first word of the page (aligned mode), the N words of the page are sequentially output. If the first data required do not correspond to the first word of the page (misaligned mode), once the end of the page is reached, first subsequent data is the first word of the same page (wrapped mode). The dimension of the burst page, constituted of N words, does not necessarily equal the dimension of the physical page of the device, as we will see later on.

- Synchronous N-Word Not Wrapped Burst Read Mode (see Fig. 11.21). Differently from the previous mode, the N words of the page are all consecutive to the first requested word.

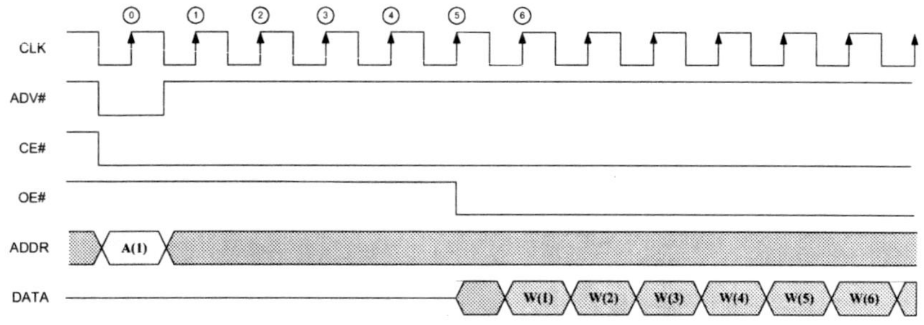

Fig. 11.18. Synchronous Continuous-Word Burst Mode Read with six latency cycles (6,1,...,1)

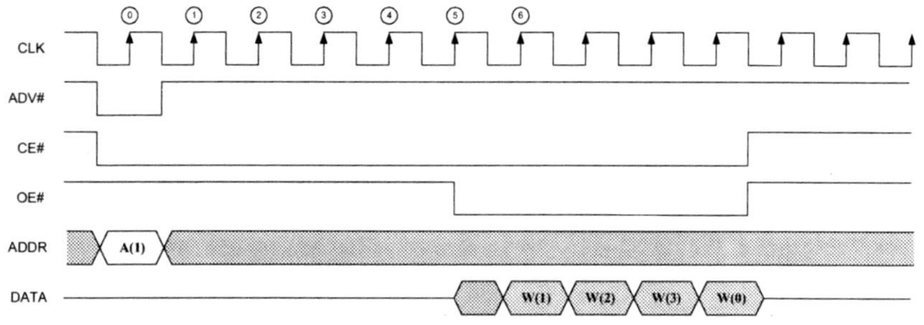

Fig. 11.19. Synchronous 4-Word Wrapped Burst Mode Read with six latency cycles (6,1,1,1)

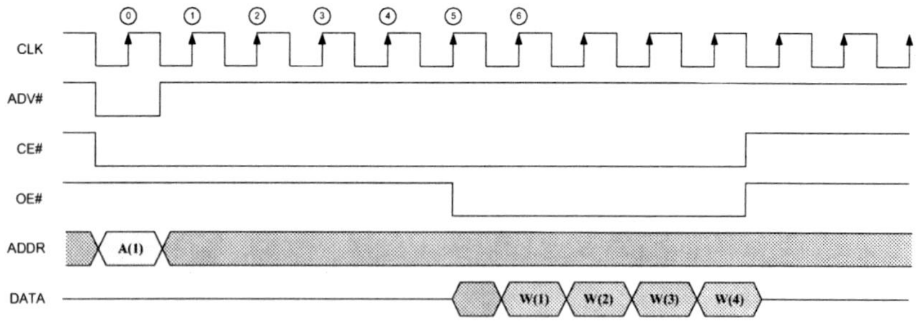

Fig. 11.20. Synchronous 4-Word Not Wrapped Burst Mode Read with six latency cycles (6,1,1,1)

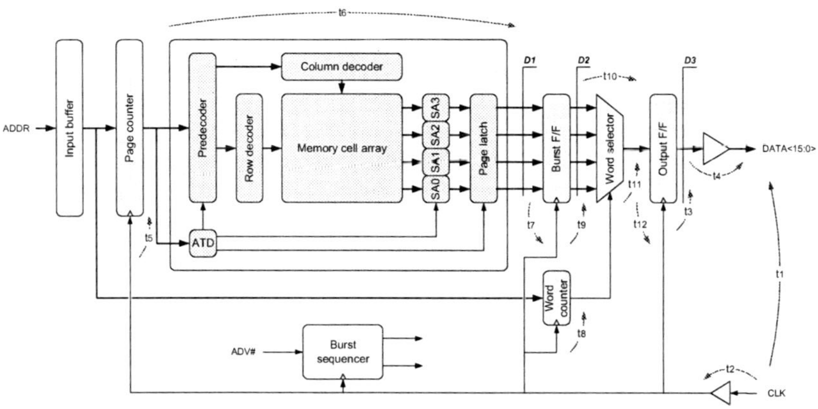

Fig. 11.21. Representation of the memory and basic blocks involved in the burst read

To illustrate an architecture that can realize these operation modes and show the basic parameters involved in these designs, it is better to start from the analysis of a simple solution to the Synchronous Continuous-Word Burst Mode Read. In view of the high throughput required with respect to the basic mode, it is easy to figure out the necessity for a parallel structure similar to the one used for the Asynchronous Page Mode Read.

Since the operations of the memory must by synchronized with the external processor, the external clock must be plugged into the memory through a suitable buffer. With reference to such a clock signal (Fig. 11.22), the device specifications impose a minimum clock period (T), and a minimum time between clock edge and valid data (t1), so as to guarantee a sufficient margin for the setup of the input flip-flops of the processor and for the propagation delay of the signals on the board (ts).

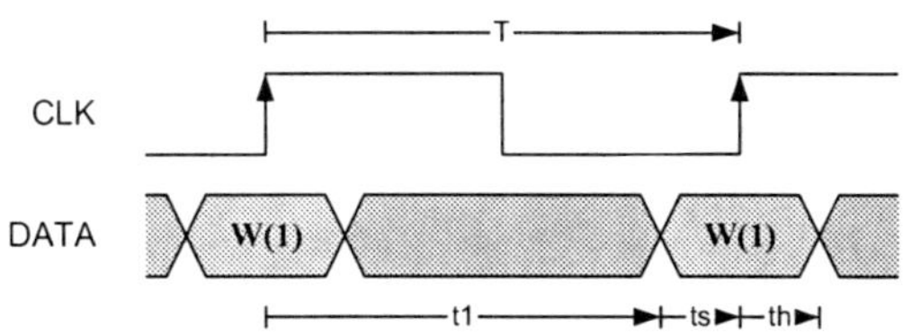

Fig. 11.22. Timing diagram of clock signal and data in Synchronous Burst Mode. T is the clock period, t1 is the maximum delay between active clock edge and valid output data, ts and th are the setup and hold time respectively, that are guaranteed to the user

The distribution of the clock inside the device must fulfill two requirements. The first is the minimization of the delay between the clock signal applied to the CLK pad and the data output on the DATA pads and sampled with respect to the system clock, so as to obtain the maximum transfer speed in burst mode. The second requirement is that the clock must reach each flip-flop of the burst control logic with the same delay. To fulfill both requirements, clock buffer and routing of the clock to the flip-flops of the output buffer are optimized so as to minimize the delay. Moreover, a reduced clock tree ensures the minimum clock skew on the flip-flops of the burst control logic. The delay between the two clock distributions at the interface between control logic and output buffers must be accurately considered. The output buffer represents a very important element of the device in terms of access time, area occupation, and power consumption. In order to optimize such a block, it is necessary to assign a major part of the cycle time to it, by defining an architecture with a first level of pipeline included in the output buffer.

With reference to Fig. 11.21, we find t2, t3, and t4 on the path between CLK and DATA, where t2 represents the propagation delay of the clock from its pad to the farthest flip-flop of the output buffers, t3 is the time to present the data on the flip-flop output calculated with respect to the active clock edge, and, finally, t4 is the propagation delay of the output buffer with the maximum output load allowed by the specification. Named T the clock period and t1 the time between CLK and valid DATA, the following relations must be fulfilled:

$$t1 > t2 + t3 + t4 \tag{11.1}$$

where

$$t1 + Tsetup\ \mu < T \tag{11.2}$$

Let's now focus on the input of the addresses applied by the user. These addresses must be latched on the active clock edge when ADV# is active, and incremented during the execution of the burst read. The address counter, made of flip-flops, is divided in two parts: page and word counter. The page counter, initialized by ADV#, provides the selection address to the decoder and ATD. Each time this counter is updated, the reading of the selected page is automatically triggered. The word counter, which is also initialized by ADV#, provides the selection address of one of the words contained in the burst buffer. After the address input and data output section of the memory has been established, the next area of focus is on the memory core. In order to guarantee constant data flow during the continuous burst with period T, and implementing a core with access time t6 > T, it is necessary to use a parallel architecture of sense amplifiers and latches, to be able to read and store an entire page of data containing N words. The data that are output by the memory core at the end of the read operation are latched in a bank of flip-flops (burst FFs) to allow the execution of the subsequent read operation. The degree of parallelism that is necessary for a continuous burst is given by the following relation:

$$N \cdot T > t5 + t6 + t7 \tag{11.3}$$

where t5 is the delay in updating the address applied to the core with respect to the clock edge, t6 is the access time to the memory core from the application of the address, t7 is the setup time of the bank of flip-flops, and N is the number of words that must be read simultaneously.

If N is the number of words that must be read simultaneously (N = 4 in the example of Fig. 11.21), it is possible to size the Word Counter that, by means of a selector, addresses one of the N words to send to the flip-flops of the output buffers.

Problem 11.3: Try to identify what and how many time relationships involving the burst buffer and the Word Counter must be fulfilled for a correct burst operation.

The operation flow in the pipeline, i.e. the control of the latency time, and the increment of Page and Word Counter, are managed by a Finite State Machine (FSM) called "Burst Sequencer".

The Burst Sequencer is a synchronous FSM whose clock is CLK (Fig. 11.23). The GOSEQ signal, derived from ADV#, as we will see later on, operates as an asynchronous reset for the FSM and as an asynchronous load for the page and word latency counters. The combinatory logic of the FSM receives the MAX_WORD signal, active when the word counter reaches its maximum value, and the MAX_LAT signal, active when the latency counter reaches zero. In the described structure, all the outputs of the FSM are latched and the following control signals are generated: DEC_LAT to decrement the latency counter LAT, INC_PG_CT to increment the page address and, at the same time, recharge the LAT latency counter, INC_WR_CNT to increment the Word Counter. The functioning of the FSM is completely described through its state representation, as depicted in Fig. 11.24.

The timing diagram (Fig. 11.25) of the main signals of the system helps demonstrate the functioning of the burst machine and the progressing of the operations.

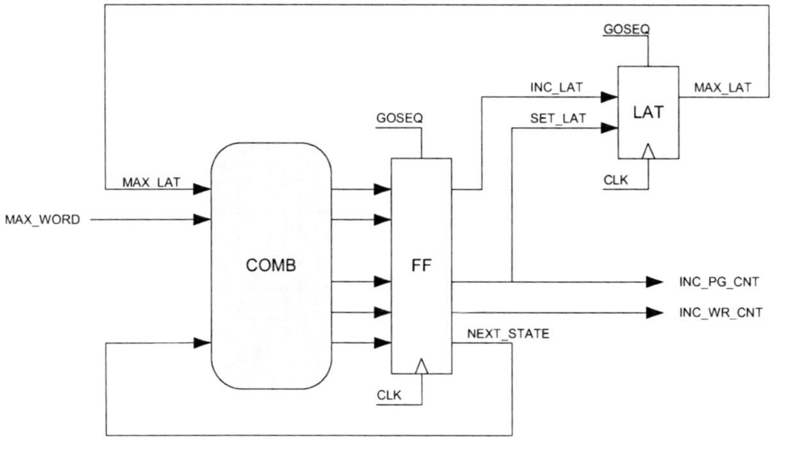

Fig. 11.23. Block diagram of the Burst Sequencer

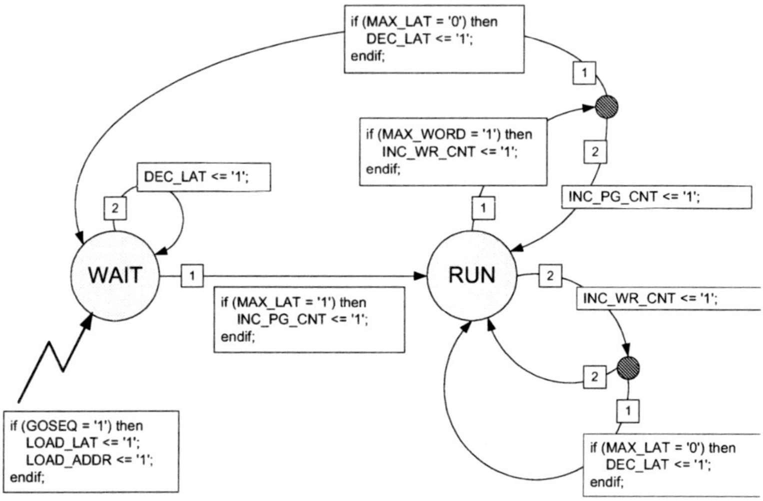

Fig. 11.24. Representation of the functioning of the sequencer state machine

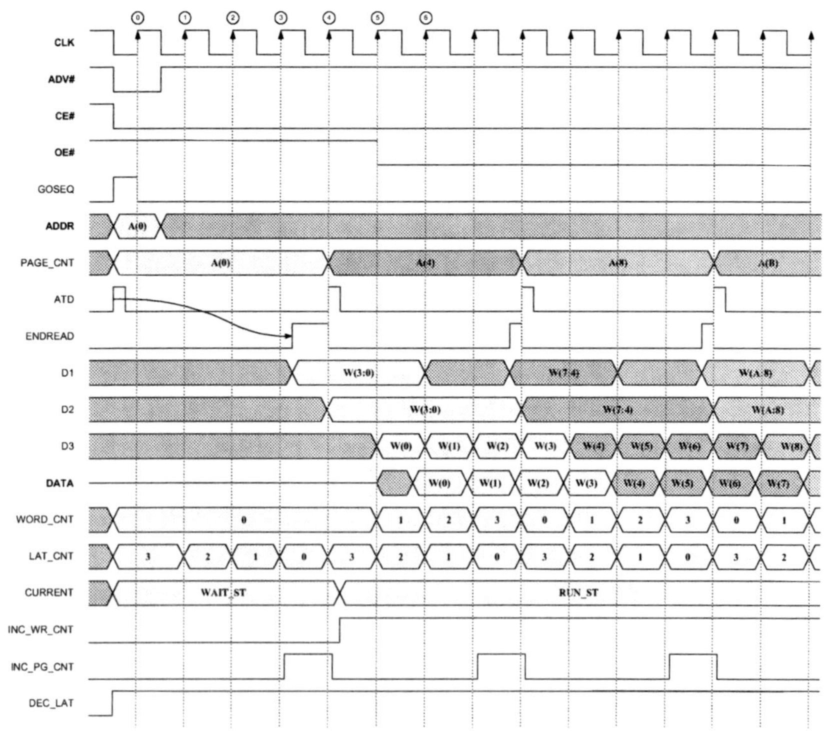

Fig. 11.25. Representation of the main signals involved in the Burst Read Mode

The burst sequence starts with the activation of the ADV# signal that triggers the internal GOSEQ signal. This signal, which controls the preset of the counters and the start of all burst operations, is activated as soon as ADV# becomes active and remains active until the first active clock edge. In this way, when ADV# is activated, the memory core can receive the new address without waiting the active clock edge, thus avoiding unnecessary delays.

Problem 11.4: Try to realize the circuit that generates GOSEQ.

An update of the Page Counter triggers the ATD signal that regulates the entire read operation. After the t6 time necessary to read the array has elapsed, the Page Latches are updated by the asynchronous ENDREAD signal (D1). At the same time, the latency counter, LAT, initialized to 3 by GOSEQ, is decremented on each clock edge, and, thus, it becomes zero after the third clock edge. The sequencer increments the Page Counter (INC_PG_CNT) on the subsequent clock edge (4) and recharges the latency counter (SET_LAT). An update of the page address disables the page latch, activates a new ATD pulse, and, hence, starts a new read cycle. On the forth clock edge, the data of the read page are stored in the burst flip-flops (D2), and the Word Counter is activated. During the forth clock pulse, the first word of the page is selected and each bit is sent to the corresponding output buffer that stores it on the fifth clock edge (D3). The data are propagated through the output buffer to the pad, and the controller must sample them on the sixth edge. At this point, the pipeline is completely full and can output data on each clock pulse. If the first word that is to be read is not aligned to the page (i.e. it is not the first word of the page), the buffer of the pipeline, constituted of the burst flip-flops, is enabled to request the update of the content after only two cycle, when the core of the memory is not yet ready. To correctly manage this event, the sequencer must be able to suspend the sampling activity of the external processor by means of the WAIT signal (Fig. 11.26), which the processor samples on the same clock edge used to sample the data or on the previous one.

Since the WAIT signal is an output that has its own flip-flop and output buffer, the INT_WAIT control signal must be generated sufficiently in advance by the FSM.

Problem 11.5: Try to modify the FSM (Figs. 11.23 and 11.24) to correctly generate the INT_WAIT signal.

In some cases, this reduction in performance can be avoided by introducing a modification in the structure of the column decoder.

In the case where a row contains eight words, four of which are simultaneously selected to be read, typically (Fig. 11.27) the least significant bit of the page address discriminates between even and odd pages. In our example, AX<2:1> select the word in a page and AX<3> discriminates between even and odd pages. With this organization, the device presents all the words of either even or odd pages to the sense amplifiers, creating, in the case of non-aligned burst, the problem described above.

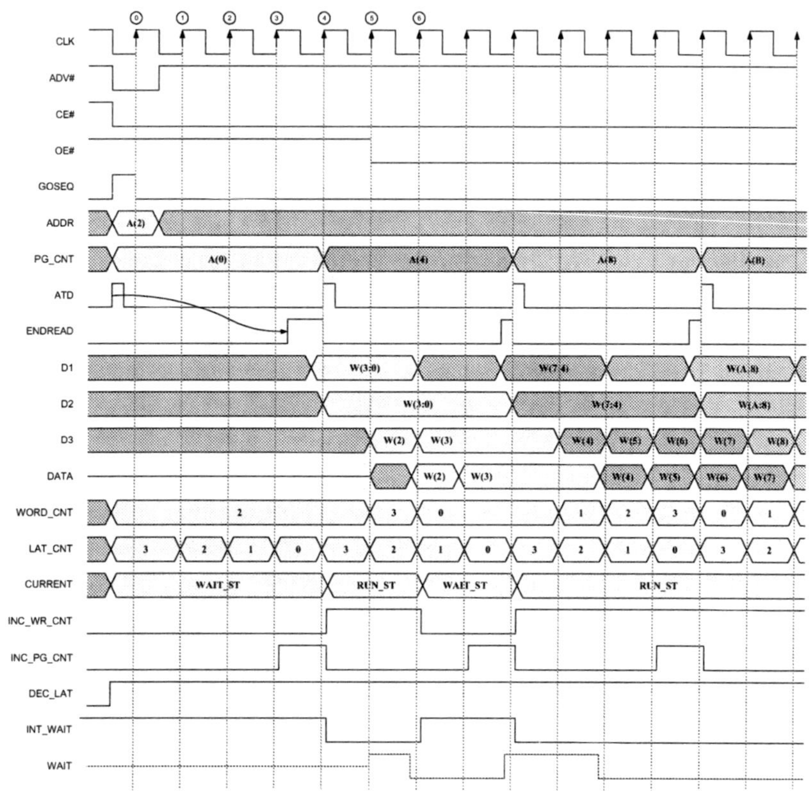

Fig. 11.26. Representation of the main signals involved in a Continuous Burst Read Mode with non-aligned starting address

The decoding structure called "barrel shift" (Fig. 11.28) reduces and, in some cases, overcomes this problem by using a predecoding logic that takes into account also AX<2:1> beside AX<3> latched in a register (AXS<2:1>) at the beginning of the burst sequence. The decoder is arranged so that a contiguous group of word is enabled to be presented to the sense amplifiers. The output selectors continue operating with AX<2:1> coming from the word counter. Such a method avoids the introduction of wait cycles as long as the sequence does not require a change of row. In this case, the WAIT signal must be handled by the sequencer, and AXS<2:1> must be reset.

The advantage of this new type of decoder is much more evident in the Synchronous N-Word Wrapped Burst Mode Read, in which the dimension of the page of the user is a multiple of the device page but a portion of the row length.

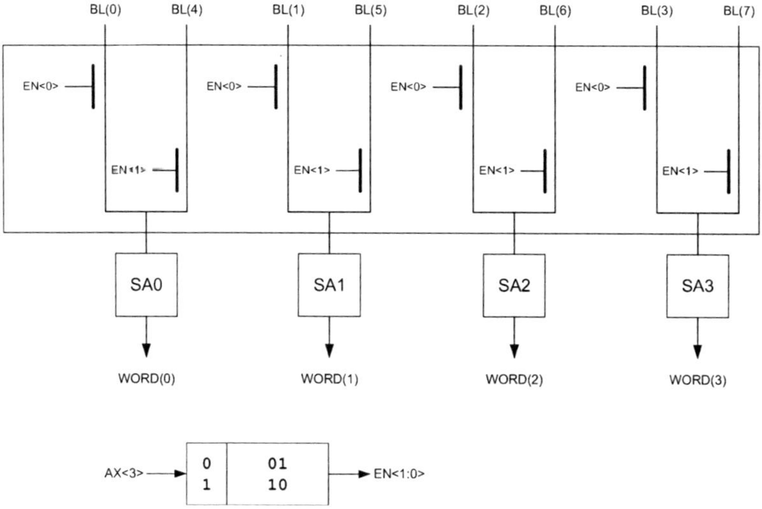

Fig. 11.27. Standard column decoder

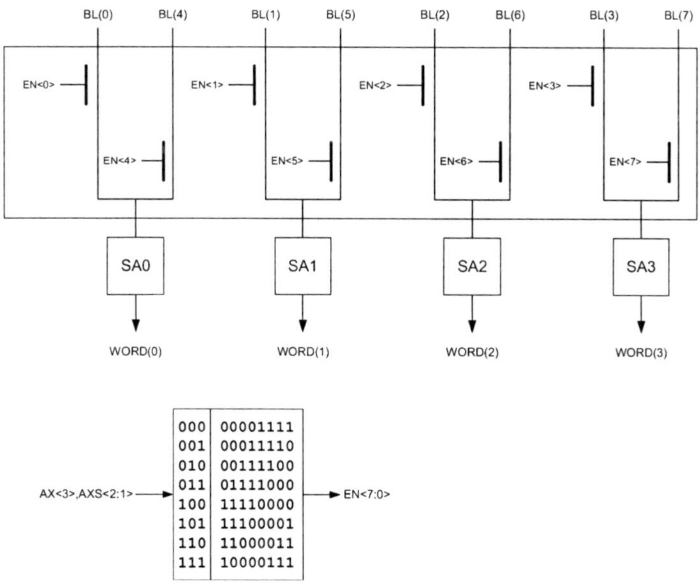

Fig. 11.28. Column decoder with barrel shift

The structure of the memory described implements the Synchronous Continuous-Word Burst Read Mode. The Synchronous N-Word Wrapped/Not Wrapped Burst Read Mode can easily be obtained by simply operating on the burst sequencer.

For the Burst Page Read Mode, in the hypothesis that the length of the burst is K, the sequencer must only increment the counters up to a maximum of K words. The Wrapped Page Mode must be handled similarly, the only difference being that the page counter must be contained in the user page.

On average, the Synchronous Burst Read Mode offers better performances than the Asynchronous Page Mode, though some points need to be clarified. In order to realize a synchronous structure, it is necessary to use a parallel structure and synchronize the operation with respect to an external clock. Having introduced a pipeline, it is necessary to subdivide the entire read operation (Dl in Fig. 11.29) into basic steps that must be separated by means of flip-flops. On the contrary, in the Asynchronous Page Mode, reading is accomplished in a single step that takes the t1 time. The blocks that are to be isolated, as shown in Fig. 11.21, are the memory core, the output buffer, and the word selector. Once the minimum clock period has been fixed according to the output buffers, and the latency of the memory core has been determined, the quantization imposed by the clock on each single operation causes an increase in the time required for the single read (D2), which is largely recovered by the transfer speed of the subsequent words (D3). Such an increment in the execution of a single reading is typical of pipeline structures that require that the operation is divided into its basic steps and sampled by means of a clock.

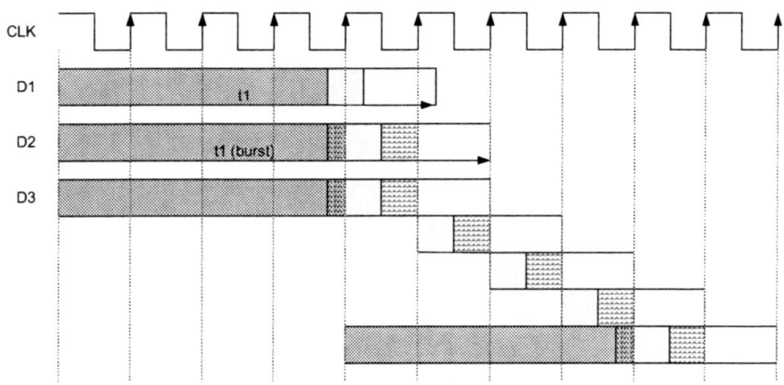

Fig. 11.29. Effect of the quantization of the single steps of the read operation: core reading, word selection, and output buffer (D1). The three phases of the sequential reading executed in Asynchronous Page Mode (D2). Effect of the quantization of the single steps of the reading by means of a clock in Synchronous Burst Mode (D3). Result of a continuous burst: after the initial latency necessary to fill the pipeline, the system outputs a word at each clock pulse

The reduction in performance becomes even more evident if the user, for any reason, cannot use a clock signal having the minimum period allowed (Fig. 11.30), since the impact of the quantization becomes more relevant. In particular, for short burst sequences, the Asynchronous Page Mode provides better system performance than the Synchronous Burst Mode. For this reason and also because of the similarities of the internal parallel structures, usually the Asynchronous Page Mode is also available in those devices in which the Synchronous Burst Mode is implemented.

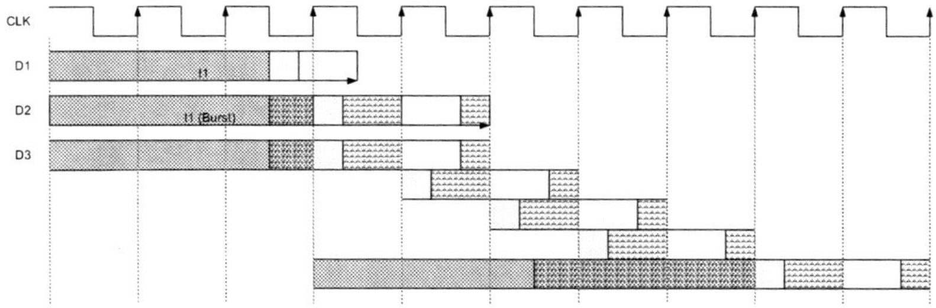

Fig. 11.30. Effect of the quantization in Synchronous Burst Mode when it is not possible to use the minimum clock period. In this case, for short burst sequences, the Asynchronous Page Read Mode guarantees better performances

Bibliography

G. Campardo et al., "Method and circuit for generating an ATD signal to regulate the access to a non-volatile memory", USA patent No. 6,075,750, (June 13, 2000).

G. Campardo et al., "Method and circuit for regualting the lenght of an ATD pulse signal", USA patent No. 6,169,423, (January 2, 2001).

H.H. Chao et al. "On the synchronization of a microprocessor", IEEE Custom Integrated Circuits Conference, pp. 447-450, (1986).

S.T. Flannagan et al., "Two 13-ns 64k CMOS SRAM's with very low active power and improved asynchronous circuit techniques", IEEE Journal of Solid-State Circuits, vol. SC-11, No. 5, pp. 692-703, (October 1986).

R. Micheloni et al., "Method and a related circuit for adjusting the duration of a synchronization signal ATD for timing the access to a non-volatile memory", USA patent No. 6,075,750, (May 22, 2001).

12 Reading Circuits

This chapter deals with the circuits used to read the information stored in the cells. The historical evolution from the EPROM to the Flash cells will be presented.

12.1 The Inverter Approach

Reading the charge stored on the floating gate of a memory cell is one of the most critical operations in a non-volatile memory device.

During the read operation, the cell source is grounded, the gate is driven by the row decoder and the drain is connected to the power rail through a load, as shown in Fig. 12.1. C_{BL} is the parasitic capacitance of the bit line; the transistors of the column decoder (M2 and M3) are dimensioned to have a negligible voltage drop on the corresponding channels. The voltage level of the cell drain during the read phase must be defined in order to avoid the phenomenon of the drain stress and, at the same time, provide a suitable current for a correct and fast read.

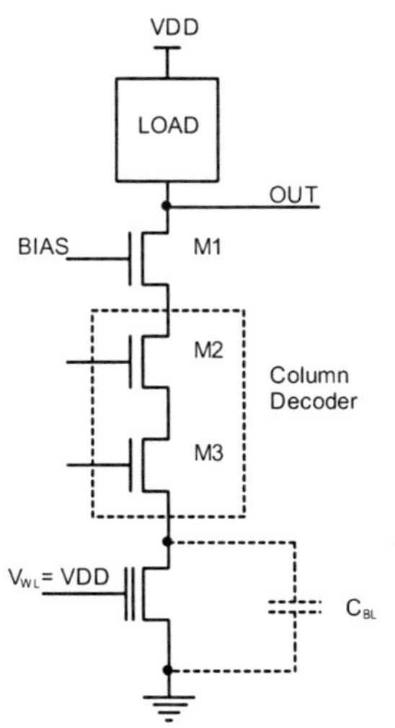

Fig. 12.1. Reading bias with a cascode stage (M1)

The drain is generally biased around 1 V by means of transistor M1. During the read phase, the gate voltage equals VDD and, hence, is greater than 1 V; as a consequence, the cells operates in linear region. As detailed in Sect. 3.4, under the same bias condition, the characteristics of cells with different threshold voltages are parallel and shifted by a quantity equal to the voltage step ΔV_T with respect to an UV erased virgin cell.

Supposing that the V_{BIAS} voltage that drives M1 is constant and equal to 2 V, let's now examine what happens in the case of a virgin cell, i.e. with threshold voltage around 2 V. Once M2 and M3 are on, there is an initial current surge that charges the C_{BL} capacitance to the value of 1 V. After reaching this value, the V_{GS} voltage of M1 is lower than its threshold voltage, also because of the body effect. As a consequence, M1 is off. In the meantime, the word line has reached and passed the threshold voltage of the cell that starts sinking current, thus discharging the parasitic capacitance and lowering the voltage of the OUT node. The written cell, instead, is not able to sink current since the gate is at VDD and the voltage of the OUT node is consequently tied to the supply voltage.

In Fig. 12.2 the voltage-current characteristics of virgin and programmed cells are reported; I_{REF} is an ideal current generator that operates as the load shown in Fig. 12.1.

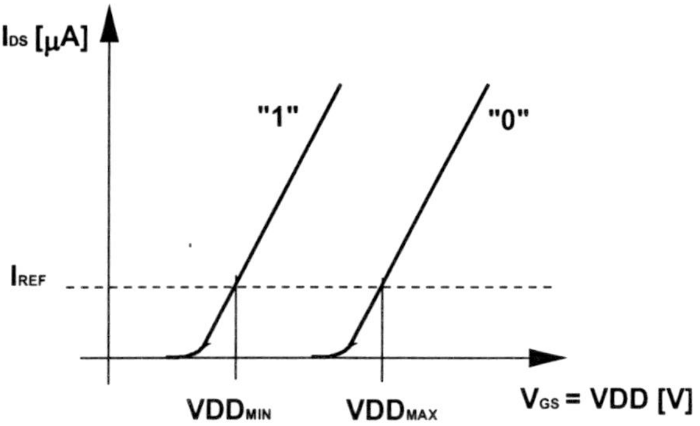

Fig. 12.2. Voltage-current characteristics of memory cells. I_{REF} is the load current generator.

In the region in which VDD < VDD_{MIN}, the erased cell is not able to sink all the current sourced by the load. V_{OUT} goes high and the cell appears as programmed. After VDD_{MIN} has been reached, the cell is biased and can sink more current than I_{REF}, causing the voltage of the OUT node to go low. The cell is then correctly read as virgin. In the case of a programmed cell, the situation is the opposite. At low gate voltages (i.e. at VDD) the cell sinks little current and V_{OUT} is high. When VDD > VDD_{MAX}, the programmed cell is able to sink more current than I_{REF}, appearing as erased.

The correct behavior of the circuit in Fig. 12.1 is thus limited to the range $VDD_{MIN} \div VDD_{MAX}$, where VDD_{MIN} is the minimum VDD, whereas VDD_{MAX} is the maximum VDD. Due to the parallelism of the characteristics of the cell in the linear region, this range is equivalent to the voltage gap, ΔV_T. The determination of the load current is not based only on the available voltage: I_{REF} must be large enough to charge C_{BL} quickly and small enough to allow the virgin cell to drive the output node low.

Let's now analyze what is the impact of having V_{BIAS} constant during the transient of the circuit described above.

We have already seen that, when the column selectors are switched on, it is necessary to charge the parasitic capacitance of the bit line, which is around 1.5 pF for 1,024 cells, before starting reading. Supposing that the initial difference of potential at the terminals of C_{BL} is zero, we have that, at the initial instant, the V_{GS} voltage of M1 equals V_{BIAS}.

In Fig. 12.3 the solution commonly used to improve the charging of C_{BL} is shown. The constant bias of M1 is substituted by a feedback loop. The SAEN# signal enables the sense amplifier. When the circuit is switched on (SAEN# low), the B node is at ground and the output of the NOR gate equals VDD. Thus, M1 has V_{GS} that equals VDD, which is the maximum voltage available to the circuit. The charging of the bit line can take place at the maximum current and in the minimum time. In order to avoid that the potential of B overcomes the limit of 1 V, it is necessary to unbalance the switching threshold voltage of the NOR gate so as to toggle as soon as the NMOS threshold voltage is reached ($V_{T,N}$ ~ 1 V). The following relation must therefore be satisfied:

$$\left(\frac{W}{L}\right)_{NMOS} >> \left(\frac{W}{L}\right)_{PMOS} \tag{12.1}$$

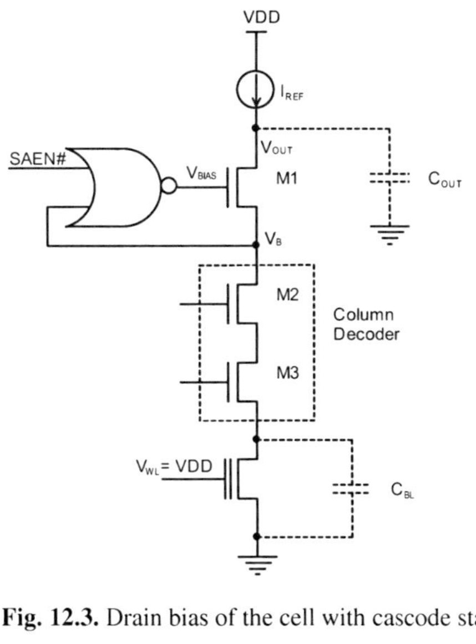

Fig. 12.3. Drain bias of the cell with cascode stage

Considering that the NOR gate in the feedback loop operates around its switching threshold voltage, in this circuit configuration DC current consumption is present. For this reason, an enable signal, SAEN, is introduced to keep the sense amplifiers "on" for the shortest time needed.

Problem 12.1: Calculate the loop gain of the feedback circuit in Fig. 12.3 and estimate its stability.

Let's finally analyze how the M1 transistor operates as a cascode between nodes OUT and B in Fig. 12.3.

M1 separates two capacitors, i.e. C_{BL}, the value of which is around a few picoFarads, and the parasitic capacitor of the OUT node. The OUT node is usually connected to a buffer that is able to drive V_{OUT} at the CMOS levels (VDD, GND), so as to allow propagation also in the case of long distances[1]. The buffer generally used is, for sake of simplicity, an inverter, and the resulting parasitic capacitance that is associated to the output node of the sense amplifier is around some tens of femtoFarads. The B node is therefore connected to a much greater capacitance than the output node. A small variation of the voltage of B, which is equivalent to a small variation of the charge of C_{BL}, results in a larger variation of the charge of C_{OUT}[2]. In this way, a sort of "amplification" of the charge is obtained, as shown in Fig. 12.4.

The basic schematic of the sensing of an EPROM cell (Erasable Programmable ROM) is based on the inverter approach previously discussed: tipically, a load transistor is used as a current source, while the memory cell operates as a pull-down.

This kind of architecture has three serious drawbacks:

1. the charge current, I_{REF}, must be dimensioned with great attention, so as to allow an erased cell to pull down the inverter output; at the same time, I_{REF} must be large enough to pull up the inverter output in a small time, so as to reduce the access time as much as possible;
2. the cell sinks little current at low VDD values and, hence, the operating range is limited. The erased cell must have a suitable V_{GS} (i.e. the word line voltage) to sink more current than the load transistor;
3. the operating range is limited at high VDD values since the programmed cell starts conducting current ($\Delta V_T = 3$ V).

Such constraints have lead to the use of the differential structures that will be presented in the following sections.

[1] The propagation of the signal is basically a charging and discharging process of the parasitic capacitance associated to the transmission line.

[2] Verify the effect that results from decoupling two capacitors of different size. Calculate the variation of the voltage of one of them as a consequence of the variation of the charge of the other.

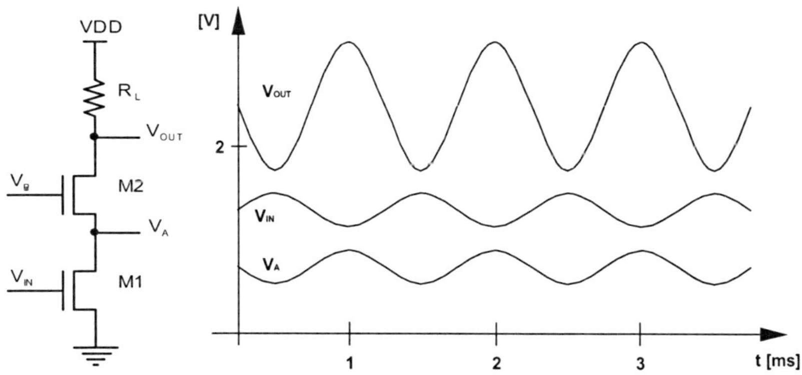

Fig. 12.4. Cascode effect

12.2 Differential Read with Unbalanced Load

The differential approach is based on the comparison between reference and real array cells. The reference cell shares the gate voltage with the array and the two resulting currents are compared to determine the status of the array cell. The advantage of this solution is that the accurate selection of the current source, with which the current of the cell must be compared, is no longer necessary. In fact, if the load transistors have variations in their conductivity, the resulting effect is common (not differential) mode, thus eliminating the dependence on the load.

In Fig. 12.5, the main schematic of a differential architecture is reported. Here we have a differential branch where the F_{REF} cell is present, whose threshold voltage is known.

In the case of an EPROM device, F_{REF} is a UV erased virgin cell, with threshold voltage V_{TUV} = 2 V. On the array side, instead, a F_{MAT} cell is present whose threshold voltage is a function of the charge stored on the floating gate.

The two cells are driven by a common gate and have the same parasitic elements. In an EPROM device, the distribution of the threshold of the virgin cells (logic "1") is very narrow, approximately ± 150 mV around the value of V_{TUV}. The distribution of the threshold voltage of programmed cells (logic "0") is wider. The most important parameter in this case is V_{TW}, the threshold of the "least written" cell, and it can be tuned by operating on the parameters that can modify the program (time, gate and drain voltages). Generally, V_{TW} is about 3 V above V_{TUV} so as to guarantee good separation between the two distributions.

The two branches of the current-to-voltage conversion shown in Fig. 12.5 are equal in terms of layout and size and, finally, the M5 and M6 transistors balance M2 and M3 of the column decoder, but are always on, since their gate voltage is VDD. Due to the previous considerations, let's initially suppose that the loads are simple resistors. The sense amplifier is completed by the comparator shown in Fig. 12.6 that allows translating the result of the current-to-voltage conversion into digital information.

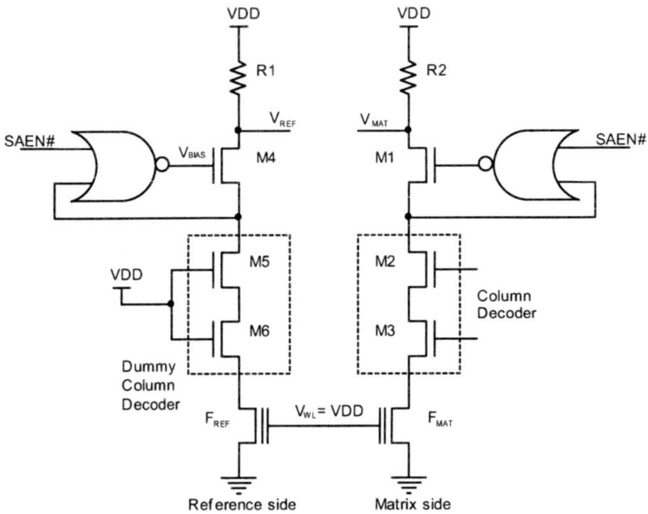

Fig. 12.5. Differential architecture

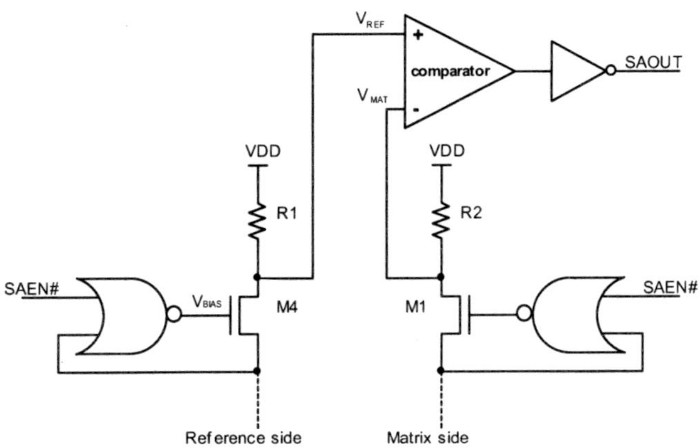

Fig. 12.6. Output structure of the sense amplifier: the result of the current-to-voltage conversion is translated into digital format.

Let's start from reading a written cell. On the array side no current is sunk, and the V_{MAT} potential is therefore tied to the supply voltage. The V_{REF} potential is instead low due to the fact that F_{REF} has been UV erased and the comparator has an input difference of potential that is sufficient for correct operation. On the contrary, if we read a virgin cell, the cell of the array has the same threshold voltage as the reference cell, apart from the statistical spread. At this point, the MAT and REF nodes have the same potential and the comparator output has a random value. It is therefore evident that the introduction of an element of asymmetry that allows distinguishing a virgin from a programmed cell is necessary. The only element on which it is possible to operate in the configuration of Fig. 12.5 is the value of the load. The V_{REF} potential has always the same value, independently of the kind of cell on the array side. Hence, the V_{MAT} potential must have higher or lower values than V_{REF} when a written and a virgin cell are read, respectively. Without asymmetry, a written cell causes V_{MAT} to have the maximum possible value, whereas in the case of a virgin cell the potential equals V_{REF}, as previously stated.

V_{MW} and V_{MV} are the names of the potentials of the array branch corresponding to a written and an erased cell respectively, and these voltages can be shifted around the reference voltage in two possible ways:

1. We may change the value of the reference load, therefore increasing the value of V_{REF} that is in between V_{MW} and V_{MV}. To do so, it is sufficient to decrease the value of the load resistance on the reference side. In this way, for the same current sunk, the V_{REF} potential is higher that V_{MV}, since the voltage drop on R1 is less than the drop on R2, being R1 < R2.
2. We may increase the value of R2, therefore diminishing the V_{MAT} potential in the case of virgin cell.

The solution that is generally adopted is the first one, since it allows limiting the resistive value of the loads. Lower load resistance means higher available current and, hence, higher speed to charge the parasitic capacitance of the column.

The load resistors can be substituted for diode-connected p-channel transistors, as indicated in Fig. 12.7. The reference side has two p-channels with the same size, the array side just one. It is preferable to realize two identical p-channels, instead of a single MOS having double size, in order to reduce possible offsets caused by dimensional variations. For example, even if the transistor length varies by 0.1 μm, the size ratio between the two branches, array and reference, does not vary. The p-channel transistors on the two conversion branches can be connected to form a mirror by having a common gate connection, to increase the voltage swing of the output signal.

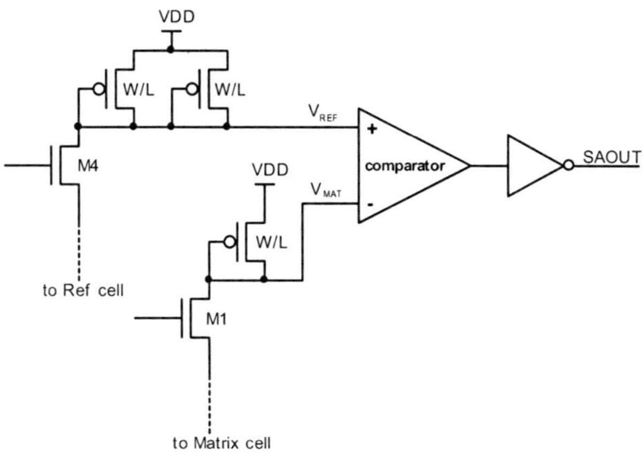

Fig. 12.7. Active loads for differential reading

At this point it is necessary to introduce a way to model the cell characteristics that will be very useful in later discussions. We would like to obtain a diagram that displays V_{GS}, or equivalently VDD, and the cell current, which could clearly demonstrate the asymmetry introduced.

In the case of a virgin cell, the reference side sinks the same current as the array side, but the REF node has higher potential than the MAT node. We could imagine that the two branches have the same load but the reference side has a cell with half the size of the corresponding cell in the array. In this way, the functionality is preserved and the asymmetry can be attributed to the reference cell. The result is represented in the graph of Fig. 12.8.

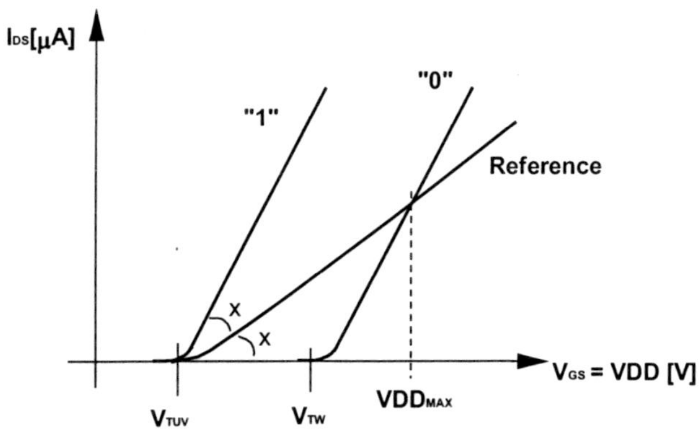

Fig. 12.8. Reading with unbalanced load

It is important to state that the reference cell does not sink half the current of the array side, but rather effectively half the current with respect to the load device. The diagram of Fig. 12.8 is only a useful representation. Modifying the size of the reference cells would be undesired. The possibility of comparing cells with the same size allows associating all the process and geometric variations to a common mode variation, due to the differential architecture.

The asymmetry introduced causes the characteristic of the reference and written cell to have an intersection for a given value of VDD. This happens when the current of a written cell equals half the current sunk by the UV erased virgin cell. The point called VDD_{MAX} in Fig. 12.8 represents the maximum value of the supply voltage for which the device correctly reads the written cell, whereas, theoretically, the minimum value of VDD corresponds to V_{TUV}.

Defining η the ratio between the load size on the reference and the array side, and recalling that $V_{TW} = V_{TUV} + \Delta V_T$, we can express VDD_{MAX} as follows:

$$VDD_{MAX} = V_{TUV} + \frac{\eta}{\eta + 1} \Delta V_T \tag{12.2}$$

Problem 12.2: What kind of issues may derive from a load ratio greater or lower than two?

12.3 Differential Reading with Current Offset

The main problem of a current-to-voltage converter with unbalanced load is the limit of VDD_{MAX}, which can be around 7 V. The VDD voltage of memory device operating at 5 V ranges between 4.5 and 5.5 V. Why then worry if VDD_{MAX} equals 7 V? The first answer is the available margin. When a device is tested, after fabrication and before being sold, the values of some operating parameters (VDD, temperature, frequency and so on) are tested to guarantee the correct behavior beyond the specification limits.

Moreover, a written cell tends to lose charge over time, due to gate and drain stress, and temperature. As a consequence, the value of VDD_{MAX} tends to diminish over the time, reducing the functioning range.

The first solution to solve the problem of VDD_{MAX} in non-volatile NMOS memory devices was the introduction of a parallel reference, i.e. with a characteristic that is parallel to the one of the virgin and the written cell of the array.

In order to obtain such a result, we must remove the mismatch of the load shown in Fig. 12.7. If we suppose a fixed contribution, I_{OFF}, to the characteristics of the cell on the array side, we obtain the situation represented in Fig. 12.9, where the characteristic of the virgin cell lays completely at the left of the reference, while the one of the written cell crosses the characteristic of the reference at low VDD. At VDD > VDD_{MIN}, the characteristic of the reference is in between the virgin and the written cell. The problem of VDD_{MAX}, caused by the mismatch of the characteristic of the reference is then solved. However, a higher VDD_{MIN} than the previous case has been introduced.

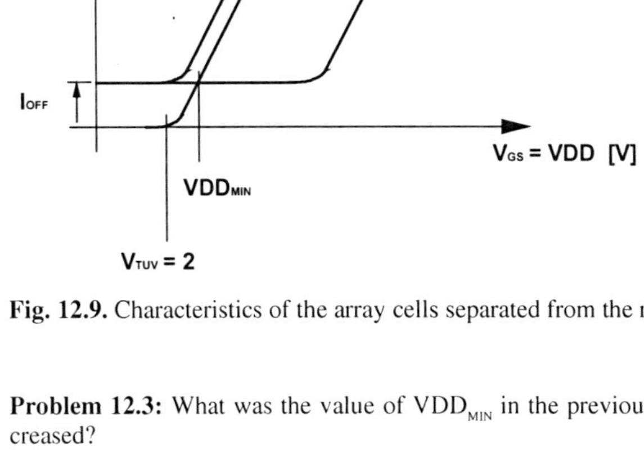

Fig. 12.9. Characteristics of the array cells separated from the reference by a current offset

Problem 12.3: What was the value of VDD_{MIN} in the previous case? How much has it increased?

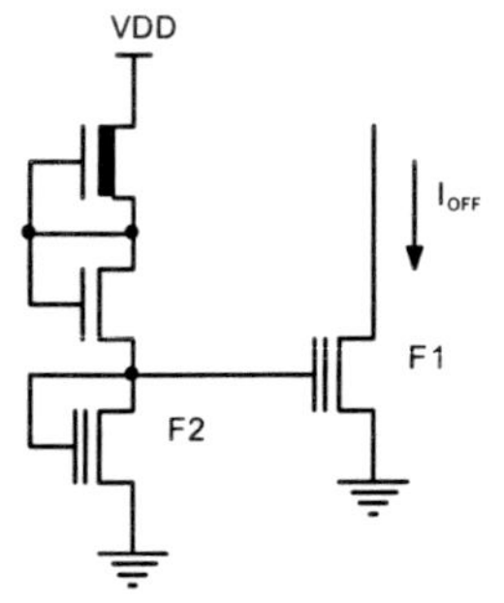

Fig. 12.10. Generation of the I_{OFF} offset current

This kind of solution was appropriate when there was no interest in low voltage operations and, hence, a reduction of the operating margin of VDD at low voltages was acceptable. The margin at high VDD values was much more important. The variation of VDD_{MIN} is proportional to the current offset (I_{OFF}) that shifts the characteristics upwards. This current has a value of some tens of microAmperes (20 to 30 μA), which is a good compromise between the value of VDD_{MIN} and the necessity of separating the characteristics of the virgin and the reference cells.

Figure 12.10 shows how the current source, I_{OFF}, can be realized. A UV virgin cell (F1) is biased with a controlled and stable gate voltage that equals the threshold voltage of another UV virgin cell (F2). Thus, F1 can sink constant current, I_{OFF}, in the operating range of interest, independently of the value of VDD. Obviously, the described circuit works properly only at bias voltage greater that the UV cell threshold voltage (2 V).

Problem 12.4: Design a current source I_{OFF} that operates also at bias voltage lower than 2 V.

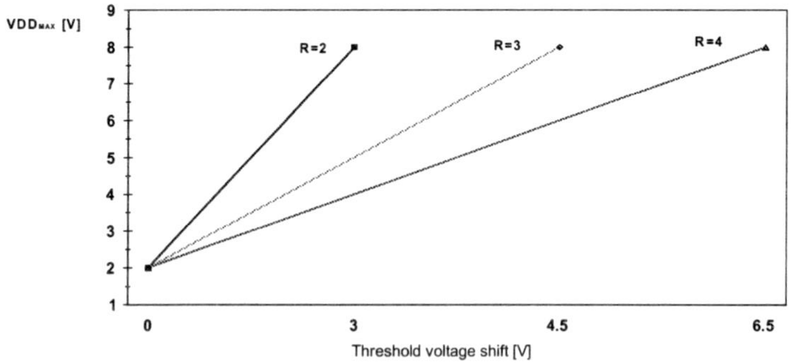

Fig. 12.11. Characteristic of VDD_{MAX} with reference to the voltage gap, ΔV_T. R is the ratio between the load of reference and array side

It is interesting to evaluate the relationship between the voltage gap and the value of VDD_{MAX} for the two types of converter presented, the one with unbalanced loads and the other with parallel characteristics. As we detailed in Chap. 3, with the bias used in read (V_D = 1 V), the cell has the following linear current-voltage characteristic:

$$I_D = \beta[(V_{GS} - V_{TUV} - \Delta V_T)] \tag{12.3}$$

The current-voltage characteristic of the reference cell in the case of unbalanced loads is:

$$I_D = \frac{\beta}{2}[V_{GS} - V_{TUV}] \tag{12.4}$$

Therefore, the point of intersection between the two characteristics is given by the solution of the system composed of the current equations of the reference and the written cell.

Thus we have:

$$V_{GS} = V_{TUV} + 2\Delta V_T = VDD_{MAX} \tag{12.5}$$

while for a parallel converter there is no crossing point between characteristics, at least theoretically. In Fig. 12.11, the foregoing relationships are shown when the mismatch ratio between array and reference varies.

12.4 Semi-Parallel Reference Current

The current-to-voltage converter described in the foregoing section, is the solution adopted for devices in NMOS technology. The memory realized only with n-channel transistors has some advantages; for example it does not suffer the latch-up effect even though some remarkable drawbacks exist. For example it is not possible to realize efficient current mirrors referred to the supply voltage but only to ground. Furthermore, for the converter with the current offset some problems may arise in case of shrink, simply because the offset current is a difficult parameter to control. The introduction of CMOS devices allowed realizing efficient current-to-voltage parallel converters to read non-volatile memory cells[3].

The basic idea is the realization of a suitable circuit to generate a reference characteristic like the one shown in Fig. 12.12. The reference has unbalanced load in the voltage range between V_{TUV} and V_S, and parallel characteristic for voltages greater than V_S.

Let's start designing an I_{REF1} current as indicated in Fig. 12.13. Notice how I_{REF1} realizes the mismatch of the load up to the V_S voltage, while for higher voltages it basically produces an offset on the reference branch.

[3] The analysis that will be present was initially conceived for EPROM devices and these concepts were later applied to Flash memories.

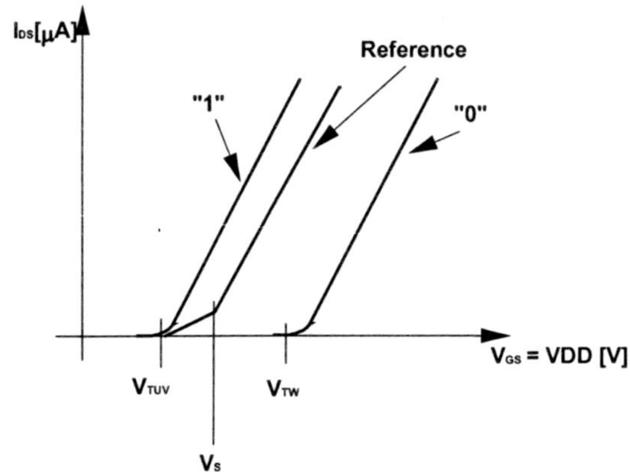

Fig. 12.12. Ideal characteristic of the reference cell

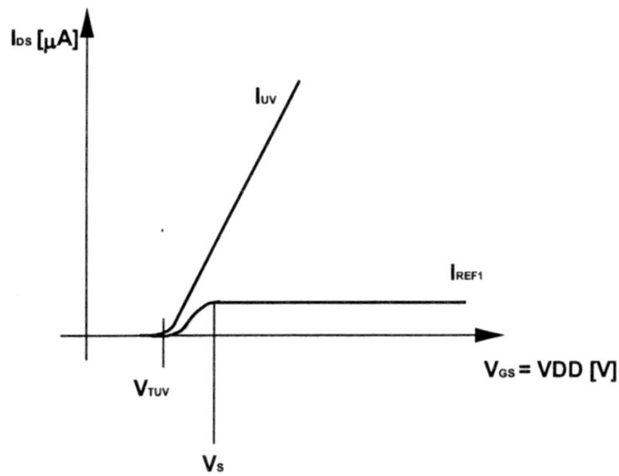

Fig. 12.13. Reference current I_{REF1}

In Fig. 12.14 the circuit that generates I_{REF1} is shown. The F6 cell is a UV virgin cell having gate directly connected to VDD. The circuit composed of M1 and M2 allows applying a voltage equal to VDD – 1 V to the gate of F3 (UV virgin cell), considering the threshold voltage of the p-channel transistor roughly equals to 1 V in magnitude. It is like F3 had a threshold voltage greater than the threshold voltage of F6 by 1 V: it is the characteristic of I_{UVS} in Fig. 12.15.

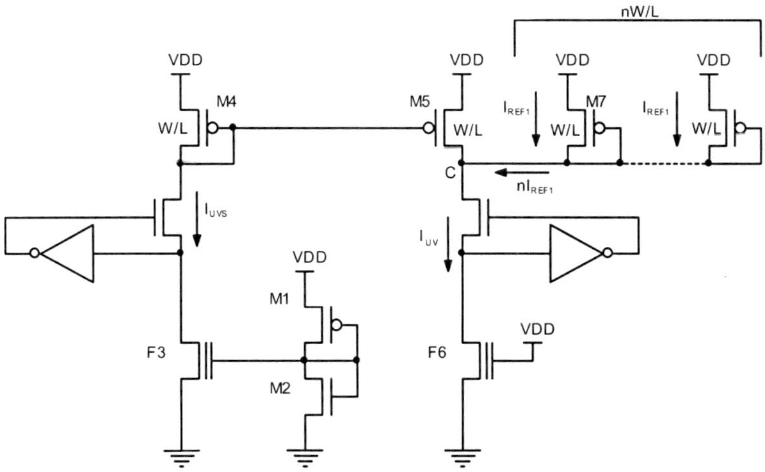

Fig. 12.14. Circuit that generates the I_{REF1} current

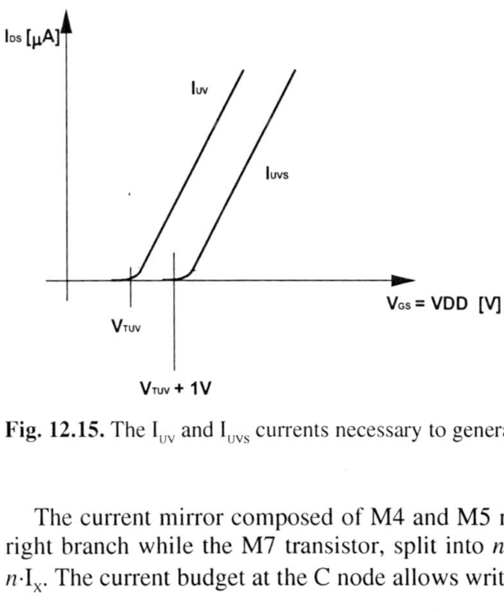

Fig. 12.15. The I_{UV} and I_{UVS} currents necessary to generate I_{REF1}

The current mirror composed of M4 and M5 reproduces the I_{UVS} current on the right branch while the M7 transistor, split into n transistors, conducts the current $n \cdot I_X$. The current budget at the C node allows writing the following relationship:

$$I_{UVS} + nIx = I_{UV} \tag{12.6}$$

that is:

$$I_{REF1} = \frac{I_{UV} - I_{UVS}}{n} \tag{12.7}$$

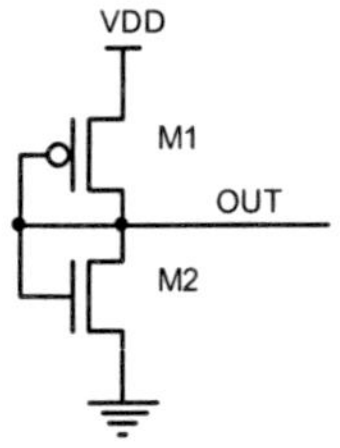

Fig. 12.16. Level shifter circuit

The term *n* as denominator is necessary to produce the portion of characteristic that reproduces the load mismatch.

Before examining the overall characteristic of the reference, it is interesting to analyze the structure constituted by the M1 and M2 transistors in Fig. 12.14, also reported in Fig. 12.16 for sake of clarity. This circuit must realize the voltage shift so that the resultant output voltage is equal to VDD plus a p-channel threshold voltage. The n-channel transistor is very resistive, e.g. 2 μm/120 μm, and its gate is connected to VDD, while M1 is very conductive, e.g. 50 μm/2 μm, so as to source all the necessary current to the n-channel with a minimum V_{SD}. Since M1 is a diode-connected transistor, $V_{SD} = V_{SG}$, the voltage of the output node basically reaches VDD - $|V_{T,p}|$.

Problem 12.5: Using the equivalent circuit for the M1 and M2 transistors, derive the algebraic relationship between the variation of the output node and the variation of the supply voltage. Discuss the result with reference to the transistor size.

After generating the I_{REF1} current, the circuit that is used to obtain the final I_{REF} is reported in Fig. 12.17. Basically, I_{REF1} does not have the parallel component above the V_S voltage. Thus, it suffices to add I_{UVS} to I_{REF1} and the required reference current can be obtained. The analysis of the schematic of Fig. 12.17 shows how the result is obtained by means of simple current mirrors that allow adding or subtracting currents at will.

In any current-to-voltage converter, the feedback circuit discussed at the beginning of Sect. 12.1 is present so as to keep the drain voltage of the cells around 1 V, preventing drain stress effects.

Problem 12.6: It is interesting to realize I_{REF} using NMOS devices only, which should demonstrate the advantage of CMOS technology.

The solution discussed in this section gave origin to the first semi-parallel reference and was implemented in a EPROM memory device several years ago. At that time, it was not possible to exploit the concept of reference cell with V_T that can be tuned during the testing phase, which became common later. Today, the I_{REF} current discussed above is obtained by suitably summing the characteristic of two written cells with the required value of the threshold voltage.

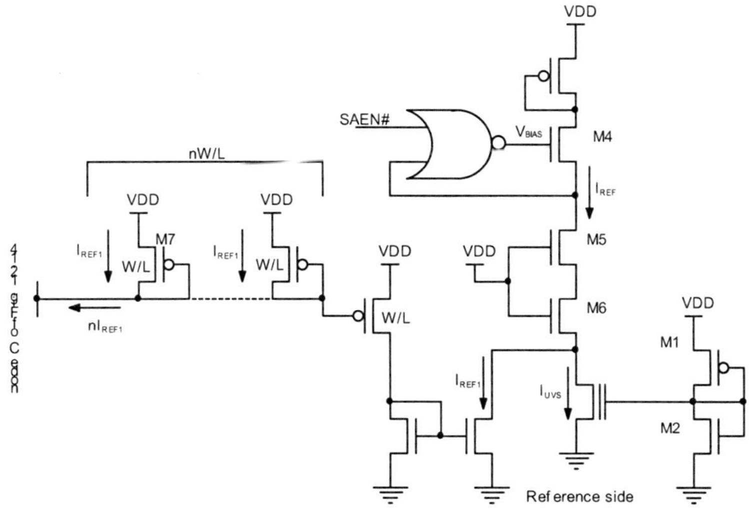

Fig. 12.17. Circuit schematic to generate I_{REF}

12.5 Techniques to Speed Up Read

In Sect. 12.4, an overview of the basic structures that allow reading the current sunk by a non-volatile cell and, hence, decoding the information stored has been presented. In this section, we will delve into some of the techniques used to improve speed and reliability of the read operation. Each concept we will introduce corresponds to a new tool that can be used to solve not only one specific problem but also other issues apparently unrelated with each other.

12.5.1 Equalization

The concept of equalization is used to reduce the voltage swing of the signals, speeding up the toggling. With reference to the converter with unbalanced load shown in Fig 12.7, let's suppose that a virgin and a written cell are read one after the other. In Fig. 12.18, the V_{MAT} and V_{REF} potentials are reported for the two cases.

For sake of simplicity, the V_{REF} potential is supposed to be constant. The V_{MAT} potential, instead, goes from the minimum of the voltage range in the case of a virgin cell to the maximum in the case of a written cell. The voltage swing amounts to about 4 V for a device biased at 5 V. This means that the parasitic capacitances of the MAT and REF nodes must be charged and discharged, which implies a longer transient time. A very limited voltage oscillation of all nodes around a fixed value of bias voltage would be the ideal situation. In this way, the transient time would be reduced to the minimum, with the consequent reduction of all the delays. To do so, the schematic of the converter is modified as in Fig. 12.19, in which a natural transistor, N1, is introduced between the MAT and REF nodes, and it is driven by the SAEQ signal, which is a 20 ÷ 30 ns wide pulse generated at each address transition.

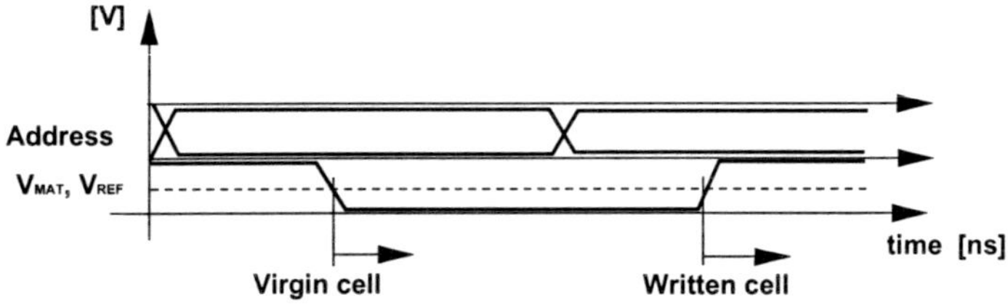

Fig. 12.18. A virgin and a written cell are read in sequence: the V_{MAT} and V_{REF} potentials in the sense amplifier are shown.

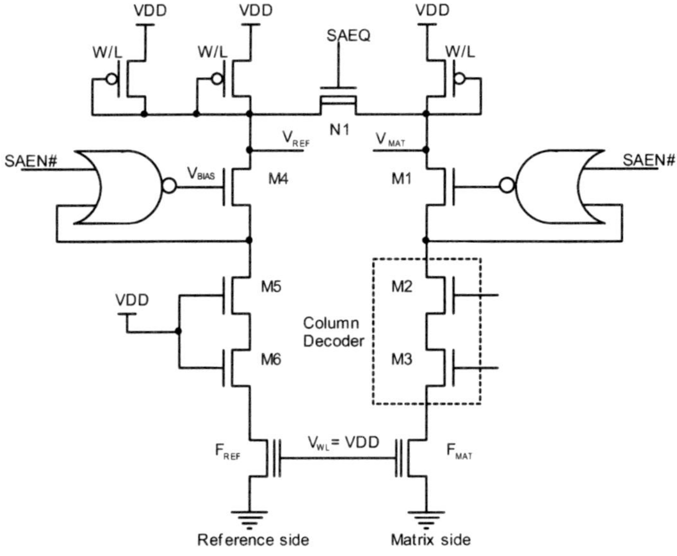

Fig. 12.19. Sense amplifier with unbalanced load with a natural transistor (N1) for the equalization

During the active phase of SAEQ, N1 is on and the MAT and REF nodes are equalized, i.e. they are driven to the same potential, the value of which is predetermined by acting on the sizes of the devices. Generally, the equalization potential is fixed to half the VDD voltage. Supposing that the bias voltage is 5 V, V_{MAT} has a voltage swing of 2.5 V in the case of virgin cell, while V_{REF} is 0.5 V. In the case of a virgin cell V_{REF} varies of 0.5 V, while V_{MAT} of about 1.5 V (Fig. 12.20).

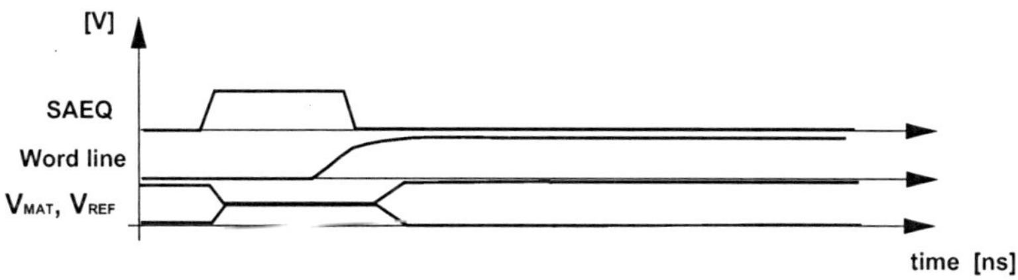

Fig. 12.20. MAT and REF node potentials when the natural transistor for the equalization is present (N1 in Fig. 12.19)

The equalization can involve also other nodes of the sense amplifier as shown in Fig. 12.21. Transistors M7 and M8 are dimensioned in order to define the value of the voltage of the nodes to equalize during the active phase of SAEQ, by acting on the ratio of the respective loads.

Problem 12.7: Discuss in detail the function of transistor M7 and M8.

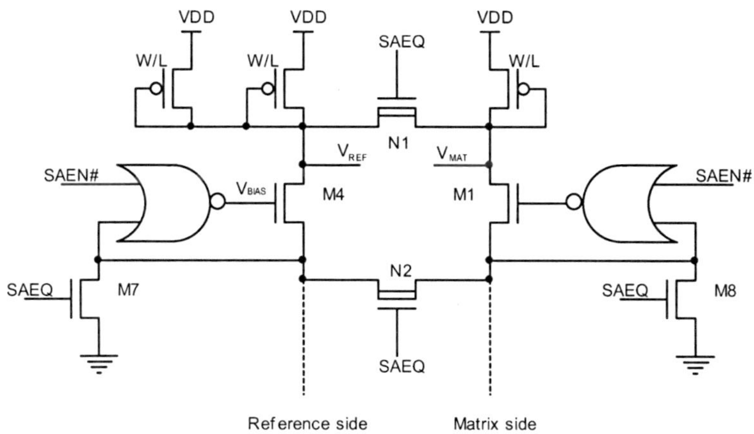

Fig. 12.21. Schematic of the sense amplifier with double equalization

One of the classical problems of equalization is that the drain potential of the reference and array cells must reach the asymptotic value before the active phase of SAEQ has finished. This kind of synchronicity is not easy to obtain, preventing at the same time waste of time. In practice, it might happen that, when SAEQ returns to the low logic state, the parasitic capacitance of the bit line has not fully been charged. Obviously, this might be paid in terms of read time.

The second problem is related to the intrinsic noise of the circuits with equalization transistors. The REF and MAT nodes are designed in order to add the minimum parasitic capacitance and be charged and discharged in the minimum time as possible. The size of N1 and N2 is limited to the minimum, contrarily to what is needed to obtain a good equalization. However, N1 and N2 couple the most important analog nodes of the sense amplifier to the logic signal SAEQ through the source-gate and drain-gate parasitic capacitors. It is even possible that the fall of SAEQ causes a temporary inversion of the relative position of V_{REF} and V_{MAT}, forcing the sense amplifier to a difficult recovery with the consequent waste of time.

12.5.2 Precharge

Another technique that is extensively used to speed up the toggling of the converter is the precharge. In Fig. 12.22, the circuit schematic for the precharge of the usual converter with unbalanced load is reported. Transistors M9 and M10 are on during the active phase of the SAEQ signal and participate in the precharge of the parasitic capacitance by sourcing more current than the diode-connected p-channels, used for the current-to-voltage conversion, could.

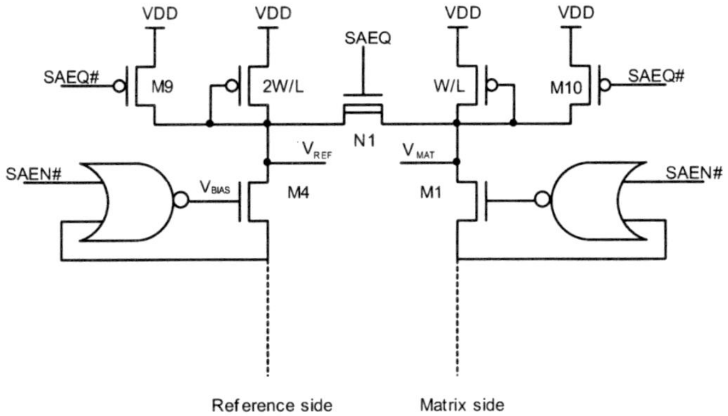

Fig. 12.22. Precharge of the MAT and REF nodes

The precharge phase is simultaneous to the equalization phase. Subsequently, M9 and M10 are switched off so as to restore the conditions needed to read in the case of unbalanced load.

12.5.3 Clamping of the MAT and REF Nodes

It is possible to further speed up the reading by modifying the converter as indicated in Fig. 12.23. Transistors M11 and M12, which are LVS-type and diode-connected, are placed between the MAT and REF nodes. In this way, the difference between the V_{MAT} and V_{REF} potentials cannot be greater than twice the threshold voltage of an LVS transistor. Let's analyze the functioning of this circuit in detail.

The equalization phase has driven the MAT and REF nodes to the same voltage value, say 2.5 V. The read phase of a written cell pulls V_{MAT} toward VDD – $V_{T,N}$ while, at the same time, V_{REF} goes low.

Transistor M12 is off because $V_{MAT} > V_{REF}$. V_{MAT} cannot go above V_{REF} plus the threshold voltage of an n-channel, since, in this case, the M11 transistor would switch on, clamping the V_{MAT} voltage.

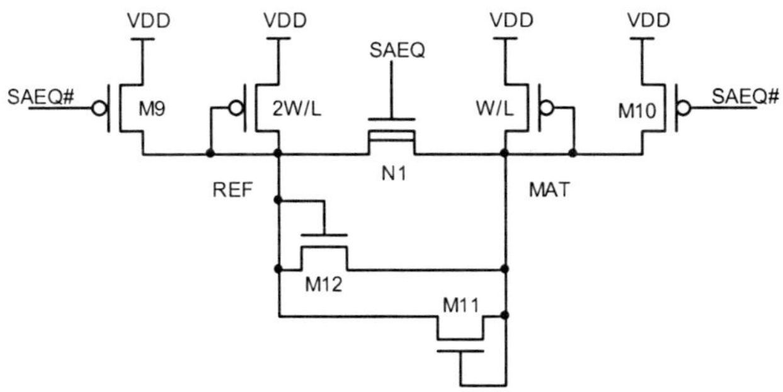

Fig. 12.23. Sense amplifier with clamping diodes for the V_{MAT} and V_{REF} voltages

12.6 Differential Read with Current Mirror

Figure 12.24a shows the schematic of a sense amplifier in which the transistors used to convert current into voltage form a current mirror. In all the previous configurations, both the load transistors are diode-connected. During the read phase, V_{REF} reaches a voltage value greater than 1 V, depending on the size ratio between the load and the corresponding virgin cell. The voltage swing of V_{MAT}, instead, is within 1 V in the case of a virgin cell, and 4 V (i.e. VDD + $V_{T,p}$)[4] in the case of a written cell. Using the current mirror allows increasing the swing of the MAT node, thus making the reading of the comparator faster and more reliable.

Problem 12.8: In the foregoing case, the voltage swing is increased whereas in the case of clamping diodes it is diminished: what is the best choice?

Let's consider the reading of a written cell. V_{MAT} tends to raise toward the supply voltage and is no longer clamped by a diode-connected transistor. V_{MAT} can therefore reach the supply voltage. The reading of a virgin cell, instead, is based on a comparison of current, like the one we discussed in Sect. 12.1. The array side

[4] It is important to bear in mind that the value of the threshold voltage of the load transistor depends also on the bias of the substrate.

has a load, M13, that sources a current equal to $I_{REF}/2$. The current mirror reproduces the current of the reference side to the array side. On the contrary, the cell sinks the I_{MAT} current that equals I_{REF}. We could say that the cell tends to sink this current but the real amount of current sunk depends only upon the load. Such a comparison between the current that the cell is supposed to sink and the actual capability of sourcing of the load is the basic principle of the circuit. As a consequence, the potential of the MAT node is pulled down to ground.

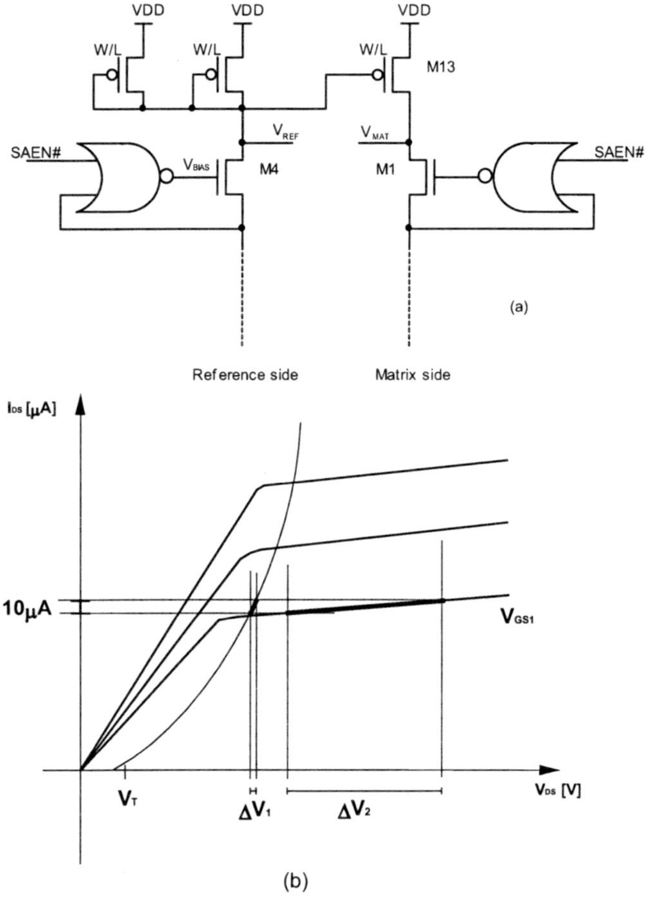

Fig. 12.24. **(a)** Differential read with current mirror **(b)** Output characteristics of a diode-connected NMOS and of an NMOS with fixed gate voltage

The use of a current mirror directly impacts on the sensitivity of the sense amplifier. To this purpose, let's consider Fig. 12.24b, in which the output characteristics of two NMOS transistors, one diode-connected and the other with fixed V_{GS}, are reported. Once the operating point has been fixed, let's consider a variation of the current, ΔI, that the transistors must sink. It is evident that the difference of potential that results is greater in the case of the transistor with fixed gate voltage, since it operates in the region with high output resistance. Under the hypothesis of linear behavior around the operating point, and considering the following relationship:

$$V_{REF} - V_{MAT} = \gamma \cdot \Delta I \tag{12.8}$$

we obtain that the proportional factor, γ, is greater in the case of mirror-connected load.

The comparator located downstream the sense amplifier is capable of discriminating between the input signal, V_{MAT} and V_{REF}, if these potentials differ at least by ΔV_{MIN}. Considering that the γ factor is greater in the case of current mirror, it descends that such circuit configuration allows discriminating smaller currents than the case with split loads.

12.7 The Flash Cell

We know that the Flash cell is basically an EPROM cell with the additional possibility of removing the charge stored during the write operation from the floating gate by applying suitable electrical potentials. In the case of the EPROM cell, the erase is carried out by exposing the device to UV radiation. The charge stored on the floating gate absorbs the radiation acquiring enough energy to cross the energy barrier due to the thin oxide, thus flowing back to the substrate. This kind of erasing automatically stops when all the charge on the floating gate has been removed. At this point the charge neutrality condition is verified and the UV value of the threshold voltage of all the cells of the array has been restored. In the case of the Flash cell, geometrical parameters, source resistance, implant variations, possible misalignment and so on, contribute to the spread of the distribution of threshold voltages of the erased cells that is quite large, as indicated in Fig. 12.25. On top of that, the presence of cells that are electrically erased faster than others can cause some cells to be depleted, or, in other words, to have negative threshold voltage.

A depleted cell is a cell that sinks current from the bit line even when its control gate is grounded. It is basically an offset current source, with all the issues related to the reading of written cells we described in Sect. 12.3. It is thus necessary to prevent the cells from being depleted during the erase phase. As we will see in the chapter dedicated to the algorithms, one of the techniques currently used to recover depleted cells is the so-called *soft-programming* algorithm that consists in a controlled program operation at the end of each electrical erasing.

The read of a Flash cell can be carried out by means of one of the techniques presented in this chapter. It is possible to adopt an approach with either unbalanced or parallel loads. Let's consider the case of the reading with unbalanced load carried out with gate voltage of 3 V. The maximum threshold voltage for an erased cell, E_R, equals 2.5 V. If the threshold of a UV virgin cell is 2 V, and the distribution of erased cells is typically 2 V wide, the distribution must be placed as shown in Fig.12.25, i.e. centered on the UV threshold voltage, guaranteeing half a volt margin to the depleted cells.

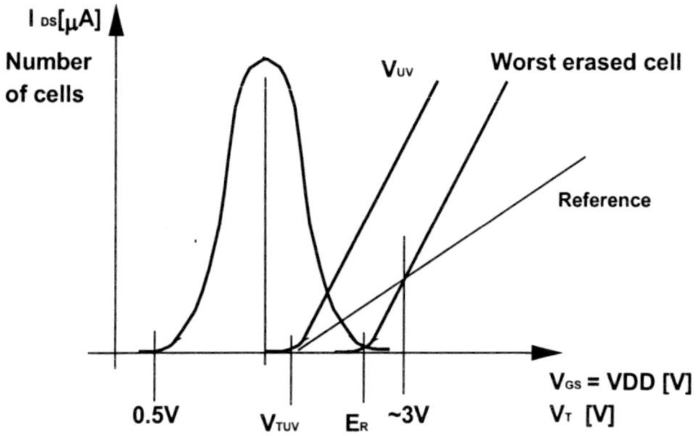

Fig. 12.25. Distribution of the V_T of Flash cells with respect to the reference (unbalanced load)

Problem 12.9: Prove that the voltage of the "less erased cell" in Fig. 12.25 is around 2.5 V.

Problem 12.10: Analyze the possible mechanism of error that a depleted cell may generate during read. Consider a read circuit with unbalanced load first, and with parallel load then, Suggestion: a current offset should be included so as to simulate the contribution of the depleted cell on the array side. Verify the immunity of the various approaches to the reading as a function of the different value of the current offset introduced. Finally, define which approach is the best to the read in presence of depleted cells.

12.8 Reading at Low VDD

The supply voltage of the devices currently required on the market tends to decrease over the time, since lower consumption and bias voltage are demanded to meet integration constraints. In future years, the devices will be biased at 1.8 V ± 10%, or maybe less. The reduction of the supply voltage implies a large number of problems due to the fact that the requirements on device functionality, speed, consumption and so on do not vary but, on the contrary, device performances must continuously be improved. The reading of an EPROM cell, and especially of a Flash cell, is difficult for a number of issues recalled hereafter.

We already know that the erasing of a Flash cell array causes a distribution of V_T and currents and, as a consequence, it is necessary to shift the distribution toward positive voltages as much as possible to avoid the inconvenience related to the depleted cell. Basically, the "less erased" cell has a threshold voltage of about 2.5 V.

With reference to the usual converter with unbalanced load, in Fig. 12.26 the relative position of the characteristics of the UV cell along with the reference and the "less erased" cell, E_R, is shown. The minimum value of functioning of VDD is defined, at least theoretically, as the intersection of the characteristics of the reference with the one of E_R. The crossing point is around 3 V.

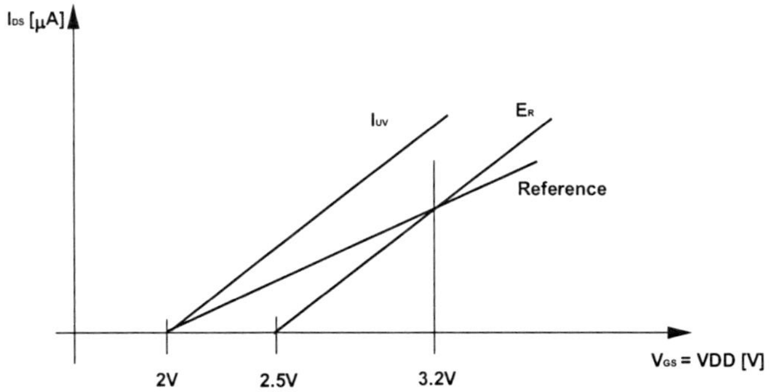

Fig. 12.26. In the case of a converter with unbalanced load, the minimum VDD is given by the intersection of the characteristics of the "less erased" cell and the reference one

The value of the bias voltage is 3 V ± 10%, which means that VDD ranges between 2.7 and 3.3 V. When VDD is 2.7 V, we have only a 0.2 V margin and the cell sinks little current, toggling slowly, but it is still recognized as written, its current being lower than the current of the reference. Moreover, the comparator located downstream the I-V converter has a indetermination region of few millivolts around the ideal toggling value, thus worsening the value of the current available to the cell. We will now examine the idea of exploiting the VDD voltage range between 2.5 and 3 V.

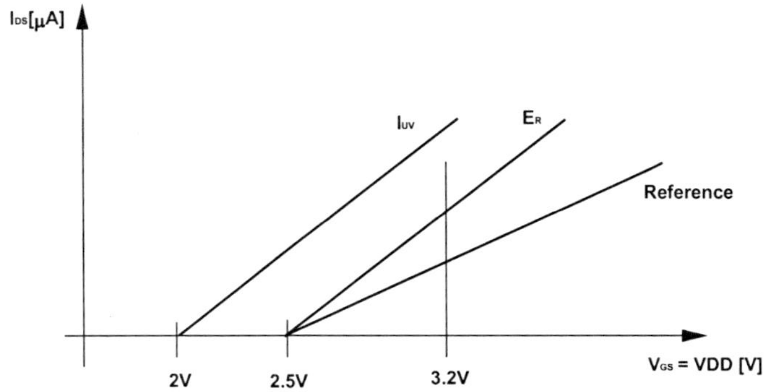

Fig. 12.27. The threshold voltage of the reference cell is shifted to 2.5 V

In Fig. 12.27 a different organization of the reference is shown, obtained by simply shifting the old reference characteristic from 2 V to 2.5 V. In this way the E_R cell theoretically sinks more current than the reference one at minimum value of the supply voltage, (i.e. 2.7 V) and, hence, the comparator can perform the reading correctly.

In practice, a 200 mV margin between the threshold voltage of E_R (2.5 V) and the minimum VDD (2.7 V) is too limited and, thus, it is convenient to "steal" a few hundreds of millivolts to the erasing, and shift the threshold voltage of E_R to 2.3 V. In Fig. 12.28 the final result in terms of characteristics is reported, when the reference has a semi-unbalanced characteristic in order to improve the performance at high VDD.

Problem 12.11: Using parts of the circuits described in the previous paragraphs, design the suitable electrical schematic to obtain the reference characteristic discussed above.

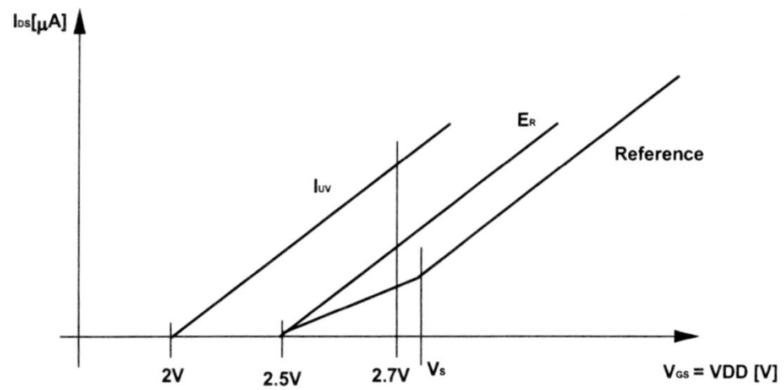

Fig. 12.28. Semi-unbalanced reference to read Flash cells

Problem 12.12: The most direct approach to the problem of the low supply voltage is the boost. Analyze the problem, highlighting drawbacks and advantages, and try to design the suitable circuitry to achieve this goal.

We can now state that the fundamental problem for operations at low voltage is not the (ideal) value of VDD_{MIN}, but the available current. The difference of current at 2.7 V between reference and erased cells we want to read is so reduced (a few microamperes) that the switching times of the converter increase since the order of magnitude of the capacitance that the cell is due to charge/discharge is some picofarads. Supposing that the cell sinks 2 μA from a 1 pF load, the time needed to discharge the capacitance of 100 mV, i.e. the value necessary for the comparator to switch, is:

$$\Delta t = C\frac{\Delta V}{I} = 50ns \tag{12.9}$$

In the read phase, this capacitance must be charged during the switching-on transient of the decoder and then, in the case of erased cell, discharged through the cell itself (if the cell is able to sink little current, the discharge will be slow). On the contrary, in order to be fast during the reading of a written cell, one must be able to source a large current on the reference side.

12.9 Amplified I/V Converter

The solution to the problem described in the previous section is to amplify the current of the array cell and compare it with the current of the reference cell. In Fig. 12.29 the circuit configuration in principle is reported.

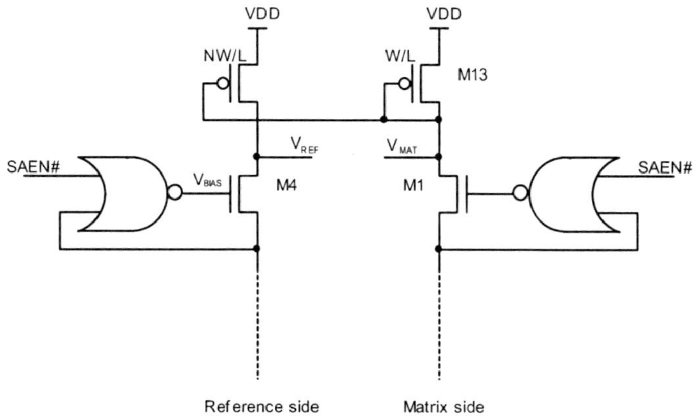

Fig. 12.29. Schematic of a converter to amplify the cell current (N > 1)

The fundamental difference with respect to the previous architectures is in the connection of the load transistors of the mirror. In this case, the cell current that is mirrored is not the reference one. This allows multiplying the array cell current, overcoming the problem of functioning at low VDD for what concerns the erased cell. The characteristics of the array and reference branches are reported in Fig. 12.30.

When the bias voltage is lower than the threshold voltage of the reference cell, V_{REF} and V_{MAT} have the same value in the case of written cell, since no current is sunk by the two branches. When VDD reaches the threshold voltage of the reference cell, the V_{REF} potential is pulled to ground and the REF node has a very different voltage from the MAT node. Once the value of the threshold voltage of the written cell has been reached, the written cell itself on the array branch starts sinking current comparable with the reference. This mechanism determines the value of VDD_{MAX}.

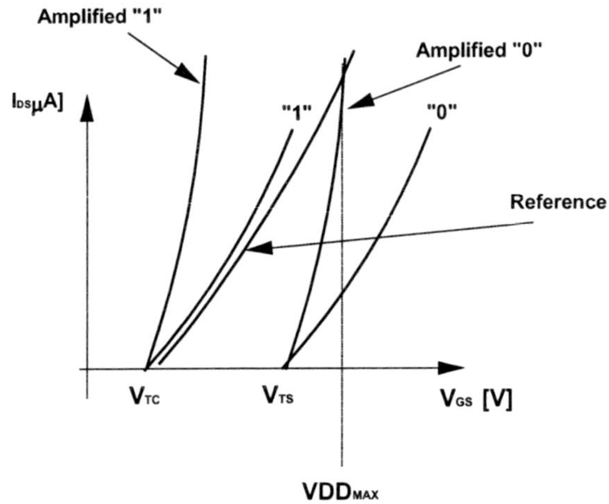

Fig. 12.30. The limit of the amplified converter is VDD_{MAX}. The characteristic of the reference and erased cells are reported separately for sake of simplicity. In the reality they overlap.

In the case of a written cell, the solution with unbalanced loads allows the separation of the characteristics over the entire VDD functioning range.

12.10 Amplified Semi-Parallel Reference

Although the solution to amplify the cell current has been found, some comments are necessary. First, if we suppose the "less erased" cell has a 2.5 V threshold voltage and the minimum bias voltage of the circuit is 2.7 V, we also suppose that a V_{GS} that is 200 mV above the threshold voltage, which means a few microAmperes (3 to 5), suffices to guarantee the proper functioning of the device. It is evident that the margin is really narrow and, hence, we might use a row booster to improve it. However, the limit on the maximum VDD shown in Fig. 12.30 persists.

We have to adapt the characteristic of the reference with respect to the bias voltage, to prevent the intersection with the characteristic of the array cells. The behavior of the amplified converter in the VDD-I_{DS} reference plane is indicated also in Fig. 12.31, where the characteristic of the cells that operate at 1 V drain voltage during reading is supposed to be linear.

The threshold voltage of the reference is named V_{TR}, and it is set to the required value during the testing phase; V_{TC} and V_{TW} are the threshold voltages of the "less erased" and the "less written" cell, respectively. Let's suppose that a boost equal to V_B is applied to the row, which pulls the row to VDD + V_B. This is equivalent to a shift to the left by V_B of the characteristics of the array cells, so that $V_{TC} - V_B$, and $V_{TW} - V_B$ are the new threshold voltage values (the cells can sink more current than before at the same VDD). Named G the cell transconductance and N the amplification gain of the array cell current, the characteristic of the cell has the following expression:

$$I_{DS} = G(VDD - V_T) \tag{12.10}$$

because when the drain is at 1 V such cells are always in the linear region. Under the foregoing hypothesis, the VDD_{MIN} and VDD_{MAX} voltages can be calculated in the case of unbalanced load:

$$VDD_{MIN} = \frac{NV_{TC} - NV_B - V_{TR}}{N-1} \tag{12.11}$$

$$VDD_{MAX} = \frac{NV_{TW} - NV_B - V_{TR}}{N-1} \tag{12.12}$$

As it can be noted, reading the written cell significantly reduces the functioning region of the memory with respect to the bias voltage. In order to remove the limitation on VDD_{MAX}, the characteristic of the reference must have a semi-parallel form as in Fig. 12.31.

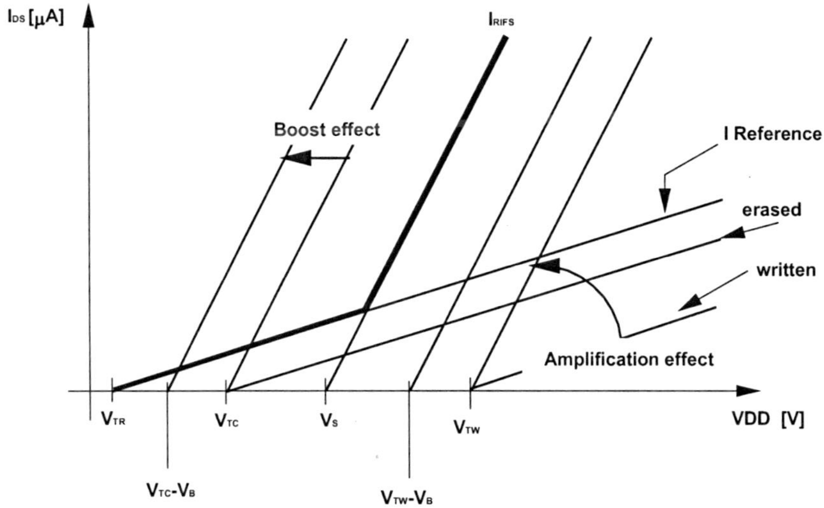

Fig. 12.31. Semi-parallel amplified sense amplifier. This configuration solves the problem of the maximum VDD and, at the same time, has the ability to amplify the cell current

At VDD < V_S the reference characteristic is the same as before, whereas it becomes parallel to the array characteristic when V_S > VDD.

In Fig. 12.32, the electrical schematic of the reference is shown; two cells, F1 and F2, which are programmed to the required V_T value during EWS, are used.

Acting on the current of F1 and F2, by means of mirrors, the current generated can bias a MOS that is used as output current mirror for all the sense amplifiers of the device. The enable signal, SAEN, is necessary only to guarantee null power consumption during the stand-by phase.

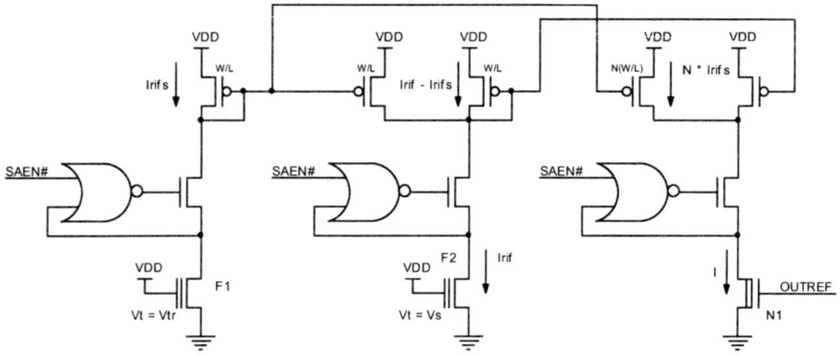

Fig. 12.32. Electrical schematic for the realization of the amplified semi-parallel reference

12.11 Sizing of the Main Mirror

Let's now determine the value of the N parameter for the amplification of the array current. With reference to Eq. (12.11) that expresses the minimum VDD, V_{TC} equals 2.5 V. This is a limit imposed by the distribution width on which it is not possible to act, V_B is the boost voltage, say 0.8 V[5], and, finally, V_{TR} is the threshold voltage of the reference cell that is set to 1.25 V, i.e. as low as possible even though not depleted.

With such values, if a minimum VDD lower than 1.8 V is required, we have:

$$\frac{2.5N - 0.8N - 1.25}{N - 1} \leq 1.8 \tag{12.13}$$

and, hence, we obtain N > 5.5, while for the maximum VDD, if we admit that the threshold voltage of the "less written" cell is 4.5 V, but, at the same time, a maximum VDD greater than 4 V is required, we have:

$$\frac{4.5N - 0.8N - 1.25}{N - 1} > 4 \tag{12.14}$$

and, thus, N < 9.1.

[5] It is not a rough estimation but it nearly equals the V_T of a HV p-channel transistor at ambient temperature (see Sect. 10.3).

We recall that VDD_{MAX} is no longer a limitation owing to the solution of the semi-parallel reference, but, supposing we have no parallelism, the main schematic of the sense amplifier is defined without any modification of the reference. In conclusion, the value of N must be within the following range:

$$5.5 < N < 9.1 \tag{12.15}$$

The choice of the value of N cannot neglect the evaluation of the possible leakage current that might flow through the array column and pollute the reading of a written cell. This current can be ascribed to the characteristics of the cell whose sub-threshold current may be not negligible. In the case of a column with many cells, a small contribution from each cell might result in an overall current of some microAmperes.

Figure 12.33 shows how a current offset shifts the characteristic of the cell upwards, increasing the minimum VDD. Assuming N = 8, in order to read up to a 2 V bias voltage, we cannot admit a leakage current greater than 20 μA. The problem is that, obviously, the converter does not distinguish the contribution of the cell from the contribution of the offset. Therefore, the real leakage that can be admitted must not be greater than:

$$(20/8)\mu A = 2.5\mu A \tag{12.16}$$

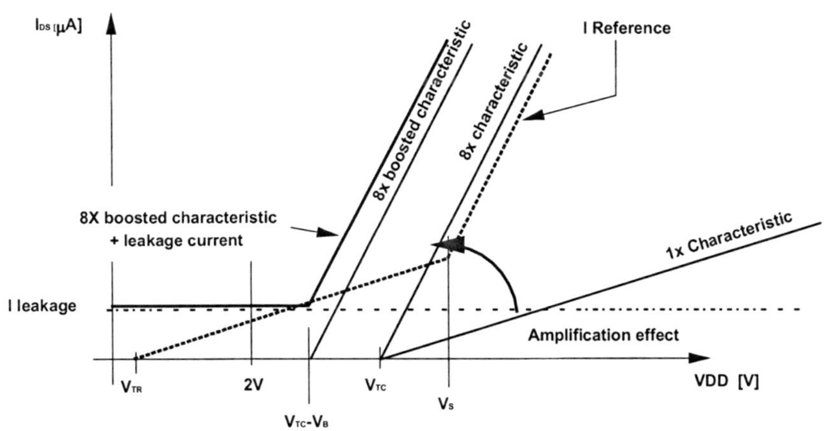

Fig. 12.33. Analysis of the functioning of the converter in the VDD-I_{DS} reference plane, considering the presence of leakage current on the array branch that is amplified and added to the current of the addressed cell

If, for the process that we use, the leakage current of a standard column with 1024 cells is 10 μA, we can accept a 256 cell column.

Finally, it is important to evaluate whether this kind of sense amplifier is robust with respect to the gain variation the memory cells might suffer during their life.

In fact, charge trapping at the silicon-oxide interface modifies the electrical characteristics of the cell resulting in gain variation. Such a gain variation of the cell can be expressed by modifying Eq. (12.10) as follows:

$$I_{DS} = \alpha G(VDD - V_T) \tag{12.17}$$

where $\alpha < 1$.

Under the hypothesis that the gain variation involves only the reference cell that, differently from the array cells, does not undergo cycling, Eq. (12.11) can be rewritten as follows:

$$VDD_{MIN} = \frac{N\alpha\left(V_{TC} - V_B\right) - V_{TR}}{N\alpha - 1} \tag{12.18}$$

where N = 8, and VDD_{MIN} ranges from 1.76 V to 1.85 V, passing from $\alpha = 1$ to $\alpha = 0.5$ due to the amplification. Thus, the variation is still very limited.

Considering that the gain variation usually involves only the array, we find a limitation in terms of VDD_{MAX} even though the reference is semi-parallel.

After having reached the V_S voltage, the reference current can be written as:

$$I_{REF} = G(VDD - V_{TR}) - G(VDD - V_S) + NG(VDD - V_S) \tag{12.19}$$

from which the expression of the maximum VDD derives:

$$VDD_{MAX} = \frac{N\alpha(V_B - V_{TC}) + (N-1)V_S + V_{TR}}{N(1-\alpha)} \tag{12.20}$$

Accepting the limitation of a VDD_{MAX} not lower that 4 V, and supposing $V_S = 3$ V, from Eq. (12.20) it descends that the sense amplifier is able to read cells with α up to 0.53.

12.12 Dynamic Analysis of the Sense Amplifier

The first step to realize a dynamic design from the static one is the application of the equalization technique that consists in short-circuiting the "critical nodes" so as to drive them not only to the same potential but also to the required voltage value, thus improving the performance. In Fig. 12.34 the classical equalization network is shown. N1, N2, and M3 are used for short-circuiting the MAT and REF nodes of the sense amplifier, operating on the transistors of the cascode and on the drain nodes of the column decoders of both the reference and the array. N1 and N2 are natural transistors, whereas M3 is LVS high voltage type because the node of the bit line is connected to the program load, i.e. to the transistor that transfers the programming voltage to the cells.

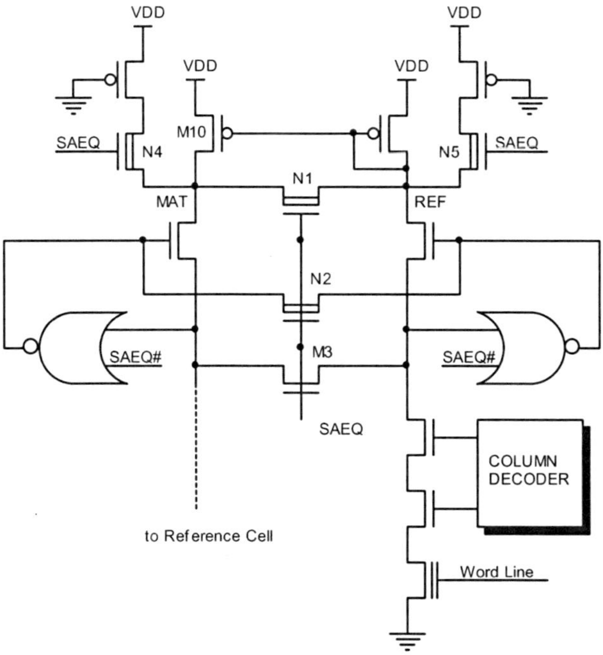

Fig. 12.34. The classical equalization network that drives the critical nodes of the comparator to intermediate voltage values to speed up toggling

Such transistors have to short-circuit the critical nodes, i.e. those nodes that have the most relevant voltage swing, during the active phase of the SAEQ pulse. The equalization restores the node voltage at each reading, driving them to a suitable voltage value. M4 and M5 have to source more current during the SAEQ active phases in order to charge the parasitic capacitance of the bit line. Finally, the p-channels that separate the M4 and M5 transistors from the supply voltage operate as a protection against electrostatic discharges. The equalization network shown does not suffice to meet our demands. If the voltage value to which the MAT and REF nodes are equalized is not suitable to keep the current mirror "on" when the equalization phase finishes, no current is sunk on the array side (since the mirror has not yet been switched on) and, hence, the two node voltages start evolving as the cell that is to be read were written. In the case the addressed cell is erased, the sense amplifier is obliged to invert the trend when, once the converter has been switched on, higher current is sunk. Of course, this behavior causes a non-acceptable waste of time. If a lower equalization level is chosen, one comes across the opposite problem. In fact, in this case, the driving voltage of the p-channels of the I/V converter of the sense amplifier is so high that the voltage of the MAT and REF nodes is pulled toward the direction of the reading of a written cell. The presence of the parasitic capacitance of the columns that sink current during the start-up phase simulating an erased cell further contributes to the problem. In this case, before reading a written cell, it is necessary to wait for the voltage levels to reset to the correct values, with the consequent increase in reading time.

The solution is the introduction of a path of current toward ground on the array side, which is still driven by the reference circuit, so as to reproduce the same current on the two sides of the converter (Fig. 12.35). Particular attention must be paid to the size of transistor N7 to balance the current of the reference branch. Transistor M8, instead, operates only as a switch. In practice the size of M7 is 1/8 of the size of transistor N1 shown in Fig. 12.32, so the two branches conduct the same current during the equalization phase. The size of N7 is determined under the hypothesis that no parasitic load is present on the branches of the sense amplifier.

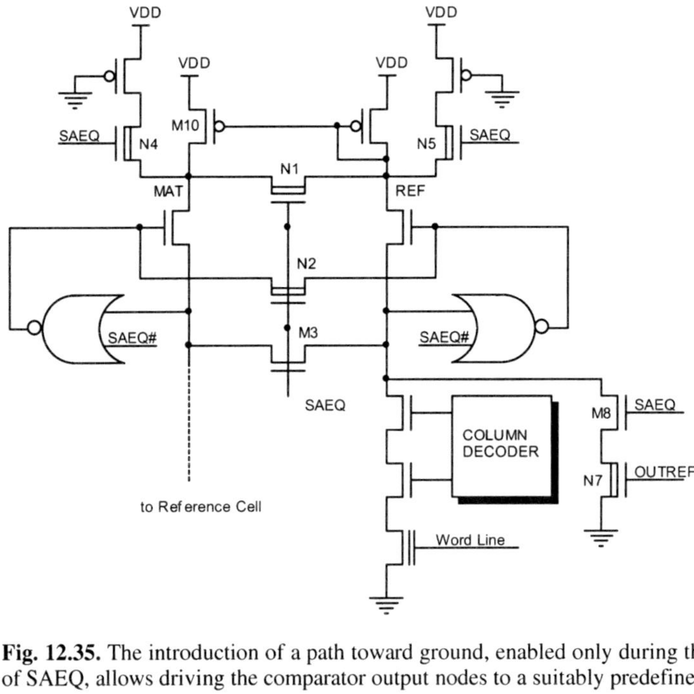

Fig. 12.35. The introduction of a path toward ground, enabled only during the active phase of SAEQ, allows driving the comparator output nodes to a suitably predefined potential

Unfortunately, there are parasitic loads, and, moreover they are large and unbalanced. On the reference branch it is important to pay attention to the layout, and, thus, to minimize the parasitic load, so that the switching-on of the reference side is very fast. The array side has capacitance around some picofarads. The results of the simulations are shown in Fig. 12.36a and 12.36b. The behavior of the voltages shows that the equalization voltage value is too high.

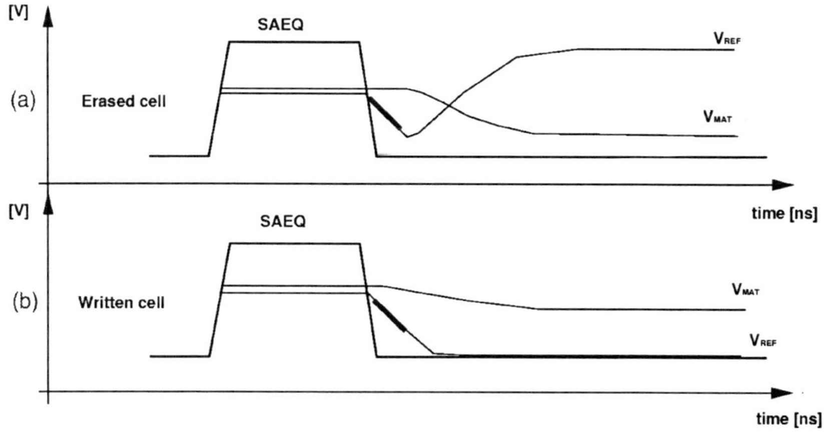

Fig. 12.36. Behavior of the V_{MAT} and V_{REF} potential as a function of the selected equalization point

At the end of the equalization, the V_{MAT} and V_{REF} potential are free to evolve. The mirror is still off, transistor M10 does not source any current and the comparator reads a written cell. We are in the correspondence of the concavity of the graph in Fig. 12.36a. Such a condition is then overcome when the mirror switches on and starts functioning properly. The bold lines have the same slope, demonstrating what just said.

In order to reduce the concavity, one has to reduce the equalization level, leaving the mirror "on" in a broader voltage range, paying attention not to exaggerate and not to come across the opposite problem. Therefore, the p-channels in Fig. 12.35, used only for ESD protections, are diode-connected, which reduces the concavity but slows down the separation of V_{MAT} and V_{REF} in the case of written cell.

12.13 Precharge of the Output Stage of the Comparator

After verifying that, in this case, the sensing time for the written cell is less than the one for the erased cell, it is possible to introduce a circuit that unbalances the output of the comparator and favors the reading of an erased cell. A circuitry like the one in Fig. 12.37 speeds up the toggling of the sense amplifier. The SAEQdelay signal switches on a two diodes network: thanks to the unbalanced triggering voltages of P1 and P2, the comparator output (DATA#) is forced low. After some delay, in order to guarantee a margin to settle V_{MAT} and V_{REF} voltages, the diodes are driven to the three-state condition, restoring the control of the DATA# signal to the sense amplifier comparator. In this way, useless toggles of the sense amplifier, which cost in terms of both power consumption and time, are avoided.

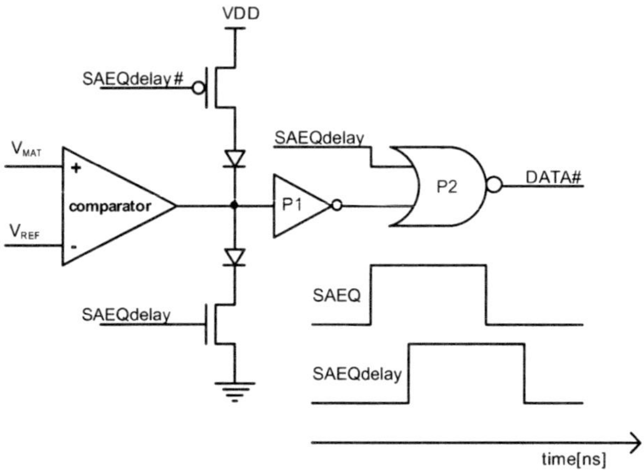

Fig. 12.37. Precharge of the comparator output to favor the toggling of the erased cell. The diodes are realized by means of MOS transistors.

12.4 Issues of the Reference

Let's now open a parenthesis to analyze the issues related to the realization of the reference to read. During the read phase of a memory cell, the discrimination of the status depends mainly on the correct functioning of the reference cell. In fact, the sense amplifier determines whether the bit to store is "0" or "1" by comparing the drain current of the addressed and the reference cell. The importance of this comparison emphasizes the necessity of having cells the characteristics of which are as similar to each other as possible. The necessity of minimizing the variations comes into conflict with other project requirements. Two different approaches to the design of the reference can be distinguished.

12.14.1 EPROM-Like Reference

In this case an array column is used for each output as reference column (Fig. 12.38). In practice, we try to place the reference cell as close as possible to the array.

Such a method has several advantages. First of all, the spread is minimized due to the proximity of the two cells. The reference is switched on together with the cell to read, since the row of the reference cell is in common to the array cell, which eliminates issues of timing. Furthermore, the loads on the two branches of the sense amplifier are the same. The drawbacks are: the stress of the reference since it undergoes the operations of the array cells and, moreover, to realize the reference, an array column is lost for each output. In the case of the Flash memories, further drawbacks exist. First of all, the reference columns must have separated ground from the rest of the sector where they are placed, since, otherwise, they might be depleted during erase. Secondly, the stress the reference column suffers might provoke cycling problems. Finally, there is little possibility of writing a large number of reference cells during EWS because it is a time consuming procedure.

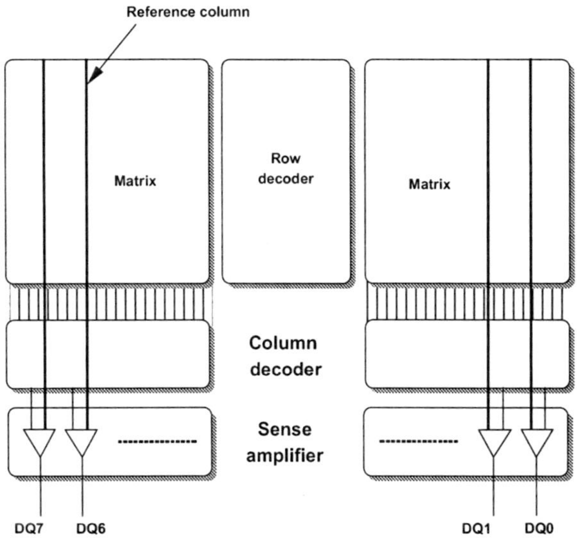

Fig. 12.38. Architecture of the EPROM-like reference: a reference column is dedicated to each output

12.14.2 Mini-Matrix

In this case, the reference cells are placed in a small dedicated array and the reference can be unique for the entire device. Therefore, in the testing phase, it is possible to write or erase the cells so as to obtain the best possible reference. The problem is that the reference and the matrix cells to read are located in different arrays, with the consequent issues of matching of the characteristics. In such a condition, a possible solution consists in using only one branch of the reference for all the sense amplifiers of the device, as indicated in Fig. 12.39. The drawback is related to the capacitive coupling of the circuit, mainly due to the bit line capacitance and the parasitic C_P capacitance. When the output switches, the dynamic voltage variation of the array branches of the respective sense amplifiers perturbs the reference branch which may cause problems during reading. Such a drawback is overcome by restoring the configuration of the diode-connected p-channels of the current-to-voltage converter. On the other hand, in this way, nodes MAT1, MAT2 and so forth, can reach at most the bias voltage minus the value of

MAT2 and so forth, can reach at most the bias voltage minus the value of one threshold voltage, i.e. with a reduced signal swing. This effect is undesired and has a non-negligible importance when the bias voltage is low.

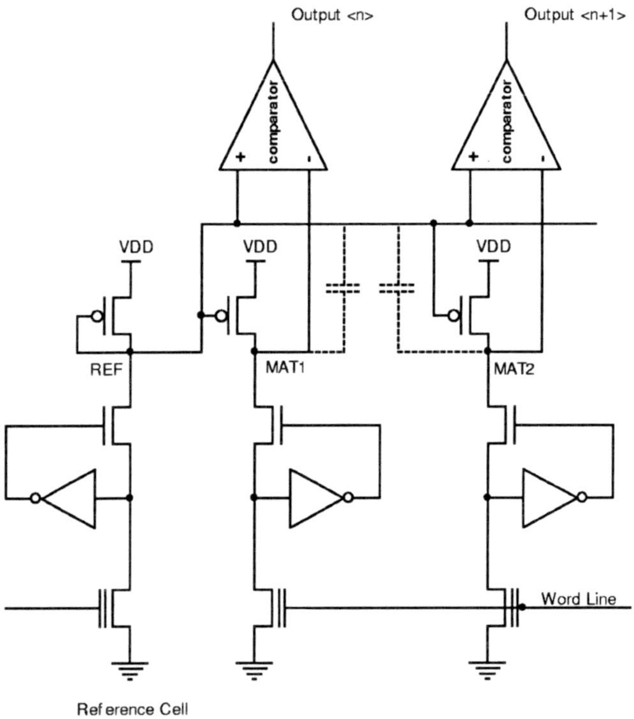

Fig. 12.39. Parasitic coupling that perturbs the reference voltage

12.15 Mirrored Reference Current

In order to overcome the foregoing drawbacks, it is possible to use one single reference cell, external to the array, but with completely separated sense amplifiers, so that they do not share directly the reference branch as indicated in Fig. 12.40.

The current used as a reference is the one that flows in the N1 diode-connected natural MOS of Fig. 12.32, that mirrors its current on the other natural MOS's (N2, N3, N4) that are connected to the reference branches of all the other sense amplifiers[6].

[6] It is not possible to substitute N1 for a Flash cell, since it should be erased in order to sink enough current. This means that also for N2, N3 and so on some cells whose threshold voltage should be corrected during the testing phase should be used, with the consequent great overload.

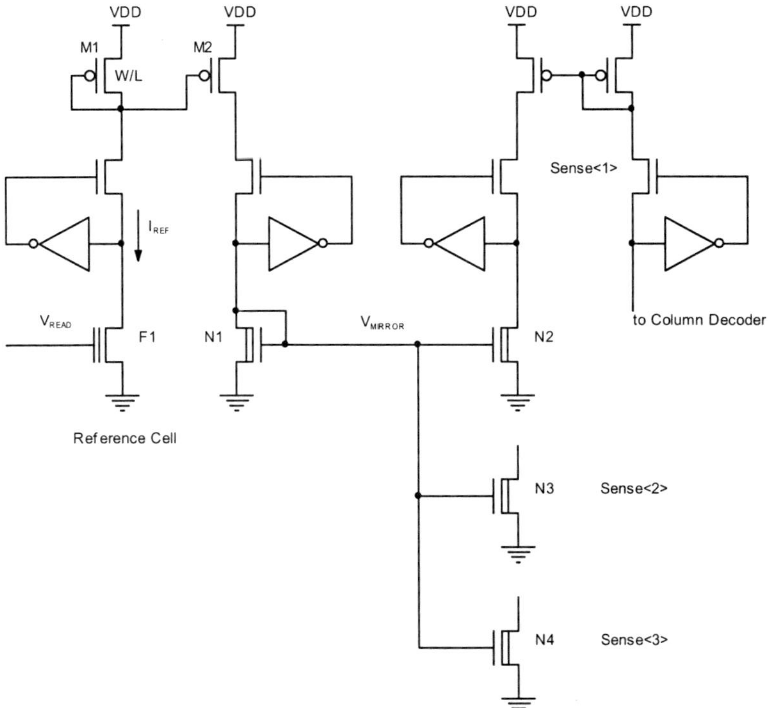

Fig. 12.40. The current of the reference cell, F1, is mirrored in the reference branches of the sense amplifiers that read the current of the array cell directly

An important point that should be considered is the speed with which the V_{MIRROR} potential reaches its steady-state value, to allow the various sense amplifiers to properly function. A parasitic capacitance is associated with the MIRROR node, due to the following contributions:

- the C_{GATE} input capacitance of the sense amplifier;
- the C_{ROUT} capacitance due to the routing connection of the MIRROR node to the sense amplifier;
- the gate capacitance of M1 and M2.

Once the memory architecture, i.e. the number of sense amplifiers present and their layout disposition has been defined, C_{GATE} and C_{ROUT} are fixed. At a first analysis, it may appear that the settling time of V_{MIRROR} can be reduced by simply multiplying the I_{REF} current by a factor $\eta > 1$. From the circuit point of view, such an operation consists in sizing the M2 transistor η times greater than M1. In this way, however, the gate capacitance of M2 increases, which slows down the switching on of the M1-M2 mirror. Referring to the charging of a capacitor at constant current, we can write the following relationship to express the settling time, T_{SET}:

$$T_{SET} \approx \left(\alpha \frac{C_{ROUT} + C_{GATE}}{\eta I_{REF}} + \beta(\eta + 1) \frac{W \cdot L \cdot C_{ox}}{I_{REF}} \right) \quad (12.21)$$

where C_{OX} is the capacitance of the gate oxide and α and β are factors of proportionality.

Considering that Eq. (12.21) contents two addends, one directly proportional, the other inversely proportional to η, it descends that, once C_{PAR}, C_{ROUT}, W and L of M1, and I_{REF} have been defined, there is a value of η that minimizes the settling time of V_{MIRROR}. Generally, such a value ranges between three and five.

As we have seen, having one single current reference for many comparators implies the use of circuits based on current mirrors. The problem that might pose is that such mirrors work properly only if the components are identical toward VDD as well as toward ground. Furthermore, also the bias conditions must be the same. Great attention must then be paid to the realization of the layout, so as to limit geometrical and process mismatch.

What just said implies that each sense amplifier has a different current reference. The sense amplifier that requires more current to read an erased cell limits the operations during erase, the one that requires less current suffers a penalty in terms of access time since it has to discriminate a lower current difference.

12.16 The Verify Operation

The program and erase procedures are followed by the verify operation, which attests that the modification of the content of the cell has actually been accomplished. In order to guarantee the functioning during read with a wide margin, such a verify operation is not a simple reading operation, but a reading "with margin", i.e. carried out by setting the cell in the worst case condition with respect to the standard reading. The way in which the verify operation is performed depends upon the type of sense amplifier used. Let's assume that we use a read circuitry with unbalanced loads.

12.16.1 Erase

The verify operation implies a reading "with margin" of the matrix cells after the erase pulse. This means that the loads of the sense amplifier must be modified so as to recognize the cell as erased only if it has a threshold voltage below a predefined value. In Fig. 12.41 the method applied in the case of the sense amplifier with unbalanceds load is shown. The read reference is substituted for another reference with threshold voltage V_T so that, when a gate voltage equal to the one indicated in the figure is applied to both the array and the reference cells, a cell is recognized as erased only if its threshold voltage is lower than V_{EV}. The cell is recognized as erased if it sinks more current than the reference one. This is one of the possible approaches and was used in the first generation of Flash memories.

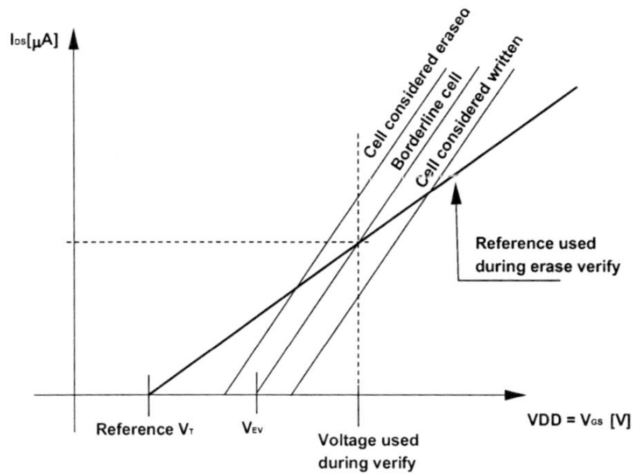

Fig. 12.41. Erase verify using the unbalanced load approach

In the most recent Flash memories, a parallel reference is instead used (Fig. 12.42). The characteristic of the reference is parallel to the characteristic of the matrix and, thus, the load of the two branches of the sense amplifier must be equal and not unbalanced as during reading.

The use of this kind of comparison has the advantage that a non-specific voltage can be applied to the cell gate; the current difference does not depend upon the biasing gate voltage since the characteristics are parallel. This fact is very important in a single supply device, in which each voltage greater than the minimum voltage value must be generated by means of charge pumps and suitable regulators. As usual, the V_{EV} threshold voltage of the reference cell used during erase is defined during the testing phase (EWS).

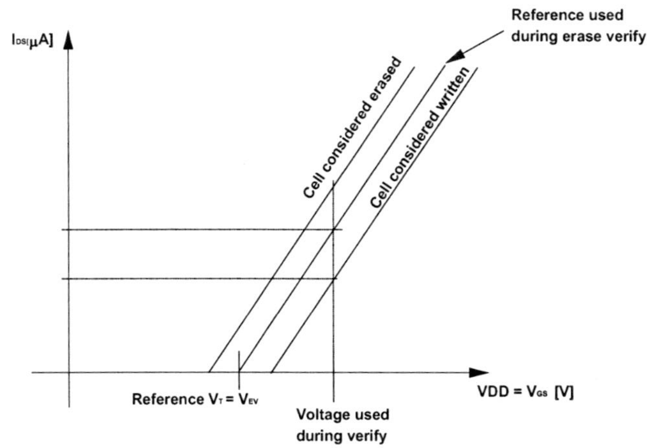

Fig. 12.42. Erase verify using the parallel approach. The V_{GS} voltage can vary in a very wide range.

12.16.2 Program

Program verify is carried out after each write pulse, to check the status of each single cell of the word and allow the control logic to block the subsequent pulse for the cel that has already reached the required V_T step. As for erase verify, the read reference is modified also in this case, so as to make the verify condition harder to be satisfied than during a standard read. In the first generation devices, the unbalanced reference was often used during reading as well as during the verify operations, which imposed the suitable gate voltage.

In the hypothesis that the characteristics are linearized and a reference with UV threshold voltage (around 2 V) is used, with a cell gain of 30 μA/V and a threshold voltage value of the "less written" cell of 4.8 V, the Program Verify voltage (Fig. 12.43) must be around 7.5 V.

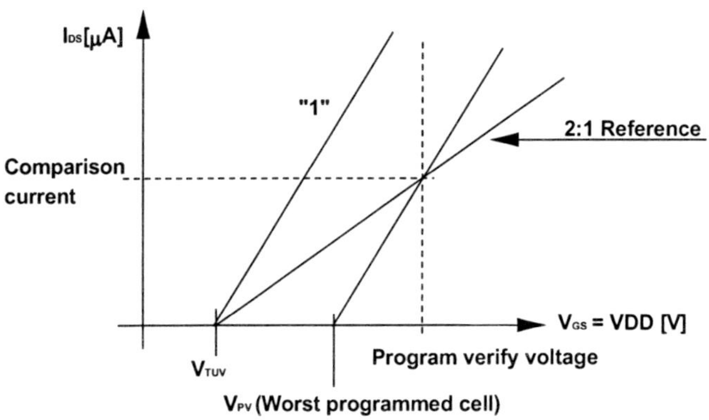

Fig. 12.43. Program verify using the unbalanced load approach

This value of the gate voltage defines the intersection of the reference characteristic with the characteristic of a cell having threshold voltage of 5 V. The advantage of this approach is that we use always the same reference just applying different gate voltages. The drawback, in general, is that high voltages must be applied, which induces undesired stress to the cells. In the case of single supply device, all the issues related to the generation of voltages above VDD are still present.

Also in this case it is easer to use a parallel reference, as for erase verify, which, on one hand obliges us to complicate the structure of the sense amplifier but, on the other hand simplifies the generation of the gate voltage. In Fig. 12.44 the schematic of the I-V converter is shown. It is an amplified converter during read and parallel converter during verify. The VER signal is grounded during read and activated only during verify. The N1 and M2 transistors guarantee the mirror-connection of M4 and M5 during read. The size of M5 is eight times larger than M4. During verify, M6 and M7, which have the same size, are activated, thus allowing a parallel approach. M3, M8, M9, and M10 are used to switch on and off the converters.

The value of the gate voltage at which program verify is carried out using the foregoing schematic is usually around 6 V.

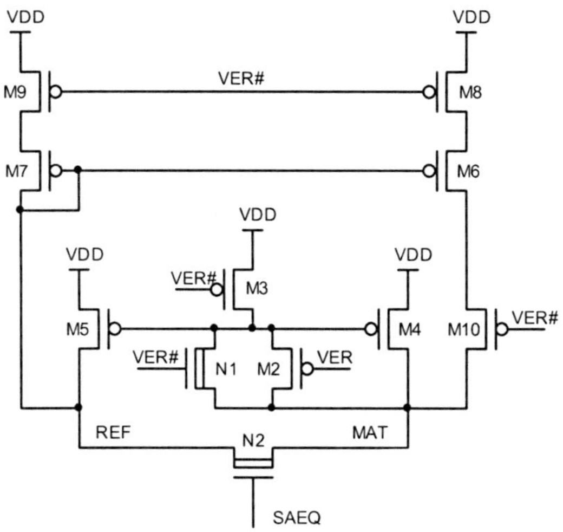

Fig. 12.44. Current/Voltage converter with a different configuration during read and erase

Bibliography

S. Atsumi, A. Umezawa, T. Tanzawa, T, Taura, H. Shiga, Y. Takano, T. Miyaba, M. Matsui, H, Watanabe, K. Isobe, S. Kitamura, S. Yamada, M. Saito, S. Mori, and T. Watanabe, "A channel-erasing 1.8V-Only 32-Mb NOR flash EEPROM with a bitline direct sensing scheme", IEEE J. Solid-State Circuits, vol. SC-35, pp. 1648–1654, (Nov. 2000).

R. Bez et al., "Depletion Mechanism of Flash cell induced by parasitic drain stress condition", VLSI Technology Symposium, (1994).

A. Brand et al., "Novel Read Disturb Failure Mechanism Induced By FLASH Cycling", IEEE IRPS, (1993).

G. Campardo, "Circuit for sensing the state of matrix cell in NMOS EPROM memories including an offset current generator", USA patent No. 4,949,307, (August 14, 1990).

G. Campardo, "Sense amplifier having capacitively coupled input for offset compensation", USA Patent No 5,729, 492, (March 17, 1998).

G. Campardo, "Read Path: Sensing technique", in Flash memory, P. Cappelletti et al., Ed Norwell, MA: Kluwer, (1999).

G. Campardo, M Dallabora, D. Novosel, "L'amplificatore di lettura nei dispositivi di memoria EPROM", Alta Frequenza, Vol. LVII, No. 6, pp. 143-154, (Luglio-Agosto 1988).

G. Campardo, R. Micheloni, S. Commodaro, "Method and circuit for reading low-supply-voltage nonvolatile memory cells", USA patent No. 6,128,225, (October 3, 20).

G. Campardo, R. Micheloni, M. Maccarrone, "Circuit and method for generating a read reference signal for nonvolatile memory cells", USA Patent No. 5,805,500, (September 8, 1998).

G. Campardo, R. Micheloni, M. Maccarrone, "Read circuit and method for nonvolatile memory cells with an equalizing structure", USA Patent No 5,886,925, (March 23, 1999).

G. Campardo, et al., "40-mm^2 3-V-only 50-MHz 64-Mb 2-b/cell CHE NOR Flash memory", IEEE J. Solid-State Circuits, vol. SC-35, no. 11, pp.1655-1667, (Nov. 2000).T. Jnbo et al., "A 5V only 16Mb Flash memory with sector Erase Mode", IEEE Journal of Solid State Circuits, Vol. 27, N 11, (November 1992).

G. Campardo et al. "A 40mm^2 3V 50MHz 64Mb 4-level Cell NOR Type Flash Memory", 2000 ISSCC, San Francisco.

G. Campardo et al., "Method and circuit for dynamic reading of a memory cell at low supply voltage and with low output dynamics", USA patent No. 6,639,833, (October 28, 2003).

M. Dallabora, et al., "A 20MB/s data rate 2.5V Flash memory with current controlled field erasing for 1M cycle endurance", 1997 IEEE International Solid-State Circuits Conference (ISSCC) Dig. Tech. Papers, San Francisco (California, U.S.A.), pp.396, (Feb.1997).

S. D'arrigo et al., "A 5V-Only 256K Bit CMOS Flash EEPROM", ISSCC 89, pp. 132-133.

S.H. Dhong et al., "High Speed Sensing Scheme for CMOS DRAM's", IEEE Journal of Solid-State Circuits, vol. 23, no. 1, pp. 34-40, (February 1988).

C. Golla et al. "A 30M Samples/s Programmable Filter Processor", ISSCC90, High-Speed Signal Processors.

H. Hidaka et al., "Twisted Bit-Line Architectures for Multi- Megabit DRAM's", IEEE Journal of Solid-State Circuits, vol. 24, no. 1, pp. 21-27, (February 1989).

IEEE 1995 Nonvolatile Semiconductor Memory Workshop, "Flash Memory Tutorial", Monterey, California, August 14, (1995).

S. Kato et al., "Read Disturb Degradation Mechanism due to Electron Trapping in the Tunnel Oxide for Low-Voltage Flash Memories", IEEE/IEDM, (1994).

A. Kramer et al., "Flash -Based Programmable Nonlinear Capacitor for Switched-Capacitor Implementations of Neural Networks", IEEE/IEDM Technical Digest, (1994).

Y. Konishi et al., "Analysis of Coupling Noise Between Adjacent Bit Lines in Megabit DRAM's", IEEE Journal of Solid-State Circuits, vol. 24, no. 1, (February 1989).

H. Kume et al., "A Flash-Erase EEPROM Cell with An Asymmetric Source and Drain Structure", IEEE IEDM, 25.8, (1987).

M. Lenzilinger, E. H. Snow, "Fowler-Nordheim Tunneling into Thermally Grown SiO_2", Journal of Applied Physics, Vol. 40, No. 1, pp 278-283, (January 1969).

R. Micheloni, et al., "The Flash memory read path building blocks and critical aspects", IEEE Proceeding of the, Vol. 91, No. 4, pp. 537-553, (April 2003).

A. Montalvo et al., "Flash EEPROM array with paged erase architecture", USA Patent N 5,126,808, (June 30, 1992).

D. Novosel, G. Campardo, "Sense circuit for the state of matrix cells in MOS EPROM memories", EP Patent, No. 0270750, (June 15, 1988).IEDM NVRAM, technology and application Short Course, (1995).

T.C. Ong et al., "ERRATIC ERASE IN ETOX™ FLAH MEMORY ARRAY", VLSI Symposium on Technology, 7A-2, pp. 83-84, (1993).

Cheng-Sheng Pan et al., "Physical Origin of Long-Term Charge Loss in Floating-Gate EPROM with an Interpoly Oxide-Nitride-Oxide Stacked Dielectric", IEEE Electron Device Letters, Vol. 12, No. 2, (February 1991).

A. Pierin, S. Gregori, O. Khouri, R. Micheloni, G. Torelli, "High-Speed Low-Power Sense Comparator for Multilevel Flash Memories" in Proc. 7th Int. Conf. Electronics, Circuits and Systems, vol. II, pp. 759–762, (Dec. 2000).

Betty Prince, "Semiconductor Memories. A Handbook of Design Manufacture and Application", Wiley & Sons, (1993).

A. Silvagni et al., "Modular Architecture For a Family of Multilevel 256/192/128/64MBIT 2-Bit/Cell 3V Only NOR FLASH Memory Devices", ICECS 2001 The 8th IEEE international Conference on Electronics, Circuits and System, September 2-6, Malta, (2001).

G. Verma & N. Mielke, "Reliability Performance of ETOX Flash Memories", IEEE IRPS, pp. 158-166, (1998).

S. Yamada, "A Self-Convergence Erasing Scheme for a Simple Stacked Gate FLASH EEPROM", IEEE IEDM, 11.4.1, (1991).

13 Multilevel Read

13.1 Multilevel Storage

The necessity of non-volatile memories to have higher and higher densities leads to the consideration of multilevel memories. In this case, it is possible to generate more than two logic levels by controlling the amount of charge stored on the floating gate.

Figure 13.1 shows the characteristics ($V_{GS,}$ I_{DS}) of a bi-level Flash cell in reading condition. The characteristic with the V_{TV} threshold voltage, generally ranging between 0.5 and 2.5 V, is assigned to logic value "1", whereas the characteristic with V_{TW}, typically greater than 5 V, is assigned to value "0". As stated in Chap. 12, reading consists of the conversion of the current drawn by the cell at a defined V_{GS} into a voltage that is then translated into the corresponding logic level.

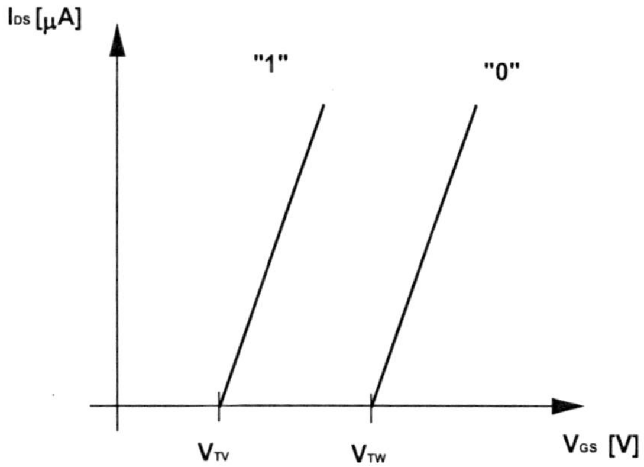

Fig. 13.1. Bi-level cell

In the case of multilevel cells, the (V_{GS}, I_{DS}) plane is divided by several characteristics. For example, in the case of 2 bits/cell, there are four resulting characteristics (Fig. 13.2) to which the 11, 10, 01, and 00 logic values are associated. In reality, we obviously deal with threshold voltage distributions spaced with respect to each other in a way that allows the determination of the logic value by means of the sense amplifiers.

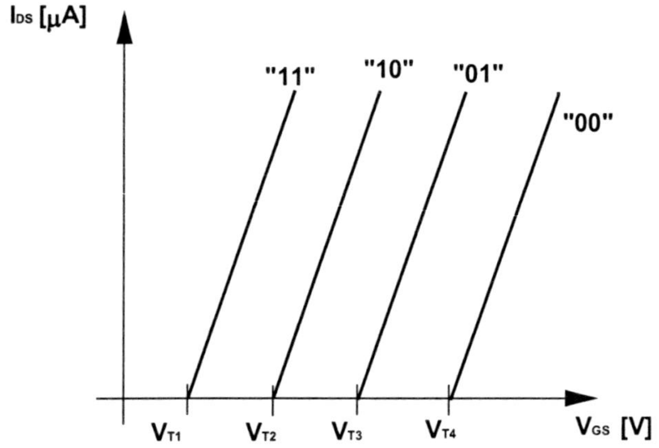

Fig. 13.2. Binary encoding of the threshold voltage in the multilevel case

One of the problems in reading the cell is related to the V_{GS} voltage. Since reading is a current-to-voltage conversion, the voltage that is applied to the cell gate must be at least greater than V_{T3}. Figure 13.3 shows one possible placement of the distribution; for sake of simplicity, the distributions are reported in the same scale (except the 11 value which has a larger V_{GS} range) and equally spaced. The read voltage, for example, can be fixed to 6 V.

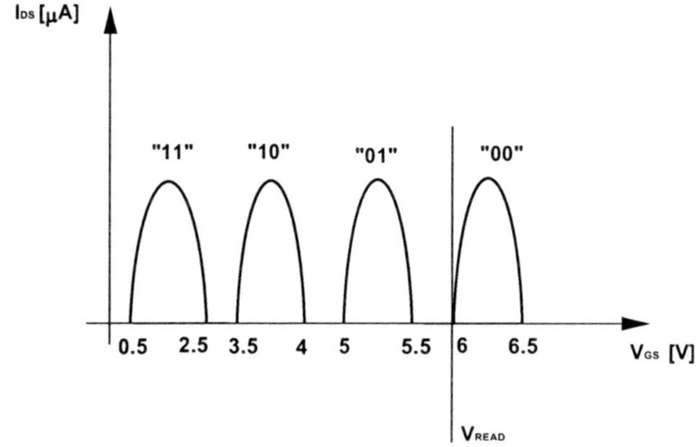

Fig. 13.3. Position of the threshold voltages for a cell containing two bits

13.2 Current Sensing Method

One way to discriminate the cell current is the usage of three reference cells, with V_T programmable during the testing phase and located between the various characteristics (Fig. 13.4). For sake of simplicity, also in this case the three references are centered between the distributions of cell currents representing the logic states.

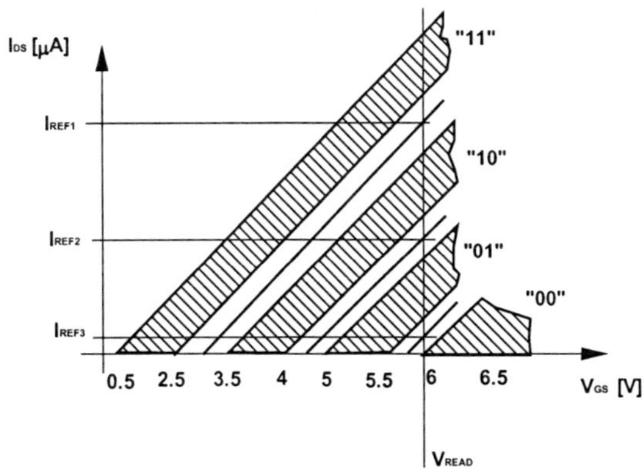

Fig. 13.4. Current sensing approach: the cell is read at constant V_{GS}, exploiting the three different reference cells

The read operation is accomplished by applying a gate voltage to the memory cells (V_{READ}) and comparing the current drawn by the cell to the three reference currents, I_{REF1}, I_{REF2}, I_{REF3}. A circuit structure, composed of several I-V converters, compares the conversion results and provides the logic value to the output, as shown in Fig. 13.5. The minimum working window of the threshold voltage is determined by reliability requirements rather than the number of bits that the cell stores. It is then evident that the multilevel structure causes a reduction in the distance between the various V_T distributions. This reduction increases the difficulty of sensing, since there is less differential current between logic states. Due to the limited gain of the cells (between 20 and 25 μA/V), the sense circuitry must discriminate currents on the order of 10 μA, also in the case of only four levels.

Figure 13.6 shows the real position of the three reference cells. The characteristics of the first two, R1 and R2, are parallel to the characteristics of the array cell. The threshold voltage of the third cell, R3, is instead set to the upper value of the "01" threshold voltage distribution. To understand the particular position of R3 it is necessary to recall that, while V_{READ} remains practically fixed as bias voltage and temperature vary (it is derived from a bandgap reference), the cell threshold voltage drifts with the temperature. Thus, there is the risk of a too low reference current when the temperature decreases. The unbalanced read technique is just for lowering the threshold voltage of R3.

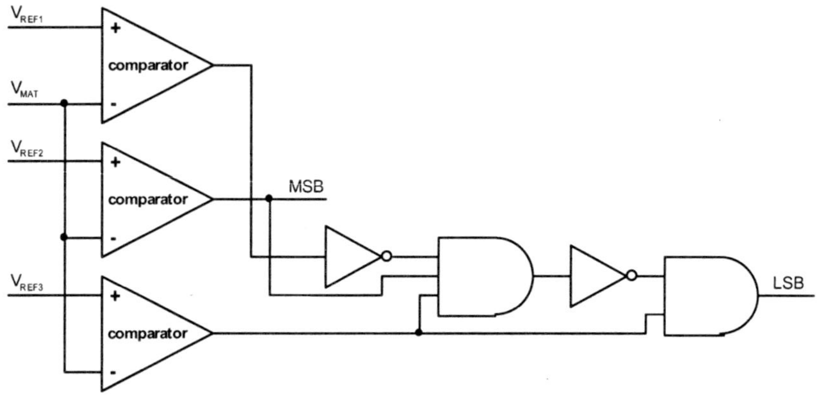

Fig. 13.5. Analog-to-digital converter for parallel multilevel sensing

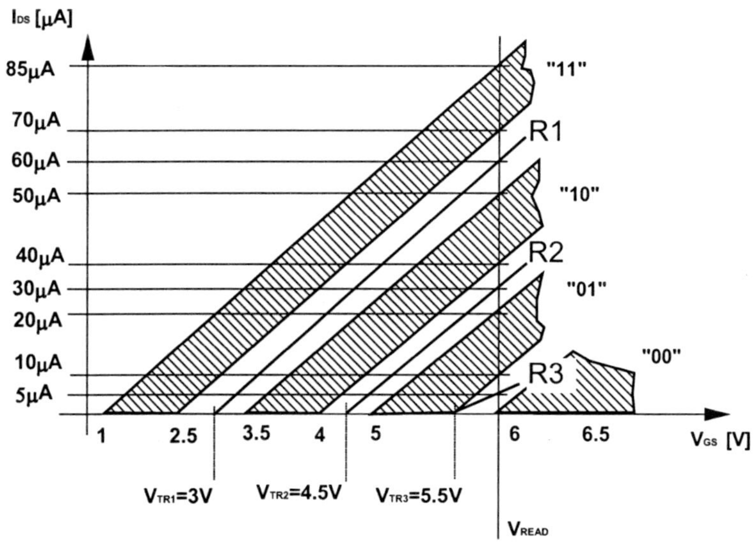

Fig. 13.6. Actual position of the reference cells: R3 is used in the unbalanced configuration

In order to fix the problem of the limited available current, it is possible to use the amplified I-V converter, already analyzed in Sect. 12.9. It is preferable to amplify the current of the cell corresponding to the "01" level because this is the one that sinks less current during read due to the high V_T. For example we could decide to multiply the current of the "01" cell by three; while the "11" and "10" distributions are only doubled so as to limit power consumption. Figure 13.7 summarizes the ratios defined between the currents of the various distributions.

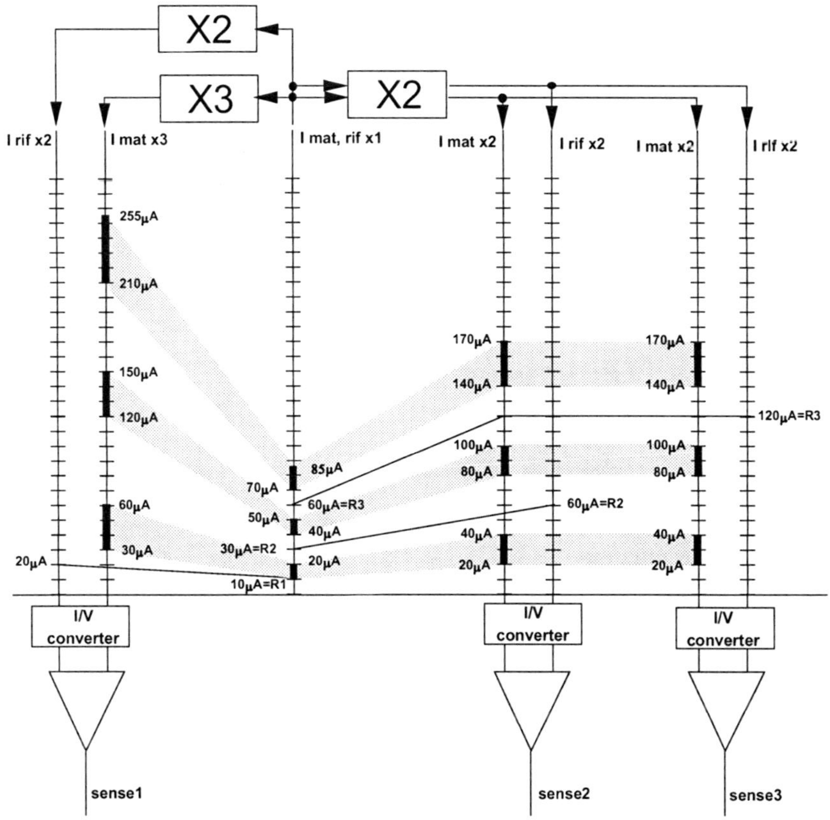

Fig. 13.7. Amplified I/V converter for multilevel reading

The vertical axes represent matrix and reference currents. The "Imat, rif x1" axis indicates the current values referred to the tails of the different distributions, as shown in Fig. 13.4. The first distribution is in the range between 100 and 70 μA, the second between 50 and 40 μA, the third between 20 and 10 μA. The three references sink 60 μA (R1), 30 μA (R2), and 10 μA (R3), respectively. The different currents are collected by means of current mirrors (the blocks located above) and transferred to the current/voltage converters.

The current of the array cell is sensed and multiplied, in one case by three, in the other cases by two. The reference currents are always multiplied by two and, thus, in the case of the converters connected to sense2 and sense3, the overall effect is the doubling of the dynamic signal range. In the case of sense1, with the lowest reference current R3, we must introduce a different multiplication factor for the array, to be able to distinguish the array from the reference. In this specific case, R3 must always be lower than the current drawn by the cell and, hence, the cell current must have a higher multiplication factor. Figure 13.7 shows the different currents that reach the current/voltage converters: for each reference, the different vertical axes represent the possible position of the current of the various distributions. Basically, the three reference currents are collected by the small reference array, amplified by a factor 2 and presented as inputs of the respective I/V converters. The cell current is instead collected and multiplied by two or three. The current multiplied by two is input to the converters of sense2 and sense3, while the current amplified by three is input to sense1. In this way, we obtain the desired result of increasing the dynamic range of the available current, limiting, at the same time, the threshold voltage of reference R3 at the upper value of the "01" distribution.

13.3 Multilevel Programming

The precision required in the multilevel case is not limited to the read operation; it is also necessary to design a write mechanism to program the cells within the different distributions. The write mechanism usually employed in Flash memories with NOR architecture is the hot electron technique presented in Chap. 3: by applying a voltage of about 10 V on the control gate, a voltage of about 5 V to the drain and leaving the source connected to ground, it is possible to generate an intense electric field between isolated gate and channel and, at the same time, a high current between source and drain. This permits the electrons flowing between source and drain to acquire high kinetic energy and reach the floating gate from the channel, overcoming the energy barrier to cross the tunnel oxide.

The advantages offered by the implementation of a multilevel memory are accompanied by formidable problems related to the reduction of the difference between the different threshold voltages corresponding to the various charge quantities stored on the floating gate and, hence, between the different cell current flow.

The programming of a multilevel memory cell requires very precise control of the charge stored on the floating gate.

It is possible to define a linear relationship between the variation of the voltage applied to the gate during the program phase, ΔV_G, and the threshold voltage step, ΔV_T, obtained at fixed V_D and V_S, as shown in Fig. 13.8.

The memory cell is, therefore, programmed by applying a series of linearly increasing step voltages to the gate: each program pulse has to differ from the previous and the subsequent one by a constant value, ΔV_G. The program gate voltage has therefore the appearance of a staircase with constant step. In particular, a significant result obtained by the use of this method is that, neglecting all the effects related to retention, temperature variation and so forth, the width of the threshold voltage distribution equals the step of the staircase voltage applied.

The method described above is inherently slow since it requires the application of a sequence of pulses to the cell gate. In order to obtain a single byte program time comparable with a standard bi-level cell it is therefore necessary to program several cells in parallel. Let's suppose that the single byte program time is 6 μs in the bi-level case; if we need 200 μs to apply the entire program stair, it is necessary to program 256 bits simultaneously to obtain a program time of 6 μs for each single byte.

The reader can refer to the next chapter for a more detailed description of the entire program algorithm.

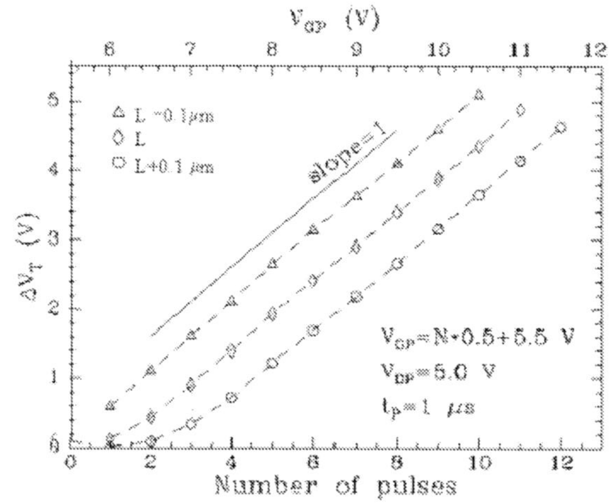

Fig. 13.8. Staircase programming: ΔV_T as a function of the number of pulses for three different channel lengths. The upper axis shows the gate voltage at each program step

13.4 Current/Voltage Reference Network

As stated in Sect. 12.15, the number of reference cells typically used in Flash memories is lower than the number of sense amplifiers present; as a consequence, the V_{REF} reference voltage is derived from one cell and shared among several sense amplifiers. Generally, two or four reference voltages can be obtained from one reference cell, and each of them is shared among a group of sense amplifiers, as sketched in Fig. 13.9. For example, the V_{REF} reference voltage can be the V_{MIRROR} voltage in Fig. 12.40; by replying I_{REF} several times (M1, M2 mirror), several V_{MIRROR} can be obtained.

The conversion of array current into voltage is carried out locally inside the sense amplifier. The reference current is instead converted into a voltage in a location away from the sense amplifiers since the reference cells belong to the dedicated small array. The reference voltages are obviously referred to a local ground of their I/V converters and, hence, it is important that the offset in the ground connections be minimized between the reference circuitry and the sense amplifiers.

In Fig. 13.10 the layout of the power grids inside a generic device is shown. The ground potential is routed by means of a ring of metal lines, whose width is the result of a tradeoff between area occupation and parasitic resistance. The voltage drop on the ground lines depends on the line layout while the current consumption of the sense amplifiers depends on the current of the array cells that are read. Thus, the current consumption and the consequent voltage drop on the ground lines depend on the pattern that is read.

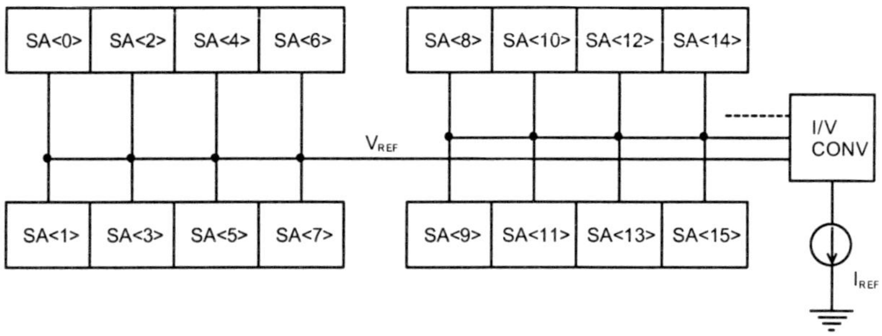

Fig. 13.9. V_{REF} reference voltage shared among several sense amplifiers

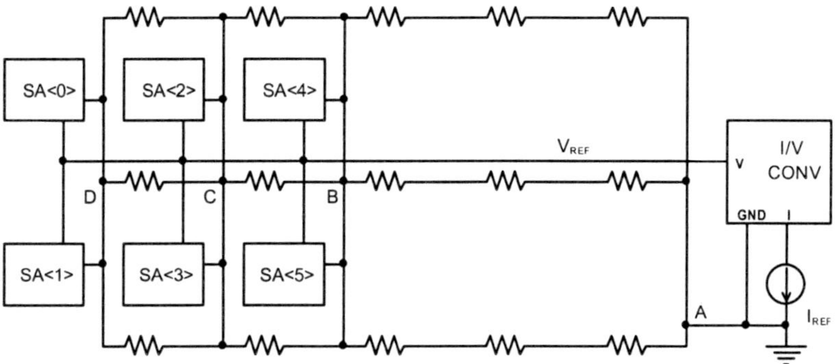

Fig. 13.10. Layout of the ground lines and V_{REF} in the area of the sense amplifiers

In order to better understand the importance of the effect described above, let's consider the case of a multilevel memory.

Programming is performed by increasing the cell threshold voltage by small steps (typically 300 mV for a multilevel memory with 2 bits/cell). Starting from the distribution of erased cells ("11"), a verify operation is performed at the end of each single step, i.e. an actual read operation is performed. The cells that have reached the required distribution are not subjected to further programming pulses. During the various verifications, the current that flows in the sense amplifiers depends on several cell currents, which are changing as the cells are programmed. The verifications are thus performed with different ground currents and can have variation in the ground offset as the cells are programmed to different threshold voltages.

The problem arises when one or more cells reach the required distribution. Although they are not involved (because they have been verified correctly), further program pulses will be applied to the other cells. In practice, at the end of the program phase, the voltage drop on the ground lines does not equal the drop during the first cell verification.

Let's examine Fig. 13.10 and suppose that we want to program the cell connected to SA<0> to level "11" and all the others to "00".

The I/V conversion takes place according to the local ground of the converter. For the reference it is node A in Fig. 13.10, for the SA<0> sense amplifier, it is node D, for SA<2> it is node C and so forth.

The first verification is successful for SA<0>, while the other cells have not yet reached the programmed state so these cells do not verify correctly. The first verification is executed with the sense amplifiers connected to the cells that sink the highest current and, therefore, the voltage drop between nodes D and A is the highest.

As the threshold voltage of the cells is increased, the current consumption diminishes; and therefore the voltage drop between D and A decreases. This variation of V_{DA} is transformed into a differential voltage input signal for the voltage comparators of the sense amplifiers. This means that the distributions are enlarged and consequentially the device has lower noise immunity and increased access time.

Problem 13.1: Due to the effect described above, do the distributions of the threshold voltage enlarge only in one direction or do both tails move?

The solution generally adopted in such a case is to bring the generation of V_{REF} closer to the group of sense amplifiers that share the reference, as indicated in Fig. 13.11. In this case, the reference signal transmitted inside the memory is the I_{REF} current which implies advantages in terms of noise immunity.

Problem 13.2: What may happen when a current is reproduced by means of transistors in mirror configuration that are located in very different areas of the chip? Recalling the PMOS case, consider that a small voltage drop takes place between the sources of the two p-channels.

In Fig. 13.11, the ground for the I/V conversion of the reference current is node B. In this way we eliminate the contribution of V_{BA} because this drop is now a common mode signal for the voltage comparator of the sense amplifiers. However, the variation of the voltage drops has not been overcome since the problem of V_{DB} is still present.

A good solution in such cases is the adoption of a different ground line for each I/V converter. The various ground lines are connected to a single node (node B in Fig. 13.11) to which the I/V conversion of the reference current must be referred. In this way, the voltage drop between the ground of the single converter and node B is no longer influenced by the current variations of the other sense amplifiers.

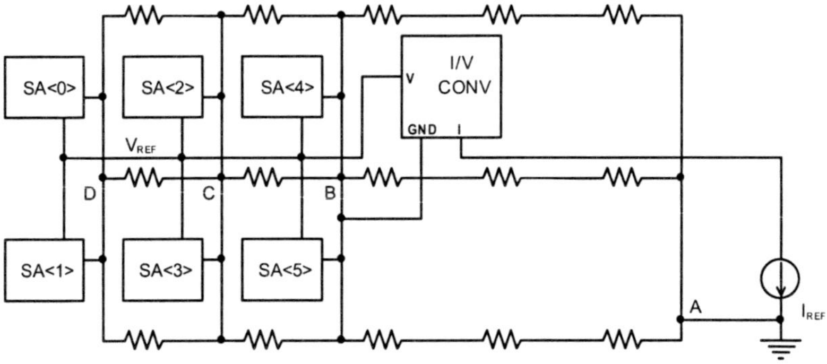

Fig. 13.11. The reference signal is the I_{REF} current and the I/V conversion is performed locally, close to the group of sense amplifiers that share the same V_{REF}

Problem 13.3: Draw the layout of ground and reference voltages in the case of dedicated ground line for each sense amplifier.

Problem 13.4: Evaluate the impact of the realized layout in terms of area occupation.

13.5 Voltage Sensing Method

Basically, the reading of a non-volatile cell can be accomplished in two ways: with either constant gate voltage or constant drain current, as shown in Fig. 13.12. In both cases, the drain and body voltages are kept constant. If the gate voltage is kept fixed, we talk about current sensing, since the value of the current indicates the program status. On the other hand, if the drain current is kept constant, voltage sensing is utilized since V_{GS} provides the means to determine the data stored. While the current sensing method and its variations are the conventional techniques used to read the traditional bi-level and four-level memories, the voltage sensing method has been developed to suitably read the more complex multilevel devices.

Figure 13.13 shows the eight logic levels that are necessary to store three bits in a single cell. In this case the voltage sensing is appropriate. Instead of biasing the word line directly to 6 V, the V_{GS} voltage of the cells increases in a staircase fashion. For example, the first and the final value of such a staircase could be 2 and 6 V respectively. The number of steps of the row voltage equals the number of levels minus one.

The operating principle of this technique is based upon the comparison between the current drawn by the cell and an I_{REF} current generator. In Fig. 13.14 the "101" distribution is shown and the reference current is 10 μA. The transient of the word line voltage is reported in Fig. 13.15. At each step, the sense amplifier verifies whether the cell sinks more current than 10 μA. With reference to the case of the "101" level, the comparator of the sense amplifier toggles during the third step of the V_{GS} voltage. This operation requires a single sense amplifier for each cell, with remarkable saving in terms of area since the result of the comparison is "1" or "0", like in the case of bi-level reading.

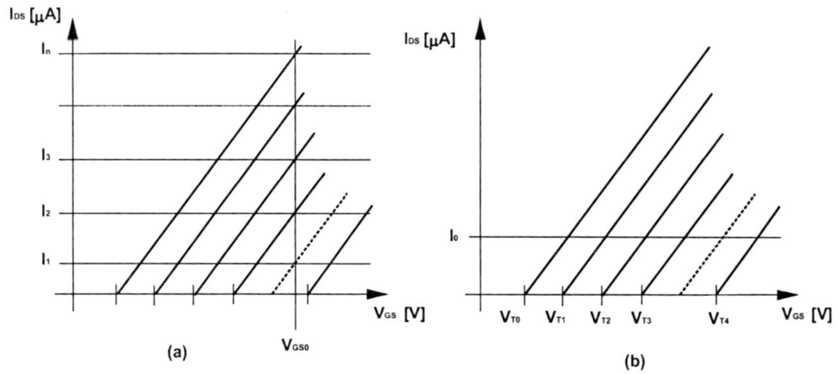

Fig. 13.12. Comparison of read methods: **(a)** constant gate voltage (V_{GS0}), **(b)** constant drain current (I_0)

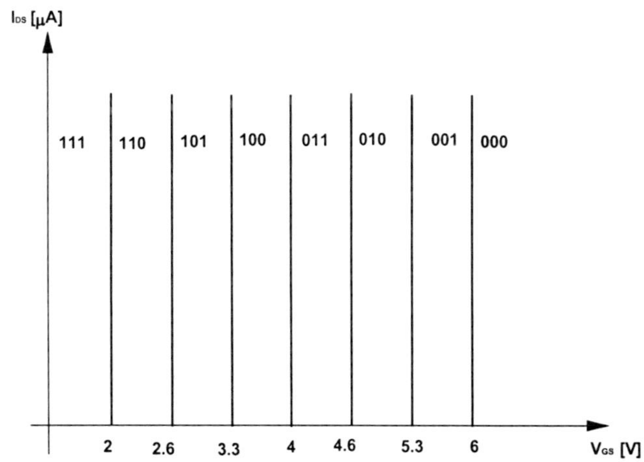

Fig. 13.13. The eight distributions that are necessary to store 3 bits in a single cell

The steps of the row voltage can have a 15 ns time duration, leading the overall time to charge the row to 105 ns. This delay significantly impacts the access time. By reading more cells in parallel and exploiting the fact that each cell stores three bits, it is possible to improve the read throughput by means of an architecture that allows burst readings (see Chap. 11).

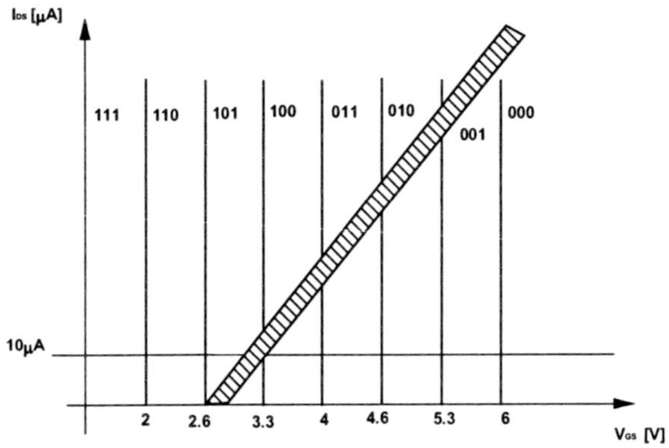

Fig. 13.14. Current/voltage characteristics of the cells corresponding to the "101" logic level

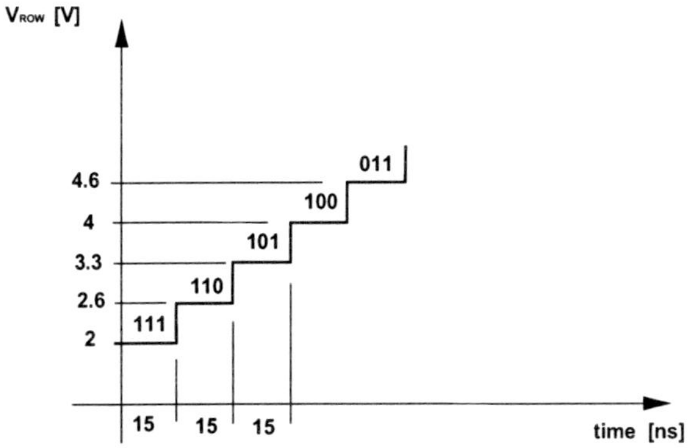

Fig. 13.15. Word line staircase voltage

The comparison between the cell current and the current of the reference generator can be carried out with a classic sense amplifier structure, as sketched in Fig. 13.16. The I/V converter has to pull down the V_{MAT} potential so as to be lower than V_{REF} when the cell sinks more than 10 μA. The toggling of the comparator, together with the voltage value the stairs has reached identifies the set of bits stored in the memory cell.

Problem 13.5: Considering what discussed in Chap. 12, define the details of the sense amplifier in Fig. 13.16.

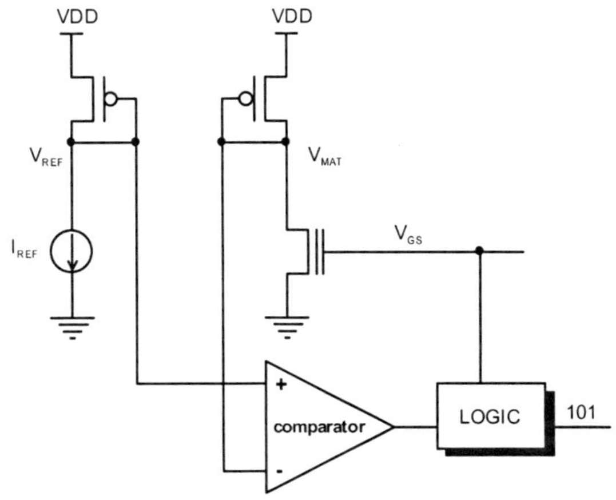

Fig. 13.16. Schematic of the sense amplifier that performs the comparison between the cell current and the reference generator (for example 10 μA)

13.6 Sample & Hold Sense Amplifier

The foregoing solution is not easily employed when more than three bits are stored in the same cell. In fact, the necessity of higher and higher number of steps for the word line voltage has an increasingly negative impact on the access time. In this section, a sense amplifier that performs a sort of sample & hold operation and also holds the value of the row voltage that caused the toggling is presented.

As we saw in Chap. 9, row and column decoders allow the unambiguous selection of the cells to read or program. The byte (or word) is made of cells that are located on the same row but on different columns. In this way, the current drawn by each cell is converted into a voltage. This kind of architecture does not permit independent behavior of the word line for each cell unless a serial read is carried out.

In order to separate the eight cells that compose the byte it is possible to adopt a hierarchical row decoding. In Fig. 13.17 the typical structure of a hierarchical selection of the word line is indicated. The local decoders separate the sectors, i.e. the portions of the array that are simultaneously erased.

Figure 13.18 shows a different kind of hierarchical organization, in which the same sector is further divided into blocks by means of local row decoders. In this way, bits belonging to different bytes (or words) are grouped in the same block and, therefore, each cell of the byte has its own local row. Using such a technique, it is easy to design the sample & hold circuit which stores the row voltage. Figure 13.19 shows the sample & hold circuit of two cells belonging to the same byte but located on different rows. For the sake of simplicity, in the representation of the local decoder only the PMOS transistor (PY0, PY1) used to transfer the read voltage has been highlighted. The reader can refer to Sect. 9.12 for further details on the structure of the local decoders.

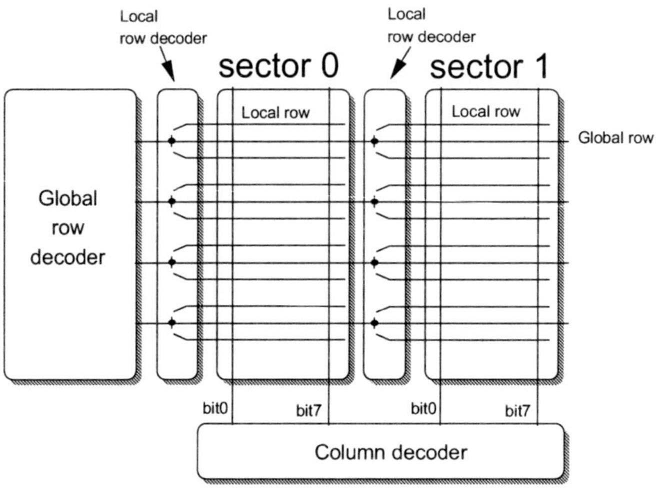

Fig. 13.17. Hierarchical row decoding

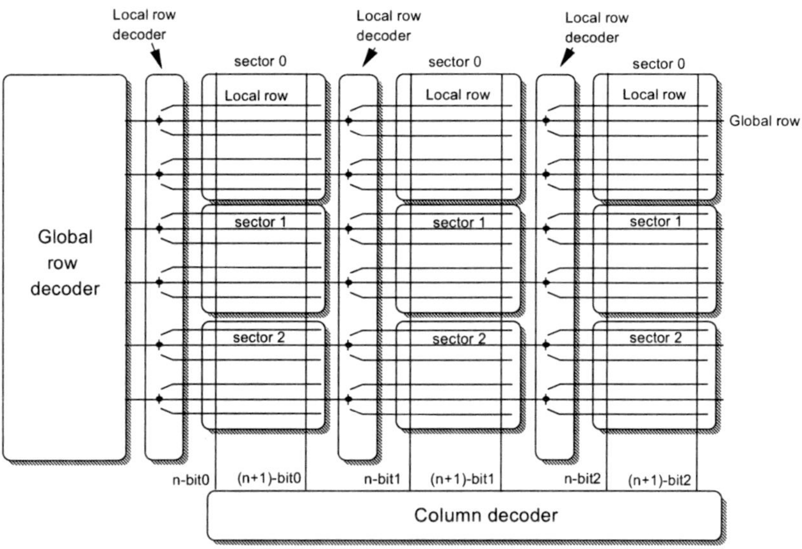

Fig. 13.18. Row decoder for sample & hold sense amplifiers

The voltage of the global rows increases linearly with time. When the comparator signals that the cell sinks more than 10 μA (I_{REF}), the P1 p-channel transistor is switched off. The local row voltage is then stored on the parasitic capacitance (0.5

÷ 1 pF), and is approximately the voltage reached by the main word line (MWL) at the time the comparator commutes. Obviously, each local row (LWL) is able to reach a voltage value independently of the other rows. We could imagine a voltage ramp applied to the global row having 100 ÷ 150 ns time duration, depending on the number of levels to discriminate. The principle is to take the access time of a standard 3 V bi-level flash memory as a reference, i.e. 70 ns typically. In the case of 16 levels, 8 cells produce 32 output bits, which means 4 times the number of bits of the bi-level memory that works by bytes.

The access time to read 4 bytes of a bi-level memory is therefore 280 ns, which is the figure we refer to as a benchmark.

Let's now determine how to read the voltage of the word line. In Fig. 13.20 the circuitry of two cells is shown: let's concentrate on the cell marked with the circle. The voltage ramp is applied to MWL<1>, while YO<0> is active. In such conditions, M1, M3, M5, and M7 are on while M4 and M8 are off. The only local row that is high, as a result of the combination of the high main word line and YO<0>, is the local row to which the indicated cell belongs. The MWL starts going up, driven by the global row decoder (not indicated in the figure). When this cell sinks 10 μA, the comparator related to cell 0 triggers, turning M1 off and, hence, sampling the local row. Meanwhile, M2 is on and M6 is off. The voltage of the addressed local row is transferred to the A/D converter through the wl-bit line. The conversions of all the other cells that form the byte (word) are performed in parallel. The local rows have transistor M2 in common, which is also connected to M6.

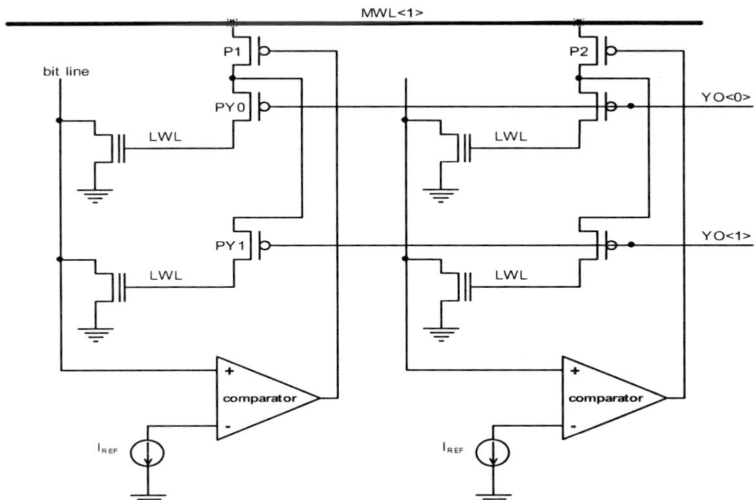

Fig. 13.19. Sample & hold sense amplifier for two bits: the voltage value reached by the word line is hold in the parasitic capacitor of the row

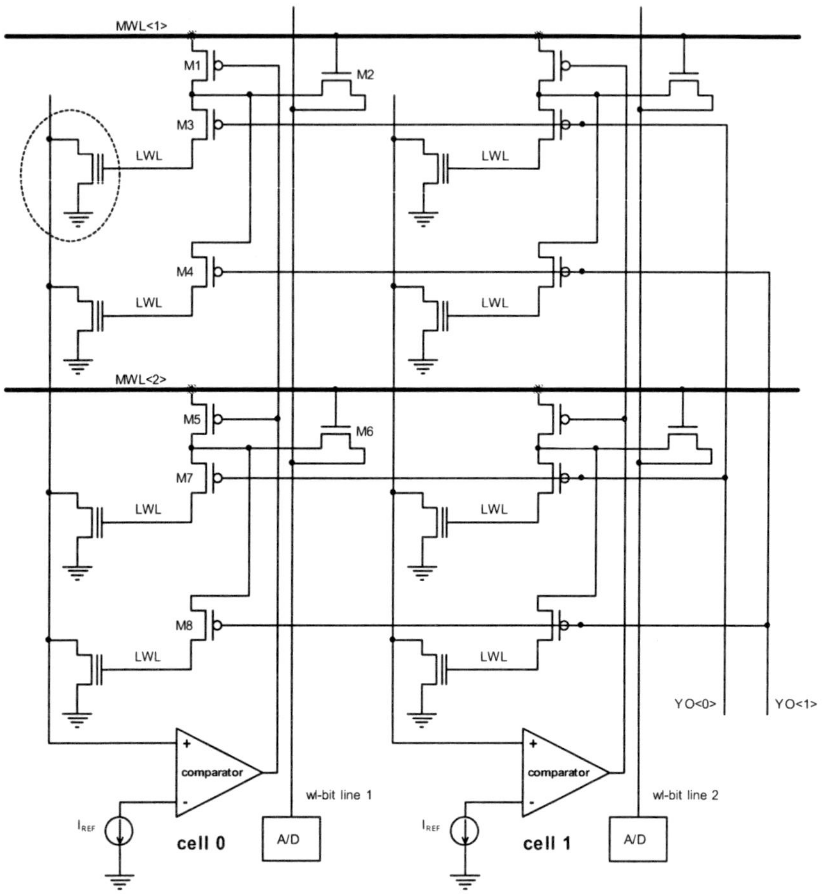

Fig. 13.20. Sample & hold sense amplifier of the line called "wl-bit line" that allows the voltage of the word line to be transferred on M2 and M6 though the A/D converter

The problem is that M2 has a positive threshold voltage and, therefore, the parasitic capacitor of the wl-bit line starts being charged only when main word line 1 has passed the V_T of M2. Given the V_{SH} voltage of the local word line, we can calculate the V_{AD} voltage transferred to the A/D converter taking into account the parasitic network shown in Fig. 13.21

$$V_{SH} \cdot C_{WL} + \left(V_{SH} - V_{T,M2}\right) \cdot C_B = V_{AD} \cdot \left(C_{WL} + C_B\right) \tag{13.1}$$

$$V_{AD} = V_{SH} - V_{T,M2} \frac{C_B}{C_{WL} + C_B} \tag{13.2}$$

Considering the case in which $C_B >> C_{WL}$, we obtain:

$$V_{AD} = V_{SH} - V_{T,M2} \tag{13.3}$$

The above relationship assumes the body effect is negligible. In order to actually eliminate the contribution due to body effect, transistors M2 and M6 must be fabricated in a triple well structure. Then the voltage that is input to the converter is that of the word line minus a constant term.

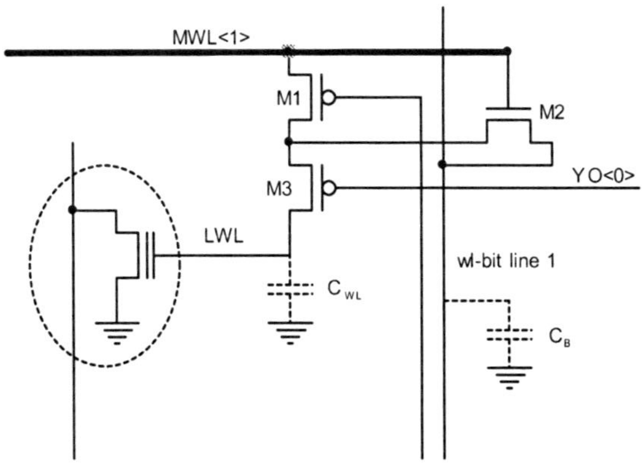

Fig. 13.21. The parasitic capacitor limits and modifies the voltage transferred from the local word line to the A/D converter

The solution proposed above suffers from the delay introduced by the circuit used for the feedback (switching P1 off in Fig. 13.19). If the signal that drives the loop opening does not operate properly, V_{GS} will be incorrect since it will contain a ΔV term proportional to the delay. In fact, the ramp applied to the word line continues regardless whether the local word line is connected. An error due to the loop delay will be more significant if the ramp rises quickly. The ΔV term can be regarded as an offset, and therefore eliminated if it is independent of operating conditions and of cell status. Another limit is due to the nature of the sample circuits: once the loop is open, the V_{SH} voltage is fixed and cannot be changed. Any disturb that causes the comparator to toggle will result in an incorrect reading of the cell contents.

13.7 Closed-Loop Voltage Sensing

Another way to implement voltage sensing is to use a real closed-loop feedback. The difference between the current drawn by the selected and the reference cell drives (more or less directly) a voltage source connected to the cell gate. Equilibrium is reached when the two currents are equal. The gate is regulated to the voltage value that maintains a state of equilibrium. The block diagram of the closed-loop voltage sensing is reported in Fig. 13.22.

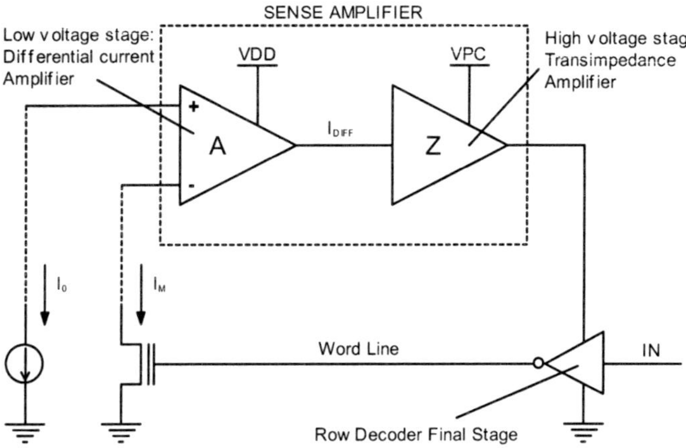

Fig. 13.22. Closed-loop voltage sensing approach

The advantage of this solution is the fact that the V_{GS} voltage settles only when the current of the cell reaches the required value. Possible delays in the signal propagation do not impact the final value but only the settling time. The main problem of the closed-loop reading method is that high speed reading increases power consumption. High speed means that the voltages involved have to vary quickly, and the parasitic capacitors of the system has to be charged and discharged with high current.

In order to realize the circuitry, it is possible to use a class AB amplifier with two stages, as indicated in Fig. 13.23.

The class AB amplifier has good performances during both the word line charging and discharging phase. This solution also minimizes the overshoots caused by delays in the feedback. The choice to use a two-stage structure is mainly due to the power consumption of the charge pumps. In fact, as one can see in Fig. 13.23, only the second stage is biased by means of the VPCX (> VDD) high voltage.

Let's now move on to the analysis of the circuit. The bit line limiter is a conventional common gate device, with M1 biased by inverter INV. The cell current is mirrored by the first stage and compared with the reference current source. The resulting current is amplified and buffered by the second stage. The bias circuit of the HV stage, made up of transistors M4, M5, M10, and M11, is worth noting. The operation is based on the principle of the current mirror, with a mirroring factor of unity. The input branch is made of two diode connected MOS's (M10 and M11), together with the I_{HV} source so as to control the current that flows in the branch. The voltage drop at the diode terminals is the same as the one between the gates of M4 and M5. If a current is injected into the input branch, it is mirrored in the output branch. Thus, due to the symmetry, node B has the same potential as node F. In this way, the DC currents and voltages of the high voltage stage can be controlled, without limiting the dynamic range.

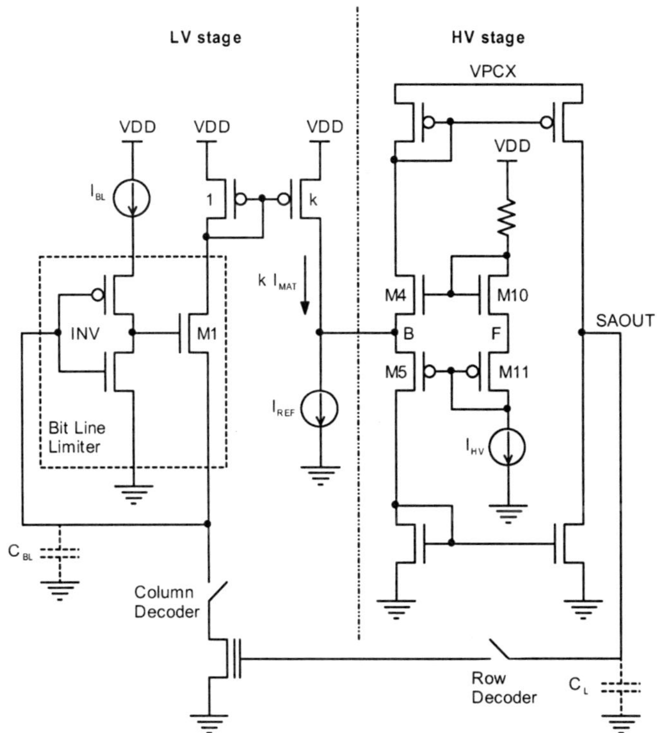

Fig. 13.23. Sense amplifier for closed loop voltage sensing

In order to speed up the sensing operation, the SAOUT output node can be set to an intermediate potential during the precharging of the bit line. When the loop is enabled, two transients can be obtained as shown in Fig. 13.24. If V_{SAOUT} reaches a final value greater that the starting one, we have a constant current charging until the cell starts sinking current, then a settling exponential behavior follows. In the case when the steady-state potential value of the word line is lower than the pre-charging voltage, we have an exponential discharge.

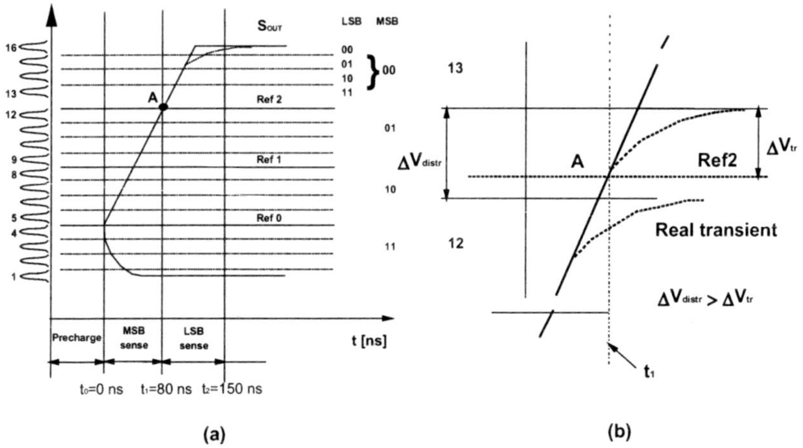

Fig. 13.24. **(a)** Transient of the sensing loop output voltage (the reference levels required for the A/D conversion are also shown; solid lines: MSB references; dashed lines: LSB references); **(b)** detail of the transient around point A for level 12 and 13

With this kind of read operation, it is possible to obtain access times around 200 ns in the case of 4-bits/cell memory. The consumption in terms of current sourced by node VPCX is around some tens of microAmperes for each sense amplifier.

13.8 Hierarchical Row Decoding for Multiple Sensing Loops

In order to close the loop it is necessary to find a way to apply the output voltage of the regulator to the cell gate. The structure of the decoder is still the one shown in Fig. 13.18, where a single sensing circuit has to be used for all the array blocks that are vertically arranged. Before delving into the circuit details, it is important to understand what kind of voltages we have to apply to the cell gate. During the read phase, the minimum voltage that is read is typically around 1.5 V, and the maximum voltage is around 7 V. In reality, the feedback loop represents a direct path to also transfer the programming voltages to the cell gate , which typically vary from 1 V to 9 V.

The method to allow both modes of operation in the same row circuitry is accomplished by the insertion of an inverter in the schematic of the hierarchical decoder in Fig. 9.34, thus obtaining the schematic in Fig. 13.25. Such an inverter must be driven by the MWL and, on the other hand, has to drive a metal row that is realized within the block but performs the same functions as the MWL in the hierarchical row decoder. For this reason it is named SUBMWL.

As for the hierarchical row decoder, SUBMWL is connected to 4 local word lines. At this point, it is necessary to clarify that the V_{SAOUT} voltage represents the sense amplifier output voltage during the read operation, whereas, during the program phase it represents the output voltage of the program voltage stairs. Therefore, the M3 p-channel, operating as a pass transistor, closes the loop. Transistor M5 is necessary to prevent non-addressed SUBMWL from being in a floating condition.

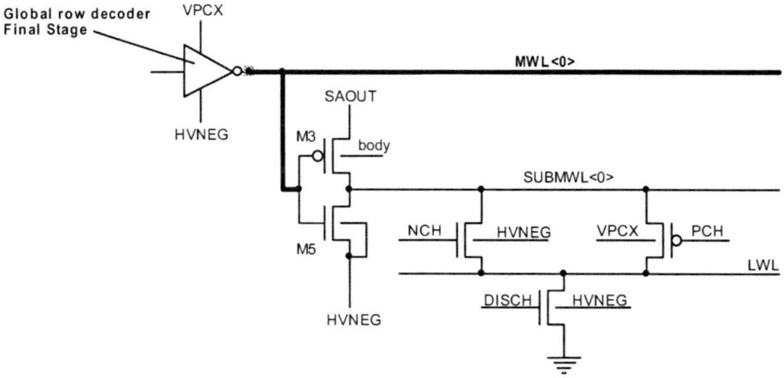

Fig. 13.25. Introduction of SUBMWL in the hierarchical decoder

Actually, in this implementation, the structure has some drawbacks. In fact, it is important to note that it is necessary to drive transistor M3 with low voltage in order to transfer the output voltage of the sense amplifier to the cell gate, whereas transistor M3 must be driven with high voltage to tie SUBMWL to ground. While during read and program, the value of V_{SAOUT} is not necessarily known, and M3 can be switched off only by applying to its gate the highest available voltage during each specific operation. This means that only the selected MWL must be tied to ground while all the unselected MWL's must be at high voltage.

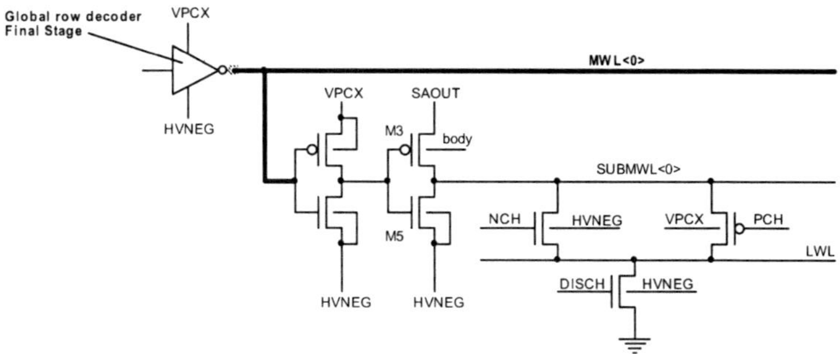

Fig. 13.26. Introduction of a second inverter not to have selected MWL's at low voltage and unselected MWL's at high voltage

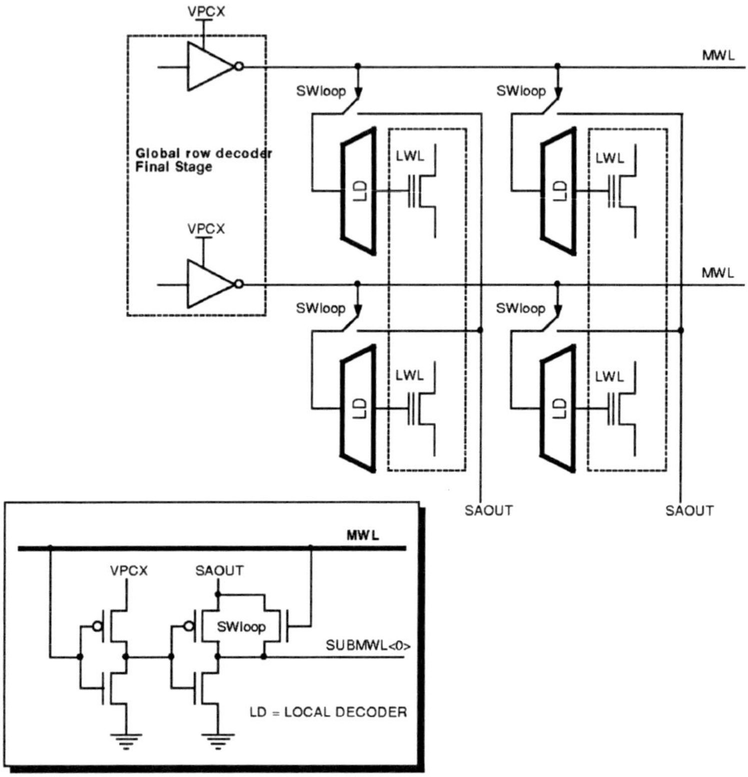

Fig. 13.27. Hierarchical row selection with separate addresses and analog voltage transmission functions

The problem is that the high voltages are generated internally by means of charge pumps and the charging of considerable capacitance may significantly impact the time needed to reach the required voltage. Thus, it would be better to be in the opposite situation: a single MWL selected at high voltage and all the unselected word lines tied to ground. The solution is the local insertion of an inverter biased between VPCX and HVNEG, as indicated in Fig. 13.26.

At this point, we still have the problem of where to connect the bulk of M3. Connecting it to the sense amplifier output would slow down the settling of the output voltage since node SAOUT would have a very high capacitance especially if a single sensing circuit is used for a row of vertical sectors, as previously described. In fact, all the parasitic capacitors between the n-well and the p-substrate of these transistors should be biased at V_{SAOUT}.

Therefore, we have to connect the bulk of M3 to the VPCX potential. However, in this way, due to the body effect, the threshold voltage of M3 (9 V in the case of VPCX) reaches 1.8 V and, hence, it is not sufficient to use only this transistor to transfer the sense output voltage to the SUBMWL's, since it would not be possible to transfer voltages around 1 V. It is then necessary to add an n-channel transistor in parallel to M3, directly driven by the MWL.

With such a structure, the area occupation increases even though all the transistors introduced have minimum size, but they still have to fit in the row pitch.

In conclusion the memory is organized as in Fig. 13.27.

13.9 A/D Conversion

The analog-to-digital conversion phase is the one in which the analog value of a cell operating parameter (gate voltage or drain current) is translated into a digital value.

The voltage (or the current sourced by the cell) is compared with a set of references and, depending on the result of this comparison, it is possible to obtain the value of the bit stored. Hereafter, some methods used to accomplish such a task are proposed.

The first technique, directly derived from the bi-level case, is the comparison of the analog value with respect to all the references. The comparison can be carried out in parallel, in series, or in a hybrid way.

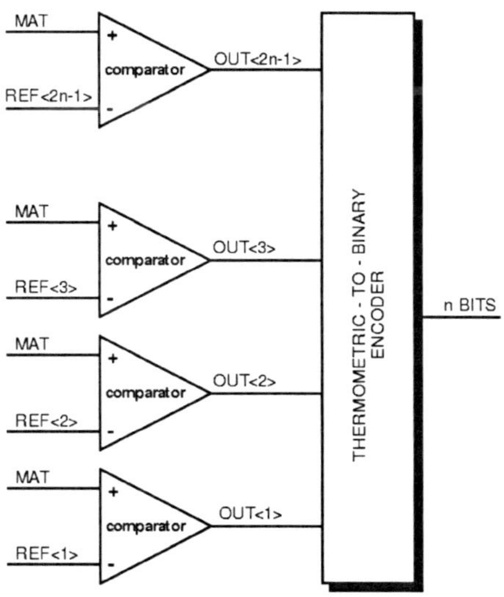

Fig. 13.28. Parallel sensing architecture

The parallel method basically uses a flash-type A/D converter as shown in Fig. 13.28. The final result is obtained during the time of a single evaluation. The drawback is that one comparator is used for each level. Such a method is therefore expensive in terms of both area and power consumption, making it practically unfeasible in the case of 16 levels. Furthermore, we would also have the problem of reproducing the analog value derived from the array cell a large number of times, with the resultant limitations in terms of precision. The decision of not using the same array signal for all the converters is chiefly due to the need for limiting the kick-back effect induced by the simultaneous toggling of a large number of comparators.

A method that does not introduce these disadvantages is the serial one: a single comparator is used with obvious saving in terms of area and power consumption. The voltage of the memory cell is applied to one of the comparator inputs, while the references are applied to the other input, one at a time. According to the reference that has caused the comparator to toggle, the data are decoded. Two implementations are possible. In the standard serial method, the references are applied to the comparator in a monotonic way. For example, starting from the lowest reference value and continuing in increasing order, the digital conversion is carried out based on the number of levels compared (Fig. 13.29). Unfortunately, the disadvantage is that the conversion time depends on the data stored. In the case of 16 levels, 15 comparisons are necessary before accomplishing the decoding of data corresponding to the last level.

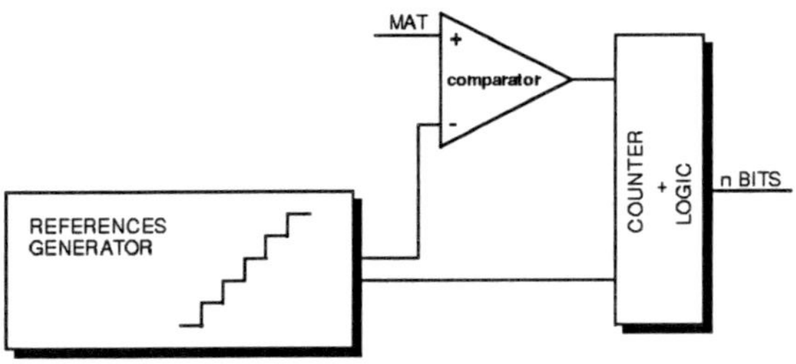

Fig. 13.29. Classic serial conversion method

If we design a more complicated control circuitry, the successive approximation method can be implemented. The read operation starts by comparing the analog value of the cell parameter with the value of the center reference level (the eighth in the case of 16 levels). Analyzing the result of this operation, it is possible to determine which half of the range the stored values belong to. At this point, the middle value of the selected range is chosen to perform the second comparison. This operation is repeated until the data are decoded (Fig. 13.30). The number of comparisons equals the number of bits stored in the single cell. Although faster than the classic one, also this method is quite slow. It is also necessary to pay attention to the area required for the additional logic so as not to waste the savings due to the serial method.

To overcome the problems related to the methods previously described (power consumption and area occupation for the parallel approach, long decoding time for the serial one), a mixed approach has been developed, which represents a hybrid solution between serial and parallel. Theoretically, it represents a generalization of the dichotomic serial method. The logic of the successive approximations is still used but several comparison are performed in parallel. The dichotomic method allows dividing the references in two groups. If the number of comparisons simultaneously performed is p, the references will be divided in $(p + 1)$ groups. The operation is then repeated on one of the $(p + 1)$ groups, and so forth. In the case of 16 levels, it is possible to carry out the conversion in only two steps, decoding two bits at a time, as shown in Fig. 13.31. In general, assuming that the same number of bit (z) is decoded at each step, n/z steps are necessary to sense n levels. The number of comparators that have to be used is $(2^z - 1)$. This method represents a very good compromise between area consumption, power consumption, and timing performance.

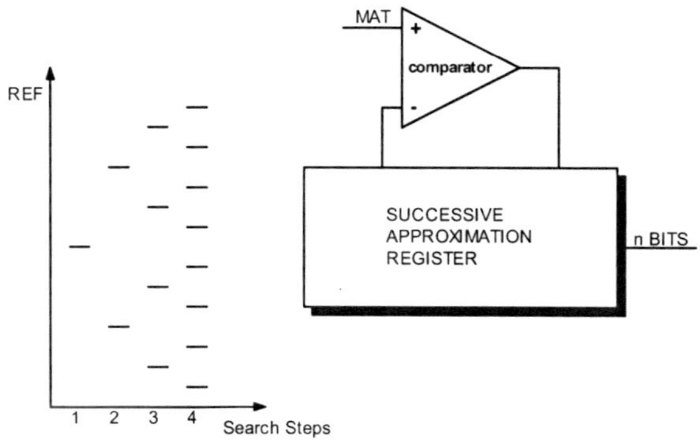

Fig. 13.30. Dichotomic serial conversion method

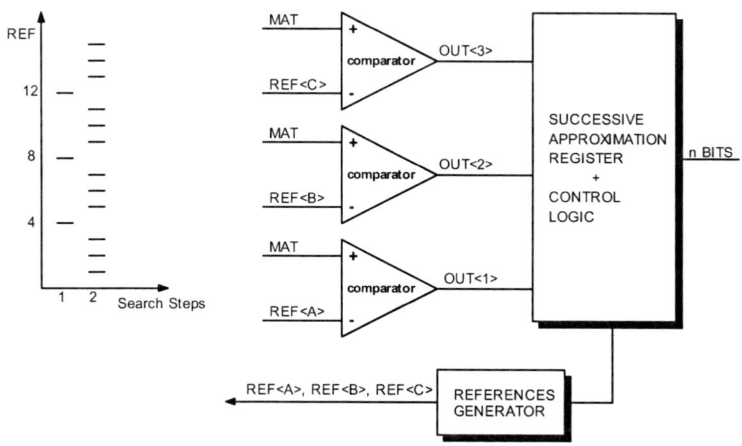

Fig. 13.31. Mixed parallel-serial sensing technique. During the first conversion step: REF<A> = REF<4>, REF<B> = REF<8>, REF<C> = REF<12>

13.10 Low Power Comparator

It should be noted that the voltage obtained as a result of signal production in a voltage mode approach is equal to the programmed threshold voltage plus an overdrive voltage ΔV which depends on the reference current drawn by the cell. Signal recognition has therefore to process voltage signals placed in a range equal to the available V_T window shifted by ΔV. This implies a high input common mode range (CMR) of the comparator, in the range of VPP (>VDD). As the power efficiency of charge pumps is low and high reading parallelism has to be provided, signal production and recognition must be carried out with limited power consumption.

As a consequence, the comparator design aims at power consumption optimization, while still ensuring the above input CMR as well as required comparison time and accuracy. To achieve adequate sensitivity and high operational speed, the comparator is made up by an input buffer followed by a regenerative stage. The required accuracy is obtained by adopting an output offset storage technique.

Figure 13.32 shows the block diagram of the comparator. Blocks G_{m1} and G_{m2} represent two fully differential stages, which share their load resistors R and R'. The first stage (powered by VPP) is devoted to input signal buffering, providing a low voltage gain ($A_V = 2$) in order to minimize the current drawn from VPP.

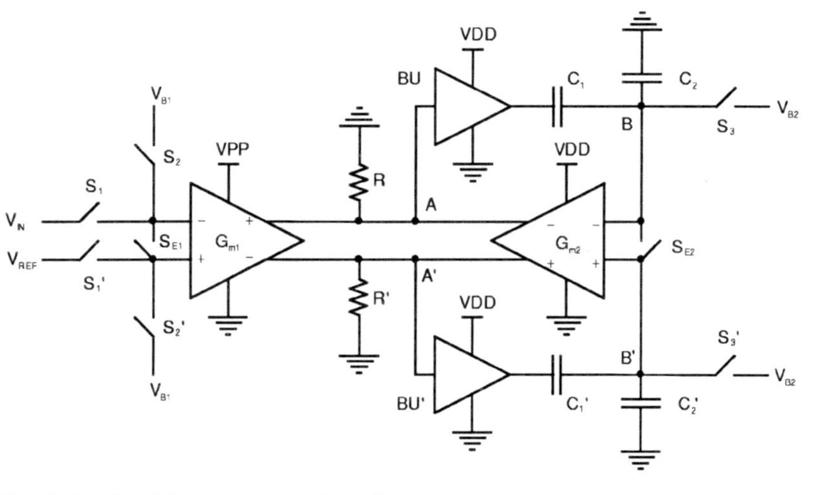

Fig. 13.32. Simplified comparator block diagram

The second stage (powered by VDD), together with feedback-connected capacitors C_1 and C_1', provides the regenerative action required to achieve high gain at high speed.

The two feedback capacitors are driven by unity-gain voltage buffers BU and BU' (which are also powered by VDD) to minimize the capacitive load of output nodes A and A'.

Offset cancellation is carried out on both differential stages, so as to minimize residual input-referred offset. Additional capacitors C_2 and C_2' are used to reduce clock feed-through effects generated when turning switches S_3, S_3' on and S_{E2} off. V_{B2} is a bias voltage.

To reduce total power dissipation, a two-phase offset cancellation technique can be used. Indeed, this approach allows the time duration of the working phases to be relaxed as compared to a three-phase approach with the same comparison time.

In conventional two-phase offset cancellation techniques, the first working phase provides input equalization and output offset storage, while input signal amplification and output signal regeneration are carried out simultaneously in the second working phase.

However, the comparator operation at the beginning of the second phase (that is, when starting the regenerative action) is affected by disturbs coming when opening the equalization switches of the input stage (S_2, S_2' and S_{E1}) and closing input switches (S_1 and S_1'). Indeed capacitive coupling with control phases, charge injection, and charge sharing between input lines and the comparator inputs take place. Moreover, the inverting input of the first differential stage suffers by larger disturb effects as compared with its non-inverting counterpart, because the input signal V_{IN} is common to all comparators in the sensing circuit. This can lead to transient inversion in the input signal polarity, which triggers the regenerative stage in the wrong direction. This effects turns out to be particularly severe in the presence of a non negligible impedance of the input signal generator.

To overcome this drawback, a different working phase sequence can be adopted. During the first phase, the input signal (V_{IN} – V_{REF}) is applied to the input stage (S_1, S_1' on; S_2, S_2', S_{E1} off), while the input nodes of the regenerative stage are equalized (S3, S3', S_{E2} on). This way, the input signal as well as offset contributions by both stage are amplified by a predetermined small factor, and the resulting voltage is stored on capacitors C_1 and C_1'.

During the second phase, the input nodes of the first stage are equalized (S_1, S_1' off; S_2, S_2', S_{E1} on), while the positive feedback around the second stage is enabled (S3, S3', S_{E2} off). As a consequence of the input node equalization, a voltage drop equal to $G_{m1}R(V_{IN} - V_{REF})$ takes place across the output terminals A and A', which, in turn, causes a voltage imbalance across the input terminals of the regenerative stage (nodes B and B '):

$$V_B - V_{B'} = G_{m1} \cdot R \cdot \frac{C_1}{C_1 + C_2} \left(V_{IN} - V_{REF} \right) \tag{13.4}$$

The above voltage imbalance, which is offset compensated, starts the regenerative action of the second stage, thus providing the desired high-speed large output swing.

As the input terminals of the comparator are equalized by S_{E1} during the regeneration phase, any disturb effect at the comparator input can be regarded as a common mode contribution and is therefore rejected by the input differential stage. Residual offset is mainly due to charge injection caused by the switches connected to the input nodes of the regenerative stage, which are turned off at the beginning of the second phase. To minimize the contribution due to mismatches between S_3 and S_3', a small delay is provided before switching S_{E2} off. Moreover, charge injection effects caused by the latter switch, are reduced by additional capacitors C_2 and C_2'.

Problem 13.6: Transform the block diagram in Fig. 13.32 in a circuit schematic.

With a comparator such as the one described above, a comparison time of about 20 nanoseconds can be achieved, assuming an input signal of 10 mV. The VPP consumption is around 10 μA.

Bibliography

M. Bauer, R. Alexis, G. Atwood, B. Baltar, A. Fazio, K. Frary, M. Hensel, M. Ishac, J. Javanifard, M. Landgraf, D. Leak, K. Loe, D. Mills, P. Ruby, R. Rozman, S. Sweha, S. Talreja, and K. Wojciechowski,, "A multilevel-cell 32Mb Flash memory", in 1995 IEEE Int. Solid-State Circuits Conf. Dig. Tech. Pap., pp. 132–133, (Feb. 1995).

G. Campardo et al., "40mm2 3V Only 50 MHz 64 Mb 2-b/cell CHE NOR Flash Memory", J. of Solid State Circuits, 35, 1665, (2000).

G. Campardo and R. Micheloni "Architecture of non volatile memory with multi-bit cells" Elsevier Science, Microelectronic Engineering, Volume 59, Issue 1-4, pagg. 173-181, (November 2001).

Y.-J. Choi, K.-D. Suh, Y.-N. Koh, J.-W. Park, K.-J. Lee, Y.-J. Cho, and B.-H. Suh, "A high speed programming scheme for multi-level NAND Flash memory", in 1996 Symp. VLSI Circuits Dig. Tech. Papers., pp. 170–171, (June 1996).

C. de Graaf, P. Young, and D. Hulsbos, "Feasibility of multilevel storage in Flash EEPROM cells", in Proc. 25th European Solid State Device Research Conf., pp. 213-216, (Sept. 1995).

M. Grossi, et al., "Program schemes for multilevel Flash Memories", IEEE Proceeding of the, Vol. 91, No. 4, pp. 594-601, (April 2003).

M. Horiguchi, M. Aoki, Y. Nakagome, S. Ikenaga, and K. Shimohigashi: "An experimental large-capacity semiconductor file memory using 16-levels/cell storage," IEEE J. Solid-State Circuits, vol. 23, no. 1, pp. 27-33, (Feb. 1988).

T.S. Jung, Y.-J. Choi, K.-D. Suh, B.-H. Suh, J.-K. Kim, Y.-H. Lim, Y.-N. Koh, J.-W. Park, K.-J. Lee, J.-H. Park, K.-T. Park, J.-R. Kim, J.-H. Lee, H.-K. Lim , "A 117-mm^2 3.3V only 128-Mb multilevel NAND Flash memory for mass storage applications", IEEE J. Solid-State Circuits, vol. SC-31, pp. 1575–1583, (November 1996).

R. Micheloni, O. Khouri, S. Gregori, A. Cabrini, G. Campardo, L. Fratin, and G. Torelli, "A 0.13-µm CMOS NOR Flash memory experimental chip for 4-b/cell storage," ESSCIRC 28th Proc. European Solid-State Circuit Conf., pp. 131-134, (Sept. 2002).

R. Micheloni, I. Motta, O. Khouri, and G. Torelli, "Stand-by low-power architecture in a 3-V only 2-bit/cell 64-Mbit Flash memory", in Proc. 8th IEEE Int. Conf. Electronics, Circuits, and Systems, vol. II, pp. 929–932, (Sept. 2001).

R. Micheloni et al., "Method for reading a multilevel nonvolatile memory and multilevel nonvolatile memory", USA patent No. 6,301,149, (October 9, 2001).

R. Micheloni et al., "Read circuit for a nonvolatile memory", USA patent No. 6,327,184, (December 4, 200).

A. Modelli, R. Bez, A.Visconti "Multi-level Flash Memory Technology", 2001 International Conference on Solid State Devices and Materials (SSDM), Tokyo, Extended Abstract, pp. 516-517, (2001).

A. Modelli, A. Manstretta, and G. Torelli, "Basic Feasibility Constraints for Multilevel CHE-Programmed Flash Memories", IEEE Trans. Electron Devices, ED-48, NO. 9, pp. 2032-2042, (September 2001).

M. Ohkawa el al., "A 9.8mm^2 Die size 3.3V 64Mb Flash memory with FN-NOR type Four level cell", IEEE, Journal of Solid-State Circuit, vol. 31, no. 11, p. 1584, (Nov. 96).

B. Riccò, G. Torelli, M. Lanzoni, A. Manstretta, H. E. Maes, D. Montanari, and A. Modelli, "Nonvolatile multilevel memories for digital applications", Proc. IEEE, vol. 86, pp. 2399-2421, (Dec. 1998).

P. L. Rolandi, et al.: "1M-cell 6b/cell analog flash memory for digital storage", IEEE ISSCC Dig. Tech. Papers, pp. 334-335, (Feb. 1998).

P.L. Rolandi et al., "A 32Mb-4b/cell analog Flash Memory Supporting Variable Density with Only Supply and serial I/O", 25th ESSCIRC '99.

R. Shirota, G.J. Hemink, K. Takeuchi, H. Nakamura, and S. Aritome, "A new programming method and cell architecture for multi-level NAND Flash memories", presented at the 14h IEEE Non-Volatile Semiconductor Memory Workshop, Monterey, CA, paper 2.7, (Aug. 1995).

14 Program and Erase Algorithms

First generation EPROM and Flash memories did not integrate "intelligence" to accomplish write operations. In modern devices, due to the complexity, variety, and precision required by write operations, controllers that autonomously perform such modify operations are always present.

14.1 Memory Architecture from the Program-Erase Functionality Point of View

The following schematic representation, shown in Fig. 14.1, aims at giving an overview of the main blocks involved in program and erase operations in a particular Flash memory.

Table 14.1. Synoptic table of the voltages generated by the regulators and applied to the cells during the read, program, and erase phases.

	Read	Program	Program Verify	Erase Pulse	Erase Verify	Soft Program Verify
VPD	---	VdCHE 4 V	---	VsFN 0 to 8 V	---	---
VPCX	VgRead 5 V	VgStair 1 – 9 V	VgVerify 5 or 6 V	0 V	VgRead	VgDvSel 3 V
VNEG	---	---	---	VgFN - 8 V	VgEvUns - 2 V	VgDvUns 2 V
VPCXLOC	6 V	10 V	6 or 7 V	2 V	6 V	6 V
VPCYLOC	4 V	6 V	4 V	0 V	4 V	4 V
VIPW	---	VbCHE - 2 V	---	---	---	---
WL_Sel	5 V	VgStair	VgVerify	VgFN	VgRead	VgDvSel
WL_Uns	0 V	0 V	0 V	0 V	VgEvUns	VgDvUns
BL_Sel	VdRead 1 V	VdCHE 4 V	VdRead 1 V	---	VdRead 1 V	VdRead 1 V
BL_Uns	0 V	0 V	0 V	0 V	0 V	0 V
S_Sel	0 V	0 V	0 V	VsFN	0 V	0 V
S_Uns	0 V	0 V	0 V	0 V	0 V	0 V
IPW_Sel	0 V	- 2 V	0 V	VsFN	0 V	0 V
IPW_Uns	0 V	0 V	0 V	0 V	0 V	0 V
NW_Sel	VDD	VDD	VDD	VsFN	VDD	VDD

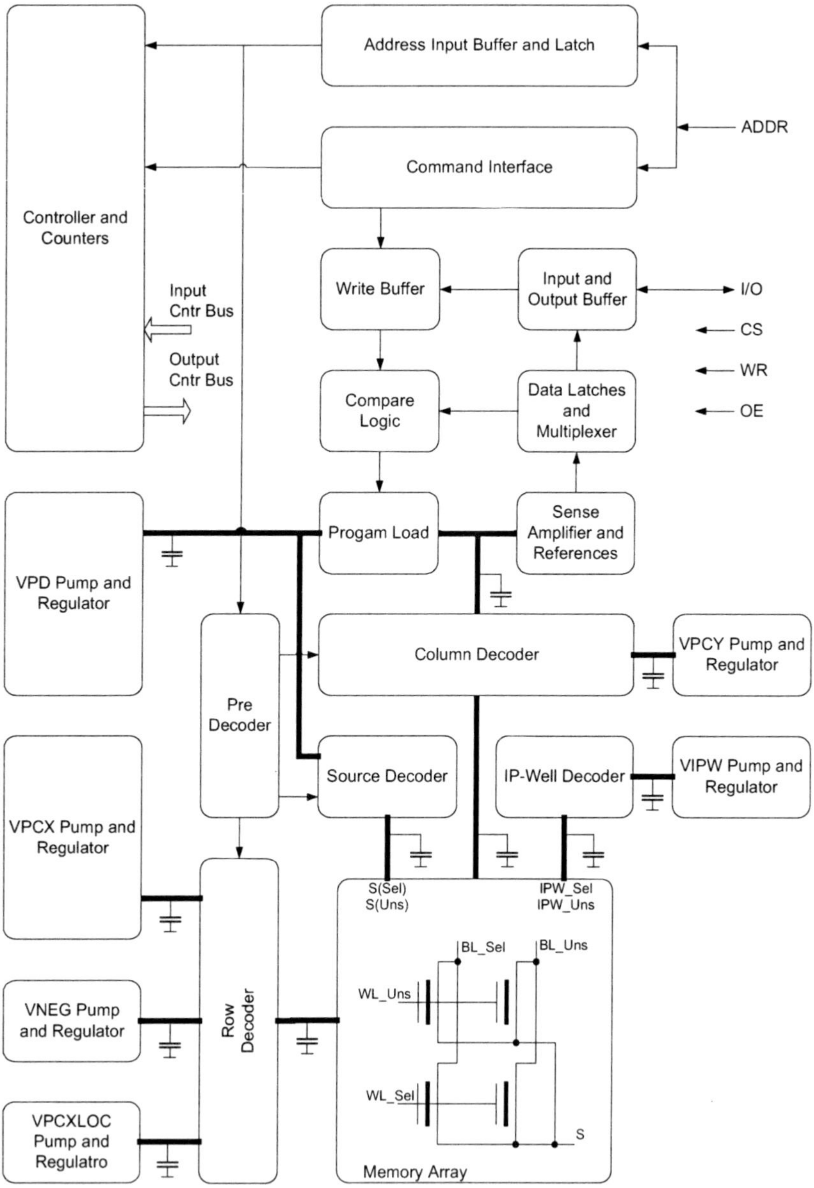

Fig. 14.1. Block diagram of a Flash memory

- Memory Array: the core of the memory is undoubtedly the cell array organized in sectors, rows, and columns. The logic size of the sector is imposed by the specification agreed with the user, while the physical size of the sector, the number of rows and columns are determined through a careful trade-off among area, performance, technological and process constraints.
- Around the array are the decoding blocks: Row, Column, Source, and Isolated p-well Decoders. Their function is to apply a suitable bias voltages to the selected sectors and memory cells. All these circuits, which, for the sake of simplicity, have been represented as a single block, are regularly distributed all around each sector. Write operations require a large number of modifications to the bias conditions of the selected cells. In order to obtain fast variations of the bias conditions, it is useful to suitably divide the load of the decoding circuits so as to move the minimum amount of charge.
- Predecoder: the predecoder block simply converts the logic selection address into control signals for the decoder.
- Pumps and Regulators: pumps and regulators must provide bias voltages and sustain the dynamic and/or static load during the different operating conditions. In Table 14.1 the voltages that must be applied to the cell during read, program, and erase are reported.
- The VPD Pump and Regulator block provides the V_{dCHE} to the drain of the Flash cell. During programming, high current consumption occurs and, as the program efficiency is strongly dependent upon the voltage applied to the cell drain, particular attention must be paid to the distribution of this voltage and to the regulator that generates it. During the erase phase, the cell drain remains floating and the same regulator is used to generate the positive V_{sFN} voltage to apply to the source and bulk of the cell. Such a voltage must follow a voltage ramp and, therefore, the regulator must be able to manage a variable value of the voltage to regulate.
- The VPCX Pump and Regulator block generates the V_{gRead} voltage during read, V_{gRead} or V_{gAll0} during program verify, and the V_{gStair} voltage, ranging between $V_{gStairStart}$ and $V_{gStairEnd}$, during the program pulses. During program and erase operations, the regulator of this block must frequently and quickly modify the regulated value on its load, which is mainly capacitive. The required precision is very high, especially in the case of multilevel devices.
- The VNEG Pump and Regulator block generates the negative V_{gFN}, V_{gEvUns}, and V_{gDvUns} voltages that are applied to the cell gates of the sector, as indicated in Table 14.1.
- The VIPW Pump and Regulator block generates the V_{bCHE} negative voltage that is applied to the cell bulk during the program pulse.
- The VPCX and VPCY Pump and Regulator blocks provide the suitable voltages to the row and column decoding circuits.

All these pumps and regulators, together with the decoders, must be managed carefully by the controller during the write phase in order to obtain fast bias voltage variations as required. This must be accomplished while avoiding any violation of the technological rules, directly or due to capacitive coupling.

- Sense Amplifiers and References: the sense amplifiers and reference generation is fundamental, not only to reading, but also to the program and erase verify operations.
- Program Load: the program load selectively provides the V_{dCHE} voltage to the cells during program. The Program Load activation is controlled by means of a group of Flip Flops (Program Load Flip Flops) present in the Compare Logic block.
- Compare Logic: this block contains the Program Load Flip Flops and the logic to update them during the different program and erase phases. Moreover, it provides the information about the program and erase completion to the controller.
- Write Buffer: during a user program command, the data that are to be programmed are loaded into the Write Buffer.
- Data Latches and Multiplexer: this block contains the latches to store data read from the array by means of the sense amplifiers, and the multiplexer to manage the page and burst read mode.
- Input and Output Buffers: this block carries out all the functionalities regarding the physical interface to external lines.
- Command Interface: it represents the section that recognizes the command sequence applied by the user; it also manages the writing of the word in the Write Buffer, and activates the Controller for program and erase operations.
- Address Input Buffer and Latch.
- Controller and Counters: the controller awakes from its stand-by state only as a consequence of a request of the Command Interface to execute program and erase algorithms, or, in Test Mode, to accomplish test sequences. It is interfaced to each block previously described, and monitors, initiates and controls the device operations. It includes other blocks involved in the algorithm execution, such as a timer, some counters to count the attempts and manage the voltage ramps that must be applied, and, moreover, a set of registers to store some parameters of the algorithms under execution, in the case the device allows the user to interrupt a write sequence.

14.2 User Command to Program and Erase

Programming a word or a sequence of words means applying a command composed of one ore more write cycles. These cycles provide the logic address of the array locations that have to be programmed and the mask. We recall that the programming, in the case of Flash memories, allows modifying the cell status only from erased to programmed and that the reverse operation must be executed by erasing the entire sector. Therefore, the data applied by the user, which accompanies the program command, must be regarded as a mask. The memory bits corresponding to '0' in the mask are actually modified to '0' (if they do not correspond to this value), whereas the bits corresponding to '1' maintain their original state.

After recognizing the program command, the Command Interface activates the memory Controller for the autonomous execution of the programming procedure.

In practice, such a procedure is too complex, critical, and specific to be managed by the user and, hence, nearly all of recent devices use internal controllers to carry out such activities. The Controller is a state machine or a small sequencer (see Chap. 17) that, by means of a clock, a set of counters, and a timer, executes a predefined algorithm.

Similarly, the sector erase command must provide the sector address before leaving the control to the Controller that executes the entire operation autonomously. The overall result of the program or erase operation carried out by the controller are usually available to the user through a state register. The information related to the execution of a modify operation by the Controller may be provided to the user also through a dedicated Ready/Busy line.

14.3 Program Algorithm for Bi-Level Memories

An example of a program algorithm for bi-level memories of the latest generation is represented in Fig. 14.2. The behavior of the voltage of the selected word line (V_{WL_Sel}), selected bit line (V_{BL_Sel}), isolated p-well of the selected sector (V_{IPW_Sel}), and the source of the selected sector (V_S) are reported in Fig. 14.3. In Fig. 14.4, the function of distribution of current and voltage for a group of cells in the erased state ('1') and another group in the programmed state ('0') are reported. Furthermore, the verify voltage applied to the gate of selected and unselected cells, the current of the reference cells for read (I_r) and verify (I_{pv}) are represented. We also assume that a current sensing scheme has been adopted and, hence, the current of the cells is compared at a fixed voltage.

Let's analyze the program algorithm for bi-level memories in detail:

1. The controller activates all the pumps and regulators involved in the program operation which are usually not active during read; therefore VPD is pulled to V_{dCHE} and VIPW is pulled to V_{bCHE}. The attempt counter, 'Att', is reset and the gate stair counter, 'Stair', is initialized to 'StartStair'. The Program Load Flip Flops are all activated.
2. In order to avoid stress to cells that do not require programming, an initial read is carried out. The gate, drain, and source voltages must be set to perform read (see READ in the bias voltage table). The comparison executed in the Compare Logic block deactivates all the Program Load Flip Flops associated with the cells that do not require programming. If all the cells already have the correct value, the algorithm exits and the read bias conditions are restored.
3. If some cells have to be programmed, the sequence of program pulses is activated. The bulk voltage of the selected sector, V_{BL_Sel} is pulled to V_{bCHE}, whereas a stair of voltage is applied to the gate of the selected cells, starting from $V_{gStairStart}$ to V_{gBlind}, and, moreover, V_{dCHE} is applied to the drain of the selected cells. The voltage ramp that is applied to the gate allows obtaining the maximum program speed, keeping the current drawn by the cells within values sustainable by the VPD pump. The last program step, having longer duration, allows the slowest cells of the '0' distribution to be programmed.

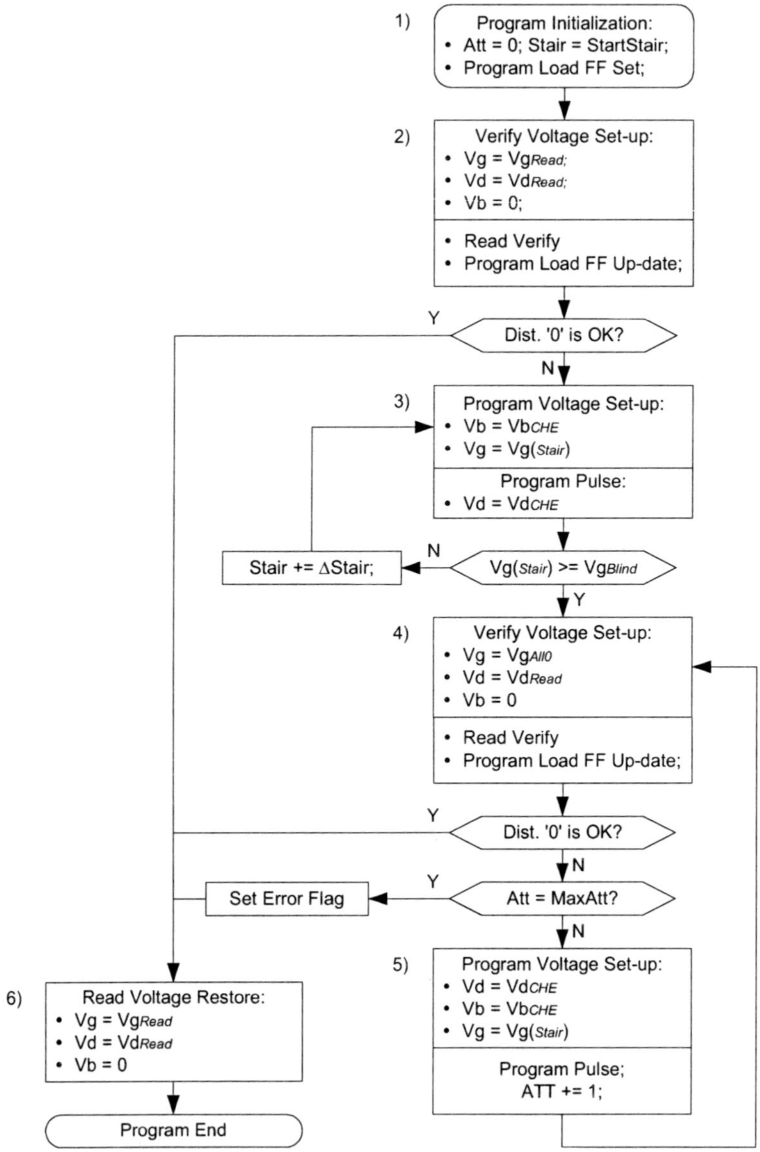

Fig. 14.2. Representation of the program algorithm for a modern bi-level memory

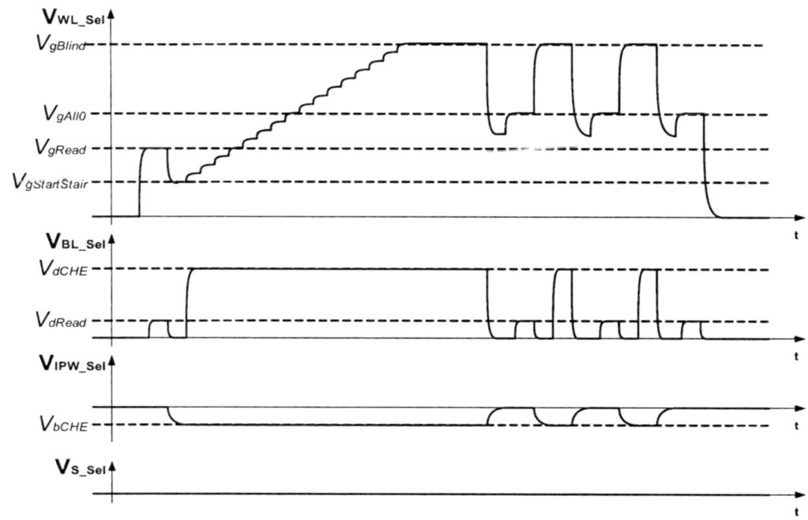

Fig. 14.3. Representation of the main voltages applied during the bi-level program algorithm

4. Once the blind part of the programming has finished, a verify is performed to determine whether all the cells are able to draw the required current. The verification is carried out with the V_{gAll0} margined gate voltage with respect to the read voltage. The current of the array cells must be lower than the current of the reference cell, I_{vp}. The Program Load Flip Flops associated with the cells that fulfill such a requirement and, hence, do not need more program pulses, are deactivated. If all the cells are programmed and, therefore, all the Program Load Flip Flops are deactivated, the algorithm exits and the read bias conditions are restored.
5. If some of the cells have not yet been programmed, the algorithm steps back to restore the program conditions for a further program pulse, at the end of which, step (4) is repeated to verify. This loop is repeated until all the cells are correctly programmed and an attempt counter is incremented at each extra-pulse. After the maximum number of attempt has been reached, the algorithm exits with an error signal and the read condition is restored.
6. Before releasing the control, the controller restores the read conditions and properly sets all the pumps and regulators to the stand-by condition, since there is no reason for them to be active during read. The end of the program operation is signaled by means of a Ready/Busy signal and an appropriate flag in the device status register.

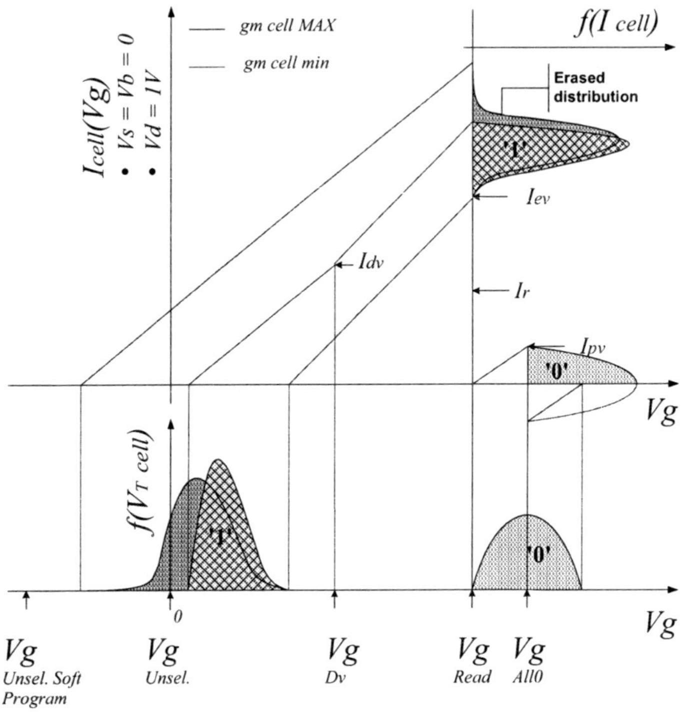

Fig. 14.4. Representation of the functions of distribution of voltage and current of a group of erased or programmed cells

During the progress of the flow, particular attention is paid to the fulfillment of the timings for the modifications of the bias conditions of the array and decoders (as shown in Fig. 14.3), in order to obtain the precision required by the different phases and prevent dangerous boost effects.

In Fig. 14.3 V_{WL_Sel} is the voltage of the selected word line. The $V_{gStartStair}$ voltage that defines the beginning of the program stair, the V_{gRead} read voltage, the V_{gAll0} verify voltage of the memory cells that must be programmed to '0', the $V_{gEndStair}$ voltage that is the maximum voltage applied to the selected gates during program, are also reported. V_{BL_Sel} is the voltage applied to one of the selected bit lines. The V_{dRead} voltage regulated by the sense amplifier at the drain terminal of the cell during the read of the verify operation, the V_{dCHE} voltage applied to the cell drain during the program pulse are also reported. V_{IPW_Sel} is the voltage of the isolated p-well of the selected sector. The V_{bCHE} voltage to apply during the program pulse is also highlighted. V_{S_Sel} is the source voltage of all the cells of the selected sector.

14.4 Program Algorithm for Multilevel Memories

Multilevel memories use the same cell studied in the case of bi-level memories and the same CHE program principle. The dynamic current range available to the Flash cell must suitably be subdivided to place N distributions. The linearity of the relationship between the voltage step applied to the Flash cell gate and the threshold voltage reached by the cell at the end of the program pulse (see Sect. 13.3), allows placing the multilevel distributions accurately.

The programming of multilevel distributions is performed by means of an alternation of program pulses and verify operations. The number of iterations is imposed by the width of the central distributions (the only real multilevel distributions) and by the sensitivity of the sense amplifiers. The program time depends on the number of iterations to perform, the duration of the program pulse, the time required to verify, and, largely on the toggling time to correctly bias the array in the two different phases. The total program time in the case of multilevel cell is remarkably longer than the program time in the case of bi-level memories in which the program algorithm progresses mainly in a blind way. To overcome such a reduction in performance, structures with high degree of parallelism are employed in multilevel memories so as to program and verify a large number of cells in parallel. The number of cells that can be programmed in parallel is limited by the capability (area) of the VPD drain pump that has to source current to the drains of the cells that are being programmed. On the other hand, also the number of sense amplifiers that are simultaneously active is limited by their area and consumption. Once the maximum capability of the pump has been fixed, the maximum number of cells that can be programmed simultaneously is also determined. Having a large number of cells that need to be programmed, the overall program time can further be reduced if the number of transitions between the program and the erase array bias condition is reduced. This requires that a larger amount of data to be programmed is available and, hence, a larger Write Buffer, a suitable number of Program Load Flip Flops to store the information related to the program status of all the cells and control logic able to manage the entire operation are necessary. This technique is extensively used in modern multilevel memory architectures and allows satisfactory programming performances to be attained. In the example of multilevel memories that we will examine, a structure with a fourfold Write Buffer size with respect to the number of cells that can actually be programmed and verified is used. Therefore, as we will see, each program and verify phase is repeated for each of the four banks of cells.

Multilevel Flash memories are regarded by the user as exactly the same as bi-level memories. The data that have to be written represent a mask of bits where a '0' indicates that the corresponding cell must actually be programmed to '0', whereas a '1' means that the corresponding cell must be left in the original state. From the multilevel program point of view, the programming mode requires that the information contained in the cells to program must be read and used as a mask with respect to the Write Buffer. This simple operation is commonly referred to as 'Bit manipulation'.

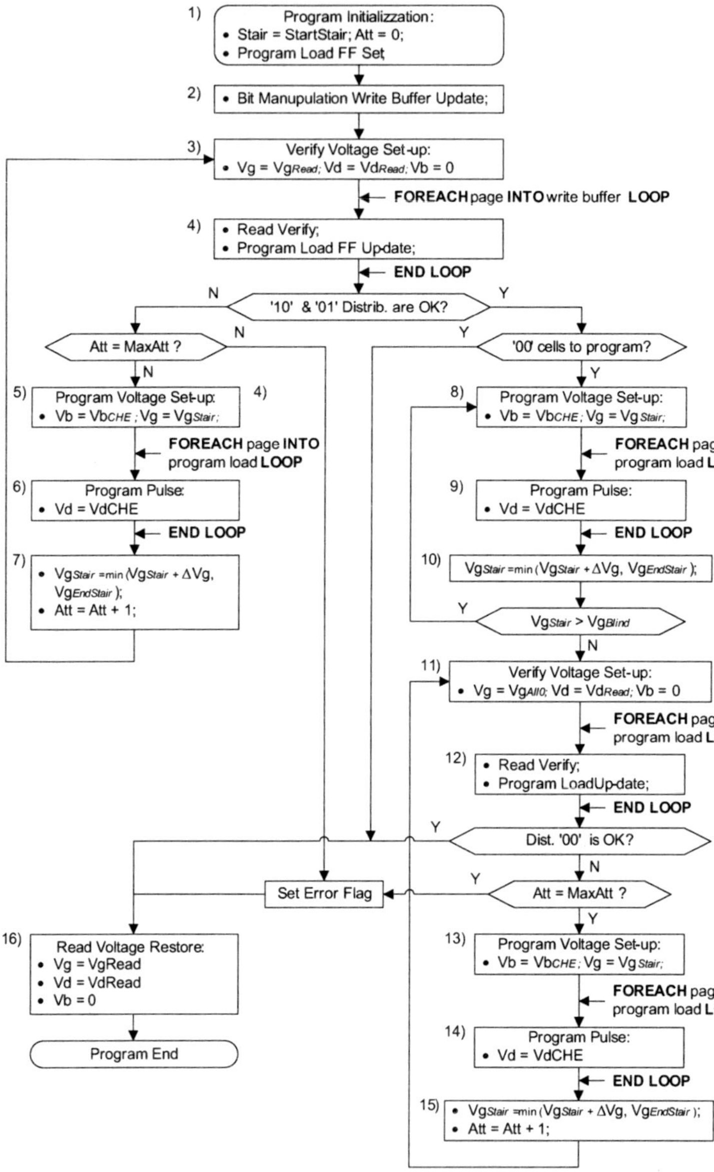

Fig. 14.5. Representation of the program algorithm of a modern multilevel memory

An example of an algorithm for multilevel memories is reported in Fig. 14.5. The behavior of the voltage of the selected word line, one selected bit line, the isolated p-well of the selected sector, and the source of the selected sector are represented in Fig. 14.6 and 14.7. V_{WL_Sel} is the voltage of the selected word line. The

$V_{gStartStair}$ voltage that defines the beginning of the program stair, the V_{gRead} read voltage, the V_{gAll0} verify voltage of the memory cells that must be programmed at '00', the $V_{gEndStair}$ voltage that is the maximum voltage applied to the selected gates during program, are also reported. V_{gBlind} is the maximum voltage applied to the gate during the blind program phase of the '00' distribution. V_{BL_Sel} is the voltage applied to one of the selected bit lines. The V_{dRead} voltage regulated by the sense amplifier at the drain terminal of the cell, and the V_{dCHE} voltage applied to cell drain during the program pulse are also reported. V_{IPW_Sel} is the voltage of the isolated p-well of the selected sector. The Vb_{CHE} voltage to apply during the program pulse is also highlighted. V_{S_Sel} is the source voltage of all the cells of the selected sector.

In Fig. 14.8 the representation of the distributions of currents and voltages for a group of cells in the '11', '10', '01', and '00' state is given. We also recall that we assume a scheme of current sensing is used and, hence, the cells are compared in terms of current with fixed gate voltage.

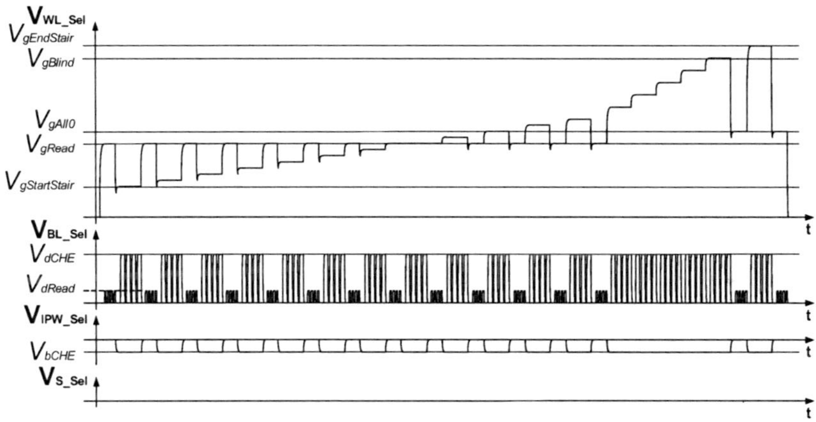

Fig. 14.6. Representation of the main voltages applied to the array during the multilevel program algorithm

Let's analyze the program algorithm for multilevel Flash memories in detail.

1. First of all, the controller activates all the pumps and regulators necessary to the program operations which are usually not active during read; therefore VPD is pulled to V_{dCHE} and V_{IPW} is set to V_{bCHE}. The attempt counter, 'Att', is initialized to 0 and the counter for the stair gate is initialized to the initial value, 'StairStart'. All the Program Load Flip Flops are activated.
2. The Write Buffer contains the information of the bits that have to be set to '0'. As previously stated, it is necessary to perform a read operation on the pages corresponding to the Write Buffer to obtain the program pattern. In this phase, the program destination pages are read from the memory and the content of the Write Buffer is updated according to the bit manipulation rules.

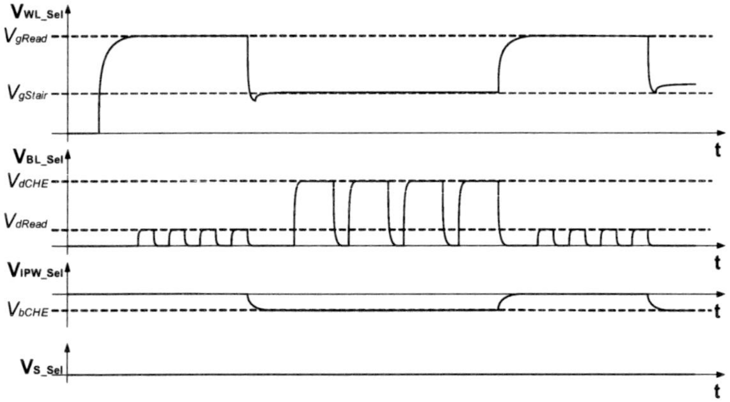

Fig. 14.7. Details referring to the four verify pulses and four program pulses of the program algorithm of a multilevel memory. In the upper part, the timings associated with each of the phases that compose the program algorithm are reported

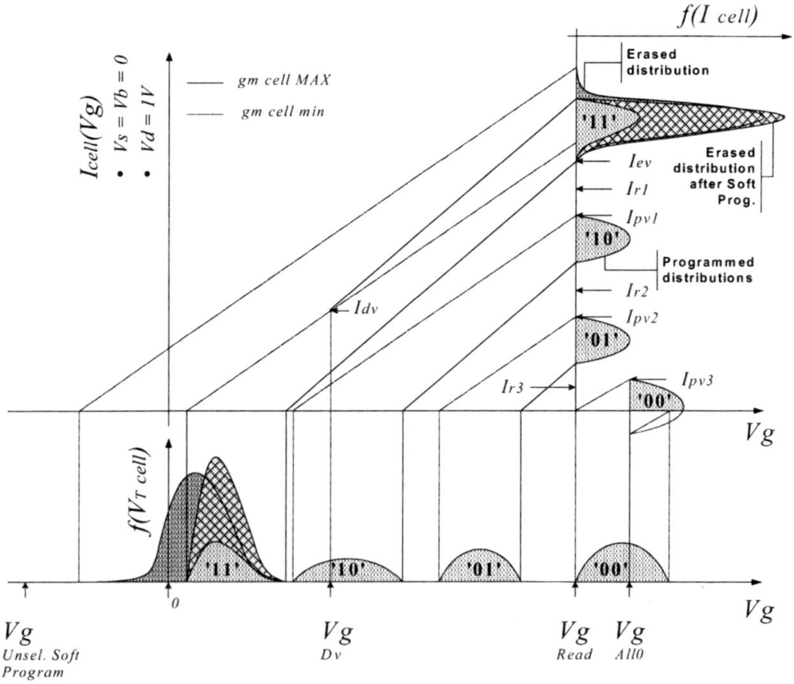

Fig. 14.8. Representation of the distributions of voltages and currents for a group of cells '11', '10', '01', and '00'

3. At this point, the actual program procedure can take place. The voltages for the verification of the multilevel distributions are set. The multilevel distributions are '10' and '01' only. The verification of such distributions is carried out by comparing the current of the cells with respect to the two reference cells, I_{pv1} and I_{pv2} respectively (Fig. 14.8), when the V_{gRead} read voltage is applied to the gate. If the V_{gRead} voltage is used for the verification of the multilevel distributions, the information obtained from the sense amplifier that compares the cell current with respect to the current of the I_{pv3} reference cell is not reliable, since the PV3 reference cell has threshold voltage greater than V_{gRead}. During the multilevel verification of the '01' and '10' cells it is thus necessary to ignore the information provided by the foregoing sense amplifiers and, on the contrary, it is convenient to keep them off as if the cells they refer to were already perfectly programmed. This prevents the current associated with the '00' cells that have not yet been programmed from disturbing the verification of the other distributions. The verify operation should be performed in the same conditions as the read operation to reduce or cancel the effects of any parasitic elements which can introduce inaccuracies.
4. After setting the verify voltages, the verification of the multilevel distributions is performed for each of the pages of the Write Buffer and the Program Load Flip Flops are updated. If all the multilevel cells (01, 10) are in the correct status, the algorithm jumps to step (8) to verify and program the cells that will assume value '00', otherwise the multilevel distributions are programmed. First of all, the number of program attempts is checked to verify whether the maximum value has been reached, in which case an error flag is set and the program algorithm jumps to step (16), and, if not, new program pulses are generated.
5. The program bias voltages are set: the bulk is biased at V_{bCHE} and the selected word line is biased at V_{gStair}.
6. A program pulse is initiated for each cell page corresponding to the Write Buffer.
7. The Stair counter and the attempt counter are incremented for the next program step. The algorithms jumps back to the verification of the multilevel distributions (4).
8. When all the multilevel distributions are OK, the verification of the existence of '00' cells to program is performed. In particular, if all the Program Load Flip Flops are deactivated, it means that no '00' cells to program and, thus, it is possible to jump to the end (16). If some '00' cells are to be programmed, a blind program is executed up to the V_{gBlind} gate voltage, and, hence, the program biases are activated. The bulk is set to V_{bCHE} and the selected word line to V_{gStair}.
9. A program pulse is initiated for each cell page corresponding to the Write Buffer.
10. The Stair counter and the attempt counter are incremented for the next program step. As in this phase it is not necessary to achieve great precision, since the only limitation in the selection of ΔV_g is imposed by the capability of the VPD pump to source current. This loop is repeated until the gate voltage reaches V_{gBlind}.

11. Typically, beyond this value, all the '00' cells are programmed. The bias to verify the '00' cells is set with a gate voltage of V_{gAll0}.
12. All the pages of the Write Buffer are verified and the Program Load Flip Flops updated. If at the end of this phase all the Program Load Flip Flops are deactivated, the algorithm jumps to the end (16). If the maximum number of program pulses has been reached, an error flag is set and the algorithm jumps to the end (16); otherwise a new program and verify cycle is started.
13. The program voltages are set.
14. Program pulses are initiated for each page that requires programming.
15. The voltage stair is incremented if the $V_{gEndStair}$ maximum value has not yet been reached and the attempt counter is incremented, jumping back to the verification phase of the '00' cells (step 11).

During the execution of the above flow, particular attention is paid to the timings to change the bias condition of the array and decoders, so as to obtain the precision required by the different phases and prevent dangerous boost effects.

14.5 Erase Algorithm

An example of an erase algorithm for bi-level memories of the latest generation is given in Fig. 14.9. The erase algorithm consists of three phases: Program All0, Erase, and Soft Program. In the "Program All 0" phase, all the cells of the selected sector are programmed. This phase is executed to guarantee homogeneous aging of the array cells, obtain a more compact distribution in erase, and prevent erased cells from being excessively depleted. The real erase phase follows, which causes all the cells to have higher current then the reference current (I_{ev}). During the erase phase, some cells may be depleted and, thus, it is necessary to recover these cells during the third phase. The latest operation is called Soft Program. In Fig. 14.8, the current and threshold voltage distributions after the Electrical Erase is shown, in which the evident tail of the distribution is due to the depleted cells, together with the final erase distribution after the Soft Program where the depleted cells have been recovered.

Let's examine in detail the erase algorithm for Flash memories.

First of all the erase algorithm requires the 'Program All0' phase. All the pages within the sector undergo a blind program, i.e. without any verification. In order to limit the current being drawn from the VPD pump and reduce the cell stress, a voltage ramp is applied to the gate for programming.

1. The controller activates all the pumps and regulators necessary for the program operations which are usually not active during read; therefore VPD is pulled to V_{dCHE} and V_{IPW} is set to V_{bCHE}. All the Program Load Flip Flops are activated, the counters that select the page within the selected sector are initialized to zero.
2. For each page of the sector the following cycle is executed:
   - The Stair counter that determines the gate voltage of the cells to program is initialized to StairStart.
   - The V_{dCHE} voltage is applied to the drains of the cells to program during the entire period of the program pulse.
   - At the end of the program pulse the Stair counter is incremented to prepare the gate voltage for the next pulse.

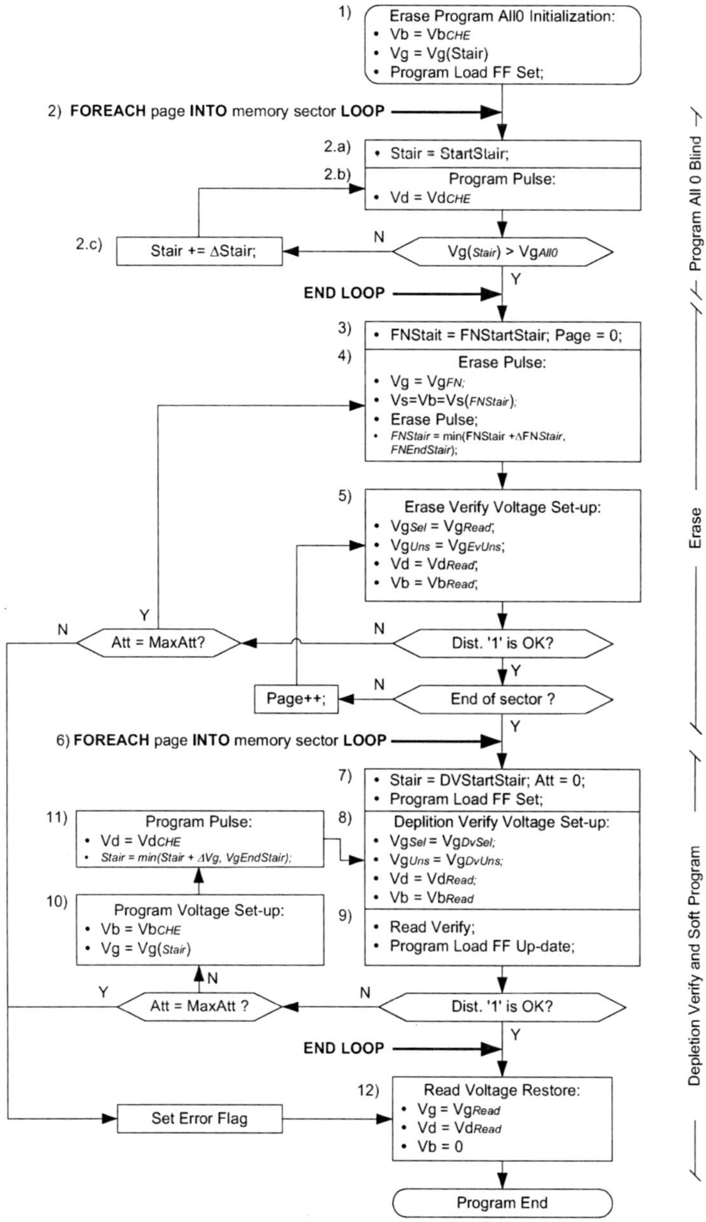

Fig. 14.9. Representation of the erase algorithm for a bi-level memory

If the Stair has finally reached the All0 final value, the procedure is applied to the subsequent page.

Once the Program All0 phase has been completed, the erase phase, which operates on all the cells of the sector, follows. Erase is carried out by FN tunneling, by applying an intense electric field (16 V) between bulk and gate. As the Flash process does not tolerate such a difference of potential outside the array, a negative voltage is applied to the gate (V_{gFN}), whereas a positive voltage is applied to the bulk (V_{sFN}). Great attention must be paid during the toggling of the signals applied to the array in order to avoid any violation of the design rules during the transitions, taking into account the large load and capacitive coupling present. Also in the erase phase, the bulk voltage of the selected sector is applied by means of a stair of pulses to limit stress effects on the tunnel oxide of the cell

3. First of all the pumps and regulators needed for erase are activated. The V_{NEG} pump and regulator are activated to generate V_{gFN}, the negative voltage that must be applied to the gate of the selected sector. VPD is prepared for generating the V_{sFN} voltage. The Stair counter is initialized to FNStartStair, i.e. the starting voltage to generate the erase stair, and the page counter is set to zero.
4. The erase potentials are applied to the array during the entire erase pulse and then the verify conditions are restored. Once the erase pulse has finished, the Stair counter is incremented to prepare the erase voltage for the subsequent pulse.
5. After each erase pulse, a verification cycle of the sector follows. The VPCX regulator is set to apply the V_{gRead} verify voltage to the selected word lines. The VNEG regulator is set to apply the V_{gUnSel} negative voltage to the unselected word lines of the sector to erase. The UnSel negative voltage applied to the word lines guarantees that the current flowing through the bit lines to the sense amplifier for comparison to the reference cell current, I_{ev}, is due only to the cell currents of the page being erased. The V_{gUnSel} value is chosen so that the voltage of the most depleted cell produces a negligible current (remember that in the NOR architecture hundreds of unselected cells are present on the same bit line). If the erase-verify operation detects that even one single cell does not conduct enough current, the algorithm repeats the erase pulse (step (3)). If the page verification is successful, the following page is examined. Once all the pages have been verified, the Soft Program phase follows. During the erase cycle, the V_b voltage is incremented at each erase pulse up to a maximum value (FNEndStair). If the number of erase pulses reaches the maximum number of attempts (MaxAtt), an error flag is set and the algorithm jumps to the end (10).

At the end of the erase phase, a part of the cells of the sector have negative threshold voltage, as it can be noted from the distribution of the threshold voltages (Fig. 14.8). These cells have to be slightly programmed to restore their threshold voltage to a positive value. This is the Soft Program phase.

6. The Soft Program phase has to examine each page of the sector and operate sequences of verify and program functions. The regulators used to Soft Program are suitably activated. VPD is activated to generate V_{dCHE} and V_{IPW} to generate V_{bCHE}. VNEG is activated to generate the V_{gUnSel} voltage that keeps the unselected word lines inactive. The voltage applied to the selected word line

and, hence, the gate of the cells being programmed, is driven, during the program pulse, by the voltage value of the Stair counter.

7. For each page, the Stair counter is initialized to DVStairStart and all the Program Load Flip Flops are activated.
8. A verify is executed to determine whether all the cells of the selected page have positive threshold voltage. The comparison is carried out between the current of the reference cell, I_{vd}, and the current of the selected cell when both are biased at V_{gDvSel}. The I_{vd} current must be small to be able to precisely detect the position of the threshold voltage, but large enough to allow the sense amplifiers to operate. The unselected word lines are biased with a negative voltage, V_{gDvUns}, to keep all the cells belonging to unselected word lines inactive, so as not to interfere with the page being verified.
9. After the reading has been carried out, the Program Load Flip Flops are updated. If all the cells of the page pass the verification, the algorithm moves on to verify the subsequent page (7).
10. If some of the cells are depleted, then they are programmed slightly, and also the proper program voltages are applied to bulk and gate.
11. The V_{dCHE} program voltage is applied to the drain of the cells selected by the active Program Load Flip Flops, and remains for the entire duration of the program pulse. At the end of the pulse, the Stair counter is incremented for the next possible pulse, and the verification is repeated (8).
12. The cycle ends when all the pages have correctly been verified. Before releasing control, the controller restores the read conditions and drives all the pumps and regulators into their stand-by condition, since there is no need for them to be active during read. The end of the erase operation is signaled to the user by means of either a busy signal or a proper flag set in the device status register.

During the execution of the flow, particular attention is paid to the timings to change the bias conditions of the array and decoders, so as to obtain the precision required by the different phases and prevent dangerous boost effects.

The erase algorithm for multilevel memories is identical to the one for bi-level memories. Better accuracy of the sense amplifiers and voltages allows obtaining a more precise placement of the erase distribution during the Erase Verify and Erase Depletion phases.

14.6 Test Algorithms

Device test represents a important portion of the device cost. Many tests can be executed by controlling each part of the device by means of an external controller but the need for carrying out a large number of program and erase operations makes it useful (or essential) to use test algorithms executed by an internal controller that communicates with the external test machine. The most common test capabilities required by the controller, in addition to the program and erase ones, are the ability to execute portions of the following algorithms: 'Program All0', 'Erase Only', and 'Soft Program Only', or, simply, the verification of patterns throughout a sector or the entire device.

Bibliography

S. Maramatsu et al. "The solution of over-erase problem controlling poly-Si grain size-Modified scaling principles for Flash memory", IEDM Tech. Dig., pp. 847-850, (1994).

T. Nakayama, "A 60ns 16Mb Flash EEPROM with Program and Erase Sequence Controller", ISSCC91, paper FA 16.1, pp. 260-261

A. Silvagni, et al., "An overview of logic architectures inside Flash Memory devices", IEEE Proceeding of the, Vol. 91, No. 4, pp. 569-580, (April 2003).

K. Yoshikawa, S. Yamada, J. Myamoto, T. Suzuki, M. Oshikiri, E. Obi, Y. Hiura, K. Yamada, Y. Ohshima and S. Atsumi, "Comparison of current Flash EEPROM erasing method: Stability and how to ocntrol", IEDM Tech. Dig., pp. 595-598, (1992).

15 Circuits Used in Program and Erase Operations

Nowadays, every Flash memory device contains integrated charge pump circuits: this feature is mandatory because power supplies have been reduced but at the same time internal high voltages are required by the memory to get the electric fields needed by both program and erase operations.

15.1 Introduction

In the past few years, the use of portable devices has considerably increased; the key feature for these applications is reduced power consumption, which calls for chips that operate from a single power supply. This is the main cause for the transition from dual voltage (VPP and VDD where VPP > VDD) to single voltage (VDD) memory design. Moreover supply voltage tends to decrease in order to limit power consumption.

One of the main issues related to voltage scaling is to insure the proper operation of the device, i.e. the correct execution of all those operation which call for voltages greater than power supply, such as erase and program of the Flash memory (Table 15.1). In fact, the physical phenomena and the electrical fields that must take place when these operations are executed are not possible with present supply voltages (from 1.8 V to 3 V), unless the gate oxide thickness is reduced. Such a solution cannot be implemented, since an excessive reduction of the oxide thickness would negatively affect the ability of the cell to retain information, thus compromising its reliability.

Table 15.1. Biasing of the terminals of the memory cell during the different operations

Operation	N-well	Ip-well	Source	Gate	Drain
Read	3 V	0 V	0 V	2 ... 6 V	0.8 V
Program	3 V	- 1/- 2 V	0 V	8 V	4.5 V
Erase	8 V	8 V	8 V	-8 V	Floating

To summarize, the single voltage supply memories require circuits which are able to generate the (high) voltages required by the various operations starting from the (low) power supply. Such circuits are usually based on the charge pump principle, and therefore they are often referred to as "charge pumps".

15.2 Dual Voltage Devices

In the case of dual voltage devices, read, write and erase operations are carried out using a high voltage, VPP, supplied to the flash memory. It is therefore unnecessary to have on-chip high voltage generators, since all the voltages greater than VDD can be obtained from VPP through an appropriate voltage partitioning.

In addition, for devices realized using technologies without p-channel transistors, the main issue is how to completely transfer such a voltage.

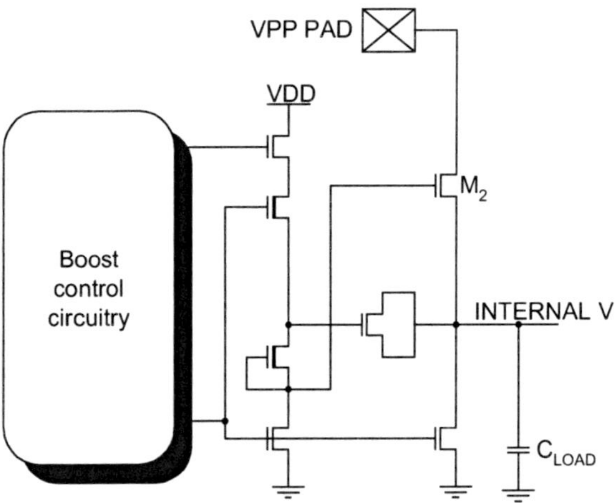

Fig. 15.1 VPP switch for an NMOS device

A circuit that can be used for this purpose is shown in Fig. 15.1. It is a circuit where a bootstrap is used to bias the gate of the M_2 pass transistor at a value greater than VPP to prevent the internal voltage from being limited by the V_T of the pass itself. The only drawback of this circuit is related to the W of M2: this transistor can be as wide as 1 mm if the load capacitances connected (to the output) are significant (e.g. both row and column capacitances) and it is necessary to charge them in a reasonable amount of time.

The externally provided voltage can be used to bias the gate or the drain of the cells during program; depending on its usage, the circuit must be properly designed and/or modified.

If the VPP voltage is used to bias the gate of the cells, then the node called INTERNAL V, i.e. the node of the circuit shown in Fig. 15.1, is connected to the supply node of the row decoder.

The switch described above can be used to transfer into the chip both VPP and VDD voltages as shown in Fig. 15.2. Again, the main issue is the size of M_2; in order to minimize this device, the circuit that requires the VDD supply during the read operation must be carefully selected.

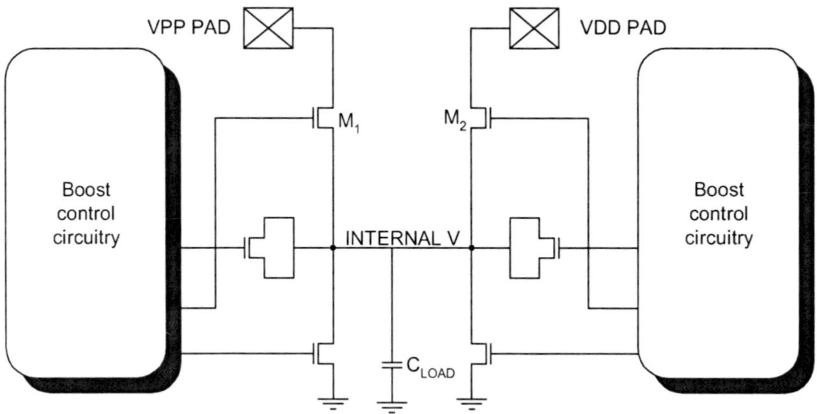

Fig. 15.2. VPC switch for the first NMOS devices

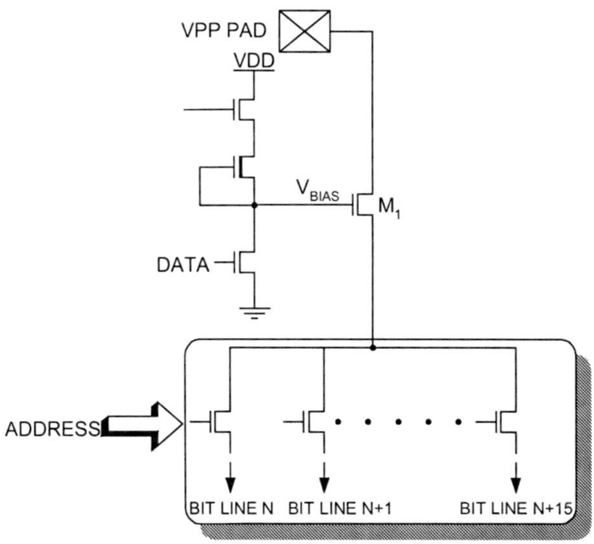

Fig. 15.3. Drain voltage without regulation

Whenever the voltage is used to bias the drain of the cell, it is necessary to employ a feedback circuit to regulate the voltage to the proper range, rather than let the voltage be determined only by the size of M_1 and by the value of V_{BIAS} of its gate. In fact, V_{BIAS} is determined by the inverter that drives transistor M_1 (Fig. 15.3). Variations of VPP, temperature and current sunk by the cell can therefore cause a wide variation on the voltage applied to the drain.

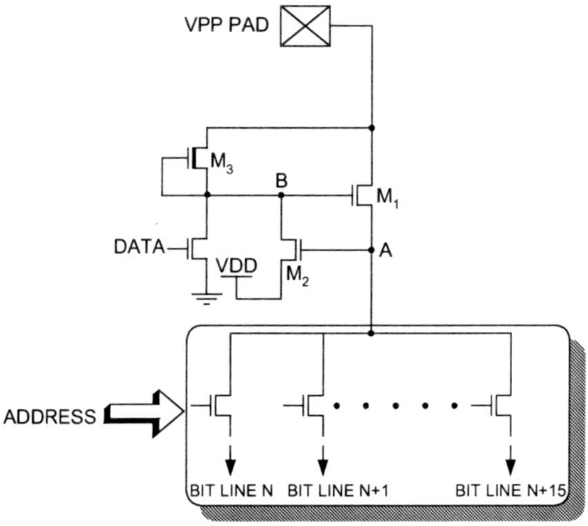

Fig. 15.4. Regulated drain voltage

The variation above described can be avoided by inserting a simple regulator as shown in Fig. 15.4. When the circuit is switched on, the voltage of node A increases until it turns on M_2. The gate of M_1 is biased at a voltage value that is determined by the drive ratio between M_3 and M_2. At this point, if the voltage of node A tried to increase, M_2 would absorb more current from VDD, thus causing the gate voltage of M_1 to decrease. Of course, the opposite would happen in case the voltage of node A tended to decrease.

15.3 Charge Pumps

Figure 15.5 shows the basic scheme of a charge pump: it is composed of a series of diodes and capacitors. Diodes are required to establish the direction of the current flow, while the main task of the capacitors is to accumulate the charge that is then transferred from one capacitor to the next capacitor, by driving the capacitors with the alternate phases CK and CK# (Fig. 15.6). A storage capacitor is placed at the output of the pump, to hold the final voltage.

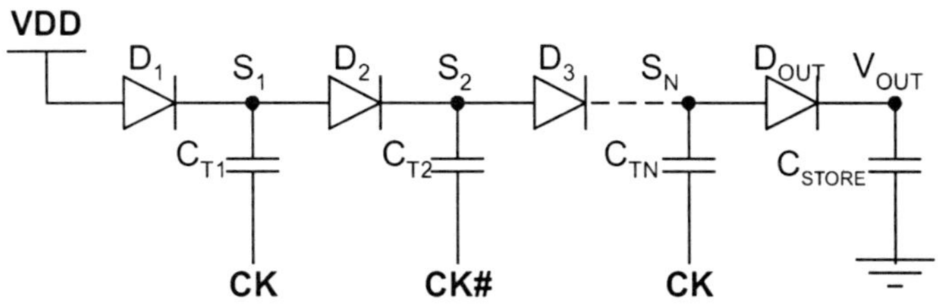

Fig. 15.5. Basic scheme of a charge pump

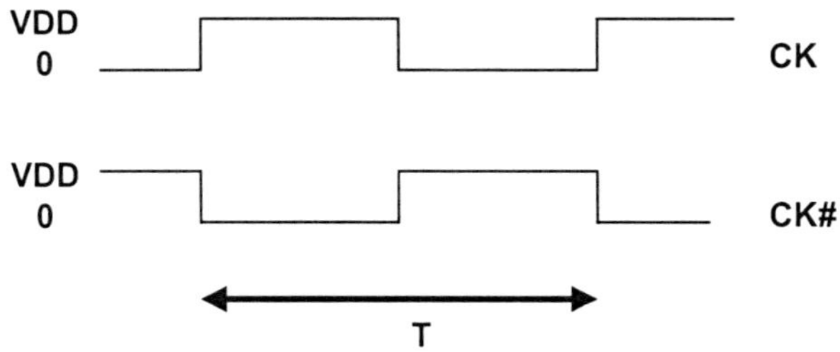

Fig. 15.6. CK and CK# phases; T is the clock period

Two important physical phenomena are the bases for the operation of a charge pump. The former is related to the basic capacitor characteristic: the voltage across the capacitor cannot change instantaneously, as shown in Fig. 15.7; the latter is the so called charge sharing phenomenon, which takes place when two capacitors precharged at different voltages are short-circuited. From Fig. 15.8 it follows that the final voltage of the short-circuited node depends on the initial voltages of the two capacitors as well as on their relative dimensions.

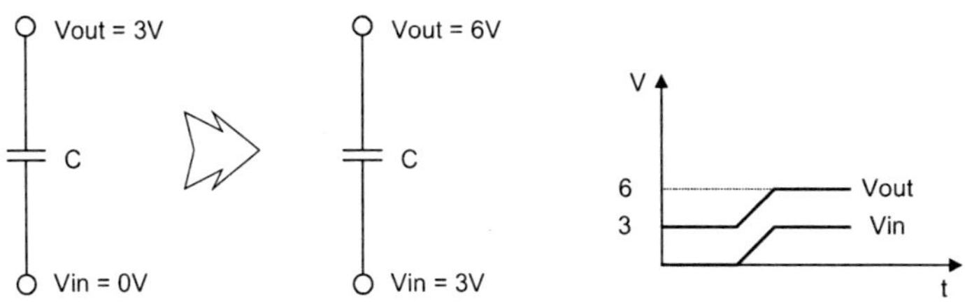

Fig. 15.7. Voltage variation at the terminals of a capacitor

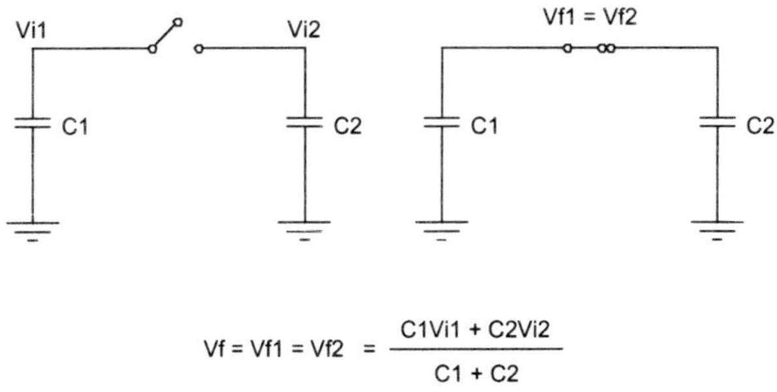

Fig. 15.8. Charge sharing principle

To understand how a charge pump works it is helpful to assume that:

- Diodes have a null threshold voltage;
- All transfer capacitors C_T are of the same size;
- CK and CK# are square waveforms (where CK# is the negated version of CK), initially at ground;
- S_i nodes are initially precharged to VDD

At the power up of the CK and CK# signals, charge storage and transfer begin:

1. when CK goes from ground to VDD, S_1 becomes equal to 2 VDD (the initial value (VDD) + VDD). At this point, a portion of the charge previously stored in capacitor C_{T1} is transferred to capacitor C_{T2}. Transfer stops when the voltage of node S_2 is equal to the voltage of node S_1. Diode D_1 prevents the charge from going back to the supply of the circuit, forcing the charge to go into capacitor C_{T2};
2. when CK goes from VDD to ground CK# rises from ground to VDD and capacitor C_{T2} then transfers a portion of its stored charge to capacitor C_{T3}. At the same time capacitor C_{T1} that partially discharged during the previous phase is recharged to the value of VDD.

The voltage of the output node continues to increase by an increment that becomes smaller and smaller until the voltage V_{OUT} reaches the value of VDD + n VDD, where n is equal to the number of stages of the pump, while the internal nodes voltage reaches a maximum value of $V(S_i)$ = VDD + iVDD. As soon as these voltages are reached, no further charge transfer across the diodes takes place.

Following an example shows how the voltages at nodes S_i and V_{OUT} of an ideal, three-stages pump vary (Fig. 15.9). A power supply of 3 V and a storage capacitor are assumed.

The initial conditions for the circuit are those previously described: CK and CK# are at ground while the intermediate stages S_i and the output node V_{OUT} are precharged at VDD.

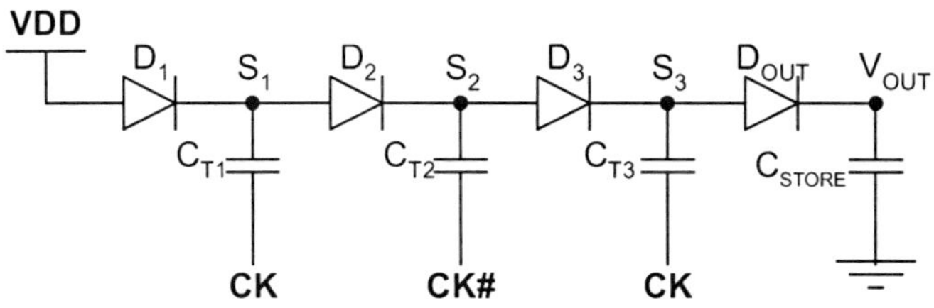

Fig. 15.9. Three-stages charge pump

As soon as the circuit is activated, CK goes to VDD, while CK# remains at ground. S_1 and S_3, thanks to the boost effect, will instantaneously charge at a voltage equal to 6 V and they will start to transfer charge to capacitor C_{T2} and to the storage capacitor C_{STORE} respectively. In this way the voltages at the terminals of C_{T1} and C_{T3} decrease, while the voltages at the terminals of C_{T2} and C_{STORE} increase. Assuming that all the capacitors have the same size and that the voltage drop on diodes is zero, it follows that the charge transfer stops as soon as the voltages at the terminals of the two pairs of capacitors are equal, i.e. when $S_1 = S_2 = 4.5$ V and $S_3 = V_{OUT} = 4.5$ V.

S_3 is initially greater than S_2 owing to the way the capacitances have been driven, i.e. by means of alternate signals CK and CK#. Diode D_3 is therefore reverse-biased and charge moves exclusively from C_{T1} to C_{T2}.

On the other hand, a boost is applied to C_{T1} and C_{T3}, and when CK goes to GND, S_1 and S_3 go to 1.5 V (4.5 V – VDD). Meanwhile CK# has reached VDD and therefore voltage S_2 becomes equal to 7.5 V (4.5 V + VDD). At this point C_{T2} starts to transfer charge to C_{T3}. But now C_{T3} is precharged at 1.5 V, therefore the final voltage on nodes S_2 and S_3 is equal to 4.5 V. At the same time capacitor C_{T1} charges to VDD through diode D_1 that allows charge flow from VDD to C_{T1} as it is forward-biased.

Summing up, at the end of a clock period we have:

$S_1 = 3$ V,
$S_2 = 1.5$ V,
$S_3 = 4.5$ V,
$V_{OUT} = 4.5$ V

At the end of the second clock period we get:

$S_1 = 3$ V,
$S_2 = 2.25$ V,
$S_3 = 5.25$ V,
$V_{OUT} = 5.25$ V

Although, at a first glance, it seems that V_{OUT} is the only voltage tending to increase, it is easy to realize that the voltage of each single node increases as the cycles continue.

Maximum output voltage that can be achievable when no current load is connected to the output is:

$$V_{MAX} = VDD + nVDD \tag{15.1}$$

In the previous example V_{MAX} is equal to 12 V (3 V + 3 · 3 V)
Since the charge transferred at each clock cycle of either CK or CK# is:

$$Q = C_T \cdot \Delta V = C_T \cdot VDD \tag{15.2}$$

the current provided on the output in a clock period T is equal to:

$$I = \frac{Q}{T} = \frac{C_T \cdot VDD}{T} \tag{15.3}$$

From Eq. (15.1) and Eq. (15.3) it is easy to obtain the output resistance of the pump:

$$R_P = \frac{V_{MAX} - V_{SUPPLY}}{I} = \frac{(n+1) \cdot VDD - VDD}{\frac{Q}{T}} = n\frac{T}{C_T} \tag{15.4}$$

As an example, Fig. 15.10 shows the plot of the output resistance of a pump implemented in a device.

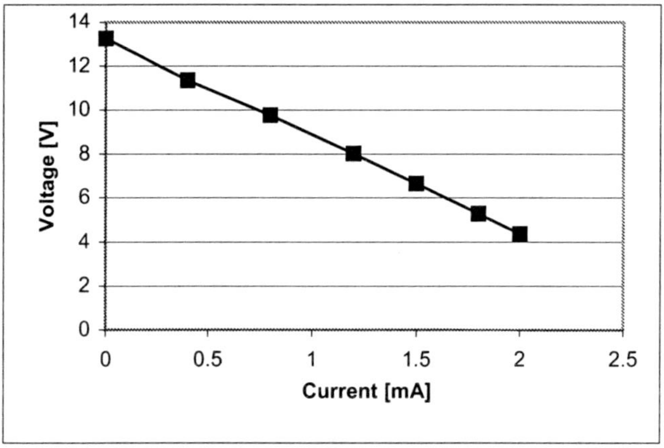

Fig. 15.10. Output resistance of a 3 V supplied charge pump made up of 4 stages.

Anyway it is necessary to point out that this equation holds true if and only if there is a complete transfer of the stored charges between a capacitor (C_{Ti}) and the subsequent one (C_{Ti+1}).

As far as we have seen until now, charge pump can be considered as a V_{MAX} voltage generator with a series resistor R_P; therefore if a load sinking a constant current I_0 were connected to the output, the general value for the voltage V_{OUT} would be:

$$V_{OUT} = V_{MAX} - n\frac{I_0 T}{C_T} = V_{MAX} - R_p I_0 \tag{15.5}$$

In this case, C_{Ti} capacitors must provide a charge equal to $q = I_0 \cdot T$ to the output at each clock period T. For this reason a reduction of the voltage value occurs for both the corresponding nodes S_i and for the output node.

Now that the operating principle of the pump should be clear, it is possible to get nearer to the "real" model, taking into account the fact that the inserted diodes have a threshold voltage not equal to zero, which we will refer to as V_D in the following.

In this case the output voltage, as well as the voltage of the intermediate stages, is smaller than the voltage we have considered so far. In fact the charge transfer from capacitor C_{Ti} to capacitor C_{Ti+1} ends as soon as the voltage of the i - *th* stage is equal to the voltage of the (i + 1) - *th* minus V_D, simply because at this point the diode between the two stages becomes an open circuit. Resulting V_{MAX} voltage is therefore:

$$V_{MAX} = VDD - V_D + n \cdot (VDD - V_D) \tag{15.6}$$

It is important to point out that the size of the diodes must be carefully chosen: if they are too big, the associated area, which equates to capacitance, is no longer negligible, while if they are too small then the resistance of the pump R_P becomes high because of the series resistance represented by the diodes themselves.

What happens if we increase the number of stages? Considering Eq. (15.4) and Eq. (15.6), two opposite effects take place: on one hand, output voltage V_{MAX}, increases; on the other hand, resistance R_P increases as well. The latter result is far from being desirable, because whenever a current absorption occurs, a high value of resistance implies a considerable decrease of the voltage V_{OUT} (which, in reality, is the available voltage) with respect to V_{MAX}.

Observing Eq. (15.4) we could think about reducing the resistance of the pump R_P by modifying either the size of the transfer capacitors or the duration of the clock period. Both solutions are not recommended.

Indeed it is a fact that there is usually a limited range of clock periods (often just one clock period) that allow us to get the maximum output voltage; such a period is often referred to as optimum. Figure 15.11 shows voltage V_{OUT} of a pump as a function of different clock periods. It is worth noting that as the current absorbed by the pump varies, V_{OUT} changes as is expected, while the optimum period does not vary.

The reason behind the existence of a limited range of values is related to the non-ideality of the components used to realize the pump: in fact, in spite of what happens in the ideal pump, charge transfer from one capacitor to the following one does not take place instantaneously, but it depends on several factors like, for instance, the threshold voltage of the diodes, the driving capability of CK and CK# phases generators, typical leakage currents of the capacitors, parasitic capacitances, etc.

To summarize, reducing the clock period while maintaining the size of the capacitors could prevent a complete transfer of the stored charge from happening: the driving of the capacitor would be less effective and the output voltage would not be at its maximum; the same undesired result would be achieved if the size of the capacitors were increased without changing clock period.

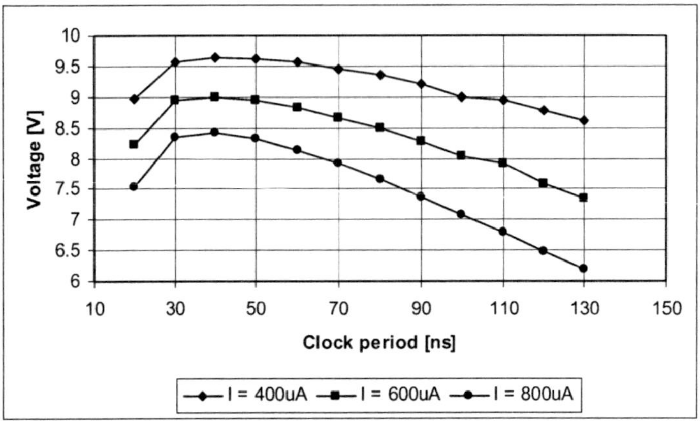

Fig. 15.11. Optimum period for a charge pump

Among the non-ideality previously listed, parasitic capacitances deserve further discussion. By "parasitic capacitances" we mean all those capacitances connected (directly or indirectly) to the transfer capacitors. They are charged during one half-cycle, and in the following half-cycle they transfer the accumulated charge towards ground, instead of transferring it to the storage capacitors.

The charge of these capacitors is therefore necessary for the proper operation of the circuit, since their presence can not be avoided but it is counterproductive to generating the output voltage because part of the charge is "stolen" from the transfer capacitors. Going further into detail, the parasitic capacitance C_P consists of:

1. capacitance between the lower plate of the capacitor and ground;
2. capacitance between the upper plate of the capacitor and ground;
3. capacitance between the anode-substrate and the cathode-substrate typical for the diode.

Resulting voltage V_{MAX} taking this effect into account is:

$$V_{MAX} = VDD - V_D + n \cdot \left(\frac{C_T}{C_T + C_P} VDD - V_D\right) \tag{15.7}$$

where C_T is the value of the transfer capacitors.

Designing a charge pump, therefore, means to explore solutions to realize an ideal voltage generator by overcoming the non ideal characteristics of components, which tend to limit the output voltage below the ideal value.

15.4 Different Types of Charge Pumps

Thanks to the continuous improvements of technologies to realize memory devices and the evolution of the electronic circuits design techniques, novel designs of charge pumps have been devised, which have improved performance, reduced R_P resistance and increased output voltage.

15.4.1 Dickson Pump Based on Bipolar Diodes

The Dickson pump is the one of the oldest type of pump, and from a structural point of view it is identical to the pump shown in Fig. 15.5: an alternating series of diodes and transfer capacitors driven by two opposite phases (CK and CK#), which in turn do not need delays or phase shift.

Even if both the pump and the phase generator are quite easy from a circuital point of view, such a structure cannot be effectively realized in those technologies where triple-well is not available. In fact, in order to design a working circuit, it is advantageous to use bipolar diodes, which are typically realized by means of n+ diffusion in a p-well tub.

15.4.2 Dickson Pump Based on Transistor-Based Diodes

In order to make the charge pump independent of a given technology, it is possible to substitute the bipolar diodes of the pump previously described with diodes realized using properly connected MOS transistors as shown in Fig. 15.12.

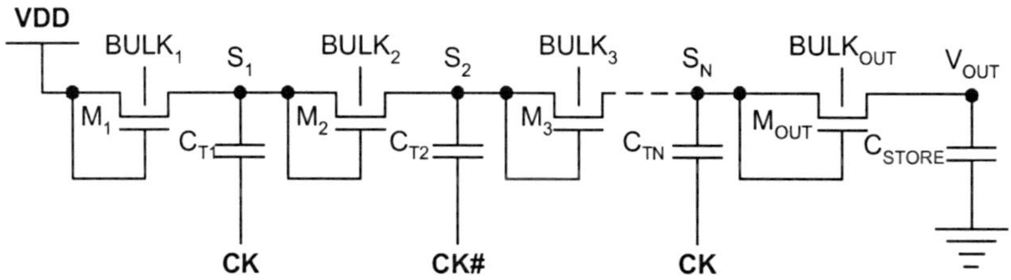

Fig. 15.12 Diodes realized using MOS transistors

One of the advantages of this solution is that the latch-up phenomenon, which is typical of pumps based on bipolar diodes, can be avoided. The first issue to cope with when designing this pump is the choice of where to connect the body terminal of the inserted transistors. The only possible solution is to connect the bulk of the NMOS (diode-connected) to ground if we want to use NMOS transistors without triple well. Indeed, in this way we are guaranteed that the body is not only connected to the lowest voltage of the circuit, but also that it is stable. Even if such a solution leads to a relatively simple circuit, the drawback is that the output voltage is considerably reduced because of body effect. In fact, the threshold voltage of the transistors tends to increase stage after stage until it reaches its maximum value on the output transistor. If we decide to replace the diodes with PMOS transistors instead of NMOS, the issue of where to connect the body must be considered, since it is difficult to say which voltage best fits, considering the behavior of the voltages at the terminals of the transistors; indeed, it is not possible to find a voltage between drain and source which is constantly higher.

There are simple solutions that overcome the body effect issue; the drawback is that in the case of an NMOS it is again necessary to use a triple well technology in order to be able to connect the bulk to a voltage that is different from the substrate voltage.

Figure 15.13 shows an example of a bulk bias circuit: it is a voltage divider circuit composed of diodes: the voltage of the intermediate nodes can be used to bias the bulk of each charge pump transistor. Diodes are used instead of resistors in order to save area with the same power consumption; moreover since the current is provided by the pump, only a minimum current can be used, without compromising the performance of the pump itself. From Fig. 15.13 it is possible to see that the intermediate nodes of the divider (P_i) are connected to the bulk through a NMOS diode-connected transistor and a capacitor. These elements are used to filter the voltage of nodes P_i since these nodes are not fixed, but they have an oscillation proportional to the output voltage oscillation (which in turn is related to the current required on the output).

Using this structure, called a bulk biaser, it is possible to reduce the body effect by connecting the bulk nodes of the pump transistors to a voltage slightly lower than the corresponding drain and source nodes.

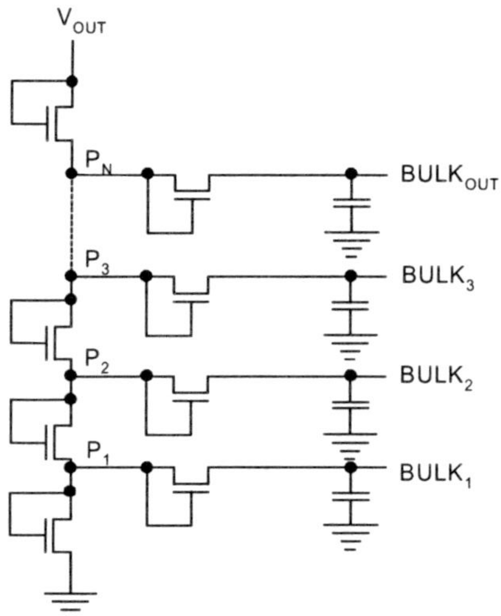

Fig. 15.13. Bulk biaser

Alternatively, a dynamic biasing of the bulk nodes could be devised. The structure shown in Fig. 15.14 allows biasing the bulk nodes of the pump transistors to the lowest voltage between drain and source of the transistor (or the highest one in case the same structure would have been realized using PMOS transistors) exploiting the stages of the pump itself. The disadvantage of this solution is that the bulk nodes of the pump might be not biased during the transient between one semi-period and the following one, because of the delay between the turning on of the two transistors M_1, M_2. This risk must be carefully evaluated in the design phase taking into account the parasitic capacitances that influence the stages, the tubs of the stages themselves, the commutation time of the phases, etc.

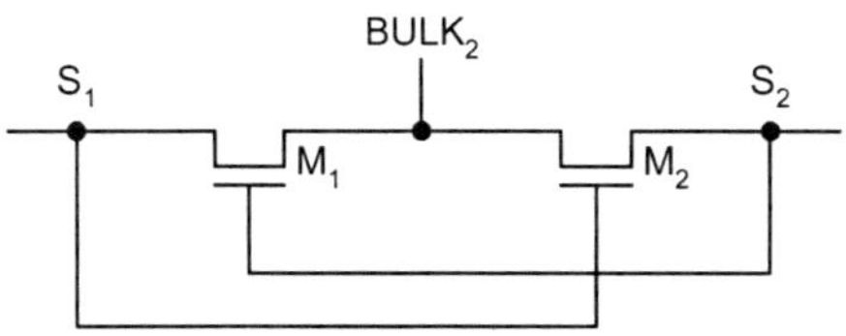

Fig. 15.14. Dynamic biasing of the bulk nodes

Problem 15.1: Analyze in detail the behavior of the circuit in Fig. 15.14.

The performance of the pump that uses the diode-connected transistors instead of the bipolar diodes is influenced by two effects:

1. parasitic resistance of the transistor, that effects the optimum period;
2. the threshold voltage of the transistors, which is typically higher than the turn on voltage of a bipolar diode, which directly effects the output voltage V_{OUT}.

Since the two solutions (NMOS/PMOS) share the same issues, it is reasonable to wonder if one is better than the other, and why.

The pumps realized with PMOS are for sure simple and reliable architectures. Anyway, in order to have a negligible parasitic resistance introduced by the transistor (in particular if supply voltage is very low, e.g. 1.8 V), the size of the transistor must be quite big, and area occupation becomes an issue.

On the other hand, the use of NMOS transistors realized in triple well is the only way to reduce the body effect. In this case it is mandatory to control the biasing of each well of the transistor, in order to prevent any forward biasing of the junctions.

15.4.3 Charge Pump Based on Pass Transistors

The diodes inside the pump are needed to establish the current direction and to prevent any charge from flowing towards the previous stage. The main drawback of the use of such components, apart from the technological complexity to realize them, is their high threshold voltage, i.e. the need of having a consistent voltage difference between drain and source, in order to have a charge transfer. It is clear that under these conditions the charge transfer between one stage and the following one cannot be complete.

The alternative to the diodes to obtain both charge transfer and node connections is represented by transistors used as switches. MOS behavior is involved, i.e.

if the voltage applied to the gate is lower than its V_T the transistor is turned off, therefore there is no charge transfer and it is equivalent to an open switch. On the other hand if the voltage applied to the gate is higher than V_T, then there is a complete current transfer with a minimum voltage drop equal to the turn-on resistance of the transistor R_{ON}, multiplied by the transferred current. In the latter case, the transistor is equivalent to a closed switch

The circuit inevitably gets more complex and it occupies a larger area. In fact, besides the transfer capacitors and the two phases already present in a simple pump circuit, it is necessary to add two boost capacitors and two additional phases to better drive the switches as shown in Fig. 15.15.

Let's suppose to transfer charge from capacitor C_{T-1} to capacitor C_T, and at the same time to block the charge transfer between the two consecutive capacitors (C_T and C_{T+1}).

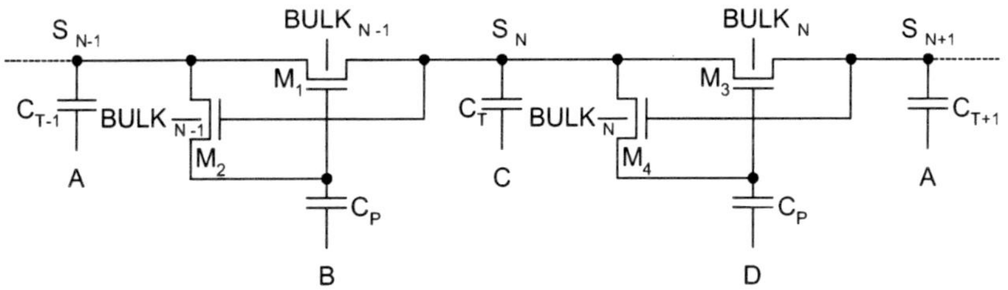

Fig. 15.15. Internal stages for a pump using pass-transistors

Let's start from the condition where no charge transfer occurs, i.e. pass transistors are switched off, and therefore B and D are at GND and C is at VDD. In this case, the gate of transistor M_1 is biased at a low potential (it is difficult to specify the value since it depends on the stage we are looking at and on the amount of discharge on C_P in the meanwhile), while node S_N is biased at VDD.

At this point, as it is possible to see from Fig. 15.16, phase A is triggered, thus applying the bootstrap to node S_{N-1} and transistor M_2 is turned off (C is still active) it is possible to precharge the upper plate of the capacitor C_P which drives the gate of the pass transistor M_1. Once this phase is over, it is necessary that C returns to GND and that the gate of M_1 reaches a value that allows charge transfer between C_{T-1} and C_T. This condition can be obtained by activating phase B for an interval corresponding to a complete charge transfer between the two capacitors. Once the transfer is over, all the operations are repeated in reverse order: first B is lowered to GND, then phase C rises. Indeed at this point capacitor C_P connected to the gate of transistor M_3 pre-charges as long as A is active. When phase D is activated, charge transfer between C_T and C_{T+1} starts.

The biasing of the bulks of the pass-transistors exhibits the same issues encountered with the pumps where diodes where implemented using transistors (see Sect. 15.4.2).

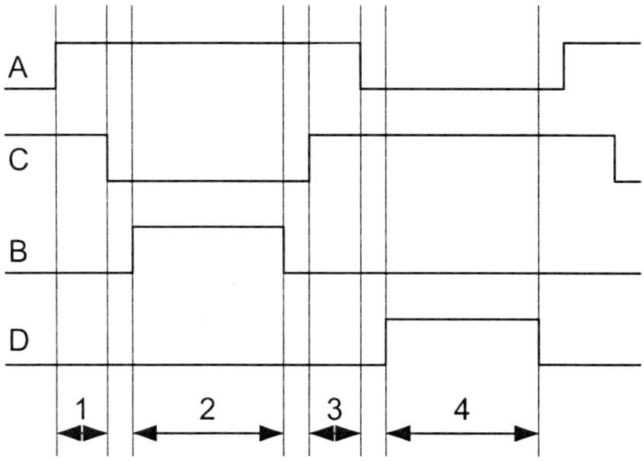

Fig. 15.16. Timing diagram of the phases for the pass-transistor pump: 1) precharge of capacitor C_P of transistor M_1, 2) charge transfer between capacitors C_{T-1} and C_T, 3) precharge of capacitor C_P of transistor M_3, 4) charge transfer between capacitors C_T and C_{T+1}

If we could use N-channel transistors in triple-well, we might think about connecting the body of the switch to its drain.

In this way, not only do we save area on silicon, because structure like bulk biaser or dynamic bulk biaser are no longer required, but we also have a positive side effect, i.e. a bipolar diode is created inside the triple well transistor (Fig. 15.17), which starts transferring charge as soon as the voltage difference between drain and source is equal to its V_T. This fact has several consequences. On one hand it is possible to start the charge transfer from one capacitor to the following one earlier and the resistance for the NMOS transistor is reduced, since the charge transfer takes place through two paths: one is the channel when the MOS is activated, the other one is the bipolar (at least until the bipolar diode is active). On the other hand, the pre-charge of capacitors C_P might be still incomplete, therefore the boost of the gate of the transistors might become inefficient. Another issue due to the presence of the bipolar is the possibility that latch-up phenomena occur.

To transfer the charges from the last transfer capacitor to the storage capacitor on the output a diode-connected transistor or a further pass-transistor can be used. Both solutions have pros and cons.

A diode-connected transistor has the advantage of being easily designed, while the disadvantage is that it requires a voltage difference at its terminal in order to operate correctly; as a consequence the voltage transferred to the output is equal to the voltage of the last stage minus the threshold voltage of the diode. On the contrary, the pass-transistor allows transferring all the voltage to the output, but it requires a boost structure for its gate (identical to the one required by the internal stages of the pump) to be properly driven, i.e. to be able to correctly transfer the voltage. Power consumption increases when this pump structure is compared to the pumps previously seen. In fact the phases must be created "ad hoc": delays must be carefully introduced so that under every operating condition of the pump simultaneous boost of the pass-transistors or of capacitors belonging to different stages are avoided. Therefore the phase generator grows in complexity and its power consumption grows as well. Furthermore part of the charge stored in the transfer capacitors is used to precharge capacitor C_P placed on the gate of the pass-transistor. As a consequence not all the charge is transferred, but part of it is lost for this operation.

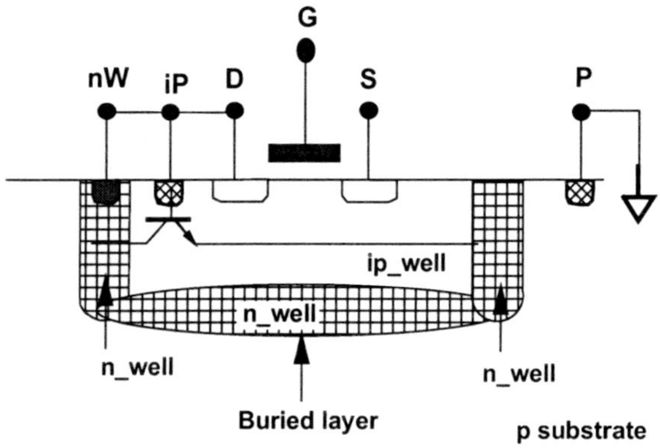

Fig. 15.17. Charge transfer using bipolar parasitic diode

15.4.4 Voltage Doubler

Another circuit whose task is to raise the voltage, also known in literature as voltage doubler, is shown in Fig. 15.18. It is a feedback system that can duplicate the power supply, and it is composed of two n-channel transistors, (M_{N1} and M_{N2}) and two capacitors (C_{T1} and C_{T2}) of the same size.

In order to understand the principle of operation of this circuit it can be assumed that, at the beginning, nodes A and B, as well as phases CK and CK#, are at GND. In this way, both transistors M_{N1} and M_{N2} are turned off.

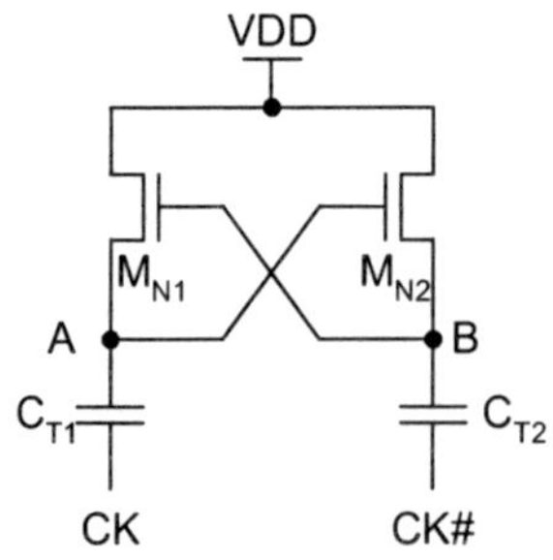

Fig. 15.18. Voltage doubler

As soon as CK phase toggles from GND to VDD, the value of node A becomes VDD, activating transistor M_{N2}. Since phase CK# remains at GND, the charge starts flowing from power supply to capacitor C_{T2} until B reaches a value equal to VDD – $V_{TH,N2}$. When phase CK goes to GND again, it brings node A along, thus turning off transistor M_{N2}. Now phase CK# gets to the value of VDD, therefore the value of node B becomes VDD – $V_{TH,N2}$ + VDD turning on transistor M_{N1}, through which C_{T1} is charged until it reaches VDD. Of course when CK# goes to GND again, node B is biased at a voltage equal to VDD – $V_{TH,N2}$. At this point the phases repeat. It is worth noting that each capacitance is boosted first, then charged, changing its value from VDD to 2 VDD. Since these operations are performed alternately on each half of the circuit, during a whole period either node A or node B is at 2 VDD (thus the reason for calling this circuit a "voltage doubler"). If, at this point, it is possible to design a circuit that can transmit on its output the 2 VDD voltage, independently of which node (either A or B) is biased at that value, then passing from the voltage doubler to the charge pump is relatively simple: in fact, it is sufficient to repeat such a structure where the output of a stage is used as the supply voltage for the following stage.

A PMOS output stage properly connected to nodes A and B, as the one shown in Fig. 15.19, is the simplest circuit to obtain an output voltage of 2 VDD: when CK is at VDD, A is at 2 VDD, while B is at VDD. Transistor M_{N1} is therefore turned off, while M_{P1} is active, transferring the voltage of node A to node OUT. In the meanwhile M_{P2} is turned off and M_{N2} is turned on, charging capacitor C_{T2}. When CK goes back to GND and CK# becomes VDD, then the circuit behaves in the opposite way: M_{N1} and M_{P2} are active (the former charges capacitor C_{T1}, the latter transfers the voltage of node B to the output) while M_{N2} and M_{P1} are turned off. The way the circuit is conceived, there are no direct paths between VDD and node OUT: these paths would jeopardize the functionality of the circuit itself.

Also in this case, the issue of the bulk biasing of the transistors exists (of course under the assumption that triple-well NMOS are used).

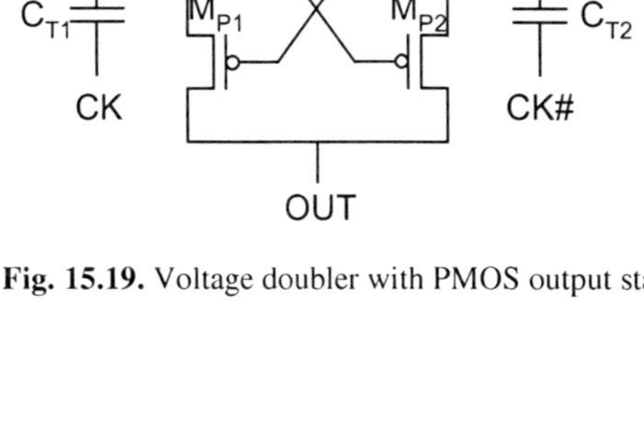

Fig. 15.19. Voltage doubler with PMOS output stage

Again, the solutions described in previous sections apply, and the same pros and cons hold true: one solution is to connect the bulk terminal to the power supply for the n-channel transistor and to the output node for the p-channel transistor, another one is the "dynamic biasing" of the bulk nodes. Figure 15.20 represents a possible circuit solution that allows keeping the body of the NMOS transistors to the lowest voltage, and the body of the PMOS transistors to the highest voltage, under any circumstances.

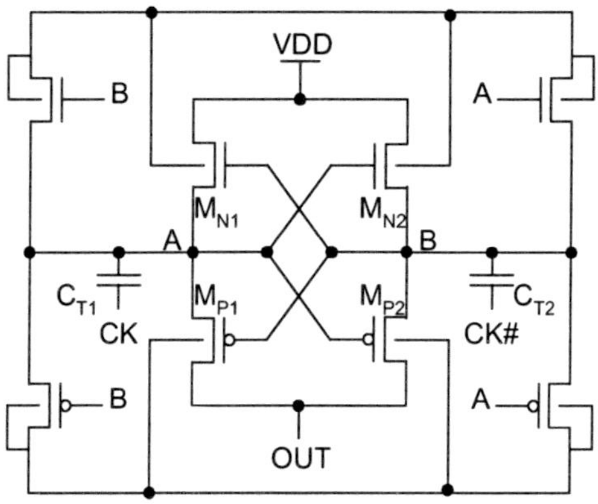

Fig. 15.20. Biasing of the bulk nodes of the transistors

A small specification is required at this point: when more stages are cascaded to realize a pump as shown in Fig. 15.21, we can decide to drive the capacitances of the stage with a signal whose swing is either GND/VDD or GND/supply of the stage. In the former case, the gain of each stage is about VDD; in the latter the gain is about 2 VDD. In any case, from an implementation point of view, the former solution is preferable.

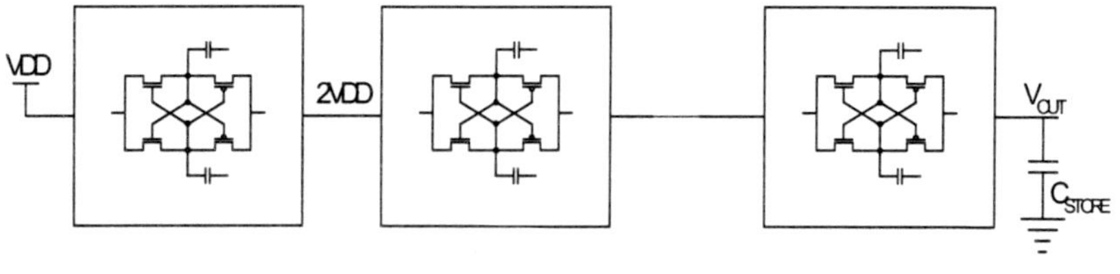

Fig. 15.21 Implementation of a charge pump by cascading a series of voltage doubler circuits; VOUT output voltage is equal to (n + 1) VDD if CK and CK# signals have a 0 – VDD swing, while it is equal to 2^NVDD if the swing of CK and CK# for the i-*th* stage is 0 – supply of the i-*th* stage

This structure has several advantages:

- the possibility of driving the capacitances using only two complemented phases. This implies a relatively simple phase generator circuit with respect to the one used in the pumps described in the previous section;
- the possibility of realizing the circuit using "low voltage" transistors (i.e. transistors that can bear a voltage difference at their terminal lower than power supply) since the voltage difference at the terminals of each transistor does not go above VDD in any operating phase of the circuit;
- no boost capacitors are required for the pass transistors (since we can use the one implemented for the charge transfer), thus achieving a reduction of the area occupation and limiting the dissipated power to the one really needed to produce the desired voltage;
- a decrease of the ripple that naturally comes from the pump.

In fact, due to the operation of the charge pump and of the capacitors in general, the output of the pump is more or less noisy depending on the phase that is toggling at a given time. Let's refer to Fig. 15.5: when the clock CK goes high, the charge stored in the capacitor C_{T3} is transferred to the storage capacitor. V_{OUT} instantaneously charges to a value equal to the previous one plus the supply voltage; the corresponding waveform will show at that time an upward peak; then V_{OUT} discharges until CK goes back to GND and CK# goes to VDD. The waveform of V_{OUT} will show an upward peak again, but its height is less than the previous case, since the charge is now shared between two capacitors (i.e. C_{T3} and C_{STORE}). Now the discharge of V_{OUT} occurs again until the new phase comes in.

When talking about "ripple", we generally refer to the height of the "peaks" that can be found in the output node waveform.

Analyzing the output of the voltage doubler as a function of different frequencies it can be seen that it is better if the clock is as short as possible. This fact can be a limitation for the circuit, since in these kinds of applications, the clock generators are often the result of a trade-off between area occupation and precision: in other words the risk is that it may be difficult to implement structures that occupy a small area and at the same time guarantee the same clock period as voltage and temperature conditions vary.

15.4.5 Voltage Tripler

This circuit is derived from the voltage doubler and provides an output voltage whose value is three times the supply voltage value.

Figure 15.22 shows the schematic of such a stage.

Since the structure is symmetric, it is possible to understand the circuit behavior by studying just one half of it. Let's start again from the condition where both CK and CK# are forced to GND, i.e. all the transistors are turned off. When CK goes to VDD, CK# remains at GND. Under these conditions, B is biased at VDD, A is biased at 2 VDD and since M_{N5} is turned on, D is biased at VDD. On the contrary, F is driven to GND by M_{N6} and transistor M_{P4} is turned off.

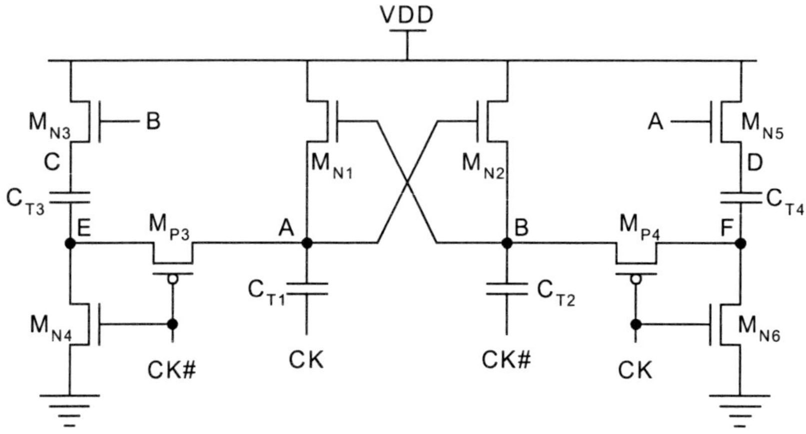

Fig. 15.22. Voltage Tripler

When CK# goes from GND to VDD, then B goes to 2 VDD while A goes to VDD, thus turning off M_{N5}. M_{N6} is turned off as well, since it is driven by the signal CK that is at GND. M_{P4} is turned on instead, and it short-circuits nodes B and F. Therefore F reaches a voltage of 2 VDD and, due to the boost effect, D is biased at 3 VDD.

The behavior of the portion of the circuit not analyzed yet is essentially complementary. In fact, when CK# is at VDD, capacitor C_{T3} is precharged at VDD: node C is kept at VDD, while E is driven to GND by M_{N4}; when CK# is at GND, then E is biased at 2 VDD and C, due to the boost effect, is biased at 3 VDD.

At this point, we just need a structure that is able to output the 3 VDD voltage at each clock cycle. To achieve this result, we can use again the structure shown in Fig. 15.19.

Also in this case we need to cascade several stages in order to build a pump, and the output of a stage must be used as the power supply for the following one. If N is the number of cascaded stages, then the maximum voltage that can be achieved is equal to:

- $V_{MAX} = (N + 2)$ VDD if the phases that drive the stages have a GND/VDD swing (it is worth noting that the $(n + 2)$ factor is due to the fact that for all the stages there is a voltage increment equal to VDD except for the first one, whose increment is 2 VDD);
- $V_{MAX} = 3^N$ VDD, if the phases that drive the stages have a swing between ground and the voltage of the driven stage.

Contrary to the two-phases pump analyzed above, this pump cannot be implemented using low voltage transistors only, since the voltage drop at the terminals of the transistors can go above VDD under certain operating conditions.

15.5 High Voltage Limiter

The voltage generated by the charge pump linearly varies as a function of the supply voltage. Taking into consideration that in certain applications the difference between the minimum and maximum supply voltage can be about a Volt (e.g. the devices that work with a power supply equal to 3 V must work at both 2.7 V and 3.6 V), the following two situations might both occur if we implement any sort of charge pump structure with a given and predefined number of stages:

1. get the voltage required by the operation that we want to perform on the cell, using a low supply voltage;
2. get a voltage that is above the maximum limit for the oxides of the cells, if we use high supply voltages.

Therefore it would be useful to cascade the charge pump with a structure that, acting on the pump itself, is able to limit the output voltage V_{OUT} to a value that cannot damage the memory cells, and acts independently from the supply voltage.

There are mainly two ways to achieve this result, i.e. applying either a continuous or an ON/OFF regulation. If the clock of the pump is realized using an oscillator (a cascade of an odd number of closed-loop inverters), it is quite simple to achieve a continuous regulation: it is sufficient to vary the number of inverters that compose the oscillator, thus obtaining a variation of the clock frequency. On the other hand, an ON/OFF regulation does not structurally change the clock generator, but the different phases that drive the capacitors of the pump are kept at ground when the pump itself has to be stopped. In general it is preferable to act on the phase generator if this structure is available, otherwise it is the clock generator that is turned on and off.

A typical circuit used for the ON/OFF regulation is the high regulator shown in Fig. 15.23.

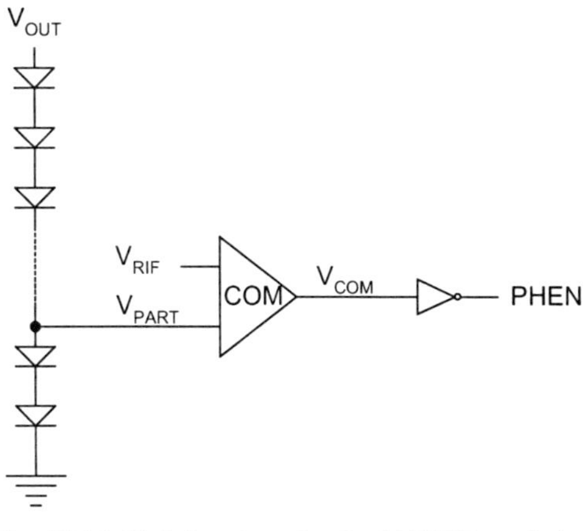

Fig. 15.23. High Regulator for the ON/OFF regulation

The output voltage of the pump, V_{OUT}, is sent to a divider circuit composed of low-conductivity diodes in order to limit as much as possible the current consumption of the charge pump. Voltage V_{PART}, whose behavior is the same as V_{OUT} in a smaller scale, is tapped from the divider circuit and compared with a reference voltage, V_{RIF}. As soon as V_{PART} becomes greater than V_{RIF}, V_{COM} becomes VDD and PHEN goes to '0': the charge pump phases are immediately forced to GND, so that no charge transfer can occur between the capacitors; in this case, V_{OUT} voltage can only decrease since the charge absorbed by the circuits connected to V_{OUT} is not "replaced". Therefore V_{PART} decreases as well and as soon as it becomes smaller than V_{RIF}, V_{COM} goes to GND and PHEN goes to VDD, thus turning on again the charge pump.

The effect of the limitation of the pump output voltage, V_{OUT}, to a precise value, is evident if the plot of the output resistance is analyzed: as shown in Fig. 15.24, the output voltage of the pump is about 8 V when the current absorption is between 0 and 1.2 mA; in the case of higher absorption, the regulator remains inactive.

From the description of the behavior of the high regulator, it should be clear that the value of the voltage which V_{OUT} is limited to can be easily changed by varying the reference voltage V_{RIF}.

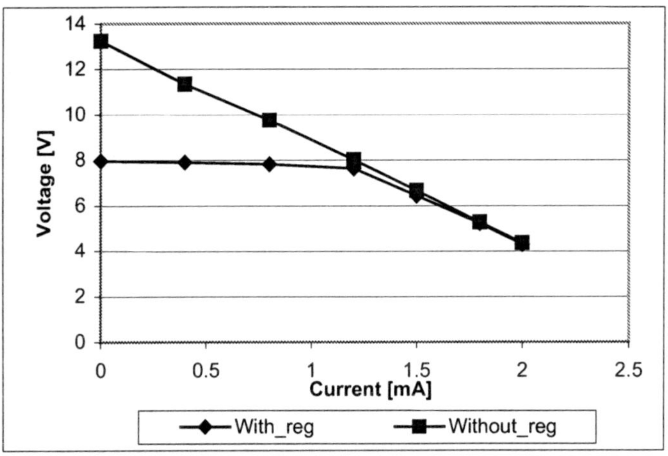

Fig. 15.24. Comparison between the output voltages of the same pump obtained with and without the use of a high regulator

15.6 Charge Pumps for Negative Voltages

Apart from the positive voltages, it is necessary to generate negative voltages inside the memory in order to bias the gate of the cells during the erase operation. Usually the negative pump is dimensioned by considering the time that is targeted to bias the gates of all the cells to be erased and therefore not taking into account a specific current demand (since no current is absorbed through the gate).

The operating principle of a negative pump is the charge transfer between one capacitor and the next one, as in the positive pumps case. The difference is that the transfer takes place from the storage capacitor toward ground, so that the output node can reach negative voltages.

In order to ease the charge transfer between one capacitor and the next one, the boost technique for the transfer capacitors is used, and either two or four phases are implemented depending on the desired type of pump.

The principle scheme of the negative charge pump is shown in Fig. 15.25.

If the triple-well technology is not available, the diodes of the pump must be implemented using P-channel transistors. The main drawback is again related to the body effect, which has a major influence primarily on the last stages. As we have already seen for the positive pumps, a possible solution to mitigate this negative effect is to use a pump realized using PMOS with dynamic biasing of the bulk nodes.

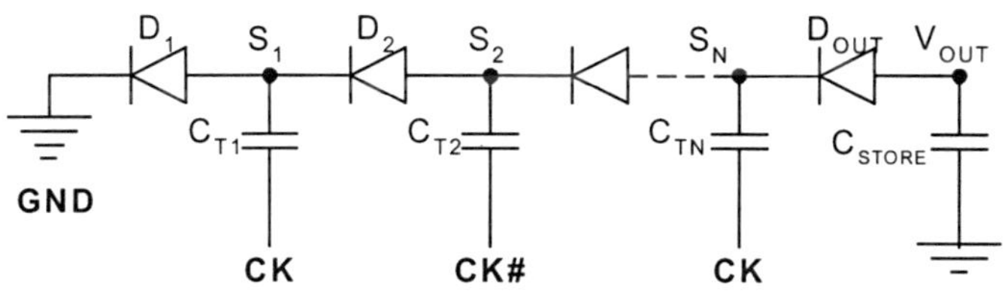

Fig. 15.25. Principle scheme of a negative charge pump

Since the output voltage of the pump depends on the number of stages, on the supply voltage and on the voltage drop on the transistors used for the charge transfer from a capacitor to the next one, the problem of guaranteeing a voltage that allows a proper erase operation (i.e. without damaging the cells) arises. It is therefore mandatory to use a high regulator that limits the output voltage V_{OUT}. The operating principle is similar to the one described in the previous section: the output voltage is compared with a reference one and, depending on the result of the comparison, the phase generator of the pump is turned either on or off. The main difference is related to the "reaction time" of this regulator: since it is only used during the erase operation, the comparator can be slower and the diodes can be smaller, thus calling for a smaller current consumption of the pump. The main drawback is that in this case the ripple on the output voltage of the pump can become large.

15.7 Voltage Regulation Principles

As it should be clear from the previous sections, the memory cells cannot be biased directly with the charge pump output voltage, which usually needs to be "filtered" to exclude its characteristic ripple and to limit it to a stable value. The circuit that performs these functions is the so-called voltage regulator[1].

The simplest voltage regulator consists of an operational amplifier, which is composed by a differential input stage, an output stage and an appropriate feedback net; the chosen topology for each of these blocks depends on the required performances. For example, depending on applications, settling time and power consumption could be important parameters; some regulators have to supply a stable output voltage with zero dc current, others have to regulate the output voltage while sourcing a huge current to the load.

Also frequency compensation is a major parameter to be taken into account: frequency compensation is defined as the way to get stability when the operational amplifier is operated in closed loop with negative feedback.

In the following sections, an overview of problems and solutions related to the principles of voltage regulation is presented, specifically referring to the gate and drain voltage. Nevertheless, many other high voltages are required to perform Flash operations; refer to Chap. 16 to have a complete overview of the high voltages required in last generation Flash memories.

15.8 Gate Voltage Regulation

15.8.1 Circuit Structure

In the earlier Flash devices, the gate of the cell was biased with high voltages during modify only, while the core power supply level was sufficient for reading. Nowadays, as the core power supply scales down to 3 V (or 1.8 V), the VDD level is not effective to perform accurate reads, i.e. to distinguish whether a cell is programmed or erased: also high-level gate voltage is required to perform readings.

Because this voltage is produced on-chip by mean of a charge pump, it needs appropriate regulation.

Since reading the cells is the most frequent operation in a Flash memory, the circuits devoted to the read mode must have a reduced static current consumption: the gate voltage regulator has to be designed to minimize its load on the charge pump, thus reducing the power consumption from the core power supply (this requirement becomes more and more important for Flash devices targeted for portable equipment).

[1] The charge pump output voltage regulation is only an aspect of a general problem: in a Flash memory, stable voltage values less than VDD are required that have to be precise even if the core power supply varies. Also in this case voltage regulators are required, which are supplied by VDD instead of a charge pump output.

Furthermore, a reduced steady-state current allows the read charge pump to reduce its setup time at power-up.

A further requirement is fast settling time of the output voltage to its steady-state value V_{READ} as a consequence of the wordline (WL) addressing.

In fact, when a WL is addressed, the output voltage V_{GROUT} shows a drop with respect to V_{READ} because of the charge sharing phenomenon between the WL parasitic capacitance, C_{WL}, and the output capacitive load of the regulator, C_L, as schematically sketched in Fig. 15.26.

The resultant voltage drop ΔV is very fast and can be excessive, i.e. it can move V_{GROUT} out of the limits that permit a proper read of the memory cell. To avoid incorrect read results, the output voltage recovery has to be fast with respect to the RC of the addressed WL, so as not to degrade the memory access time.

Finally, as the charge pump output voltage that supplies the regulator shows a ripple at its operative clock, it is necessary to take care of the positive Power Supply Rejection Ratio (PSRR), so as to minimize the amount of power supply noise injected into the output voltage.

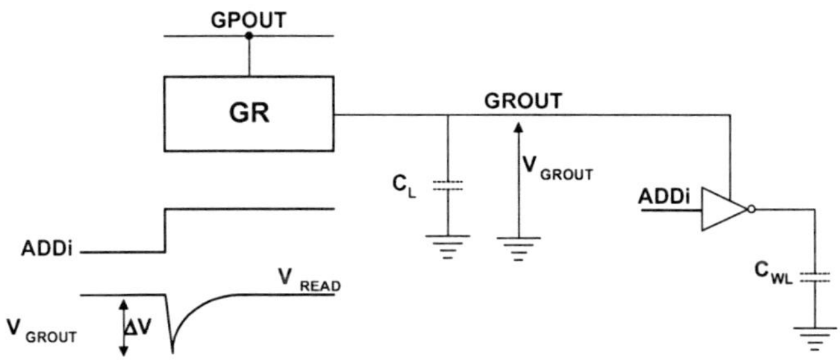

Fig. 15.26. Charge sharing between the regulator output and the addressed wordline

All of the above requirements taken into account, the most suitable topology for the read gate voltage regulator is the one depicted in Fig. 15.27. It is composed of an input differential amplifier and a single-ended PMOS output stage; a resistive divider (R_1+R_2) acts as a feedback net. The inverting input of the differential amplifier is driven by the reference voltage V_{BG}, generated by a band-gap circuit, while the non-inverting one is connected to the regulator output by the feedback net.

Assuming the circuit consists of ideal components (namely, no offset and sufficiently high loop gain), the regulator output voltage is:

$$V_{GROUT} = V_{BG}\left(1 + \frac{R_1}{R_2}\right) \tag{15.8}$$

Since integrated resistors can be made with good matching, the precision of V_{GROUT} depends essentially on the precision of V_{BG} (refer to Chap. 5).

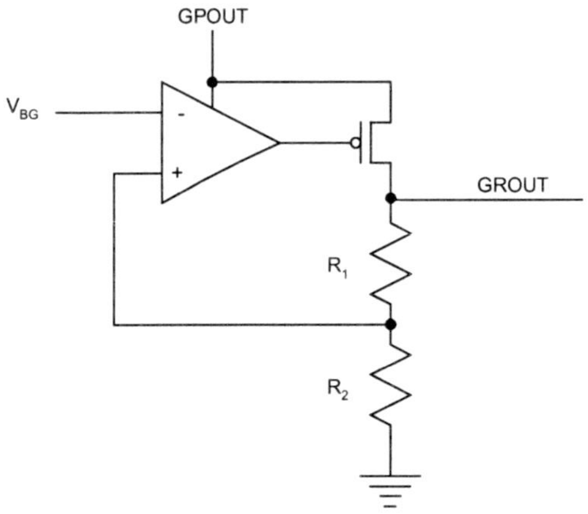

Fig. 15.27. Principle scheme of the regulator GR

The resistive feedback, which acts as a pull-down for the second gain stage, allows the regulator dc current to be kept low if R_1 and R_2 are large; in other words, the output stage steady-state current is:

$$I_{out,DC} = \frac{V_{GROUT}}{R_1 + R_2} \tag{15.9}$$

If the aspect ratio of the output p-channel transistor is kept high, the choice of a resistive feedback makes the recovery from undershoot easier. On the other hand, the recovery from an overshoot can be performed by mean of the RC network composed of the series of R_1 and R_2, and the load capacitance.

The output load of the regulator GR is a huge parasitic capacitance C_L (hundreds of pF), since V_{GROUT} biases the final stages of the row decoders, i.e. the circuits that charge the WLs to the read voltage. The value of the output load increases with the size of the memory device (more row decoders), making the design of a fast regulator difficult.

To perform the read operation, the addressed WL is connected to the regulator output through the row decoder, while the other WLs are kept at GND: in such a way, the WL parasitic capacitance is connected in parallel to the regulator output load, C_L. As the charge sharing is almost instantaneous, the output voltage drop, ΔV, is equal to:

$$\Delta V = \left(\frac{C_{WL}}{C_L + C_{WL}}\right) \cdot V_{GROUT} \cong \frac{C_{WL}}{C_L} V_{GROUT} \tag{15.10}$$

Let us make an example to determine the values: supposing that the target device is a 64Mbit Flash memory, typical values are:

$C_L = 200$ pF, $C_{WL} = 4$ pF, $V_{GR} = 6$ V, $V_{BG} = 1.3$ V, $I_{out,dc} = 50$ µA

The resistance values to be chosen to satisfy the current requirements are:

$R_1 \sim 90\ k\Omega$ $\qquad$ $R_2 \sim 30\ k\Omega$

and the voltage drop is:

$\Delta V = 120\ mV$

The circuit scheme of GR is shown in Fig. 15.28, where it is not difficult to find the blocks we talked about in the past few pages.

The n-channel transistors M1 and M2 act as a differential input stage, whose current mirror is represented by the PMOS transistors M5 and M6. M3 and M4 are inserted both to disable the regulator (when the regulator enable signal GREN is at "0"), and to decouple the internal nodes of the regulator from V_{BG} when GREN = "1", acting as cascode-connected transistors (they have to be biased in saturation region).

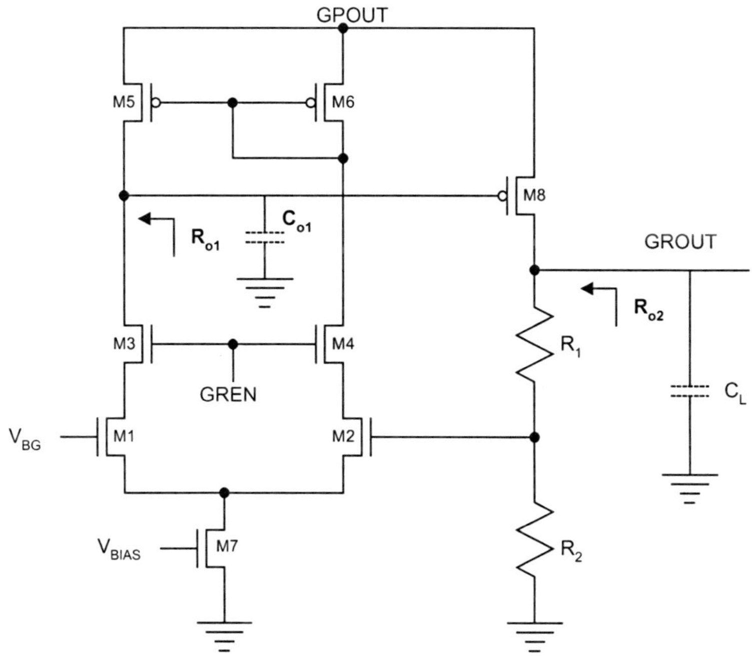

Fig. 15.28. The voltage regulator GR with its parasitic capacitances, C_{o1} and C_L

Without M3, the C_{gdM1} parasitic capacitance would couple V_{BG} with the drain of M5, i.e. the differential stage output. Since this node is a "moving" one (it is subject not only to the differential output voltage swing, but also to the charge pump ripple), M3 prevents the noise from being transferred to the bandgap output.

The differential stage current generator is represented by M7, biased in saturation region by mean of suitable gate voltage V_{BIAS} that produces the steady-state differential stage current:

$$I_{first,DC} = I_{DM7} = \frac{\mu_N C_{OX}}{2}\left(\frac{W}{L}\right)_{M7}\left(V_{BIAS} - V_{thM7}\right)^2 \tag{15.11}$$

where μ_N , C_{OX} e V_{thM7} are carriers mobility, gate oxide specific capacitance and threshold voltage of the NMOS M7, respectively.

The value of $I_{first,DC}$ cannot be set too high because GR has to meet low power consumption requirements; this poses limitations to the dynamic behaviour of the regulator, especially with respect to slew rate performance.

The output gain stage is a p-channel transistor (M8 in Fig. 15.28), which is designed with a high aspect ratio in order to minimize the time penalty when charging the parasitic capacitance of the addressed WL.

15.8.2 Frequency Compensation

Operational amplifiers have usually multiple poles: for this reason they need frequency compensation, i.e. their open-loop transfer function has to be modified for the closed-loop circuit to be stable.

Referring to Fig. 15.29, where $A=V_{out}/V_{in}$ is the open-loop gain and β the feedback factor (supposed to be frequency independent), the loop gain function G_{loop} of a negative feedback system is defined as:

$$G_{loop} = \frac{Y(s)}{X(s)} = \frac{A(s)}{1 + A(s)\beta}, s = j\omega \tag{15.12}$$

The poles locations of the closed-loop transfer function can be found by solving $1 + A(s)\beta = 0$.

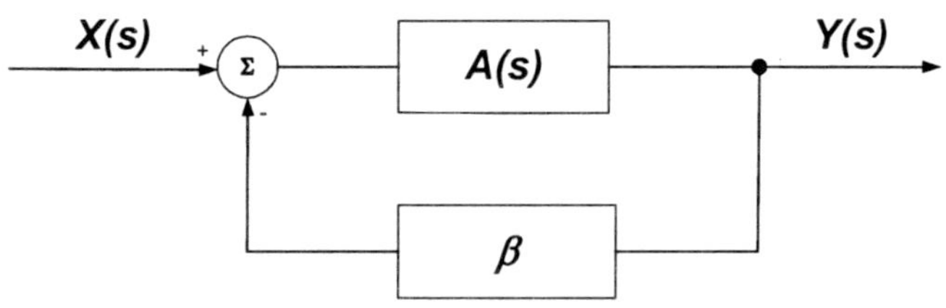

Fig. 15.29. Negative feedback schematic representation

The regulator GR needs both fast and highly stable response: so the G_{loop} function must have a high gain-bandwidth product (GB) and at least 60° phase margin.

The major obstacle to these objectives is the high value of the output parasitic capacitance C_L, as shown in the following.

Let us consider the regulator GR as shown in Fig. 15.28; its steady-state open-loop gain A_{dC} is:

$$A_{dC} = g_{mM1} R_{o1} g_{mM8} R_{o2} \qquad (15.13)$$

g_{mM1} e g_{mM8} are the transconductances of M1 and M8; R_{o1} e R_{o2} are the output resistances of the differential stage and of the PMOS output stage, respectively (the output resistances are schematically sketched in Fig. 15.28). R_{o2} is equal to the feedback network resistive series ($R_1 + R_2$), in parallel with the output resistance of M8. R_{o1} is the parallel of M5 and M3 output resistances; since M3 is connected in common-gate configuration, it shows very high output resistance:

$$R_{cascode} = r_{dsM1} g_{mM3} r_{dsM3} \qquad (15.14)$$

For this reason, R_{o1} is essentially equal to r_{dsM5}.

In order to evaluate the stability of GR, let us calculate the position of poles and zeroes.

In presence of huge output capacitive load and feedback resistance (as in our case), the dominant pole is the one associated with the regulator output:

$$|p_1| = \frac{1}{R_{o2} C_L} \qquad (15.15)$$

In a first-order analysis, the regulator is a two-poles system and the resistance and parasitic capacitance of the first stage determine the second pole. The parasitic capacitance of the differential stage output is the sum of M3 and M5 drain junction capacitances (let it be called C_1), and the gate-drain capacitance of M8. In any case, this latter capacitance is placed across a gain stage, thus exploiting the so-called Miller effect: a capacitance C placed between the input and the output of a gain stage, with gain A, can be represented as a capacitance AC connected between input and GND of said gain stage, and a capacitance C connected between its output and GND, as depicted in Fig. 15.30.

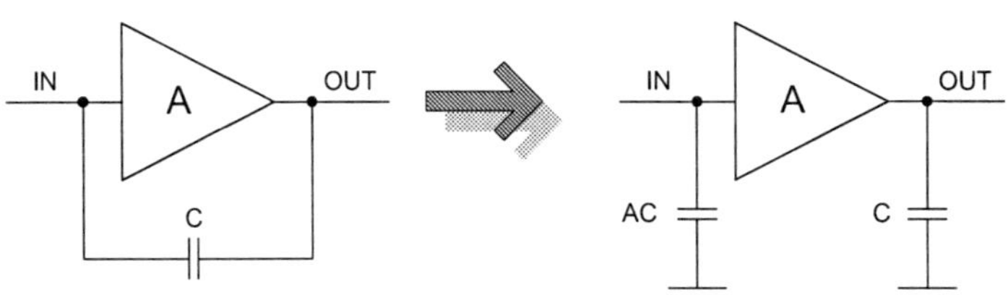

Fig. 15.30. Miller effect

For this reason, a capacitance $C_{MILL} = C_{gdM8} R_{o2} g_{mM8}$ is placed in parallel to C_1 (let us name the sum of them C_{o1}), while the output load is increased by C_{gdM8} (anyway, since $C_{gdM8} << C_L$, the position of the dominant pole does not change).

The parasitic capacitance C_{o1} leads to a pole and a zero respectively at the frequencies:

$$|p_2| = \frac{1}{R_{o1}\left(C_1 + g_{mM8} R_{o2} C_{gdM8}\right)} \cong \frac{1}{R_{o1} g_{mM8} R_{o2} C_{gdM8}} \tag{15.16}$$

$$z = \frac{g_{mM8}}{C_{gdM8}} \tag{15.17}$$

How poles and zero are placed with respect to each other depends on the parameters of the regulator. Usually, it is quite hard to design the regulator in order to have a good phase margin without a settling time penalty.

A possible solution to compensate a second pole placed in the bandwidth is to place a half-plane zero exactly at the frequency of the second pole: in such a way, the regulator acts ideally as a single-pole system (the dominant one). Anyway, this kind of compensation is usually not reliable, because a well-placed zero at a given operating condition may not "follow" the pole that it has to compensate when the operating condition varies (for example, as a consequence of temperature variations).

Another way to compensate is to use the Miller effect by placing a suitable compensating capacitor C_C across the second gain stage, as depicted in Fig. 15.31.

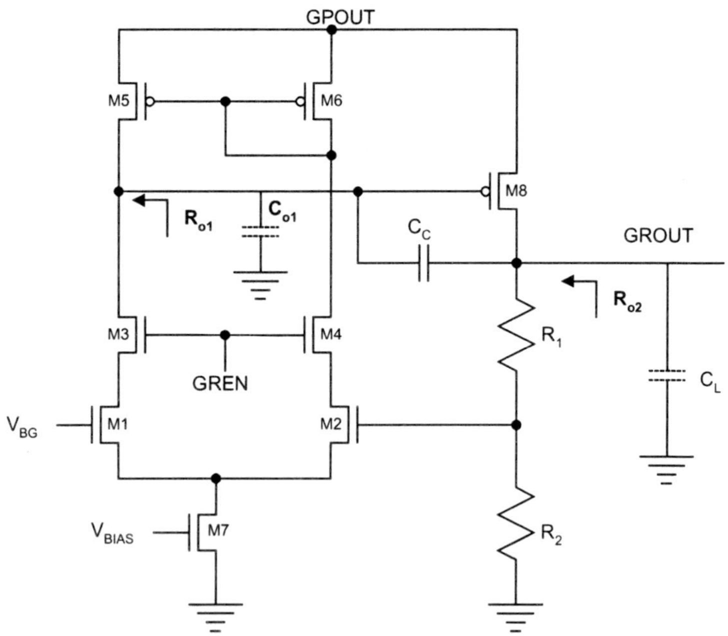

Fig. 15.31. GR with Miller compensation

Concerning the position of the poles, C_C acts in two ways: first of all, the capacitance C_{o1} is increased approximately by $g_{mM8}R_{o2}C_C$, thus reducing the frequency of the pole p_2 related to the differential stage output. Furthermore, the pole p_1 moves to higher frequency thanks to the negative feedback, which reduces the output stage resistance: this phenomenon is referred to as pole splitting.

By the effect of Miller compensation, the poles p_2 e p_1 move respectively to the frequencies:

$$|p_2| \cong \frac{1}{C_{o1}R_{o1} + C_L R_{o2} + C_C\left(R_{o1} + R_{o2} + g_{mM8}R_{o1}R_{o2}\right)} \cong \cong \frac{1}{g_{mM8}R_{o1}R_{o2}C_C} \tag{15.18}$$

$$|p_1| \cong \frac{C_C g_{mM8}}{C_{o1}C_C + C_L C_C + C_{o1}C_L} \cong \frac{g_{mM8}}{C_L} \tag{15.19}$$

The approximations in Eqs. (15.18) and (15.19) are valid when $C_L>C_C>C_{o1}$ and under the reasonable hypothesis that the second stage has an adequate gain. C_{o1} can be reduced by mean of accurate design and layout of the first stage. Depending on the value of C_C, it is possible to invert the position of the poles with respect to the scheme of Fig. 15.28, thus making the pole p_2 of the first stage dominant.

Because of the unity-gain feedback between input and output of the second stage at high frequency, C_C leads also to a right-plane zero located at the frequency:

$$z_1 = \frac{g_{mM8}}{C_C} \tag{15.20}$$

The phase shift introduced by a right-plane zero is negative as the one of a left-plane pole: for this reason, this zero degrades the phase margin, and it may become a problem if it is placed near to the unity-gain frequency.

Remembering that the unity-gain frequency in a dominant-pole approximation is given by:

$$GB = A(0)\cdot|p_{dom}| = \frac{g_{mM1}}{C_C} \tag{15.21}$$

to obtain 60° phase margin, it is necessary that:

$$\pm 180° - \tan^{-1}\left(\frac{\omega}{|p_1|}\right) - \tan^{-1}\left(\frac{\omega}{|p_2|}\right) - \tan^{-1}\left(\frac{\omega}{z}\right) = 60° \tag{15.22}$$

To satisfy Eq. (15.22), it is necessary that z>10GB; in this case we obtain $|p_2|$>2.2GB. As a result, the following relationships hold:

$$\frac{g_{mM8}}{C_C} > \frac{10\cdot g_{mM1}}{C_C} \Rightarrow g_{mM8} > 10\cdot g_{mM1} \tag{15.23}$$

$$\frac{g_{mM8}}{C_L} > \frac{2.2 \cdot g_{mM1}}{C_C} \Rightarrow C_C > 0.22C_L \tag{15.24}$$

Referring to Eq. (15.23), the zero is far away from the unity gain frequency if the transconductance of the second stage is much higher than the one of the first stage. In a CMOS design, it is difficult to achieve large differences between g_{mM1} and g_{mM8}, since the transconductance increases with the square root of the current and the aspect ratio. If the zero is near the unity-gain frequency, it modifies significantly the frequency response near to the "zero dB crossing", thus degrading the stability conditions.

Furthermore, as shown in Eq. (15.24), a huge output load C_L requires a big C_C, thus degrading the slew rate of the differential stage ($SR = I_{first,DC}/C_C$), since its biasing current $I_{first,DC}$ must meet low power consumption requirements. However, we cannot use a smaller compensation capacitance: if the phase margin is less than 60° (that is, contained between 45° and 60°), the frequency response of the system leads to damped oscillations. As a consequence of how we designed GR, it is crucial that its output voltage recovers to the steady-state value without any overshoot which could lead to an excessive increase of the gate voltage of M8, thus cutting the output stage off. If this happens, the recovery to the regulated value V_{READ} can be performed only by discharging C_L through the resistive divider $R_1 + R_2$, but the time constant of this discharge is very high by design, because the output load C_L is huge, and the sum $(R_1 + R_2)$ is great in order to reduce the static output current. Referring to the previous 64 Mbit Flash memory example, the discharge time constant of the output node is:

$$\tau_{disch} = (R_1 + R_2) * C_L = 120\ k\ \Omega * 200\ pF = 24\ \mu s \tag{15.25}$$

In practice, we cannot leave the circuit as it is, but we must find some remedy to the unsatisfactory phase margin.

The right-plane zero problem (and the related feedforward) can be solved by introducing a unity-gain buffer between GROUT and C_C, as depicted in Fig. 15.32.

By mean of the unity-gain buffer, the right plane zero is cancelled, while the position of the poles is kept unchanged with respect to Miller compensation.

Problem 15.2: Demonstrate it by mean of the small signals equivalent circuit depicted in Fig. 15.33.

If the voltage buffer has a finite output resistance, R_0, the zero is not cancelled perfectly, but a pole and a left-plane zero replace it.

With suitable modifications to the small signal equivalent circuit, one can demonstrate that the new pole and zero are located respectively at:

$$|p_3| \cong \frac{1}{R_0 C_{ol}} \tag{15.26}$$

$$z_1 \cong -\frac{1}{R_0 C_C} \tag{15.27}$$

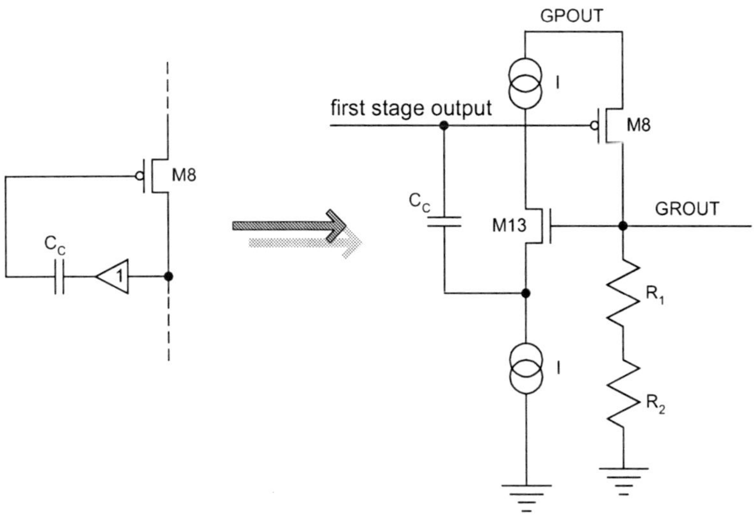

Fig. 15.32. Unity-gain buffer frequency compensation (Follower Feedback Compensation, FFC)

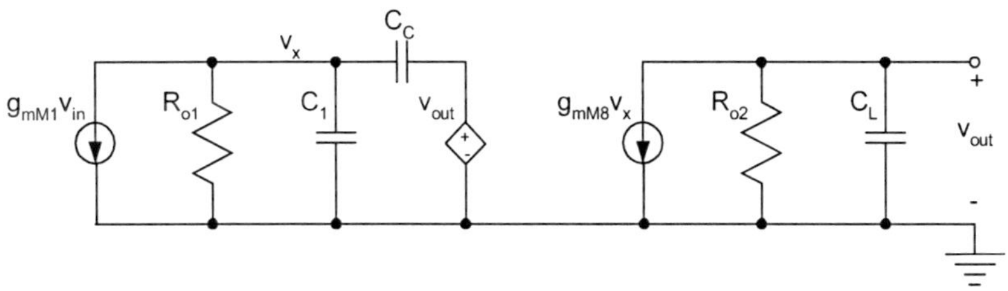

Fig. 15.33. Small signal equivalent circuit of GR with FFC

To obtain a unity-gain buffer, it is convenient to use a n-channel transistor in source follower configuration (Follower Feedback Compensation, FFC), as M13 in Fig. 15.32. The follower has to be biased with an appropriate dc current generator I; for the follower to be working properly, V_{GROUT} must never be less than the sum of the voltages required to bias in saturation region both M13 and the current generator connected towards GND. This poses limitations to the use of this kind of compensation if V_{GROUT} may assume low levels (for example <2V), as it will be explained in Sect. 15.8.4.

Instead of using a voltage buffer, we can use a current buffer in series with the compensating capacitor. Since the easiest way to obtain a current buffer is to use a transistor biased in common-gate configuration (cascode), as shown in Fig. 15.34 we can name this kind of compensation Cascode Feedback Compensation (CFC).

V_{BCASC} is chosen to keep M23 in saturation, i.e. to make it work as a cascode. The current generators can be designed as in Fig. 15.34; for this purpose, another biasing voltage V_{BIAS} is shown, not necessarily different from V_{BCASC}.

In steady state, the compensating capacitor C_C is charged to (V_{GROUT} - V_{sM23}) and the current generators provide the current I, which flows in M23. When C_C has to be charged as a consequence of an output voltage drop, the additional current ΔI is provided by the differential stage and a current $I + \Delta I$ flows in M23. The current ΔI keeps the compensating capacitor charged in such a way that the source terminal of M23 acts as a virtual ground.

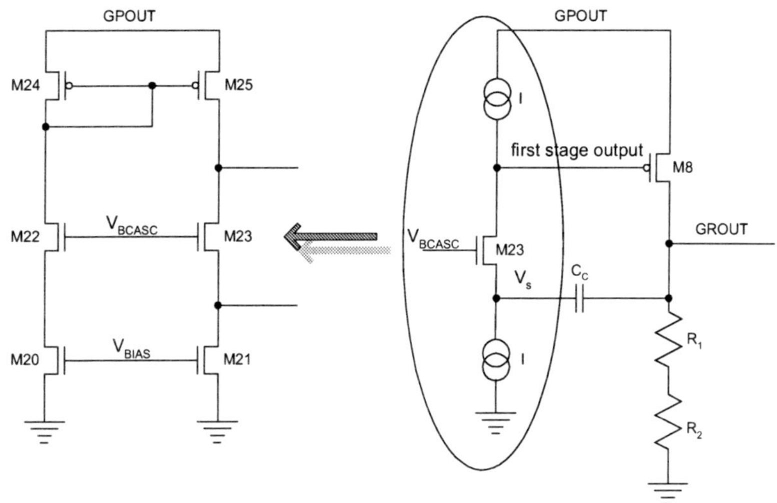

Fig. 15.34. Cascode Feedback Compensation

It is not trivial to determine the transfer function of the system; anyway, starting from the small signal equivalent circuit, with some algebra we obtain the open-loop dominant pole:

$$|p_2| \cong \frac{1}{g_{mM8} R_{o1} R_{o2} C_C} \tag{15.28}$$

Note that the expression of the dominant pole is the same as Eq. (15.25); however, since R_{o1} is smaller than in Miller compensation, the dominant pole appears at higher frequency; in fact, R_{o1} now is:

$$R_{o1} = r_{dsM3} // r_{dsM5} // r_{dsM23} // r_{dsM25} \cong r_{dsM5} // r_{dsM25} \tag{15.29}$$

g_{mM23} is the transconductance of M23, r_{dsM25} e r_{dsM23} the channel conductances of M25 and M23, respectively. Since M23 is biased in common-gate configuration, its channel conductance is very high and can be neglected.

The second pole frequency is:

$$|p_1| \cong \frac{g_{mM8} C_C}{C_{o1}(C_C + C_L)} \tag{15.30}$$

As in FFC, the system shows a third pole and a left-plane zero, located respectively at:

$$|p_3| \cong \frac{g_{mM23} C_L}{C_C^{\ 2}} \tag{15.31}$$

$$z_1 = -\frac{g_{mM23}}{C_C} \tag{15.32}$$

Usually (i.e. when C_L is the biggest capacitance in the system), the feedback cascode compensation allows the second pole high frequency shifting to be ~C_C/C_{o1} times more effective than in Miller of feedback follower compensation.

Assuming that $C_{o1} << C_C$ or $C_{o1} << C_L$, the unity gain frequency is still given by g_{mM1}/C_C; to be sure that the second pole is well beside the unity gain bandwidth, in FFC and CFC we have to satisfy, respectively:

$$\frac{g_{mM1}}{g_{mM8}} < \frac{C_C}{C_L} \tag{15.33}$$

$$\frac{g_{mM1}}{g_{mM8}} < \frac{C_C^{\ 2}}{C_{o1}(C_C + C_L)} \tag{15.34}$$

The reduced power budget limits g_{mM1} and g_{mM8}, even if the latter is chosen to provide the regulator with the required current capability. To ensure an adequate phase margin, for each set of g_{mM1}/g_{mM8}, C_L, and C_1, the compensating C_C has to be chosen to satisfy either Eq. (15.33) or Eq. (15.34). Anyway, since both FFC and CFC allow more effective compensation with respect to conventional Miller technique, a smaller C_C can be used, thus achieving better SR performances for any given value of $I_{first,DC}$.

Furthermore, even if the reduced value of g_{mM1} limits the feedback loop gain-bandwidth product, an additional feedback loop is created from GROUT to the gate of M8: through the follower M13 and C_C in case of FFC, through C_C and the cascode M23 in case of CFC. By mean of these new loops, an output voltage drop ΔV is reflected directly in a fast decrease in M8 gate voltage, which in turn leads to a fast increase in the output current. Nonetheless, the FFC additional loop is not as fast as the CFC one. When compensating with the feedback follower technique, the parasitic capacitance seen at the differential stage output is the sum of C_{o1} and C_C, while it is just C_{o1} when CFC is used. Referring to the 64Mbit example, C_{o1} is on the order of pF, while C_C is about ten pF: in case of cascode feedback compensation, the additional feedback loop will be faster.

15.8.3 Positive Power Supply Rejection Ratio (PSRR)

Good power supply rejection ratio performance is mandatory in a regulated system, especially when the regulator is supplied with a voltage produced by a charge pump instead of VDD. A characteristic of a good positive PSRR is when the noise on the output of the charge pump is not transferred to the regulated voltage; in our case we define the positive PSRR as:

$$PSRR^{+} = \frac{(V_{GROUT} / V_{in})_{V_{GPOUT}=0}}{(V_{GROUT} / V_{GPOUT})_{V_{in}=0}} \tag{15.35}$$

The PSRR performance depends on the compensation strategy used in the system; the exhaustive analisys of the positive PSRR can be performed only if we calculate the PSRR tranfer function by mean of the small signals equivalent circuits. Nonetheless, a few qualitative considerations can be deduced directly from the schematics.

From the gate of M8 (the differential stage output), the charge pump noise can be transferred through C_{gdM8}, which is usually small and can therefore transfer only high-frequency noise.

If we adopt the conventional Miller compensation, the concepts remain the same: C_C (>>C_{gdM8}) couples the positive supply directly with the regulator output, however it may happen at lower frequencies, depending on the C_L value (remember Eq. (15.24)). Also lower frequency noise can be transferred to the output and for this reason, this type of compensation should be avoided.

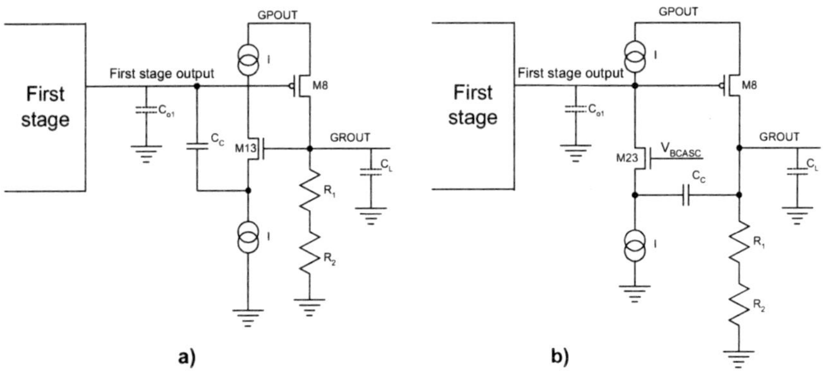

Fig. 15.35. FCC (**a**) and CFC (**b**) positive PSRR

When compensating with FFC or CFC, the direct M8 gate-to-drain connection via C_C is eliminated (of course, the connection via C_{gdM8} is still present, but typically it acts well above the charge pump noise frequency). The noise transmission mechanism is therefore different, and depends on the biasing of M8 (Fig. 15.35).

As a general rule, the differential stage output filters any signal higher than its cut-off frequency, i.e. the gate of M8 cannot "follow" a noise beyond this frequency. In this case, the noise signal V_{sgM8} is approximately equal to the noise on GPOUT (V_{sM8}), thus modulating the biasing of M8: for this reason, the drain terminal of M8, that is GROUT, varies as well, since it is forced to source a static current due to its biasing by $R_1 + R_2$.

As discussed in the previous section, since the first stage output has a heavier capacitive load in FFC than in CFC, its cut-off happens at a lower frequency, thus tranferring the charge pump noise to the output (if the charge pump operating frequency is located in the pass band). Therefore, the larger CFC noise pass band allows better positive PSRR performances.

15.8.4 Program Gate Voltage

In order to perform effective and precise program operations, the best method for biasing the WLs is with a linear "staircase" voltage; in such a way, if the drain terminal is kept to a suitable and constant voltage value V_{DP}, after an initial transient a 1:1 relation exists between gate voltage increment, ΔV_{GP}, and voltage shift after each program pulse, ΔV_T.

This method requires a small amount of current from the drain side (in single supply devices the drain program voltage is produced by a charge pump, as will be explained in the next section), since the cells are not biased with an excessive overdrive, and this permits an increase in the amount of program parallelism.

To save silicon area and power consumption, it is convenient to use the same regulator GR that produces the read voltage to provide the gate staircase program voltage as well; furthermore, this approach assures that in a Program & Verify scheme (refer to Chap. 14) the verify and the read voltages match. Ultimately, the gate voltage regulator should act as an Analog-to-Digital converter.

To obtain "programmable" gate voltage values, we can divide the feedback resistive partition into a series of smaller resistances, each one having a switch in parallel.

When adopting this solution, two major problems have to be taken into account: the first concernes the linearity of the output voltage as a function of the selection command. Figure 15.36 shows an example of the ideal staircase voltage (dashed line) with respect to the real one (continuous line). The differential linearity error ε_d is given by:

$$\varepsilon_d = \frac{|V_{ID}(n) - V_{RE}(n)|}{\Delta V_{GP}} \qquad (15.36)$$

where $V_{ID}(i)$ is the ideal voltage at the i-th step, $V_{RE}(i)$ the real one ΔV_{GP} the program step.

The second problem concerns the charge injection related to the commutation of the switches, which leads to voltage overshoot on the feedback node. In order to have a fast settling time on the program voltage, these overshoots have to be reduced as much as possible.

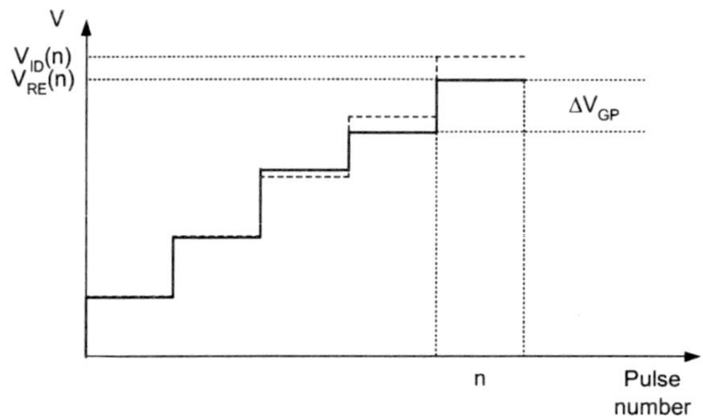

Fig. 15.36. Linearity error

To satisfy linearity requirements, a good solution consists in a "programmable" R_2 (R_2 is the resistor connected between node E and GND), composed of a parallel combination of *n* binary weighted resistors, a certain number of which are active at the same time as a function of the program voltage, as sketched in Fig. 15.37a. In this case, this solution is not feasible because the resistor divider has to assure low power consumption: the total value of the resistances is unacceptable. Furthermore, the output stage current variation, due to R_2, leads to an offset voltage in the regulator (since the differential stage current does not change).

One can choose the solution in Fig. 15.37b, where R_1 is made programmable by mean of a series of binary weighted resistors, each one with a selection switch in parallel; in this case, the number of switches simultaneously open depends on the selected program voltage level, giving arise to serious linearity problems (as much as +/-15% of the voltage step). Furthermore, when many switches commute simultaneously, the node E is subject to high charge injection, especially if the switches are designed with high aspect ratio to minimize their parasitic resistance. The best solution from the linearity point of view consists in dividing the resistor R_1 in a series of *w* equal resistors of valure ΔR, while R_2 is kept constant (Fig. 15.37c).

The voltage step ΔV_{GP} is achieved at each step by incresing the value of R_1 by a fixed amount ΔR; at the *p*-th program step the gate program voltage V_{GP} is equal to:

$$V_{GP} = V_{BG}\left(1 + \frac{R^*_1 + p\Delta R}{R_2}\right) \tag{15.37}$$

where R^*_1 is the value that R_1 assumes at the first program step. The program step voltage ΔV_{GP} is:

$$\Delta V_{GP} = V_{BG}\frac{\Delta R}{R_2} \tag{15.38}$$

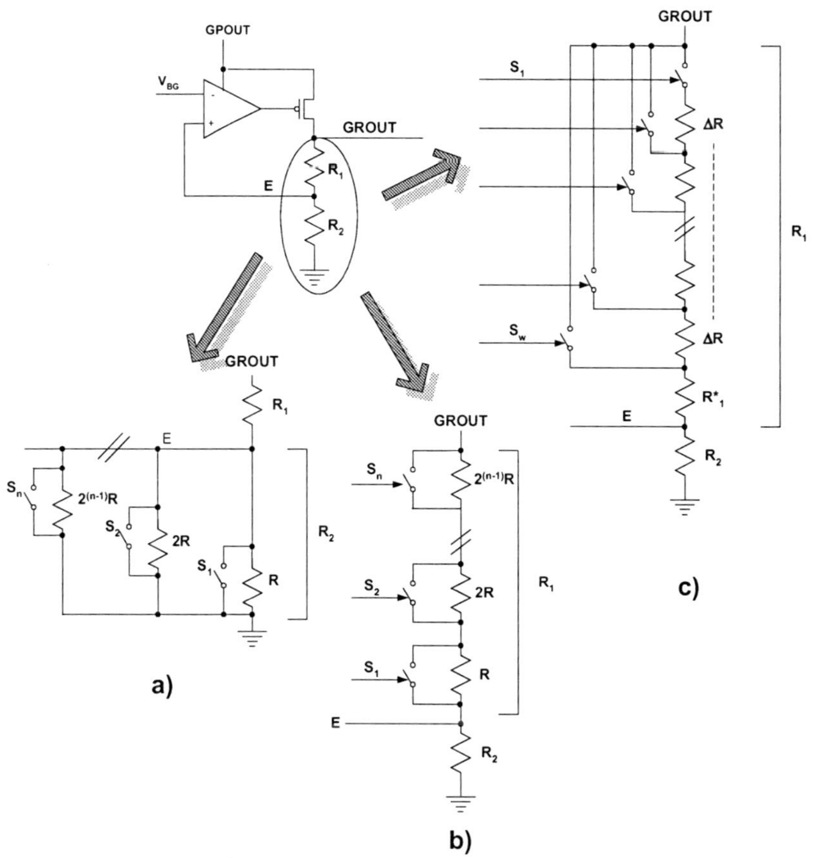

Fig. 15.37. Three different ways to make the feedback net programmable

At each p-th program pulse, a suitable value of resistor ΔR is shorted by mean of the control logic that produces the w signals to drive the switches from S_1 to S_w, thus achieving $R_1 = R^*_1 + p\Delta R$. To avoid introducing parasitic resistances that may vary from one step to the next, only one switch is active during each program pulse, while the others are off.

Usually the gate voltage ramp starts at ~1.5 V and continues until 9.5 V; the voltage step ΔV_{GP} depends on the program accuracy.

After each program pulse the cells are verified, i.e. they are read to check whether the desiderd threshold voltage has been reached or not. During verify, the gates of the cells are biased with the verify voltage V_{PV} (usually the verify gate voltage is equal to the read voltage, $V_{PV} = V_{GR}$).

To verify after each program pulse means that a highly capacitive node, i.e. the output node GROUT, must be driven to different voltage levels; as depicted in Fig. 15.38, two different situations may occur, depending on wheter $V_{GP}<V_{PV}$ or $V_{GP}>V_{PV}$. In the former case, at the end of the *k*-th program pulse C_L has to be charged to V_{PV}, while at the beginning of the *(k+1)*-th pulse C_L has to be discharged to V_{GP}; in the latter case at the end of the *k*-th program pulse C_L has to be discharged to V_{PV}, while at the beginning of the *(k+1)*-th pulse C_L has to be charged to V_{GP}.

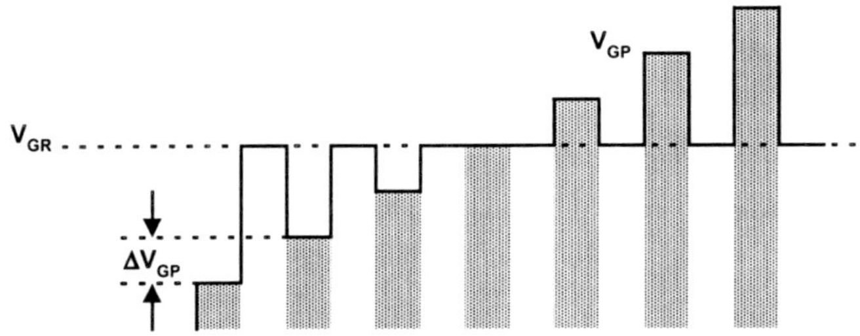

Fig. 15.38. Gate voltage qualitative behaviour during Program & Verify algorythm

As explained in the previous section, it is not difficult to charge C_L to a higher voltage level if the pull-up transistor M8 is designed with a high aspect ratio. On the contrary, the discharge operation is slow because the resistive feedback net limits the pull-down current. To get a faster discharge, the topology shown in Fig. 15.39 can be used.

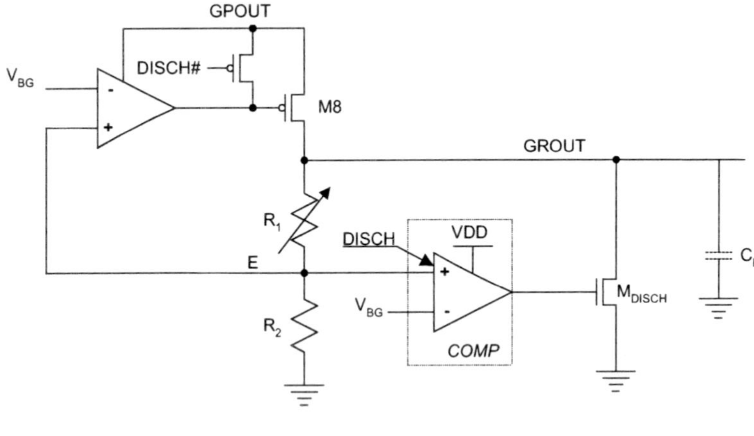

Fig. 15.39. Program gate voltage regulator with discharge circuit

When the regulator output has to be discharged from the actual value V_{PRES} to $V_{NEXT}<V_{PRES}$, the comparator COMP (that can be designed as a combination of a differential stage and an inverter) is enabled by a suitable signal DISCH, while the pull-up transistor M8 is disabled by DISCH# not to short GPOUT to GND. Meanwhile, the resistor R_1 is programmed to obtain V_{NEXT}. In such a way, since the new value of R_1 is less than the previous one and the feedback loop of the regulator is interrupted, the voltage V_E at node E increases, thus turning M_{DISCH} on and enabling the discharge path from GROUT to GND. This path remains active until $V_E > V_{BG}$, i.e. until $V_{GROUT} > V_{NEXT}$. As DISCH is switched to the low logic level and M8 is turned on, the next output voltage level V_{NEXT} can be reached quickly. This solution allows the reduction of the current requirements of the charge pump GP, thus achieving a faster settling time and reducing the total programming time.

15.9 Drain Voltage Regulation and Temperature Dependence

Now we will explore the problems related to drain voltage regulation, which is a crucial parameter affecting the programming operation. Figure 15.40 shows the program high voltages scheme in a dual supply Flash memory.

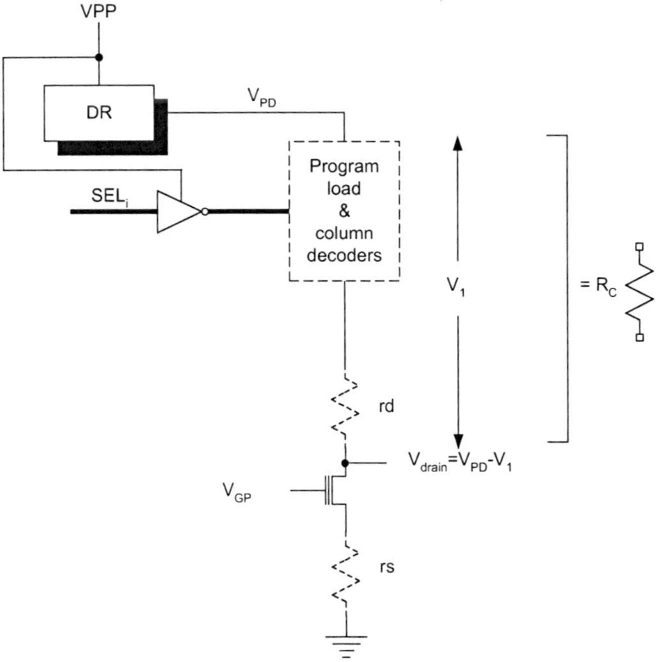

Fig. 15.40. Drain voltage regulation (V_{PD}) in a dual supply Flash memory

The drain voltage is obtained by mean of a regulator, DR, that regulates the voltage V_{PD} before the program load (refer to Sect. 16.5) and the column decoders. In such a way, the voltage on the drain is not V_{PD}, but V_{PD}-V_I, where V_I is the voltage drop across the bit line and pass transistors parasitic resistance. In order to limit this voltage drop, the column decoder pass transistors must be properly biased so they operate in linear region.

Figure 15.41 shows an example of a program load regulator, implemented in dual power supply Flash memory devices, which compensates for the voltage drop across the column decoder.

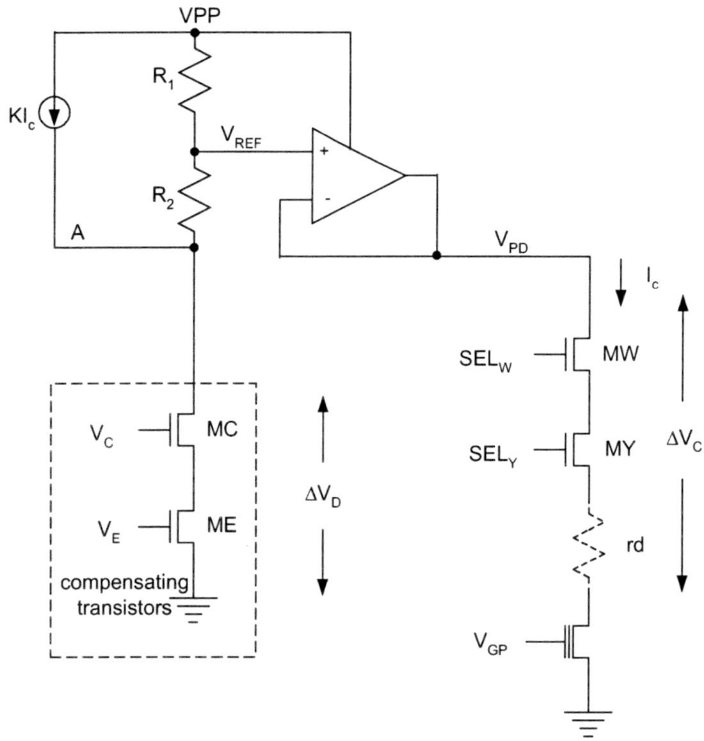

Fig. 15.41. Adaptive V_{PD} regulation

A suitable voltage V_{PD}~V_{REF} is applied to the load by mean of an operational amplifier connected in a unity-gain non-inverting configuration. V_{REF} is obtained with a resistive divider made up by the resistors R_1 and R_2, and two series-connected MOS transistors operating in the triode region (MC and ME). These are matched with the bitline select transistors MW and MY, respectively, with a suitable scaling factor, and are driven by the same control signals. Since MC and ME do not perform any selection, they will be referred to as compensating transistors.

A current-controlled current source injects a current KI_c into the node A, so as to provide an additional contribution ΔV_D to V_{REF} with respect to its quiescent value at $I_c=0$, V_{R0}.

$$V_{REF} = V_{R0} + \Delta V_D \tag{15.39}$$

The scaling factor K is chosen so as to make ΔV_D equal to the drop $\Delta V_C = I_c R_{on,s}$ across the bit line select devices ($R_{on,s} = R_{on,MW} + R_{on,MY}$). Thanks to the adaptive feedback loop, the voltage V_{PD} on the selected bitline is ideally equal to V_{R0} during the whole program operation, regardless of the variation of the fabrication process or the programming current. This architecture is also suitable for parallel programming, for example on byte or word basis. In this case, the current fed to the load is equal to the sum of the currents drawn by the *n* cells being programmed, nI_c. Each cell is addressed through its corresponding pair of bitline select transistors. For correct compensation of the voltage drop across these devices, the current KnI_c must be delivered to *n* identical branches MC_i, ME_i placed in parallel, as shown in Fig. 15.42. The enable devices ME_i activate the correct number of compensation branches, according to the data to be stored.

It is not easy to extend this solution to single supply voltage devices, especially when high program parallelism is required. In these devices the drain program voltage is provide by an on-chip charge pump, DP, which should substain not only the worst case total drain program current (and the current consumption of the regulator), but also its replica, multiplied by the scaling factor K.

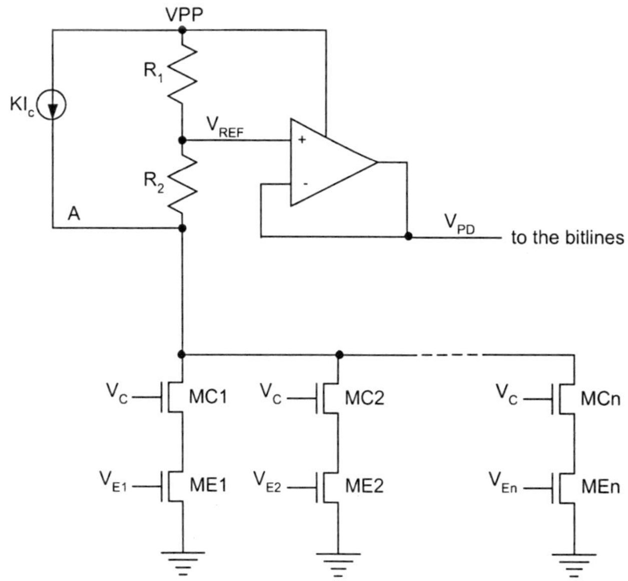

Fig. 15.42. Adaptive program load regulator for parallel programming

In single supply Flash devices we forego the ΔV_C drop compensation, and we design a regulation scheme like the one shown in Fig. 15.43, where V_{PD} is obtained from a band-gap reference. The column decoder parasitic resistance, R_C, increases with temperature, thus decreasing the program voltage applied to the drain (assuming the program current is constant). For example, if the program current is 400 μA per cell, the voltage drop ΔV_C may vary from 200 mV at –40° C to 350 mV at 120° C. It should be clear that the efficiency of the program circuits, which must force a sufficiently high drain voltage, is reduced as the temperature increases.

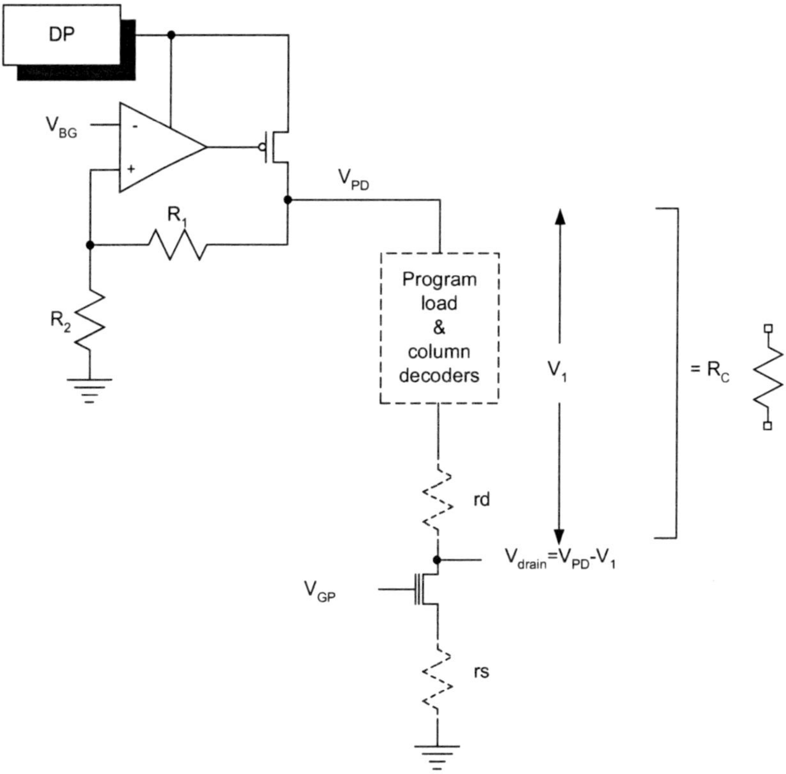

Fig. 15.43. V_{PD} regulation in single supply Flash devices (the voltage drop across the column decoder pass transistors is not compensated)

We can modify the circuit in Fig. 15.43 by substituting the input voltage on the inverting input of the differential amplifier with a suitable voltage, whose mean value is equal to the band-gap voltage V_{BG}, and linearily increases with temperature. Such a voltage can be produced by the circuit represented in Fig. 15.44. Three operational amplifiers make up the thermal tracker: the first reproduces V_{BG}, while the last one performs the difference between this voltage and a suitable amplification V_x of the emitter-base voltage of a pnp transistor.

Since the forward-biased diode voltage shows a linear decreasing of – 2mV/°C, V_{OUT} in Fig. 15.44 increases linearily with temperature.

The following relationshipss hold:

$$V_{OUT} = V_{BG} + (V_{BG} - V_{BE})\frac{R_B}{R_A} - \frac{R_B}{R_A}\frac{R_1}{R_2}V_{BE} \tag{15.40}$$

$$\frac{\delta V_{OUT}}{\delta T} = -\left(1 + \frac{R_1}{R_2}\right)\frac{R_B}{R_A}\frac{\delta V_{BE}}{\delta T} \tag{15.41}$$

If $R_1 = R_2$, from Eq. (15.40) we obtain:

$$V_{OUT} = 2V_{BG} - 2V_{EB}\frac{R_B}{R_A} \tag{15.42}$$

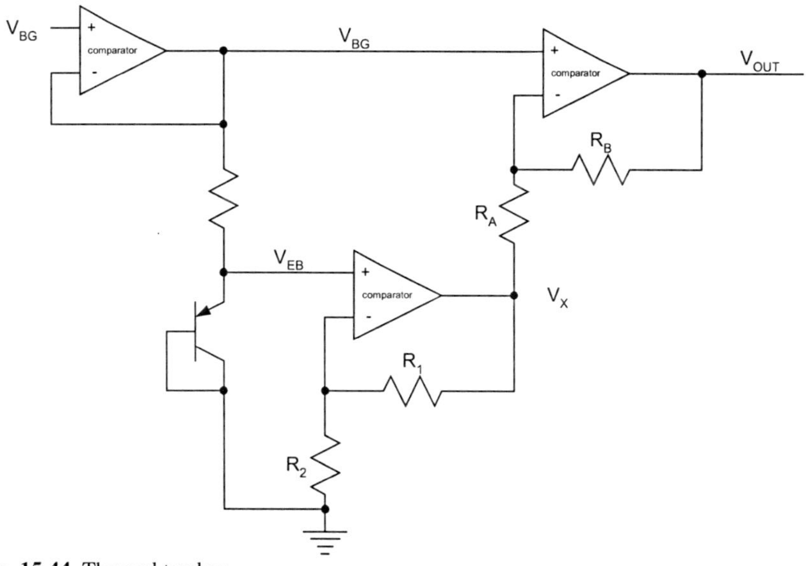

Fig. 15.44. Thermal tracker

Furthermore, Eq. (15.41) can be rewritten as:

$$\frac{\delta V_{OUT}}{\delta T} = -2\frac{R_B}{R_A}\frac{\delta V_{EB}}{\delta T} \tag{15.43}$$

i.e. the thermal derivative of V_{OUT} depends on R_B/R_A ratio.

To summarize, V_{OUT} can be applied to each regulator that uses V_{BG} without any other structural modification. Substituting V_{BG} by V_{OUT} in Fig. 15.43, we notice that $V_{PD} = K*V_{OUT}$ increases with temperature, as V_{OUT} increases. Since the program voltage applied on the cell drain is:

$$V_{drain} = V_{PD} - \Delta V_C \tag{15.44}$$

it is possible to compensate for the increase in ΔV_C by an offset in V_{PD}, so as to assure the thermal tracking of the drain voltage.

Bibliography

B.K. Ahuja, An Improved Frequency Compensation Technique for CMOS Operational Amplifiers, Journal of Solid-State Circuits, vol. SC-18, pp. 629-633, (Nov. 1983).

P.E. Allen, D.R. Holberg, CMOS Analog Circuit Design, Holt, Rinehart and Winston

C. S. Bill el al., High Voltage charge pumps with series capacitances, U.S. patent No. 5,059,815, (Oct. 22, 1991).

G. Di Cataldo and G.Palumbo, "Double and triple charge pump for power IC dynamic models which take parasitic effects into account", IEEE Trans. Circuits Syst., vol. CAS-40, pp. 92-101, (Feb. 1993).

J. Dickson, "On-chip high voltage generation in MNOS integrated circuits using an improved voltage multiplier technique", IEEE J. Solid-State Circuits, vol. SC-11, no. 3, pp. 374-378, (Jun. 1976).

P. Favrat, P. Deval, and M. J. Declercq, "A high-efficiency CMOS voltage doubler", IEEE J. Solid State Circuits, vol. SC-33, pp. 410–416, (Mar. 1998).

A. Ghilardelli, G. Campardo, J. Mulatti, "Bidirectional charge pump generating either a positive or negative voltage", USA patent No. 6,184,741, (Feb. 6, 2001).

P.R. Gray, P.J. Hurst, S.H. Lewis, and R.G. Meyer, Analysis and Design of Analog Integrated Circuits. New York, NY: Kluwer John Wiley & Sons, Inc., ch. 4, (2001).

S.S. Haddadi et al., Flash E^2PROM Array with Negative Gate Voltage Erase Operation, U.S. patent n. 5,077,691, (Oct. 23, 1989).

T. Kawahama et al., Bit line clamped sensing Multiplex and accurate high voltage generator for quarter micron Flash memory, Journal of Solid State Circuit, Vol 31, No 11, p. 1590, (Nov. 96).

T. Kawahara, T. Kobayashi, Y. Jyouno, S. Saeki, N. Miyamoto, T. Adachi, M. Kato, A. Sato, J. Yugami, H. Kume, and K. Kimura, "Bit-line clamped sensing multiplex and accurate high-voltage generator for 0.25µm flash memories", in 1996 IEEE Int. Solid-State Circuits Conf. Dig. Tech. Pap., pp. 38–39, (Feb. 1996).

O. Khouri, S. Gregori, R. Micheloni, D. Soltesz, and G. Torelli, "Low output resistance charge pump for Flash memory programming", 2001 IEEE Proc. Int. Workshop on Memory Technology, Design and Testing, San Jose, CA (USA), pp. 99–104, (Aug. 2001).

O. Khouri, S. Gregori, A. Cabrini, R. Micheloni, and G. Torelli, "Improved charge pump for Flash Memory applications in triple-well CMOS technology", in 2002 IEEE Proc. Int. Symposium on Industrial Electronics, L'Aquila (Italy), pp. 1322–1326, (Jul. 2002).

O. Khouri, R. Micheloni, and G. Torelli, "Very fast recovery word-line voltage regulator for multilevel non-volatile memories", in Proc. Third IMACS/IEEE Int. Multiconference Circuits, Communications and Computers, Athens, Greece, pp. 3781–3784, (Jun. 1999).

O. Khouri, R. Micheloni, I. Motta, A. Sacco, G. Torelli, "Capacitive boosting circuit for the regulation of the word line reading voltage in non-volatile memories", U.S. Patent No. 6.259.635, (Jul. 10, 2001).

O. Khouri, R. Micheloni, I. Motta, A. Sacco, G. Torelli, "Capacitive compensation circuit for the regulation of the wordline reading voltage in non-volatile memories", U.S. Patent No. 6.259.632, (Jul. 10, 2001).

O. Khouri, R. Micheloni, I. Motta, and G. Torelli, "Voltage regulating circuit for a capacitive load", U.S. Patent No. 6.249.112, (Jun. 19, 2001).

O. Khouri, R. Micheloni, A. Sacco, G. Campardo, and G. Torelli, "Program word-line voltage generator for multilevel Flash memories", in Proc. 7th IEEE Int. Conf. on Electronics, Circuits, and Systems, vol. II, pp. 1030–1033, (Dec. 2000)

O. Khouri, R. Micheloni, S. Gregori, and G. Torelli, "Fast Voltage Regulator for Multilevel Flash Memories", in Records 2000 IEEE Int. Workshop on Memory Technology, Design and Testing, pp. 34–38, (Aug. 2000).

M. Maccarrone et al, "Program load adaptive voltage regulator for Flash memories", Journal of Solis State Circuit, Vol. 32, No 1, p. 100, (Jan. 1997).F. Maloberti, Analog Design for CMOS VLSI Systems, 2001 Kluwer Academic Publishers, Boston

M. Mihara, Y. Terada, and M. Yamada, "Negative heap pump for low voltage operation flash memory", in 1996 Symposium VLSI Circuits Dig. Tech. Pap., pp. 76–77, (Jun. 1996).

D.B. Ribner, M.A. Copeland, Design Techniques for Cascode CMOS Op Amps with Improved PSRR and Common Mode Input Range, IEEE Journal of Solid-State Circuits, vol. SC-19, N. 6, pp. 919-925, (Dec. 1984)

G.A. Rincon-Mora and P. E. Allen, "A low-voltage, low quiescent current, low drop-out regulator", IEEE J.Solid-State Circuits, vol. SC-33, pp. 36–44, (Jan. 1998).

T. Tanzawa and T. Tanaka, "A dynamic analysis of the Dickson charge pump circuit", IEEE J. Solid-State Circuits, vol. SC-32, no. 8, pp.1231-1240,(Aug. 1997).

J.S. Witters, G. Groeseneken, and H. Maes, "Analysis and modeling of on-chip high-voltage generator circuits for use in EEPROM circuits", IEEE.J. Solid-State Circuits, vol. SC-24, pp. 1372–1380, (Oct. 1989).

C.C Wang and J. Wu, "Efficiency improvement in charge pump circuits", IEEE J. Solid-State Circuits, vol. SC-32, pp. 852–860, (Jun. 1997).

R.J. Widlar, "New developments in IC voltage regulators", IEEE J. Solid-State Circuits, vol. SC-20, pp. 816–818, (Feb. 1971).

J.T. Wu and K.L. Chang, "MOS charge pumps for low-voltage operation", IEEE J. Solid-State Circuits, vol. 33, pp.592-597, (Apr. 1998).

M. Zhang. N. Llaser, and F. Devos, "Improved voltage tripler structure with symmetrical stacking charge pump", El. Letters, vol. 37, pp. 668–669, (May 2001).

16 High-Voltage Management System

16.1 Introduction

In Chap. 15 we explained how to generate regulated high voltages (HV), describing the commonly used architecture and their working principle. In the following, we will talk about the Flash memory high voltage management system, focusing on why and how these blocks are used and the related problems.

16.2 Sectors Biasing

All operations on memory cells require adequate voltages to be passed to the selected sector, as well to the unselected ones. Table 16.1 shows the voltages to be applied to the sector to perform the various operations, as well as an example of the minimum set (C_0, C_1) of control signals which identify the sector status (read/verify, program, erase, no selection), under the hypothesis that a triple-well technology is adopted.

Sectors are organized in *m* horizontal and *n* vertical strips, so the selection of any sector is carried out by means of the respective horizontal (X) and vertical (Y) strip coordinates. Since usual row and column decoding circuits provide address signals X<*i*> (i = 1 to *m*) and Y<*j*> (j = 1 to *n*) for sector strips, the enable signal for the addressed sector can be generated from the respective signal pair X<*i*> and Y<*j*> by means of a simple combinational circuit.

Table 16.1. Voltages applied to the sector and corresponding control signal coding (C_0, C_1)

Operation	Source	iP-well	N-well	C_0	C_1
Read/verify	0 V	0 V	VDD	0	0
Program	0 V	-1 V – -2 V	VDD	1	0
Erase	~ 8 V	~ 8 V	~ 8 V	1	1
No selection	0 V	0 V	VDD	0	1

This is the so-called "flat" sector control: each sector must be provided with its own enable signal and the control signals needed to manage the operation to be carried out. Moreover, a control block HVC, made up by a combinational circuit and HV level shifters (see Chap. 5), is needed for each sector (Fig. 16.1). These

HV level shifters are necessary to drive the pass transistors that feed the appropriate biasing voltages to the source, iP-well, and N-well terminals of the sector, respectively (when the negative-substrate programming technique is not used, the iP-well terminal of each sector is short-circuited to the respective source line, which allows a HV level shifter to be saved). Biasing supply lines (BSL's), carrying the erase voltage, V_{ERASE}, and the negative-body program voltage, V_{BODY}, provide HVC blocks with the appropriate HV levels.

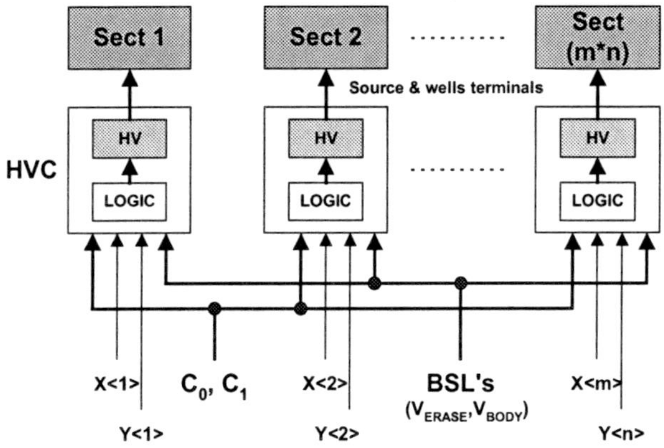

Fig. 16.1. Flat sector control

What is the impact in terms of circuits' overhead when considering a flat sector control? Supposing to have a 64Mbit Flash array composed of 64 1Mbit sectors, arranged in a 8 x 8 matrix, we deal with:

- 64 sector enable signals, obtained from the sector addresses. Any signal has to be routed towards its own sector in the area adjacent to the horizontal and vertical sector strips;
- 2 biasing lines (V_{ERASE}, V_{BODY});
- at least 2 control signals, as depicted in Table 16.1, which identify the sector status: read, program, erase, no-selection;
- 64 HVC blocks, each composed of CMOS control logic, three positive HV level shifter to provide any terminal – source, body e N-well – with the erase voltage, and a negative one for the negative-body program.

The above solution needs large silicon area, in particular due to the required signal routing towards each sector and HV level shifter count. On the other hand, routing the source, iP-well and N-well terminals of each sector towards a specific chip area where all control and decoding circuits are located is not a performing solution: a bus of (64 sectors) x (3 terminals) should be routed. Furthermore, minimum width metal lines could not be used, because the current flowing during charge/discharge of the nodes involved is in the order of mA[1].

An alternative organization is based on the hierarchical approach for sector biasing, which in sketched in Fig. 16.2. All sectors in the same horizontal strip share BSL's, while selection lines (vertical control lines, VCL's), common to the sectors owing to the same vertical strip, control the connection of the terminals of every sector to the respective biasing lines in each operation.

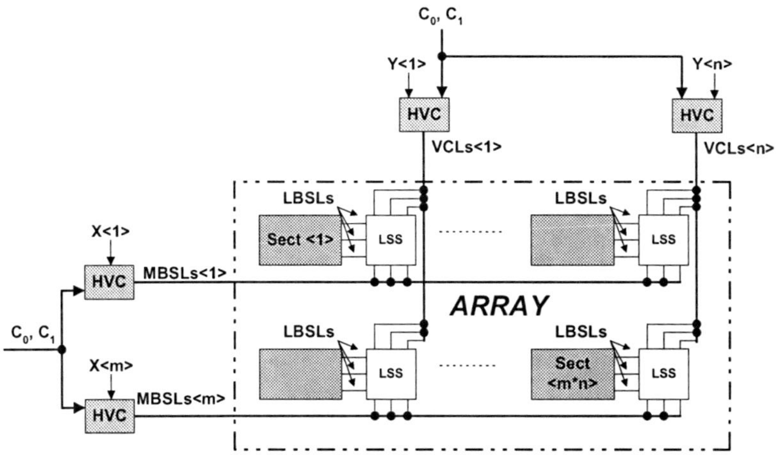

Fig. 16.2. Hierarchical sector biasing example

To provide for all possible operations, a bus of three biasing lines is needed for each horizontal sector strip, each of them carrying the voltage to be supplied to the source, iP-well and N-well terminals, respectively[2]. The three biasing lines are fed with the appropriate operating voltages by means of HV line switches, which are located outside the memory array and are driven by the corresponding X<*i*> address and control signals C_0 and C_1 through HVC blocks. Each biasing bus is routed in the horizontal direction nearby the corresponding horizontal sector strip within the memory array. According to the hierarchical approach (refer to Chap. 9), the biasing lines are therefore split in Main Biasing Supply Lines (MBSL's) and Local Biasing Supply Lines (LBSL's), the connection between them being performed by Local Sector Switches (LSS's). The LSS circuitry connects each sector terminal to the corresponding biasing line under the control of its respective

[1] Body (iP-well) parasitic capacitance, which is in the order of 500 pF or more, has to be charged/discharged in few μs. Furthermore, when a body-assisted programming strategy is chosen, a DC current flows in the iP-well that is about 20-50% the program drain current.

[2] VDD and GND lines are also routed between horizontal sectors strips. Anyway, since they are global power supply, they are not treated as MBSL's.

VCL. Each VCL runs in the vertical direction nearby the associated vertical sector strip and is driven, through a HVC block, by a combinational circuit whose inputs are the corresponding Y<*j*> address and control signals C_0 and C_1. All HVC blocks are placed outside the sector matrix, which results in a compact memory array. Figures 16.3–16.5 show how hierarchical sector biasing works during read, erase and program mode, respectively. For sake of simplicity, a matrix array composed of (2x2) sectors only has been considered.

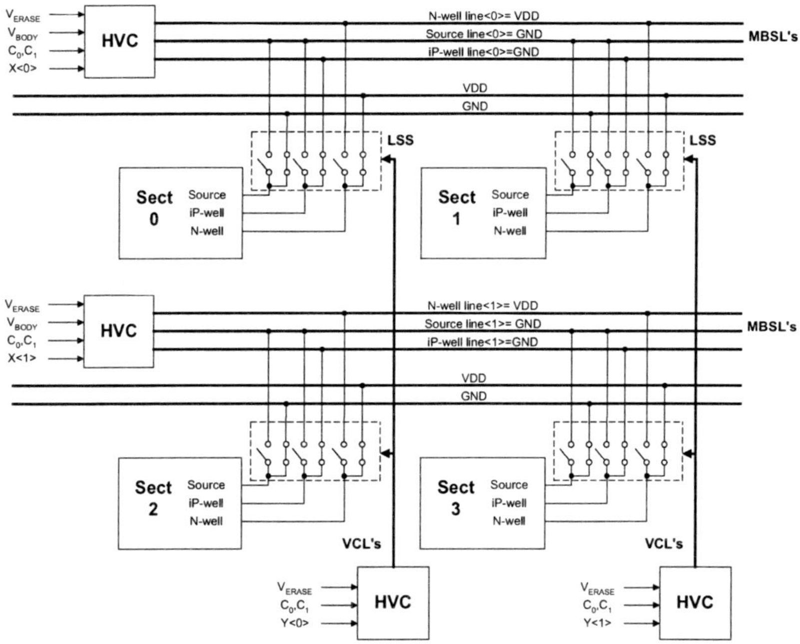

Fig. 16.3. Hierarchical sector biasing: read mode

In terms of circuits' overhead, a hierarchical sector biasing approach requires:

- one HVC block for each sector strip, containing 3 positive and 1 negative HV level shifter;
- one HCV block for each VCL, containing 3 positive HV level shifters;
- 3 MBL's for each horizontal sector strip;
- 3 VCL's for any vertical sector strip;
- 1 LSS for each sector.

The hierarchical sector biasing approach becomes a must as the sectors count increases.

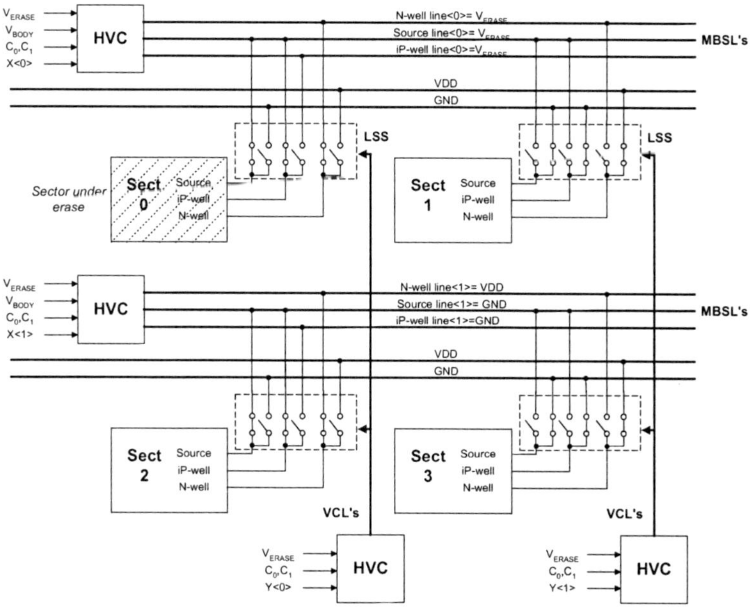

Fig. 16.4. Hierarchical sector biasing: erase

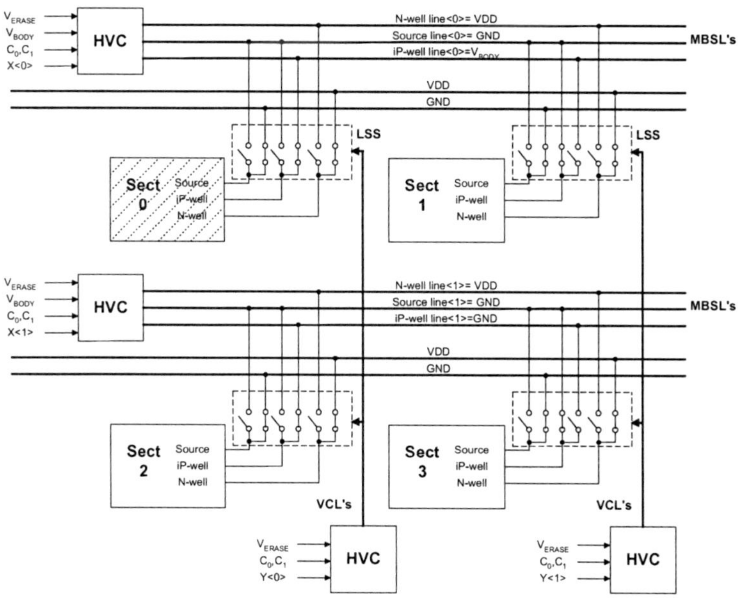

Fig. 16.5. Hierarchical sector biasing: program

16.3 Local Sector Switch

Let us see how the (Local) Sector Switch of the previous section can be implemented at transistor level.

In a conventional CMOS technology (e.g. P-substrate with N-doped well), the implementation of a sector switch does not require many transistors: only the common source terminal has to be biased with a voltage different than GND.

Figure 16.6 shows an example of the circuit that biases the source terminal. This circuit is usually referred to as source switch. Great care has to be taken when designing a source switch: the source node has to be charged/discharged slowly, in order to avoid voltage bumps due to the huge parasitic capacitance between source and bulk, and electromigration phenomena due to high current density.

ERASE enables the V_{ERASE} clamping transistor MP1 (ERASE = 1); SRCCLAMP enables the ground-clamping transistor MN1 (SRCCLAMP = 1). MN1 is designed with a high aspect ratio to keep the source node to GND during program.

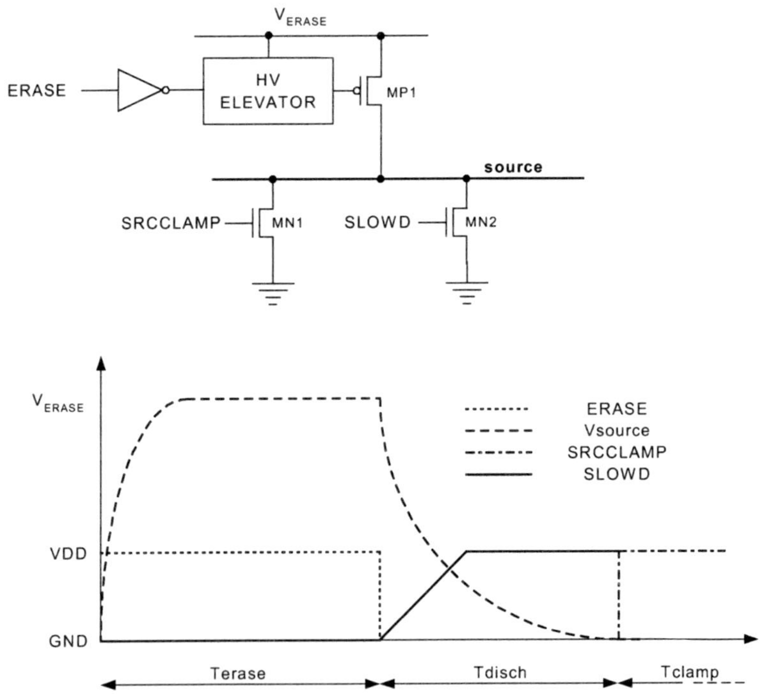

Fig. 16.6. Source switch and signal timings for single-well CMOS technology

The gate signal SLOWD has a slow rising edge to delay the activation of MN2 at the beginning of the discharge. In the meantime, MN1 is switched off by keeping SRCCLAMP low. Another reason for a slow rise of SLOWD is to avoid a snapback effect for MN2. This undesired condition takes place when high drain and gate voltage are applied; a way to avoid snapback for MN2 is to decrease its drain voltage (the source line) while its gate (SLOWD) is rising.

When a triple-well technology is available, the insulated body terminal (iP-well) of the cells can be biased at negative voltage (about –1 V to –2 V), thus increasing the channel hot electron injection efficiency during program.

In this scenario, three sector terminals – source, iP-well and N-well – have to be biased, as previously shown in Table 16.1. The timing these terminals are biased with has to be controlled carefully, in order to avoid direct biasing.

To bias a node with both positive and negative voltages with a hierarchical approach, at least three transistors are required:

1. one pull-up transistor for positive high voltage (i.e. a p-channel one, to transfer positive voltage without any threshold voltage loss);
2. one pull-down transistor for negative voltage (i.e. a n-channel one);
3. one pull-down transistor to transfer the '0' logic level if the biasing line is at low potential and the node has not to be biased at negative voltage.

Referring again to Table 16.1 and assuming that a hierarchical sector biasing approach is chosen, this results in at least 2 transistors (1 n-channel to transfer GND level, 1 p-channel to transfer V_{ERASE}) to bias the sector source, three transistors to bias the iP-well (1 n-channel to transfer GND, 1 n-channel to transfer V_{BODY}, 1 p-channel to transfer V_{ERASE}) and 2 transistors to bias the N-well (1 p-channel to transfer VDD, 1 p-channel to transfer V_{ERASE}).

Figure 16.7 shows an example of how to implement a (local) sector switch.

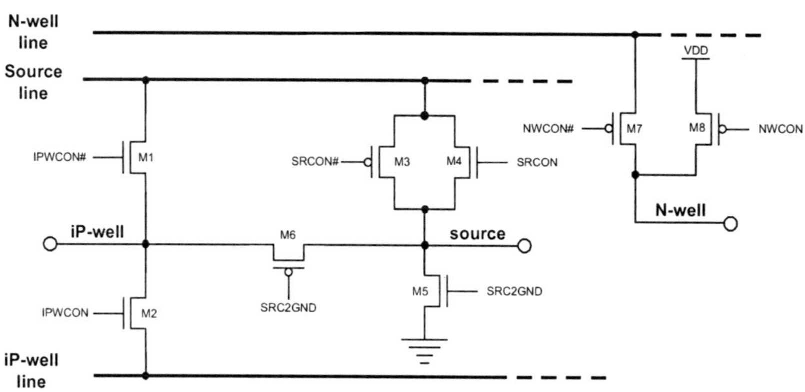

Fig. 16.7. Local sector switch implementation

Basically, 4 control signals are required:

- SRCON (and the inverted SRCON#) controls the connection of the source terminal to the source line;
- SR2GND controls the source clamping to GND;
- IPWCON (and the inverted IPWCON#) controls whether the iP-well terminal has to be biased with the voltage applied to the iP-well line or not;
- NWCON (and the inverted NWCON#) controls if the N-well terminal has either to be connected to the N-well line or to VDD.

Let us see how this sector switch works when reading, programming and erasing, with respect to both selected and unselected sectors:

Read: N-well line at VDD, source line and iP-well line at GND

M5, M8, M2 on, other transistors off

Program: iP-well line at V_{BODY}, source line at GND, N-well line at VDD

SELECTED SECTOR: M5, M2 and M8 on, other transistors off
UNSELECTED SECTORS: M5, M1, M8 on, other transistors off

Erase: N-well line and source line at V_{ERASE}, iP-well line at GND

UNSELECTED SECTORS: M8, M5, M2 on, other transistors off
SELECTED SECTOR: we distinguish between:

- charge sequence: at first, M7 is turned on to charge N-well at V_{ERASE}. M2 and M5 are turned off, while M3 and M4 are on, to charge the source to V_{ERASE} too. Finally, the iP-well is charged to V_{ERASE} by turning M6 on: in such a way the voltage on iP-well is never higher than that of source, thus avoiding any direct biasing. M1 is turned on by applying a IPWCON# level equal or lower than the power supply: in this way the iP-well is charged from the beginning of source line charge.
- discharge sequence: at first, M3 and M4 are turned off, while M2 is turned on to connect iP-well line to iP-well, thus discharging it to GND. Since M6 is on, the source is discharged to $V_{T,M6}$; after this discharge phase, M6 is turned off, and M5 is turned on to clamp the source to GND. Finally, M7 is turned off, while M8 is turned on to discharge the N-well to VDD.
  These timings ensure that any direct biasing is avoided.

The GND clamping transistor M5 is active both during read and program mode, where it has to drain a huge current. It has to be completely turned off before V_{ERASE} charges its drain terminal (matrix source) to avoid its degradation during the memory chip life.

Problem 16.1: Discuss how to connect the body terminal of the transistors that compose the local sector switch depicted in Fig. 16.7.

The high voltages involved during the cells operations require that the transistors of the local sector switch have HV oxide. Depending on the adopted technology, single/double drain extension transistors[3] should to be used.

16.4 Stand-By Management

At system level, address, data and often control busses do not owe to the Flash memory only, but also to the plurality of other integrated devices (such as microcontrollers, volatile memories, etc.). For this reason, every device is provided with a chip enable signal (CE#), so as to avoid bus contention: CE# high sets the address buffers in tristate and the output buffers in high-impedance, and makes the device insensible to any user command except than enabling. This operating mode is the so-called stand-by. As stated by the specs, no memory access time penalty is admitted when recovering from stand-by.

As the mobile applications market increases (cellular phones, digital still cameras, ...) strong efforts have to be made to reduce as much as possible system current consumption – and Flash memory's too – not only during active mode, but also during stand-by, which may represent a high percentage of the application's "Flash time".

Reducing Flash current consumption to few μA means disabling almost all the circuits with steady-state current that can be reactivated without access time penalty when recovering from stand-by. The circuits with a steady-state current (i.e. that drain current even when no cell is addressed) are charge pumps, regulators, reference voltage generators and VDD detectors. Chosing what to disable and how is not an easy task: the choice has to be made depending on power consumption, area occupation and circuit overhead.

One could disable every circuit (deep power-down), thus nulling the power consumption. Since the internal nodes are left to discharge to GND, no regulated or reference voltage are available at stand-by recovery. Furthermore, the logic circuits may be in an indeterminated state if the VDD dropped, because the disabled VDD detector was not able to detect it. For this reason, the VDD detector cannot be disabled during stand-by: as a consequence, it has to meet low power consumption requirements.

The reference voltage, usually generated by a bandgap circuit (refer to Chap. 5), is used everywhere a stable voltage is needed. The bandgap circuit may be designed to achieve different goals: low power consumption (~ μA), thus implying a slow turning-on transient (in the order of μs), or fast turn-on (~ ns), with the counterpart of higher current consumption (tens of μA). In the former case, the

[3] In a single (double) drain extension transistor a small region of drain (and source) diffusion towards the channel region is less doped. This results in an additional resistance in series to the drain (and source).

bandgap circuit can be kept on during stand-by, so as to have a stable reference voltage when recovering from stand-by; in the latter case, the bandgap circuit has to be turned off during stand-by, but during recovery its turn-on transient has to be fast even avoiding spurious overvoltages on the regulated voltages. Usually the former solution is preferred in low power applications: it is clear that maximum current count during stand-by includes the reference generator's one.

Charge pumps and regulators have of course to be disabled during stand-by, but a solution has to be found to disable them while making them settling fastly and accurately when recovering. These circuits cannot be turned off without any precaution, otherwise the leakage current would discharge their output nodes. This is particularly important for the read voltage, that would not be available when switching to active mode because of the slow settling transient of the read charge pump and regulator in presence of high capacitive load. To avoid the consequent unacceptable CE# access time penalty, an effective stand-by management system has to be implemented.

The preferred architecture depends on system requirements. Two key points have to be taken into account. First, during stand-by charge pumps and regulators have to be disabled, but their output nodes have to be kept charged so as to ensure a fast settling when re-entering active mode. Furthermore, if this kind of solution is adopted, the voltage settling when recovering from stand-by has to be analyzed carefully. In fact, when switching from a recharge situation to one where the voltage is regulated by a negative feedback loop, spurious transients may occur on the regulated node that vanify the effect of the previous recharge. This kind of transients has to be carefully avoided.

In the following, an example of the stand-by management system in a low power supply, low power consumption Flash memory is presented, where the gate read voltage is provided by a regulator, GR, supplied by a charge pump, GP. In this architecture, the charge pump and regulator outputs are kept charged to a suitable voltage value by mean of a low consumption charge pump, which is used dring stand-by only. In this memory mode, SB = 1 (corresponding to CE# = 1) disables GR and GP, disconnecting their output nodes from the nets GROUT and GPOUT respectively by mean of SW_2 and SW_1 switches (Fig. 16.8). Both nets are kept charged to a suitable voltage level V_{RSB}, proximum to the nominal read value V_{READ}, by mean of the auxiliary stand-by charge pump, SBP; GROUT and GPOUT are connected to SBPOUT by of SW_3 and SW_4 switches. A "smart" way to achieve this goal is to activate the stand-by pump only when it is necessary, as it happens to a "normal" charge pump, which is boosted only when its "high regulator" enables the boosting clock: we drive the stand-by clock with a non-continous clock generated by a regulator only if needed.

The stand-by regulator, SBR, compares a suitable partition of V_{GROUT} with the bandgap voltage (we suppose that the bandgap reference has low power consumption, thus being enabled during stand-by); the comparison result is ENOSC, which enables or disables (ENOSC = 1 or ENOSC = 0, respectively) a ring oscillator ROSC that generates the stand-by clock, SBCK. In such a way, SBP recharges GROUT a GPOUT to the stand-by voltage V_{RSB}.

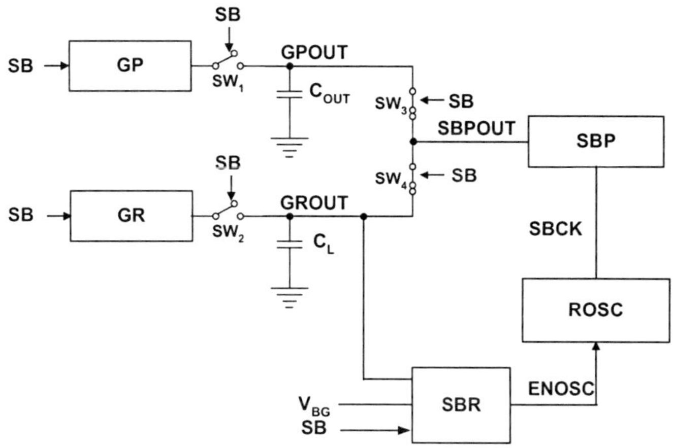

Fig. 16.8. Stand-by management architecture

Even if it would minimize the spurious transients when recovering from stand-by, it is practically impossible to set V_{RSB} equal to V_{READ}, because of unavoidable process spreads (V_{RSB} and V_{READ} are regulated by different circuits); furthermore, high current consumption is required to achieve a very precise regulation.

Moreover, when re-entering active mode the word line parasitic capacitance C_{WL} is connected in parallel to the regulator output load, C_L, the subsequent charge sharing reducing the value of V_{GROUT}: for this reason, the nominal value of V_{RSB} is chosen slightly higher than V_{READ} (e.g., $V_{RSB} = V_{READ} + \sim 100$ mV).

To detect V_{GROUT}, a voltage divider $Z_1 + Z_2$ (Fig. 16.9) is used; the voltage across Z_2 is applied to the inverting terminal of a comparator and compared to the reference voltage V_{BG}, applied to the non-inverting terminal.

To reduce the current consumption of the stand-by charge pump, the boosting clock is generated only when its output voltage (GROUT and GPOUT nodes) becomes less than V_{RSB}. To this end, the ring oscillator that provides the boosting clock is controlled by a feedback net including the voltage divider. This feedback net can be made in different ways with low power consumption: e.g., it can be made as a series of n_l equal-sized diode connected PMOS transistors, each of them performing a V_{DS} voltage drop. A low consumption comparator compares V_{DS} and V_{BG} values: the ring oscillator enabling signal, ENOSC, is driven to logic high level only if the voltage on GROUT is less than a $n_l V_{BG}$.

Due to the slow feedback net (because of the reduced current consumption allowed), the regulators sets V_{RSB}, but the voltage value effectively present on GROUT is $V_{RSB} + \Delta_{SB}$, where Δ_{SB} depends on regulator offset, stand-by current, stand-by pump driving capability and capacitive load:

$$\Delta_{SB} = f(V_{OS}, I_{SB}, I_{SBP}, C_{LOAD}) \tag{16.1}$$

This additional voltage contribution has to be taken into account when chosing V_{RSB}, so as to avoid an excessively high voltage on GROUT.

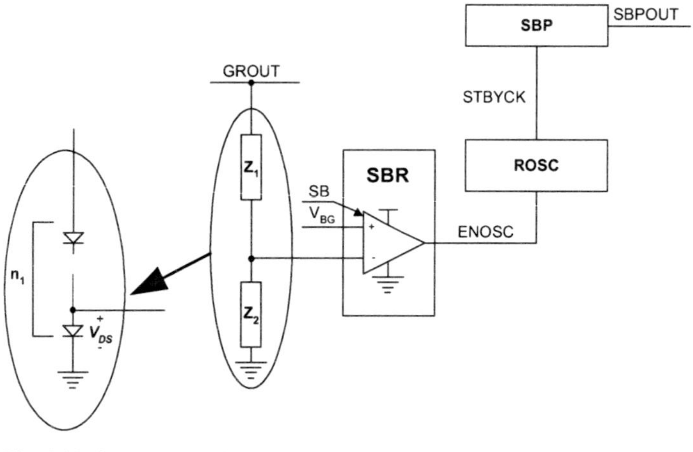

Fig. 16.9. Stand-by regulator

The diode partition is a good solution from the power consumption point of view, but it has little "programmability" performances because it is difficult to adapt the read – and stand-by – voltage to bandgap variations[4]. In this case, it could be advisable, to make the partition with high resisistivity resistors (e.g. N-well resistors), even if this solution is expensive in terms of silicon area, since we have to keep the current consumption low.

The stand-by charge pump could seem anomalous, because it has no filter capacitance to its output; this seemingly contrasts with the topologies exposed in Chap. 15. The contraddiction is only apparing, because the filter capacitance in stand-by is the parallel of the filter capacitance of GP and the output parasitic capacitance of GR. The sum of these contibutions is a huge capacitance, but the stand-by pump can be made small and with reduced driving capability because it has not to charge completely the load, but only to keep it around a given value, contrasting the leakage current (and that flowing into the series $Z_1 + Z_2$, which detects V_{GROUT}).

When the memory is not driven to stand-by, both the low-power comparator and the ring oscillator are disabled to reduce consumption and avoid any disturb on GROUT; STBYCK is forced to "0".

Let us consider the system behaviour when recovering from stand-by. Referring to Fig. 16.10, during stand-by GPOUT and GROUT are connected to the auxiliary stand-by charge pump by means of SW_3 and SW_4, respectively, and disconnected from GP and GR (SW_1, SW_2 and SW_5 open). The resisive partition is disconnected from GROUT, so as to avoid current flowing, and the feedback node E is dischared to GND by means of SW_6.

[4] The bandgap voltage value may vary from a production lot to another, depending on process parameters variations. To minimize the subsequent read voltage variations, the resistive feedback net can be made configurable by introducing resistances ΔR to be inserted at electrical testing level via CAMs/fuses.

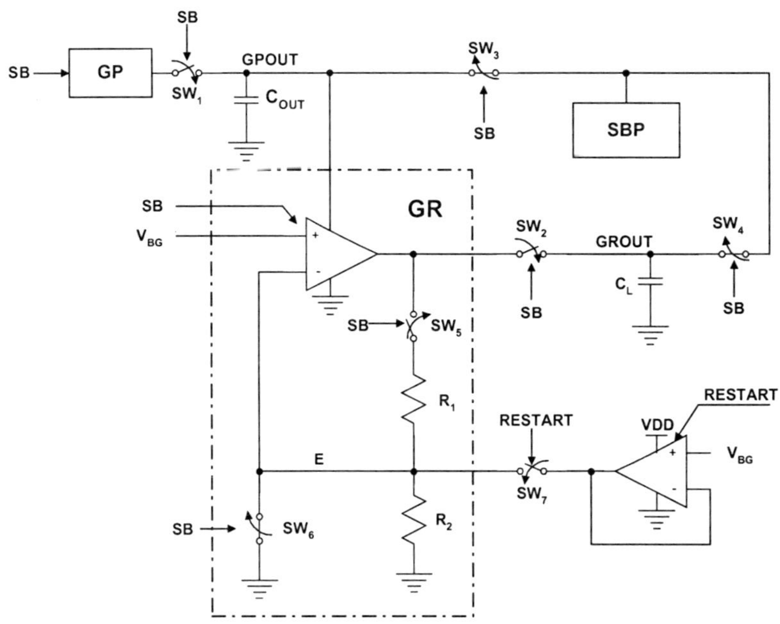

Fig. 16.10. Stand-by recovery management architecture

When the memory is switched to active mode, SW_1, SW_2 are SW_5 are turned on, while SW_3, SW_4 and SW_6 are turned off. To reach the steady state condition, the feedback node has to charge to V_{BG}: during its turn-on transient, the regulator detects a high differential input (V_{BG} is stable, while E is at GND and charges slowly), thus sourcing to the output load a huge current. Anyway, C_L is already charged to V_{RSB} (which is higher than V_{READ}), but shows further charge whose value depends on feedback net transient. The gate voltage gets higher than expected, thus preventing correct reading: until this transient is ended, any memory access is impossibile.

To solve this problem, we can force V_E to V_{BG} when recovering from stand-by for as much time as required to let the transients on the nodes of the feedback net of GR settle. To this end, a start-up regulator can be used, composed of unity-gain amplifier, supplied by VDD. The feedback node is connected to the buffer's output for the time necessary to complete the transient (the enable signal, RESTART, is active for few tens of ns). Until RESTART is high, GR ideally source no current to its load, since its differential input voltage is nominally zero, thus keeping its output to the stand-by value, V_{RSB}. Practically, the load capacitance C_L shares charge with the parasitic capacitance C_{WL} of the addressed WL; anyway, $C_{WL} << C_L$, and V_{RSB} is chosen so as to take into account this charge sharing effect.

When the recovery transient is ended, the feedback loop and the nominal behaviour of GR are restored and the start-up regulator is disabled.

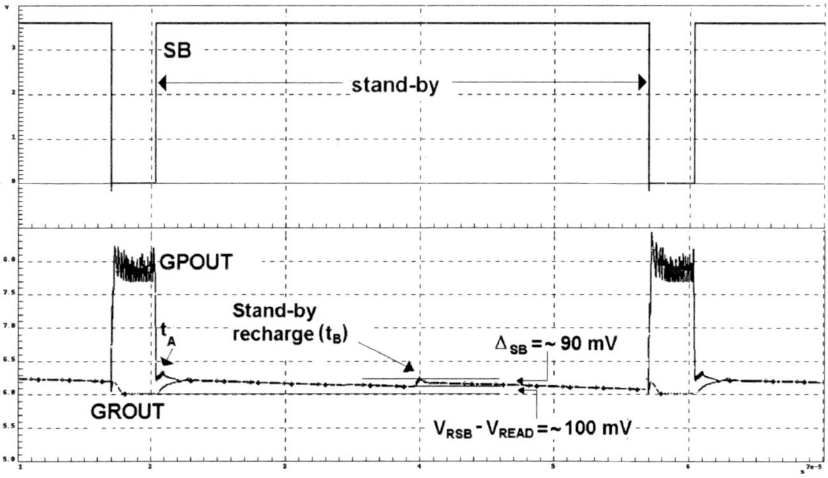

Fig. 16.11. Simulated V_{GPOUT} e V_{GROUT} waveforms

This example may be extended: more charge pump and regulator outputs may be recharged, depending on the system's requirements, simply by increasing the switches count.

Figure 16.11 shows the simulation waveforms of GP and GR output voltages during stand-by and recovery. The regulated voltage during stand-by is set about 100 mV above the read voltage V_{READ}; the overvoltage Δ_{SB} is about 90 mV. In this device, the stand-by current is less than 40 μA in every operating condition, including the bandgap circuit current (~10 μA).

As it should be clear, the power consumption of a charge pump cannot be reduced to few μA. How it is possible that the whole system current consumption is so small, taking into account that the stand-by regulator has a steady-state consumption, even if small? It all depends on how the user evaluates the power consumpion. The portable system user, i.e. a cellular phone, cannot measure instantaneous power consumption, but only mean values: setting the device to stand-by and supplying it with a battery, one can measure how much time does the Flash memory takes to discharge the battery, thus evaluating its mean consumption. In practice, short absorption peaks may be neglected with respect to long periods when the power consumption is almost zero, since the mean current during stand-by is given by the integral of the absorbed current between two subsequent recharges (t_A and t_B in Fig.16.11), divided by this time interval:

$$\overline{I_{SB}} = \frac{\int_A^B I(t)dt}{(t_B - t_A)} \tag{16.2}$$

16.5 High-Voltage Management

16.5.1 Architecture Overview

Now we are ready to analize an overview of the high-voltage management system in a single power supply Flash memory. Figure 16.12 shows an example, in which a triple-well technology is supposed to be used; the figure includes the block necessary to generate and distribute the high-voltages required in read, program and erase mode. Hierarchical row and column decoders and negative-gate erase are assumed.

A control logic (Command User Interface and State Machine, CUI & SM) generates the operating mode signals, i.e. stand-by (SB = 1), program (PROGRAM = 1), erase (ERASE = 1), read/verify (SB = PROGRAM = ERASE = 0) and erase verify (ERAVER = 1). In order to avoid performing spurious operations at power-on or when the puwer supply bumps down, a VDD detector circuit (VDD DETECT) produces suitable power-on reset (POR) and low VDD signals (VCCLOW), so as the memory to stop the algorythm and be reset.

As explained in Chap.15, the high voltages are produced on-chip by mean of charge pumps; to avoid oxide stess and to limit power supply consumption, the positive and negative charge pumps (PMP and NEG PMP, respectively) are provided with a voltage limiter, HI REG: their boosting clock is enabled (EN = 1) only if a partition of their output voltage is less than a suitable value V_{REF}.

The reference voltage $V_{REF} = V_{BG}$ for the high voltage limiter is given by a bandgap-based voltage generator (BANDGAP REFERENCE). This voltage level has to be precise and stable as the power supply and the temperature vary; the bandgap reference has low power consumption, so as to be kept on during stand-by.

16.5.2 High-Voltage Read Path

The read gate voltage V_{GR} is produced by a regulator supplied by a positive charge pump; this regulator biases the addressed Word Line (WL) through the Global Row Decoder, GRD, and the Local Row Decoder, LRD. The same concept applies also to the verify gate voltage. To disable the unaddressed Local Word lines (LWL's), the corresponding PCH signal (refer to Fig. 9.34) has to be driven at least to the read/verify voltage.

Local row decoder drivers biasing results in huge capacitive load for the read voltage; for this reason, when hierarchical row decoder approach is chosen, the biasing system can be doubled in order to reduced the parasitic load: one voltage is devoted to drive the GRD, thus biasing the addressed main and local word lines, while the other is used to supply the LRD drivers. The former voltage is named V_{GROUT}: it is produced by the regulator GR, which is supplied by the charge pump GP; the latter is named V_{LROUT}, and it is generated by the regulator LR, supplied by the charge pump LP. To ensure adequate high voltage level to PCH, GR e LR are identical, as well as GP and LP.

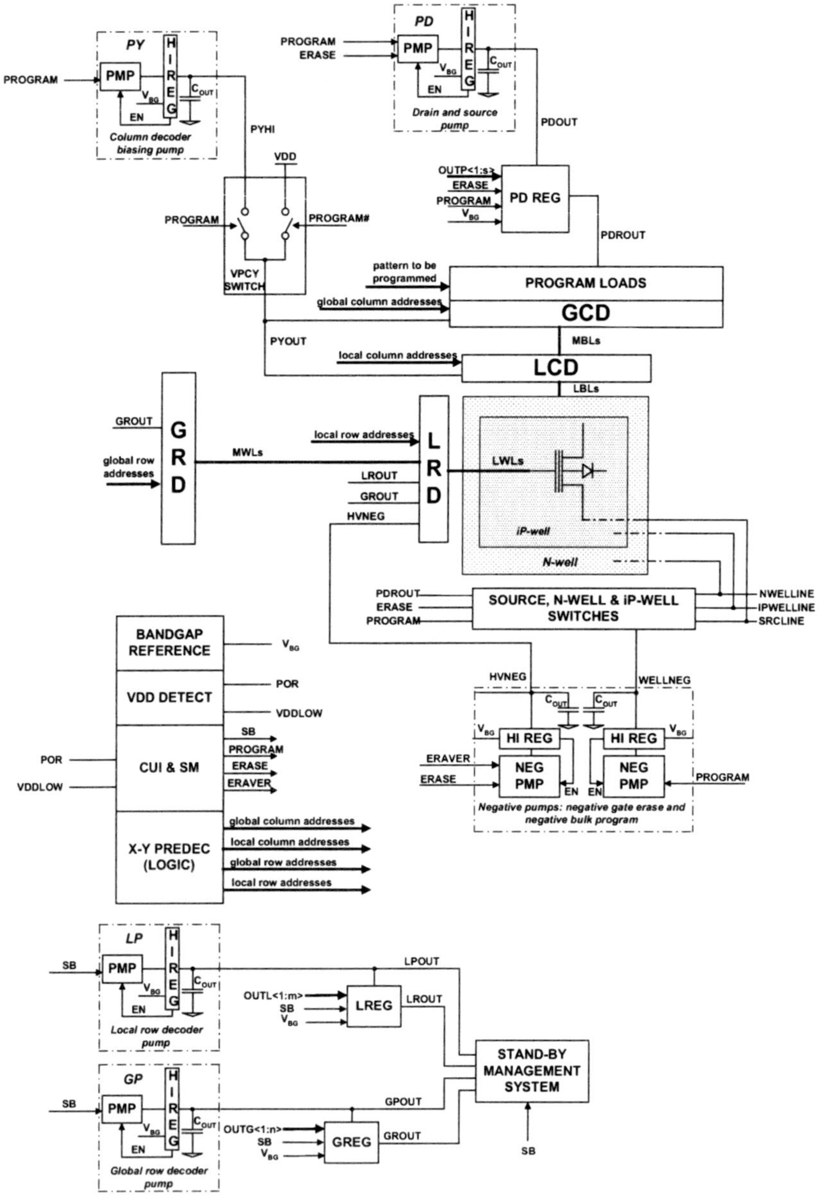

Fig. 16.12. High voltage management overview in a triple well, hierarchical decoding Flash memory

The typical architecture for GR can be that explained in Chap.15, where the output voltage is regulated by comparison with the bandgap voltage.

To achieve stand-by low power consumption, GP, LP, GR, and LR are disabled, but a stand-by management system is introduced (STAND-BY MANAGEMENT SYSTEM, for example equal to the architecture explained in the previous section) to recharge the voltages on GPOUT, LPOUT, GROUT e LROUT, so as not to degrade the CE# access time.

16.5.3 High-Voltage Program Path

The choice of the optimum values and timings for the program drain and gate voltages depends on many factors. First of all, the operating point of the memory cell has to be kept between a working window, which defines the allowed drain voltages as a function of the cell effective length, taking into account program efficiency, snap-back, drain stress and fabrication process spreads. Also the gate voltage influences important parameters such as program efficiency, drain current and snap-back triggering point: a critical gate voltage exists when the snap-back triggers also at the minimum drain voltage.

As a consequence, it is important to raise the gate voltage before connecting the drain to the high drain voltage: during the gate voltage settling the program drain voltage, V_{PD}, is not applied to the cells, in order to avoid the snap-back. Figure 16.13 shows a typical behaviour of gate and drain voltages during program; the program verify phase is shown too. According to the Program & Verify algorythm (see Chap. 14), after a program pulse the cells under program are verified, i.e. they are read with suitable references to check whether they have reached their targed distribution. This happens by comparison of the read pattern with the user pattern: only the cells that have not yet reached their target distribution are connected to V_{PD} through the program load to receive a further program pulse.

Referring to Fig. 16.12, V_{PD} is regulated by PDREG to avoid spurious phenomenon that can damage the cells or reduce their reliability. The typical topology of the regulator is similar to that of the gate regulator; the drain charge pump, PD, supplies PDREG, which provides with the program drain current. To reduce the impact on silicon area, PD and PDREG are used in erase too to produce the positive source/body voltage, V_{SE}. The output voltage value (V_{PD} or V_{SE}) is selected as a function of the operative mode (program o erase), depending on the control signals, OUTP<1:*s*>. PDREG has to be designed carefully in order to minimize its output resistance. The regulated voltage V_{PD} is applied to the cells drains through the program loads (refer to Fig. 15.14) and the column decoder. To reduce the voltage drop across these pass transistors, they are biased with a high voltage PCYHI, produced by the charge pump PCY. Since high voltage has to be tranferred to the cells drains only during program, in the other operative modes the power supply level is "high enough" to bias the column decodes: a high-voltage switch (VPCY SWITCH) is therefore necessary to select either PCYHI or VDD.

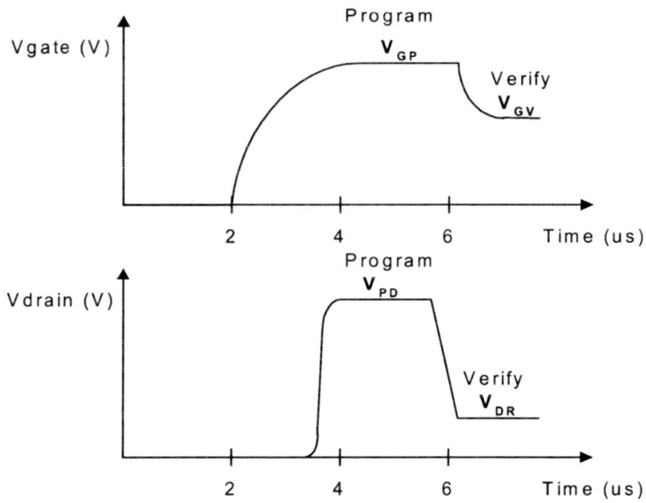

Fig. 16.13. Gate and drain voltages during Program & Verify algorythm in a Flash memory

Also the gate program voltage, V_{GP}, may be generated by GR, which provides the read voltageV_{GR} and the verify voltage V_{GV} (the latter can be equal or different from the former, depending on the verify approach); the same regulator provides also the erase-verify and depletion-verify voltages, V_{EV} e V_{DV}, respectively. The output voltage is programmable by mean of *n* control signals, OUTG<1:*n*>. Using the same regulator in read and program allows saving silicon area and minimizing power consumption.

According to the algorythm (refer to Chap. 14), the erase pulses are followed by a soft-program routine, which compacts the erased distribution by reprogramming the cells with negative threshold voltage. In order to reduce the drain current, the WL is biased with a staircase voltage, starting from very low levels (e.g. 1.5 V – 1.8 V).

As explained in the previous subsection, the voltage produced by LR has never to be less than that of GR. To simplify the structure of LR, its output voltage during soft program may be fixed to a constant potential equal to the maximum value of the gate staircase: this limits the parasitic output load to GR's, thus reducing the transient of V_{GP}. The same approach is obviously used during program. To perform a verify operation, the WL voltage has to be discharged from V_{GP} to V_{GV} (Fig. 16.13). This operation has to be sufficiently fast to minimize the impact on the program time. During program verify, the output voltage of LR has to be discharged from the highest value to the read one, in order to minimize the differences between read and verify: for this reason, also LR output voltage in programmable by mean of *m* control signals (OUTL<1:*m*>).

When the fabrication process provide for the triple-well, the bulk of the cells may be biased with a negative voltage (~ –1.5 V) to improve the program efficiency. This approach requires a negative charge pump, whose output voltage, WELLNEG, has to be provided to the iP-well of the selected sector; this charge pump has to be designed with an adequate driving capability, since the bulk current during program is about 20% to 50% of the drain current.

16.5.4 High-Voltage Erase Path

The high electric field in the tunnel oxide necessary to erase a Flash memory cell through FN tunneling can be achieved by applying a high voltage to the source terminal of the cell while connecting its control gate to ground (the cell substrate is also grounded, and its drain terminal is left floating: FN tunneling of electrons takes place from the floating gate to the source). However, the high reverse biasing condition of the source/substrate junction can degrade cell reliability. To overcome this problem, the negative-gate erase technique has been proposed: erasing is carried out by using a negative gate voltage (–8 V), which allows FN tunneling to be achieved by applying ~8 V to the source line (SRCLINE) of the sector.

The voltages used with both the above erasing schemes lead to the undesired phenomenon referred to as band-to-band tunneling (BBT). Band-to-band tunneling excessively increases the leakage current through the reverse-biased source/substrate junction and also adversely affects cell reliability due to holes injection into the tunnel oxide.

When a triple-well technology is available, the iP-well of the sector under erasing can be biased at the same voltage as its source line, thus eliminating BBT (FN tunneling takes place from the floating gate to the channel). The pump PD and its regulator PDREG provide the voltage to be applied to the source, iP-well, and N-well terminals of the sector under erasing. Only a transient current must be provided by PDREG at the beginning of any erase pulse, due to the need for charging the parasitic capacitances associated to the above regions. Erasing is generally carried out by means of a sequence of pulses (typical length ~10 ms), each followed by a verify step, and continues until all cells in the selected sector are driven into the erased state. A negative charge pump generates the negative voltage HVNEG (–8 V) to be applied to the LWL's. GROUT and LROUT values are set to their lowest voltage (e. g. 1.5 V) during any erase pulse so as to minimize transistors oxide stress. Charge pumps GP and LP are kept active because, after any erase pulse, voltages GROUT and LROUT must be driven to the level required to perform erase verify. To allow erase verify, the source line and the iP-well of the sector under erasing must be switched from the previous high level V_{SE} to GND, while the corresponding N-well has to be discharged from V_{SE} to VDD. As seen in Sect. 16.3, great care has to be taken when discharging the above nodes. Assuming a negative-gate erase scheme, during any erase pulse, the gates of the cells are (negatively) charged to –8 V, their source and iP-well regions are charged to ~8 V, and their drains are floating, which means that they are driven to the source/iP-well voltage minus the cut-in voltage of the iP-well/drain diode. If the gate of a cell is discharged to GND starting from these bias conditions, its drain terminal is boosted due to its capacitive coupling with the gate, thus causing oxide breakdown. To prevent this effect, an appropriate discharge sequence has to be provided. First, the iP-well is discharged to ground. Second, the source diffusion is driven to ground and the N-well is discharged to VDD. Then, the drain is connected to ground through its column decoder. Finally, also the gate is discharged to ground, and is now ready to be brought to the erase-verify voltage. It should be pointed out that the above discharge operations should not to be too fast, because the ground noise due to the discharge of parasitic capacitances can be excessive, thus leading to spurious switching in on-chip logic circuitry.

Cells owing to the same sector do not erase with the same speed. For instance, while most cells are erased, some have not yet reached the desired threshold voltage[5]. The erase verify is performed as a usual read, using the erase verify reference, EV; since the erase is not selective, all the cells in a sector receive erase pulses until they all pass the EV: for this reason, it is likely that most cells are depleted while the slowest have still V_T higher than the EV's. When verifying, the sense amplifiers detect the bit line current, which coincides with the selected cell one if the cells owing to the same bit line are unselected. If these cells have negative V_T, they cannot be unselected by biasing their LWLs with GND level, thus introducing a spurious subthreshold current on the bit lines. Even the single cell's current is small, the plurality of unselected cells owing to a bit line could make appear erased a slow cell that has not yet reached the EV threshold. To overcome this problem, the unselected WLs during erase verify are biased with a suitable negative voltage, e.g. –2 V, which is generated by the negative charge pump, varying the level of HVNEG by a suitable signal ERAVER. Referring to the previous discharge sequence, the gate terminals are not discharged to ground, but to the negative gate verify voltage through HVNEG; of course, after the discharge the WL to be verified is charged to the erase verify voltage. This is easy to be accomplished when the hierarchical row decoding approach is chosen. Figure 16.14 shows two main word lines, each of them owing two local word lines, for sake of simplicity. The local row decoder is the one explained in Sect. 9.12. MWL<0> contains the addressed LWL, LWL<1>, while MWL<1> and LWL<0,2,3> are unselected. According to the hierarchical approach, Fig.16.14 shows how the positive (selected) and negative (unselected) biasings are transferred to the LWLs. To prevent electrical stress and, therefore, undesired soft erase in cells outside the sector under erasing, the LWLs and the source and iP-well terminals of unselected sectors are grounded, and their N-well terminal is connected to VDD. Moreover, their drains are kept floating. A switch block (SOURCE, N-WELL & iP-WELL SWITCHES) is provided to each sector so as to bias its source and well terminals with the appropriate voltages in each operation, as explained in Sect. 16.3.

[5] Or, which is equivalent, most cells erase normally, while some are faster.

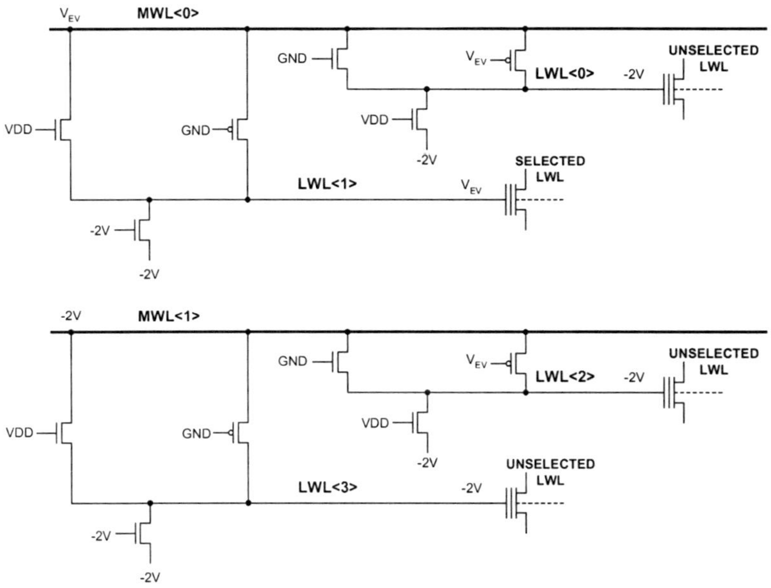

Fig. 16.14. Erase verify gate voltage in hierarchical row decoders

16.6 Modulation Effects

The "real" width of a threshold voltage distribution obtained by programming, from the read point of view is the sum of technological and circuital effects. The former are essentially related to gain and program speed spreads; the latter reside in modulations of the voltages applied to the terminals of the memory cells. But what is the cause of these modulations? Of course the circuits: variations of regulators outputs (program step, read or verify voltages modulations) or sense amplifier indetermination. For this kind of modulation, it is often possible to find circuit-level solution: for instance, referring to the program step, one should minimize the linearity error of the D/A converter. Anyway, other modulation sources are circuit-independent: usually these effects are related to the parasitic resistances that play a great role at the net in which a huge current flows.

This phenomenon is common both to single bit and to multilevel Flash memories; anyway, its effect cannot be neglected when the distributions have to be "thinner", i.e. in multilevel memories. In the following, we will refer to the case of a 2bit/cell multilevel Flash memory, showing problems and both circuital and topological solutions to reduce modulations.

16.6.1 Program Drain Voltage Modulation

As explained in Chap. 15, since the memory cells during program have to be biased with a drain voltage, V_{PD}, higher than the core power supply, this voltage is generated by mean of a charge pump, whose output voltage V_{PDOUT} is regulated by mean of a suitable voltage regulator, DR. By mean of the program load circuits, the cells are biased with V_{PD}, depending on whether they have to be programmed.

Figure 16.15 shows how the program voltage is connected to the cells' drain. Usually charge pump and regulator reside in the same layout portion so as not to introduce a parasitic resistance in series to the pump output; the output voltage of the regulator, V_{PD}, is carried to the program loads through metal lines. The program loads outputs are connected to the drains through metal lines and column decoders; for this routing to be uniform, the program loads are concentrated on a silicon area that usually is not near to where the high voltage circuits, i.e. DR, reside.

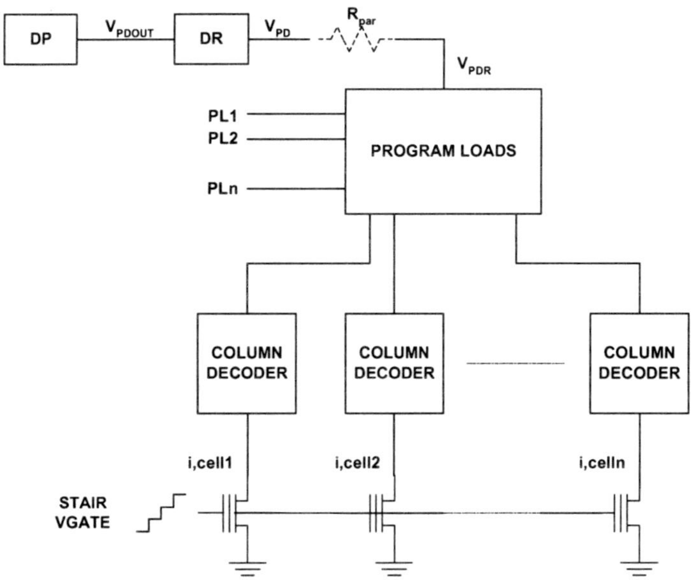

Fig. 16.15. How the cells are connected to the program drain voltage

The parasitic resistance R_{par}, due to V_{PD} routing, cannot be neglected: because of it, the program loads are not fed by V_{PD}, but by V_{PDR}, which value depends on the current flowing in R_{par}.

Let us suppose that DR has zero output resistance (this is the ideal case): its output voltage does not depend ont the ocurrent it has to source to the load. The load current is a function of how many cells have to be programmed, i.e. of how many program loads are on. Anyway, this number varies non only according to the pattern to be programmed, but also to the erased distributions width and program speed. This results in a voltage drop across R_{par} and, as a consequence, in a different drain biasing.

If *n* cells are programmed in parallel and $I_{cell,i}$ is the current sunk by the *i*-th cell, then:

$$V_{PD} - V_{PDR} = R_{par} \cdot \sum_{i=1}^{n} I_{cell,i} \tag{16.3}$$

Under ideal condition hypothesis, V_{PD} is constant even if the output current varies; of course R_{par} is constant; the current of the *i*-th cell, $I_{cell,i}$, is the program current I_{cell}, if the related program load is active, otherwise it is zero. It is clear that $V_{PD'}$ is a function of the program loads configuration, which means program pattern and step of the algorythm.

How much is this voltage drop and what is its influence on distributions broadening? Let us evaluate it with an example.

Let us suppose that $n = 64$ cells are programmed in parallel, and that the cell sinks $I_{cell} = 60$ μA during the program pulse (according to the current technologies). In this case, the total consumption from V_{PD} due to the cells to be programmed is equal to $64 \cdot 60$ μA = 3.84 mA.

The program speed is not the same for every cell, also because they start from a erased distribution. Let us suppose that after *k* program pulses *m* cells have reached the V_T of the program-verify reference, PV: their program loads are not connected anymore during the following program pulses. At the $(k+1)$-th program pulse only $(n\text{-}m)$ cells are still connected. If $m = 63$, the current required to the regulator in this phase is equal to 60 μA only.

The voltage drop between program load and cell drain is constant, because it depends on the program current of one cell only in every algorythm phase. On the contrary, the voltage drop due the parasitic resistance between regulator output and program loads may vary, because it is function of the programming pattern and of the algorythm phase (i.e. how many cells have to be connected to the program loads at each program step).

To understand the influence of this phenomenon on the distributions width, let us refer to Fig. 16.16, where the measured gate voltage versus V_T step is sketched for various drain bias. If the drain voltage is inversely proportional to the number of cells connected to the program load, the last cells to be programmed (the slowest, and/or the ones that have to reach the highest threshold voltage) move from a program characteristic to another, as shown by the arrow.

The first program steps are less effective because the drain voltage is lower: the threshold step is less than the theoretical one (i.e. the gate voltage step). Most cells are programmed and verified under this condition When these cells are programmed to the desired V_T, the drain voltage increases toward its nominal value, thus leading to a more effective programming: the last cells are subjected to higher threshold steps. This effect adds to the sense amplifier indetermination and the cells characteristics deviations, leading to unforeseen distributions width increase. To solve the drain modulation problem it is necessary to design carefully the routing of V_{PD} in order to minimize R_{par}, but this cannot be the only countermeasure adopted. First of all, the metal carrying V_{PD} cannot be designed too large because of layout constraints; furthermore, the program parallelism increase has the opposite effect, rising the current through in $R_{par.}$

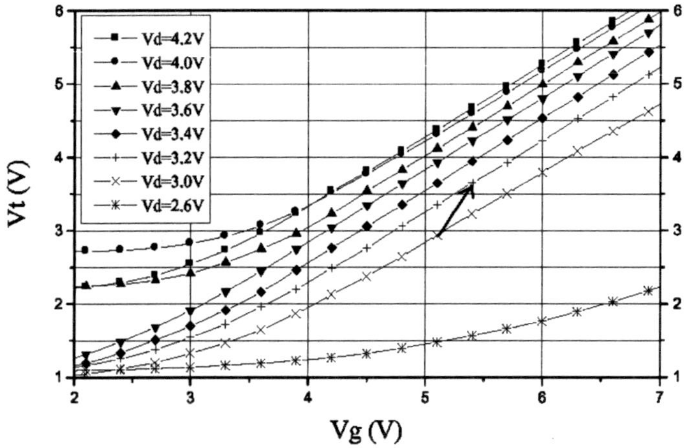

Fig. 16.16. Gate voltage vs. threshold voltage step as a function of program drain voltage

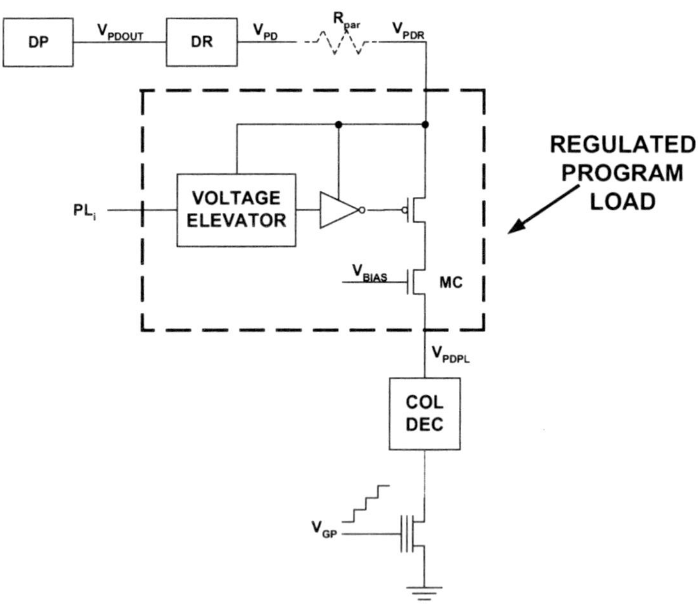

Fig. 16.17. Local drain regulation

It is possible to multiply the last stage of the drain regulator (i.e. the output stage) by the number of program loads, and locate these replicas near them; the drain regulation becomes "local", thus leading to a new program load structure with internal regulation.

This solution does not introduce routing overhead, with a minimum impact on the structure of the program load. Figure 16.17 shows the local drain regulation: in the usual structure is inserted the n-channel transistor MC after the PMOS pass transistor. MC is driven by a suitable (high) voltage V_{BIAS}, so as to keep it in common gate configuration. In such a way, R_{par} is not nulled, but its effect is minimized: even if V_{PDR} varies as a consequence of the program pattern, the voltage after the program load, V_{PDPL}, is stable.

16.6.2 Body Voltage Modulation

In the last generation Flash memories the cells bulk during program is not kept to GND, but it is driven to a suitable negative voltage: this allows creating an electric field that inhibits the hot electrons dispersion into the bulk, keeping them near to the silicon/tunnel oxide interface, thus increasing the floating gate injection rate.

The negative-body program voltage is produced by a charge pump, then applied to the bulk of the memory sector under program by mean of pass transistors, which are usually placed at the sector's boundary, as shown in Fig. 16.18.

During the program pulse, the bulk terminal sinks a constant current, thus the metal lines that feeds the negative voltage must have low resistance. As the drain side, the current required to the body charge pump varies depending on the programming pattern and algorythm status: also in this case, a voltage modulation occurs at the pass transistor output (net B in Fig. 16.18), which leads to a body voltage modulation during program. It is likely that many cells are connected in the first phase of the program algorythm: the voltage drop across R''_{par} makes the biasing of B less negative, leading to less effective program pulses, even if the drain voltage is nominal. The program pulses applied when few cells have to be programmed (i.e. at the end of the algorythm) are more effective because of the smaller voltage drop across R''_{par}: the absolute value of the voltage in B is greater. This phenomenon is analogous to the drain modulation, and has to be minimized.

To solve this problem, it is possible to regulate the body voltage locally to each sector by mean of a structure that is dual to the one of the drain voltage, as shown in Fig. 16.19.

This solution removes the modulation due to the resistive path "outside" the sector, but it is ineffective with respect to the modulation source "inside" the sector, which in program and read/verify is another time a function of the programming/programmed pattern. As shown in Fig. 16.18, the iP-well is contacted only at the sector boundary and the resistive paths inside the sector lead to the body voltage modulation: each cell under program has its bulk biased with a different voltage. As for the drain voltage, when the algorythm switches from many cells to few cells, the current injection into the bulk decreases, leading to a lower body voltage drop: the program pulses applied to the last cells are more effective.

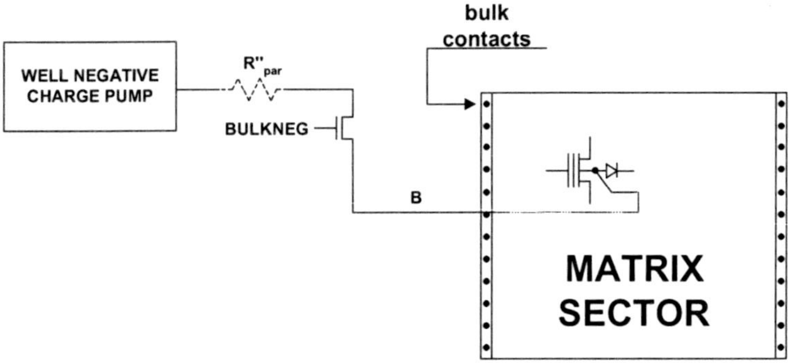

Fig. 16.18. How the negative voltage is connected to the sector bulk

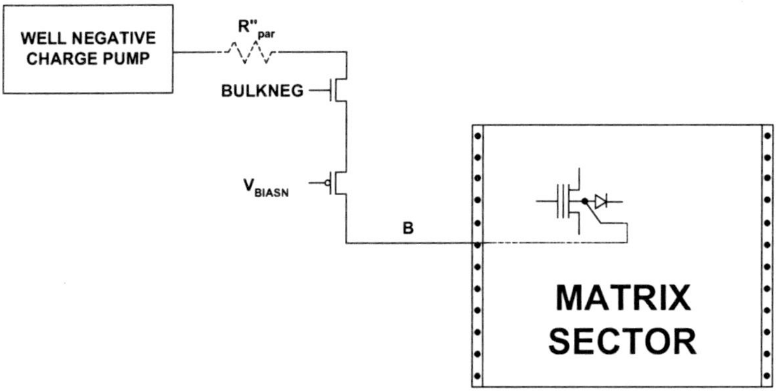

Fig. 16.19. Local bulk biasing (by sector)

Figure 16.20 shows the effect of the substrate resistivity in a 0.5 Mcells sector. The x and y axes represent the 512 rows and 1024 columns of a sector, respectively; the z axes shows the voltage drop across the bulk as a consequence of a localized current injection. The bulk is connected to the negative voltage (node B in Fig. 16.18) only on two sides of the sector (in the column direction), while the others are contactless.

Figure 16.20 a represents the situation where 64 cells are programmed in parallel at half sector: the current of 64 cells is injected into the bulk; in Fig. 16.20 b only one cell injects current. It is obvious that in these two cases the substrate voltage is quite different.

The only way to contain this problem consists in contacting the cell body not only at the boundary, but also inside the memory sector, with the counterpart of higher area occupation.

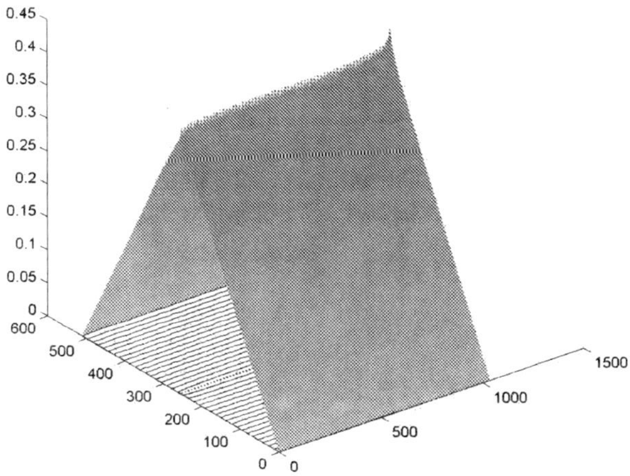

a)

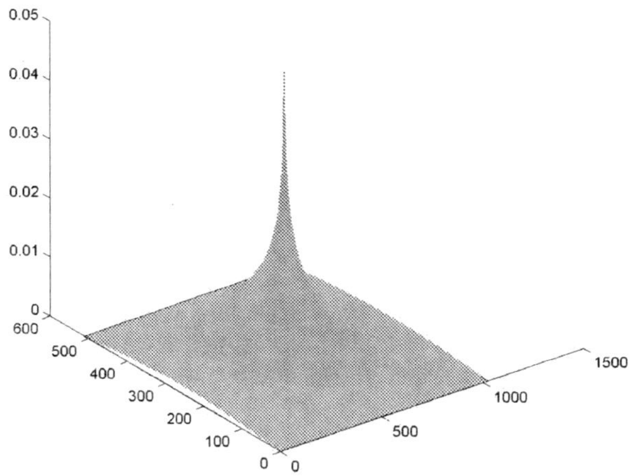

b)

Fig. 16.20. Body voltage modulation in a memory sector when 64 cells in parallel are programmed (**a**) or just one (**b**)

16.6.3 Source Voltage Modulation

Also the source voltage has to be controlled carefully during both program and read. As seen in Sect.16.3, the source terminal is connected to GND during program an read by mean of n-channel pass transistors, which are driven by suitable signals (Fig. 16.21 shows four transistors, placed at each sector's vertex, being driven by V_{SRCCON}). In fact, the source terminal cannot be connected permanently to

GND because it has to be biased with positive voltage during electrical erase. Inside the sector, the source diffusion areas are connected by mean of vertical metal strips, placed every *n* bit lines; the number *n* is chosen depending on area constraints, read/program parallelism and tolerable modulation inside the sector. If we neglect the internal modulation effect (which is not neglectable, of course, but can be solved only with thicker contacts), a modulation in the source voltage value, due the parasitic resistance of the lines and pass transistors, reflects in a different not only on V_{DS} and V_{BS}, but also on V_{GS}, that is on the overdrive.

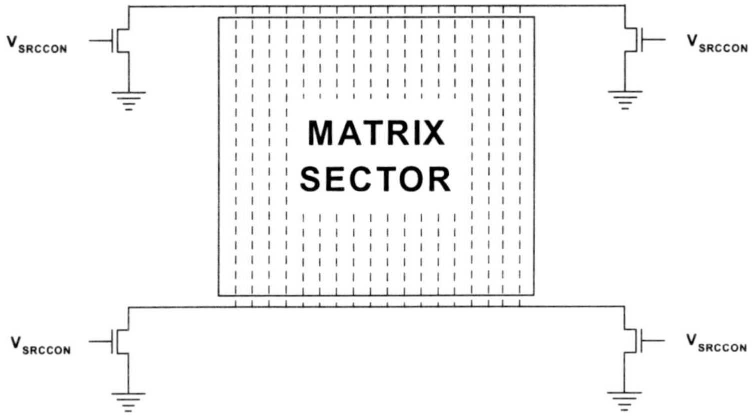

Fig. 16.21. How the source is usually connected to GND

As a consequence, the source voltage modulation influences not only program pulse effectiveness, but also verify and read mode. What may happen as a consequence of source modulation is shown in Fig. 16.22, which represents a sequence of qualitative voltage-current characteristics during various steps of the program algorythm in actual Flash cells.

STEP 1: let us suppose to start from erased cells. For a cell to be verified as erased, it must have a current higher than that of the EV reference (erase verify) when it is biased with the erase verify voltage, which is equal to the read voltage V_{READ}; the high current injection into the source node determines the folding of the high-current characteristic. Since during erase verify a current is verified, the cells V_T is slightly lower than the EV one due to this folding. Characteristic (1) represents both the *i*-th and the other cells;

STEP 2: let us suppose to program all the cells at the highest threshold distributions ("00" distribution, beyond the reference threshold PV3), except the *i*-th, which is programmed at the second distribution ("10" distribution, beyond PV1). Supposing that the program speed is almost the same for every cell, characteristic (2) may represent all the cells, including the *i*-th. The high-current folding is still present, even if it is getting lower. The *i*-th cell is verified as programmed to the desired distribution, thus its program load is permanently disconnected, while the other cells keep on programming;

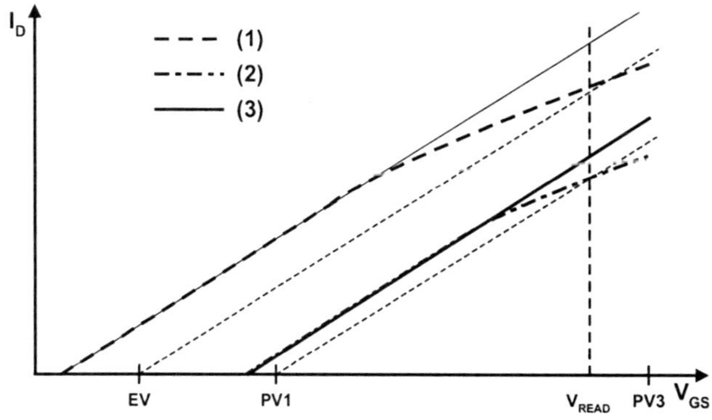

Fig. 16.22. Source voltage modulation effect

STEP 3: after all the cells reached the selected distribution, they are read. The current injection into the source terminal is almost zero, since all the cells except the *i*-th have V_T higher than the read voltage; for this reason, the *i*-th cell characteristic does not show high-current gain degradation, as shown by characteristic (3). If a supplemental margin for the source modulation has not been taken, the *i*-th cell may be read or verified in the wrong way.

This qualitative example shows that it is necessary to design a system to eliminate the effect of the source modulation, or at least to make it constant, regardeless of the charge status of the other cells it is verified/read/programmed with[6].

To regulate the source voltage, one can compare the current flowing into the source node with a suitable reference current: if the source current is less than the reference one, a voltage controlled current generator injects in the matrix source until the matrix and reference currents are equal. The reference current is set equal to the maximum current that may flow into the source node, i.e. to the current that produces the maximum modulation with respect to GND level. This solution therefore requires two paths to generate matrix voltage and reference voltage, a comparator for this two voltages and a voltage controlled current source to stabilize the source voltage.

The transistor that in Fig. 16.21 connects the matrix source to ground is splitted in the parallel of M4 e M5 (Fig. 16.23), so as to have:

$$(W/L)_{M4} = (n-1)(W/L)_{M5} \tag{16.4}$$

where n represents the program parallelism.

A suitable regulator may generate the "high" level of the selection signal, V_{SRCCON}; this solution nullifies the modulation due to VDD level variations in the specs range. Alternatively, V_{SRCCON} may be a high-voltage signal, which further reduces the parasitic resistance of the source pass transistors.

[6] This is mostly important when designing multilevel Flash memories.

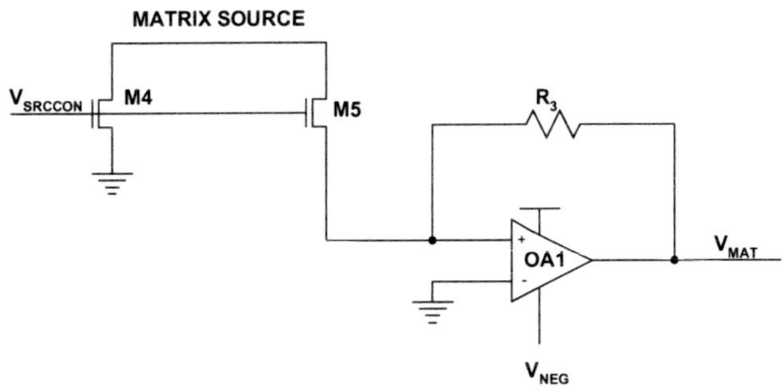

Fig. 16.23. Generation of V_{MAT}, proportional to the current flowing into the matrix source

Thanks to the virtual ground obtained by mean of OA1, M5 and M4 are equally biased, and a current $I_{DM4}/(n\text{-}1) = (V_{DSM5})/R_{eqM5}$ flows in M5. Since OA1 is connected in inverting configuration, V_{MAT} is equal to:

$$V_{MAT} = \frac{R_3 \cdot (-V_{DSM5})}{R_{eqM5}} \tag{16.5}$$

i.e. V_{MAT} is proportional to a fraction of the current flowing into the source node.

Figure 16.24 shows the circuit that generates the reference voltage. The suitable reference current is generate by $I_{REFSOURCE}$, then mirrored on the path where transistors M2 and M3 are.

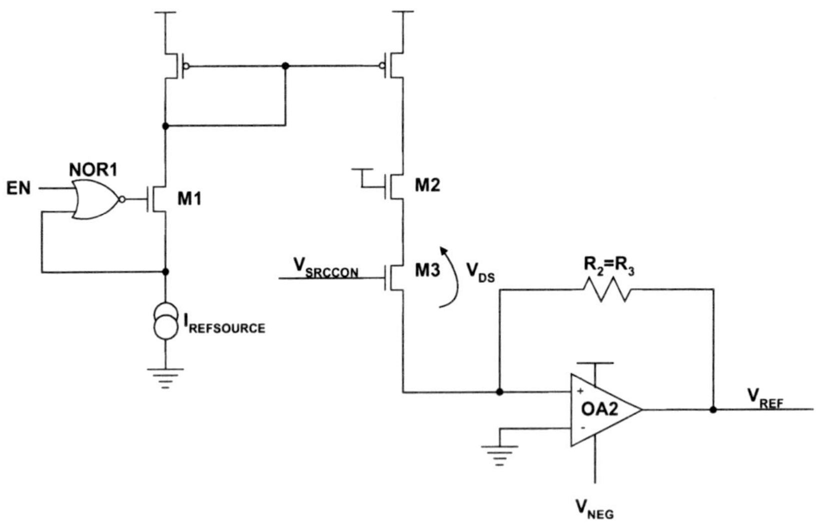

Fig. 16.24. Reference voltage generation by current mirroring

Every device depicted in Fig. 16.24 is matched with its corresponding in Fig.16.23: M3 and M5 have the same aspect ratio, R_2 has the same value of R_3 and the operational amplifiers OA2 and OA1 are equal as well. As it happens on the matrix side, the voltage V_{REF} at the operational amplifier's output is:

$$V_{REF} = \frac{R_2 \cdot (-V_{DSM3})}{R_{eqM3}} \tag{16.6}$$

M1 and NOR1 are the replicas of the drain regulator used in read to avoid cell soft writing (refer to Chap. 12).

Figure 16.25 shows the comparison between V_{MAT} and V_{REF}, performed by the comparator COMP, which is supplied by VDD and a suitable negative voltage, V_{NEG}. If $V_{MAT} < V_{REF}$, i.e. if the current flowing in the source node, $I_{DETECTED}$, is less than those that would flow if all the cells would be connected to the program load, I_{LOAD}, the comparator's output forces the p-channel transistor MFORCE to source to MATRIX SOURCE a current I_{FORCE} equal to:

$$I_{FORCE} = I_{LOAD} - I_{DETECTED} \tag{16.7}$$

thus keeping constant the voltage value at MATRIX SOURCE.

This method can be effectly used both in verify (program and erase), where the current injected into the source net is a function of the program pattern and of the point of the program/erase algorythm, and in read, where the injected current is a function of the cells status (i.e., the previously programmed pattern).

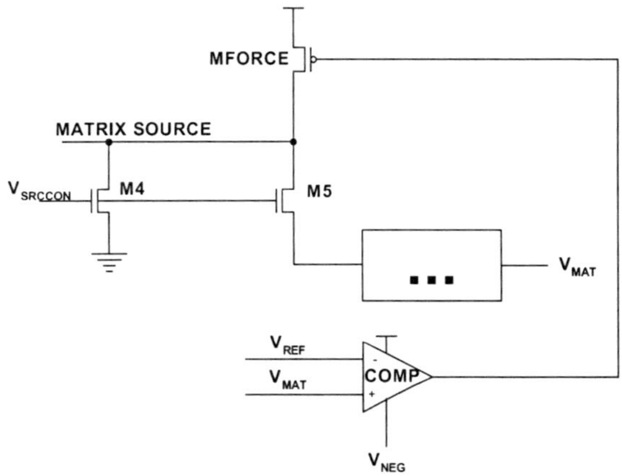

Fig. 16.25. V_{MAT}-V_{REF} comparison and matrix source injection

In read and verify, the maximum current flowing into the source is equal to that of the erased distribution, multiplied by the number of cells read or verified in parallel: to generate the reference voltage, a current equal to that of an erased cell can be used (EV reference cell).

Bibliography

M. Alam, B. Weir, and P. Silverman, "A future of function or failure?", IEEE Circuits and Device Magazine, vol. 18, pp. 2-48, (Mar. 2002).

J.C. Chen, T.H. Kuo, L.E. Cleveland, C.K. Chung, N. Leong, Y.K. Kim, T. Akaogi, and Y. Kasa, "A 2.7V only 8Mb×16 NOR Flash memory", in 1996 Symp. VLSI Circuits Dig. Tech. Pap., pp. 172-173, (Jun. 1996).

M. Dallabora et al., A 20MB/s date rate 2.5V Flash memory with current controlled field erasing for 1M cycle endurance, ISSCC, pp. 396, (1997).

S. Haddad, C. Chang, A. Wang, J. Bustillo, J. Lien, T. Montalvo, and M. Van Buskirk, "An investigation of erase-mode dependent hole trapping in flash EEPROM memory cell", IEEE Electron Device Letters, vol. EDL-11, pp. 514-516, (Nov. 1990).

L.G. Heller and W.R. Griffin, "Cascode voltage switch logic: a differential CMOS logic family", in 1984 IEEE Int. Solid-State Circuits Conf. Dig. Tech. Pap., pp. 16-17, (Feb. 1984).

G.J. Hemink, T. Tanaka, T. Endoh, S. Aritome, and R. Shirota, "Fast and accurate programming method for multi-level NAND EEPROMs", in 1995 Symp. VLSI Technology Dig. Tech. Papers., pp. 129-130, (June 1995).

S. Kenney, R. Bez, D. Cantarelli, F. Piccinini, A. Mathewson, and C. Lombardi, "Complete transient simulation of Flash EEPROM devices", IEEE Trans. Electron Devices, vol. ED-39, pp. 2750–2757, (Dec. 1992).

O. Khouri, I. Motta, R. Micheloni, G. Torelli, "Voltage regulator for low-consumption circuits", U.S. Patent No. 6,559,627, (May 6, 2003).

V.N. Kynett, M.L. Fandrich, J. Anderson, P. Dix, O. Jungroth, J.A. Kreifels, R.A. Lodenquai, B. Vajdic, S. Wells, M.D. Winston, and L. Yang, "A 90-ns one-million erase/program cycle 1-Mbit Flash memory", IEEE J. Solid-State Circuits, vol. SC-24, pp. 1259-1264, (Oct. 1989).

I. Motta, et al., "High voltage management in single-supply CHE NOR-TYPE Flash Memories", IEEE Proceeding of the, Vol. 91, No. 4, pp. 554-568, (Apr. 2003).

R. Micheloni, I. Motta, O. Khouri, and G. Torelli, "Stand-by low-power architecture in a 3-V only 2-bit/cell 64-Mbit Flash memory", in Proc. 8th IEEE Int. Conf. Electronics, Circuits, and Systems, vol. II, pp. 929–932, (Sept. 2001).

R. Micheloni, M. Zammattio, G. Campardo, "Nonvolatile memory device with hierarchical sector decoding", US Patent N. 6,456,530, (Sept. 24, 2002).

R. Micheloni, M. Zammattio, G. Campardo, O. Khouri, G. Torelli, "Hierarchical sector biasing organization for Flash memories", in Records 2000 IEEE Int. Workshop on Memory Technology, Design and Testing, pp. 29–33, (Aug. 2000).

D. Mills, M. Bauer, A. Bashir, R. Fackenthal, K. Frary, T. Gullard, C. Haid, J. Javanifard, P. Kwong, D. Leak, S. Pudar, M. Rashid, R. Rozman, S. Sambandan, S. Sweha, and J. Tsang, "A 3.3V 50MHz synchronous 16Mb flash memory", in 1995 IEEE Int. Solid-State Circuits Conf. Dig. Tech. Pap., pp. 120–121, (Feb. 1995).

A. Modelli, A. Manstretta, and G. Torelli., "Basic feasibility constraints for multilevel CHE-Programmed Flash memories", IEEE Trans. Electron Devices, vol. ED-48, pp. 2032–2041, (Sept. 2001).

K. Shimohigashi and K. Seichi, "Low-voltage ULSI design", IEEE J. Solid-State Circuits, vol. 28, pp. 408-413, (Apr. 1993).

K. Takeuchi, T. Tanaka, and H. Nakamura, "A double-level-Vth select gate array architecture for multilevel NAND Flash memories", IEEE J. Solid-State Circuits, vol. SC-31, pp. 602–609, (Apr. 1996).

T. Tanaka, T. Tanzawa, and K. Takekuchi, "A 3.4-Mbyte/sec programming 3-level NAND Flash memory saving 40% die size per bit", in 1997 Symp. VLSI Circuits Dig. Tech. Papers., pp. 65-66, (Jun. 1997).

G. Torelli and P. Lupi, "An improved method for programming a word-erasable EEPROM", Alta Frequenza, Vol. LII, pp. 487–494, (Nov./Dec. 1983).

M. Zammattio, I. Motta, R. Micheloni, C. Golla, "Low consumption voltage boost device", U.S. Patent No. 6.437.636, (Aug. 20, 2002).

17 Program and Erase Controller

Program and erase algorithms of a flash memory are very complex and the corresponding sequence of operations is critical. Modern flash memories integrate a controller that performs such operations autonomously. Such a controller can be realized through a Finite State Machine (FSM) or a microcontroller. In this chapter we will examine both the solutions, highlighting specifications, advantages, drawbacks and possible tradeoffs.

17.1 FSM Controller

The simplest way to describe the flash memory controller is a FSM. The general structure of an FSM of a flash memory is reported in Fig. 17.1, in which input and output signals that determine the boundary of the control logic are shown.

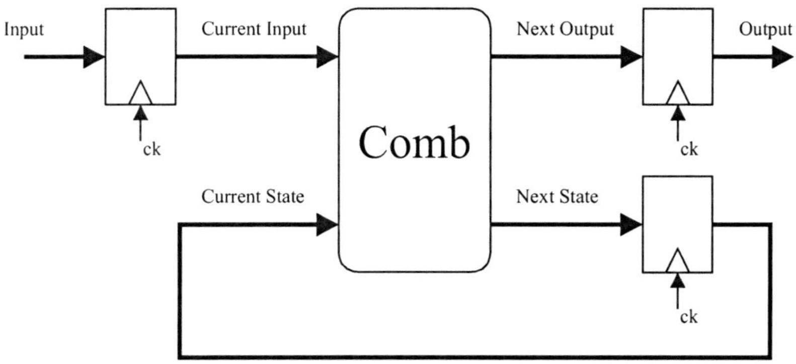

INPUT:
Command Interface Bus
Timer Bus
Program Load Status
Compare Logic Status
...

OUTPUT:
Pump and Regulator Controls
Row and Column decoder Controls
Sense Amplifier and Reference Controls
Program Load and Compare Logic Controls
...

Fig. 17.1. Generic representation of an FSM that controls the flash memory

All the input signals must be synchronized either directly or indirectly with the FSM clock, and it is good practice to synchronize also the outputs. The combinatory logic (Comb) provides the information related to the subsequent status of the FSM and outputs as a function of the present FSM and input status.

Once inputs and outputs have been defined, the behavior of the FSM can be defined completely by specifying the functionality of the combinatory logic (Comb). Due to the high number of inputs and outputs of this block, a State Diagram Representation is usually adopted.

From the implementation point of view, the flash controller described as a FSM can be realized by means of STD cells or a PLA.

17.2 STD Cell Implementation of the FSM

The logic functionalities of modern flash memories have been developed by means of STD cells and, therefore, also the FSM can be realized following the same design flow. The flash memory controller has some criticalities, in particular the algorithm definition is constantly modified during the design phase and even during debug some modifications may be necessary. Therefore, we can summarize the advantages and drawbacks of the STD cell implementation of the flash memory FSM as follows.

Advantages:

- No specific circuits are developed;
- All outputs can be updated at any time as a function of all the input signals;
- No constraints on the number of inputs and outputs;
- A direct representation of the FSM can be obtained;
- Automatic layout tools can be exploited;
- Area occupation is comparable with other solutions.

Drawbacks:

- Any modification requires re-running the optimization flow, P&R (Place & Route) and constraint verification (clock);
- After releasing the device, the STD cells cannot be modified unless the entire device is redesigned, and the routing modifications, when allowed, involve the entire back-end.

Possible actions:

- Definition of a static area to place the FSM, so as to avoid that possible modifications involve external circuitry already verified;
- Realization of the FSM using only simple STD cells that can easily be reused by the synthesis tools;
- Insertion of dummy cells that can be used for possible subsequent modifications;
- Avoiding overcrowded routing in order to carry out modifications a posteriori, without violating the constraints.

17.3 PLA Implementation of the FSM

An alternative method to implement a FSM is through a PLA. In a PLA, the combinatory block is reduced to only two levels (AND-OR) and can be implemented by means of very regular and compact structures.

A PLA is made up of the blocks sketched in Fig. 17.2.

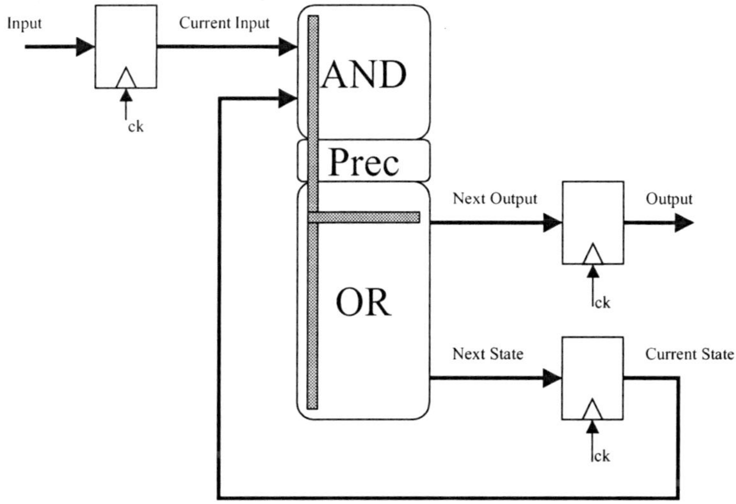

Fig. 17.2. Block diagram of a PLA

In Fig. 17.3, the section sketched in form of elementary block in the previous figure is described in detail. The AND plane is realized through a set of structures, each forming a dynamic NAND. Each column of the AND plane has the output node, B, that drives an inverter whose output, node A1, drives a distributed dynamic NOR that realizes a row of the OR plane. Signals A2, A3, are generated by a structure similar to the one that drives node A1. The event sequence is the following: the precharge and enable signals are low, therefore node B is high independently of the input values of the AND plane. The inverter output, A1, is low as well as A2 and A3 and all the other similar nodes, while node B is high. Once the precharge phase has finished, the evaluation phase starts and output B will go low only if the NAND has all the NMOS transistors turned on. Similarly, node C will go low if one of the NMOS transistors is turned on.

The precharge and, subsequently, the evaluation phases are repeated at each clock pulse. Capacitive problems due to node C for the OR plane, and to node B for the AND plane, must be taken into account when realizing the AND and OR planes. The multiplicity of the n-channel transistors connected to node C causes a large parasitic capacitance to be charged. Thus, the layout must be drawn carefully so as to minimize such a capacitance that tends to penalize the execution time of the precharge and discharge phase in the case a single transistor is turned on.

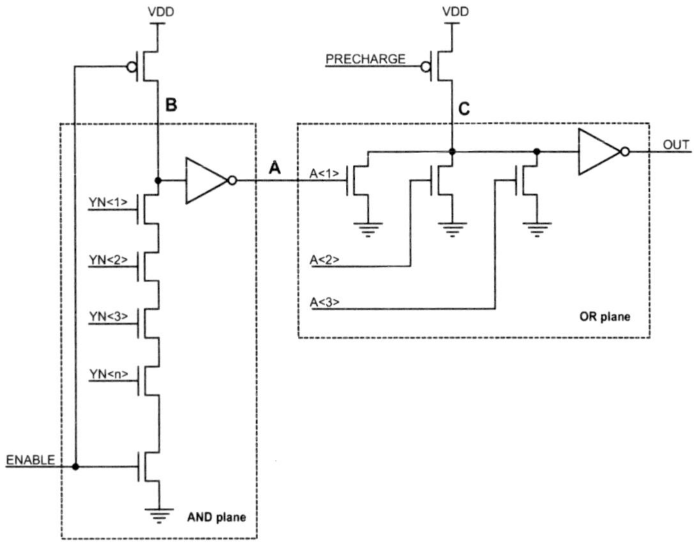

Fig. 17.3. Elementary structure of a PLA with the AND-OR planes

There is also a more subtle possible malfunctioning of the AND plane. In Fig. 17.4 a column of the AND plane is shown with the parasitic capacitance of the various internal nodes. The precharge phase pulls the evaluation signal low and, hence, node B is high. Let's suppose that an instant before the evaluation phase starts, the Y1-Yn input signals toggle but only Y1, Y2, and Y3 toggle from the low to the high state, so that the evaluation phase does not need to pull node B low. The n-channel turned on by the Y1 signal discharges a part of the parasitic capacitance of node B, Cp, charging the Cp1 parasitic capacitance, thus diminishing the voltage of B. If Y1, Y2, and Y3 were on, the charging of Cp would charge Cp1, Cp2, and Cp3. Such a charge re-distribution diminishes the voltage value of B and, although the evaluation phase does not pull node B to ground, the downstream inverter might toggle because it recognizes a low logic level in B due to the capacitive sharing only. Therefore, it is very important that the AND plane is designed so that the ratios between parasitic capacitances are appropriate to the correct functioning, and activate the evaluation signal only as a consequence of stable input signals and after the suitable setup time.

Advantages:

- The temporization of the PLA is determined in the design phase and is independent of its content;
- All the outputs can be updated at any time as a function of all the inputs;
- The PLA content can be obtained directly from the state diagram;
- The PLA content can be modified easily, usually by only one level of metallization.

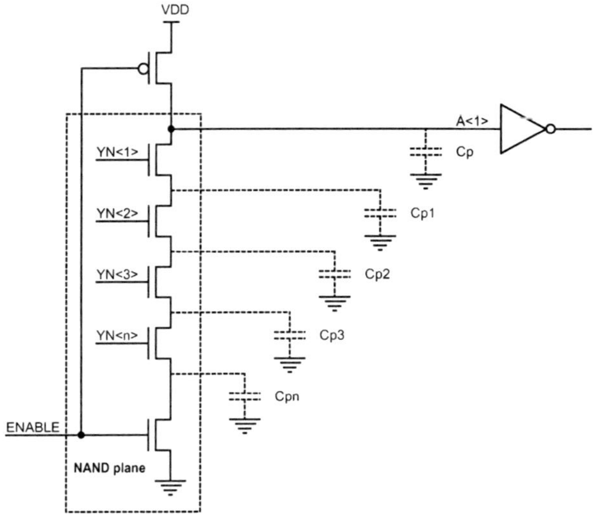

Fig. 17.4. Charge sharing problem

Shortcomings:

- The size of the PLA increases linearly as the number of inputs and outputs. As the size increases it becomes more difficult to respect the time constraints.

Solutions:

- Encoding and decoding strategies are usually adopted to limit the number of inputs and outputs and, therefore, the size of the PLA. The structure becomes less flexible, though;
- Same additional inputs, outputs and dummy PLA rows are necessary to carry out modifications;
- The structure of a PLA organized with a NAND and a NOR plane can be easily transformed into a faster structure by using a NOR organization for both planes.

17.4 Microcontroller

The low flexibility of the controller realized with either STD cells or PLA can be overcome through the implementation of a microcontroller. The controller of a flash memory executes one single program and, therefore, its structure can be reduced to the essential. In its simplest form, a microcontroller can be obtained from the FSM structure by simply dividing the combinatory logic, COMB, into two parts, the ROM and the decoder, as represented in Fig. 17.5.

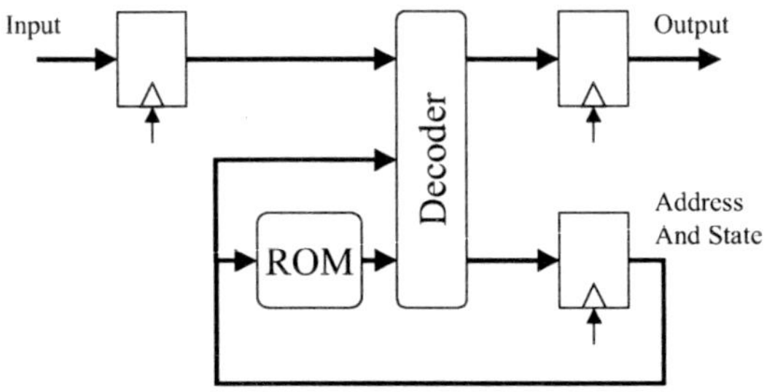

Fig. 17.5. In its most embryonic form, the structure of a microcontroller can be regarded as the evolution of the FSM structure in which the combinatory logic is divided into two parts, ROM and decoder

The decoder structure is able to activate the reading of a word from the ROM at each read cycle. The decoder determines the status of the output and the subsequent ROM address as a function of the word read and the state of the controller input. A word of the ROM, which has a limited size, does not permit specifying the status of all the controller outputs as a function of all the possible inputs. It is necessary to define a decoding for the ROM word that indicates the function to execute and the set of inputs and outputs involved in the operation. Such a code represents the microcontroller instruction set.

The impossibility of updating the outputs at each cycle as a function of all inputs is the main limitation that can be encountered when substituting the microcontroller for the FSM.

In order to determine whether the FSM of a flash memory can effectively be substituted for a microcontroller and, moreover, in order to have some useful indications about the set of functions and instructions to implement, it is important to examine the algorithm executed by the FSM and the result of its elaboration. From a first simple analysis, it is possible to obtain an indication about the number of outputs that the processor has to control during each cycle, as sketched in Fig. 17.6.

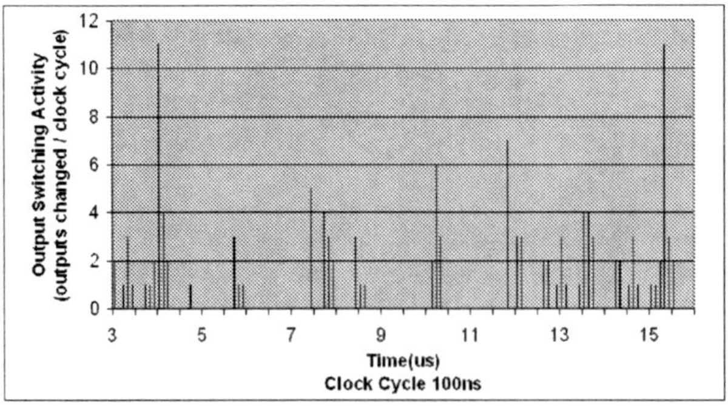

Fig. 17.6. Representation of the switching activity of the FSM during the execution of a part of the program algorithm

The average number of outputs that have to be modified simultaneously is very small compared to the hundreds of outputs of the flash FSM. The most complex transitions are rare and it is therefore reasonable, by using a microprocessor, to shift some toggles in less crowded positions so that the possible use of more cycles to carry out the transitions does not modify the overall system performance.

Let's now move on to define the architecture and the instruction set of the microcontroller. A simple representation of the microcontroller structure is reported in Fig. 17.7.

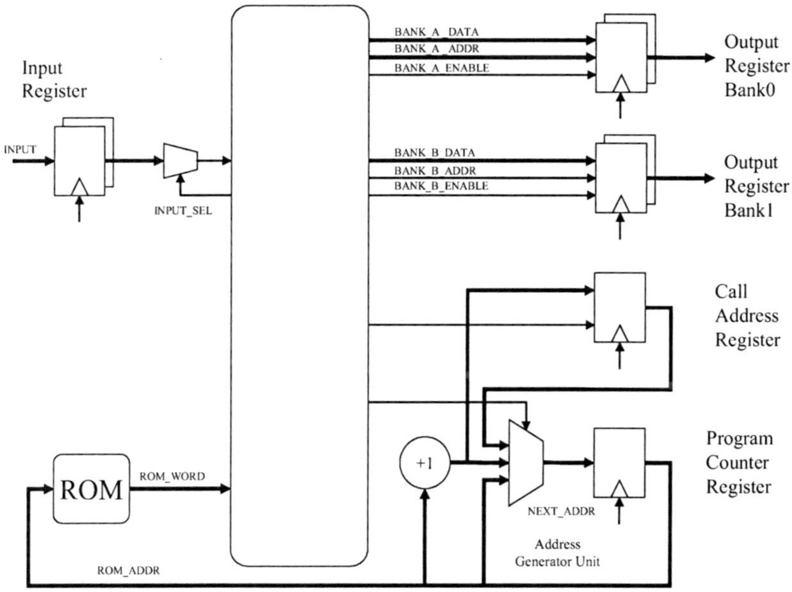

Fig. 17.7. Representation of a simple microcontroller of a flash memory

The set of instructions that are useful for the memory microcontroller can be grouped into the following categories:

- Instructions to force the outputs. As the ability to configure the outputs is a basic operation for such a microcontroller, a double I/O port is used. I/Os are grouped into four outputs. Each output group requires to be selected for configuration (BANK_A_ADDR, BANK_A_EN for port A) and it must receive the output configuration (BANK_A_DATA).
- Instructions of absolute or conditioned jump. As the useful cycles in such a microcontroller are only those that operate on the outputs, it is necessary to guarantee that the controller is able to drive at least two of the I/O ports also during verify and jump operations.

- Instructions to jump to and return from subroutines. The presence in the algorithms of cycles and often repeated routines makes the call to subroutines more efficient. A rough form of subroutine call is implemented by which it is possible to execute a single call and, thus, a single return address is stored in the Call Address Register (CAR). Also in the case of call and return instructions it is necessary to drive at least one output.
- Halt instructions. As during the algorithm execution the microcontroller has often to wait for an event, it is useful to have an instruction to freeze the functioning.

Without specifying the complete format of the instruction set, let's examine how the proposed structure can execute the required operations.

The heart of the system is the ROM that stores the program in form of sequence of instructions.

At each clock edge (see Fig. 17.8), a new address (ROM_ADDR) is applied to the ROM, and a new word (WORD_ROM) is read from the ROM and executed. In order to allow the algorithm (and therefore the operations) to proceed, a new address (NEXT_ADDR) must be generated by the Address Generation Unit (AGU), stored in the Program Counter Register (PCR) and applied to the ROM at the subsequent clock edge. For the instructions that do not require any jump, the decoder forces the AGU to simply increment the previous address by one unit. In the case of absolute jump, call to subroutines, or conditioned jump instructions, the decoder forces the AGU to provide the PCR with the address specified by the current instruction. For the instructions that require conditioned jumps, the decoder has to store the return address in the CAR, and for the instructions that require a return from subroutines, the decoder has to force the AGU to provide the PCR with the return address previously stored in the CAR. For the instructions that require the evaluation of an input signal, the decoder activates the required input signal selection (SEL_INPUT), which is internally driven (T) to the comparison. The instructions that require that a given state is forced in one of the two output banks, activate the selected port (BANK_A_ADDR), put the data that must be transferred to the output (BANK_A_DATA), and activate the port (BANK_A_EN). The data are transferred to the selected outputs on the subsequent clock edge.

Besides the controller architecture, also the instruction set must be defined. Despite the constraints imposed to the controller structure, the number of possible instructions is very large. The selection of the instruction set must take into account the minimum set that guarantees the best performances and the minimum area of decoder and ROM. A very simple set of instructions allows obtaining a reduced ROM word size and a simple decoding even though a large number of instructions is necessary to execute and store. A large number of instructions allows obtaining a compressed code with, on the other hand, a larger word and a higher complexity of the decoding. Some attempts and a little of experience help find the best solution. Before examining the complete set of instructions, let's analyze a single instruction and its possible representations (Fig. 17.9 and Fig. 17.10).

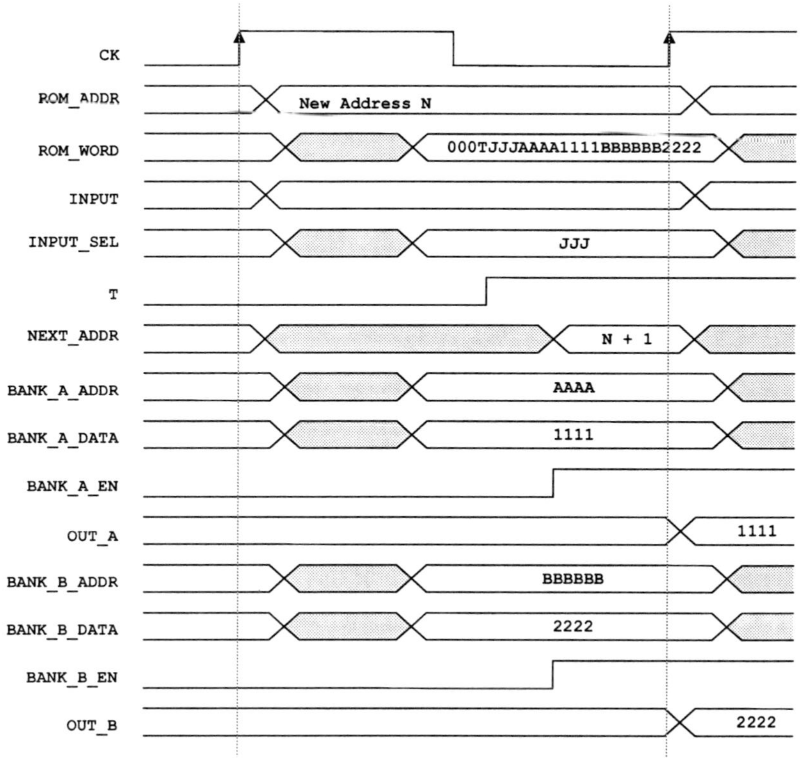

Fig. 17.8. Representation of the timings of a microprocessor instruction

0	1	0	1	1	0	0	0	0	0	1	1	0	1	1	1	0	1	0	0	0	0	0	0	1
INST. CODE			T	J			A				1				B						2			
			OPERANDS																					

Fig. 17.9. Binary representation of a test instruction and double output command

1	0	0	1	1	0	0	0	0	0	1	1	0	1	1	1	0	1	0	0	0	0	0	0	1
INST. CODE			T	J			A				1				ADDR									
			OPERANDS																					

Fig. 17.10. Binary representation of a test instruction, jump and single output command

An instruction is basically a word written in the ROM as a sequence of bits and, therefore, has a binary representation. A part of the word bits univocally identify the instruction (Instruction Code). The remaining part of the instruction contains the fields that specify the operands. Different instructions can have different number of fields for the operands and also the size of the instruction code can vary as long as the non-ambiguity of the instructions is guaranteed. In the two examined instructions, field T specifies the condition that the selected input must fulfill. Field J is made of three bits and specifies which of the eight possible inputs must be selected. Field A selects one output register out of 16 and field "1" contains the configuration that must be applied to the outputs. Similarly, field B selects one output register out of 64 and field "2" contains the configuration that must be applied to the outputs. In the jump instruction, field "addr" contains the absolute ROM address in the case of jump. The decoder executes the first instruction stored in the \$000h address as the first action after exiting the reset state. The algorithm is completely specified by the ROM content that, in the binary format, has the representation of Fig. 17.11.

ROM ADDR	WORD
\$000	0100010001100001010010010
\$001	00100000110000110000001100
\$002	...

Fig. 17.11. Binary representation of the algorithm stored in the ROM

The binary representation, although adequate to describe the algorithm, is scarcely readable to develop and maintain a program. In order to ease the program development, a symbolic representation of the instructions is usually adopted (assembler) that can be understood easily and converted into the corresponding binary representation. Since the program size is usually quite limited and each operation must be controlled precisely, no high level compiler is used.

```
IF (J = T) THEN OUT_A = OP_1 ELSE OUT_B = OP_2;
```

Representation of the instruction of Fig. 17.9 in meta-language.

```
IF (J = T) THEN OUT_A = OP_1 ELSE GOTO aaaa;
```

Representation of the instruction of Fig. 17.10 in meta-language.

A pre-processing program, together with a configuration file, allows assigning an explicative name to fields J, OUT_A, OUT_B, etc. The program that describes the algorithm has therefore the following form:

```
PGVER:IF (ZERO_FL = 0) THEN (SENSEON = 1, SMVEREQYEN = 1)
      ELSE GOTO $INCPGP;
      IF (1 = 1) THEN (SENSEON = 1); (RST_TIME = 1);
      .........
```

BINARY RAPPRESNTATION	META LANGUAGE RAPPRESNTATION	FUNCTION

000TJJJAAAA1111BBBBBB2222	IF (J = T) THEN A = 1, B = 2;	IF (J = T) THEN
		A = 1; B = 2;
		END IF;

001TJJJAAAA1111BBBBBB2222	IF (J = T) THEN A = 1; B = 2;	IF (J = T) THEN
		A = 1;
		END IF;
		B = 2;

010TJJJAAAA1111BBBBBB2222	IF (J = T) THEN A = 1 ELSE B = 2;	IF (J = T) THEN
		A = 1;
		ELSE
		B = 2;
		END IF;

011TJJJAAAA1111aaaaaaaaaa	IF (J = T) THEN GOTO addr, A = 1;	IF (J = T) THEN
		A = 1;
		GOTO addr;
		END IF;

100TJJJAAAA1111aaaaaaaaaa	IF (J = T) THEN GOTO addr ELSE A = 1;	IF (J = T) THEN
		GOTO addr;
		ELSE
		A = 1;
		END IF;

101T0JJAAAA1111aaaaaaaaaa	IF (J = T) THEN GOTO addr; A = 1;	A = 1;
		IF (J = T) THEN
		GOTO addr;
		END IF;

101T1JJAAAA1111aaaaaaaaaa	IF (J = T) THEN CALL addr, A = 1;	IF (J = T) THEN
		A = 1;
		CALL addr;
		END IF;

110T0JJAAAA1111aaaaaaaaaa	IF (J = T) THEN CALL addr ELSE B = 2;	IF (J = T) THEN
		CALL addr;
		ELSE
		B = 2;
		END IF;

110T1JJAAAA1111BBBBBB2222	IF (J = T) THEN RET, A = 1 ELSE B = 2;	IF (J = T) THEN
		A = 1;
		RET;
		ELSE
		B = 2;
		END IF;

111T0JJAAAA1111BBBBBB2222	IF (J = T) THEN RET, A = 1, B = 2;	IF (J = T) THEN
		A = 1, B = 2;
		RETURN;
		END IF;

111T1JJAAAA1111BBBBBB2222	IF (J = T) THEN HALT ELSE A = 1, B = 2;	IF (J = T) THEN
		HALT
		ELSE
		A = 1, B = 2;
		END IF;

Fig. 17.12. Representation of a possible instruction set of the microcontroller of a flash memory. All the bits of the operands have been reported with the corresponding character associated in the binary representation

Among the various instructions that can be implemented, those that guarantee better performances and code reduction have been selected (Fig. 17.12).

The selected instruction set reflects the necessity of executing as many fast test operations and output configurations as possible by this very special controller. Each instruction always contains the possibility of executing test and configuration of an output bank. The HALT instruction allows suspending the functioning of most of the controller, permitting, at the same time, reduction in device consumption.

The controller structure examined so far requires that the ROM acts as a sequential circuit and that the read time of the ROM, summed to the decoding time of the logic, is less than the cycle time. If the complexity of the operands demands one or more levels of pipeline, the decoding structure becomes even more complicated. The analysis of such structures is beyond the purpose of this brief section.

Advantages:
- The flexibility in defining the system architecture and the algorithms;
- The modification of the ROM requires modifying only one photo-mask.

Shortcomings:
- Development of the core and all the CAD tools to generate the ROM;
- Impossibility of controlling all the outputs as a function of all the inputs.

Bibliography

C. Clare, Designing logic system using state machines, Mc Graw Hill, (1973).

C.R. Clare, Designing logic system using state machine, McGraw Hill Book Company, 1973. [2] T. Nakayama, A 60ns 16Mb Flash EEPROM with Program and Erase Sequence Controller, ISSCC91, paper FA 16.1, pag. 260-261.

R. Sasagawa, I. Fukushi, M. Hamaminato, and S.Kawashima, "High-speed cascode sensing scheme for 1.0-V contact-programming Mask ROM", in Symp. VLSI Circuits Dig. Tech. Papers, pp. 95-96, 1999.

Jinn-Shyan Wang, Ching-Rong Chang, Chingwei Yeh, "Analysis and design of high-speed and low-power CMOS PLAs", IEEE Journal of Solid-State Circuits, vol. 36, pp. 1250-1262, (August 2001).

Byung-Do Yang, Lee-Sup Kim, "A low-power ROM using charge recycling and charge sharing techniques", IEEE Journal of Solid-State Circuits, vol. 38, pp. 641-653, April 2003

18 Redundancy and Error Correction Codes

Redundancy is one of the most complex items related to Flash memories, and it is also one of the less known, since the chip maker does not advertise its use, lest he should admit that the quality of the technological process is poor. As we will see, redundancy is activated in factory, before the device starts working on the field. Error correction systems allow, on the other hand, to correct the errors that may occur during the lifetime of the memory, at least to a certain extent.

18.1 Redundancy

One issue related to the fabrication of a memory device embedding millions of cells is its reproducibility, because of the defectivity that may affect every single element that composes the device itself.

The probability that one or more memory cells are corrupted, marginal or generally non-operating is not negligible, because fabrication processes are very complex and several parameters call for accurate control. In fact, a Flash process may require more than 300 separate steps: depositions, attacks, etc. In order to be able to fabricate a product on industrial scale, it is mandatory that any memory whose size is bigger than 64 Kbit contains more memory cells than those normally addressable. The additional cells, also known as redundancy cells, are used to "replace" the defective cells, if any, whose presence is detected during the test phases that follow the fabrication of the device.

Let's consider, for instance, a 1 Mbit memory with 8 outputs, composed of 1,024 rows and 1,024 columns, i.e. 1,048,576 cells. Assuming that 8 redundancy columns are present for each output, the total number of cells becomes

$$1{,}024 \cdot \left(1{,}024 + 64\right) = 1{,}114{,}112 \tag{18.1}$$

The addresses required to decode a 1 Mbit memory whose output parallelism is 8 are 17 (131,072 byte), therefore the redundancy cells cannot be decoded using the normal addresses.

In order to overcome this issue, there are non-volatile registers, called UPROM. The name UPROM is an historical heritage that comes from EPROM memories. EPROM memories can be erased by exposing them to UV radiation, which pass through a quartz window that exists on the device package. The problem is that the EPROM cells that compose the non-volatile registers are erased too; therefore the part is no longer able to activate redundancy as required. Thus these cells are protected by a metal layer that prevents the light from reaching them (and their layout is done in such a way to avoid incidental wave guides composed by the metal layers), hence the UPROM acronym (Un-Erasable Programmable Read Only Memory). Flash memories do not require the UV light to be erased, therefore there is no window on the package and there is no risk for the UPROM to lose their contents. In reality the UPROM inside the Flash memories can be electrically erased in order to change their threshold voltage and therefore increase the available current at low voltage; as we will see in the next sections, the current sunk by the Flash cells that compose the UPROM registers directly influences the time required to read them.

The task of the UPROM cells is to store the addresses of the failing bits; these registers are programmed in factory (EWS).

Inside the device, whenever an address is input to the memory, a comparison takes place between the address itself and the content of all the UPROM registers. If the comparison shows a match, and therefore the address is relative to a failing location, normal matrix decoding is disabled while decoding of the redundancy cells is activated. An address transformation takes place, and a new memory location that usually falls outside the address space is accessed (Fig. 18.1).

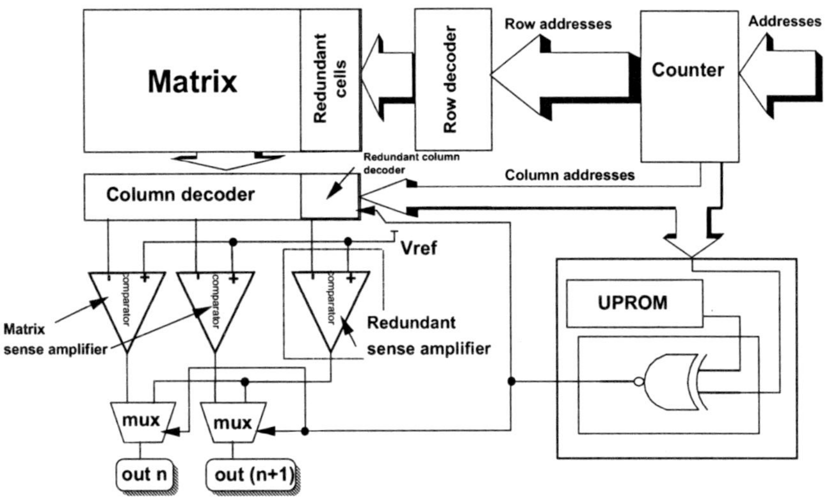

Fig. 18.1. Block diagram for a memory featuring column redundancy. Redundancy matrix is read by a dedicated sense amplifier not to negatively affect access time

Two different redundancy schemes exist: row and column redundancy. Nowadays, for very high-density memories, sector redundancy schemes start to appear. It is obvious that more redundancy means better failure coverage, even if it is important to highlight the fact that redundancy cells have the same probability as the memory cells of being defective.

Row redundancy requires additional rows in the matrix. It is worth recalling that, owing to physical arrangement of the cells, each cell shares its drain contact and source diffusion with other cells. It means that we must always use a pair of redundancy rows even if just a row is defective. Row redundancy also has a remarkable impact on the algorithm for modifying the content of the cells, since it is not possible to forget about a failing row when it has been substituted. Let's assume that a pair of rows has been replaced and that we want to erase the sector they belong to. Since source voltage is applied at the same time to all the cells inside the sectors, failing rows are erased too, and they can become deplete. Of course it is not possible to have depleted cells, otherwise written cells cannot be reliably read. In order to overcome this drawback, whenever a sector is erased it is necessary to extend soft program operation to the replaced rows too.

In case of column redundancy, the problem just mentioned does not exist, since a matrix column is entirely replaced by a redundancy one. In other words, the failing column can be completely neglected. Implementing a column redundancy is therefore usually easier than implementing a row redundancy.

18.2 Redundancy & Read Path

Before showing the selection criteria related to the amount of redundancy to be implemented, i.e. how many columns (or rows) should be added to the matrix, let's see how redundancy fits in the read path. Let's assume that only column redundancy is used.

It is worth recalling at this point the structure of column decoding. As described in Chap. 9, we know that bit line decoding is done using n-channel transistors used as pass transistors; read is performed in parallel on several cells, e.g. eight in case the memory is working by byte. Figure 18.2 shows column decoding for a matrix output composed of 128 columns where, for sake of simplicity, YO<3:0> decoding has been omitted. We can observe that byte composition is not achieved by considering 8 adjacent bit lines: inside each byte, each bit is 127 columns far from another.

Let's start considering the case of sectors organized by column recalling Fig. 7.2. Redundancy columns are inserted inside every single output and they are decoded together with matrix columns. The advantage of this solution lies in the symmetry of the arrangement, that allows using a simplified decoding for the redundancy column. The drawback is that if we want to replace a bit, e.g. on output 1, using the first redundancy column present in the output, all the first redundancy columns for each output are activated as well. Let's assume to have a 1 Mbit, with 16 outputs, organized as 1,024 rows by 1,024 columns: there are 64 columns for each output with 8 redundancy columns: it means 8 additional columns for each output, i.e. 128 additional columns for the whole matrix. The drawbacks of this kind of organization are the considerable amount of added redundancy and its poor efficiency; furthermore both normal and redundancy columns share the same sense amplifier to provide the output data.

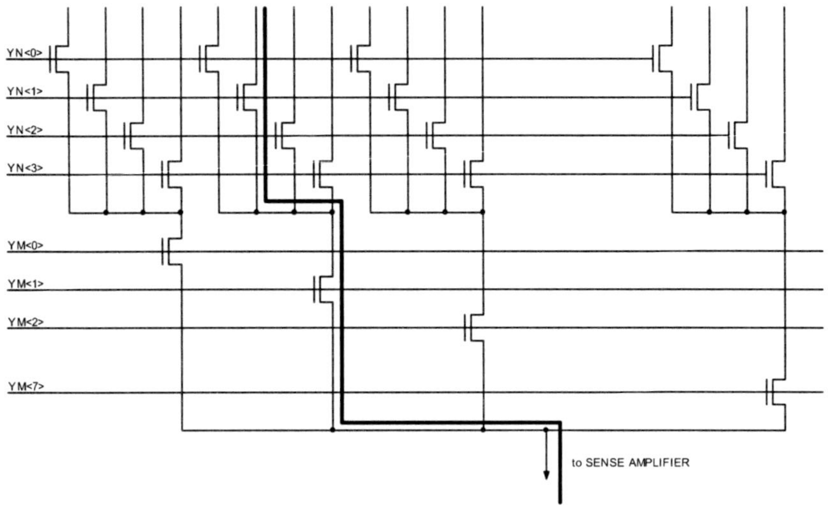

Fig. 18.2. Column decoding for an entire output. Enabling of YM<1> and YN<1> allows access to the shown local bit line. YO level of decoding has been omitted for sake of simplicity

This architecture slows down access time because redundancy column is activated only as soon as the address is detected as a failing one. In fact normal matrix decoding remains active until the comparison between the input addresses and the content of the UPROM is in progress, and therefore the sense amplifier starts its read transient on the wrong bit line.

Another possible solution is to implement a dedicated sense amplifier for redundancy, with a dedicated column decoding, and with adjacent redundancy column instead of being distributed for each output. In this way the space used for redundancy columns is reduced. If we want to repair 8 defects, as in previous example, we only need 8 columns, instead of 64; the drawback with this architecture is that the required circuitry is more complex, as sketched in Fig. 18.3.

A second advantage of this topology is the speed of read. Thanks to the dedicated sense amplifier for redundancy, normal and redundancy paths can go in parallel, and it is only after the read has been performed by the sense amplifier that the redundancy data is sent to the proper output. The former organization chooses the column to be connected to the sense amplifier, while the latter chooses the output where the redundancy data must be sent.

Column redundancy featuring the dedicated sense amplifier allows reading redundancy cells at the same time as matrix cells. In case of row redundancy, on the contrary, it is always necessary to wait until normal matrix row is deselected before addressing and read redundancy row. In fact, let's assume, for sake of simplicity, that redundancy rows are always placed inside the sector so that they can be erased together with normal ones; in other words, redundancy cells share the bit lines with replaced ones. In this case, it is necessary to wait for de-selection of failing cell before starting to read redundancy one, thus wasting time.

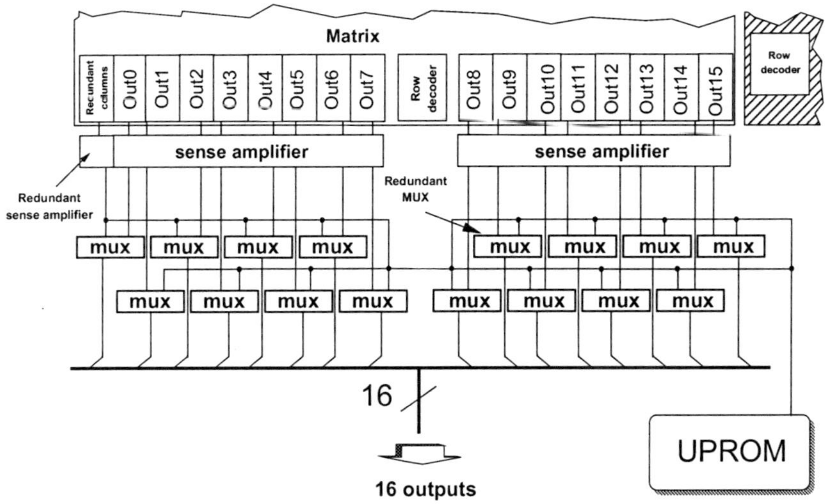

Fig. 18.3. Read path with a dedicated sense amplifier to read redundancy columns

Problem 18.1: Study a solution for redundancy row addressing that minimizes the impact on access time. Assume to sue a small sector exclusively dedicated to redundancy rows.

18.3 Yield

In this section we want to calculate the amount of the additional cells for a device where only column redundancy is present. The final aim is to maximize the yield Y, defined as the ratio between the number of operating devices and the total amount of fabricated ones.

The amount Y_0, also known as prime yield, is the yield achieved without the use of redundancy, and it is a sign of the quality and reproducibility of the process. Y_0 is composed of two yield contributions, Y_p and Y_{m0} that refer to the yield of the "peripheral" circuit and to the yield of the matrix cells respectively. Such a distinction is necessary because it is only the latter part that benefits from the introduction of redundancy rows and/or columns.

$$Y_0 = Y_p Y_{m0} \tag{18.2}$$

Assuming that failure events on distinct cells are independent, we can use the prime yield of the matrix to determine probability p of having a failing cell. The yield of the matrix can be defined as the probability of having no failing cells; if we identify with N_{cel} the number of cells that compose the array, we can write that

$$Y_{m0} = (1-p)^{N_{cel}} \tag{18.3}$$

and therefore

$$p = 1 - (Y_{m0})^{1/N_{cel}} \tag{18.4}$$

If we know such probability and the number of rows N_R that compose each sector, it is possible to calculate the probability of having at least one error on a column in the following way:

$$p_0 = 1 - (1 - p)^{N_R} \tag{18.5}$$

The term $(1-p_0)$ represents the probability that no failing cells are present on the column; its one complement automatically includes all the possible failure combinations, both as physical position and as number of defects.

In order to obtain the probability of having m columns with at least one error, where $0 \leq m \leq N_C$ (N_C = number of columns inside the sector), we assume that the event characterized by probability p_0 occurs m on N_C times

$$P_c = \binom{N_C}{m} \cdot p_0^m (1 - p_0)^{N_C - m} \tag{18.6}$$

The probability of having j error-free redundancy columns, where $0 \leq j \leq N_{RED}$ (N_{RED} = number of redundancy columns inside the sector), can be written as

$$P_{red} = \binom{N_{RED}}{j} \cdot p_0^{N_{RED} - j} (1 - p_0)^j \tag{18.7}$$

The failure can be corrected every time the number of available (i.e. error-free) redundancy columns is greater than or equal to the number of columns that contains at least one error. The probability of having a working sector, considering the possible introduction of column redundancy, can be expressed as:

$$p_{tot} = \sum_{m=0}^{N_{RED}} p_c(m, p_0) \cdot \sum_{j=m}^{N_{RED}} p_{red}(j, p_0) \tag{18.8}$$

Assuming that each sector is statistically independent from the others, final expression of the matrix yield is

$$Y_m = (p_{tot})^{N_B} \tag{18.9}$$

where N_B is the number of sectors the memory is partitioned in.

The model that has just been described does not take into account a characteristic feature of large size memory design. Let's consider the case of hierarchical column decoding with common mail bit lines for a certain number of sectors N_S. In order to minimize area occupation due to the control circuitry required to re-address the failing columns, it is possible to apply redundancy directly at main bit line level, as shown in Fig. 18.4. In other words, an error on a column of a given sector automatically causes the activation of the redundancy on N_S sectors.

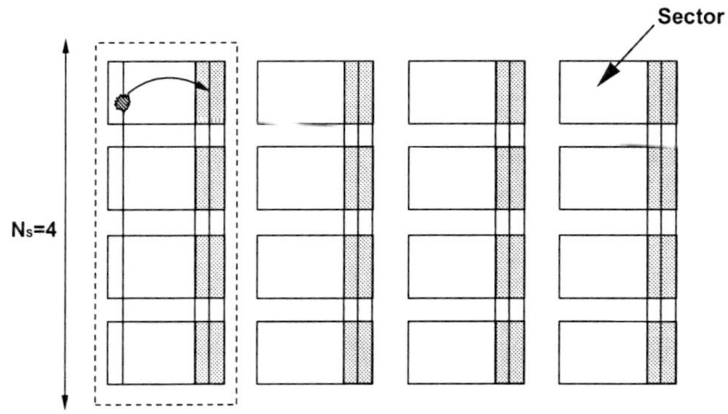

Fig. 18.4. Organization of column redundancy

In order to take into account these facts, let's assume that the N_{RED} redundancy columns are organized in N_{GR} groups of N_{CGR} columns each (e.g. four) such that

$$N_{GR} = \frac{N_{RED}}{N_{CGR}} \tag{18.10}$$

The probability that a group of N_{CGR} redundancy columns has *j* working columns, where $0 \le j \le N_{CGR}$ is:

$$p_{cgr}(j) = \binom{N_{CGR}}{j} \cdot \left(p_0^{'}\right)^{N_{CGR}-j} (1 - p_0^{'})^{j} \tag{18.11}$$

where, this time, the probability of having a failure on a column should be modified to take into account the fact that the redundancy main bit line is common to N_S sectors.

$$p_0^{'} = 1 - (1 - p)^{N_R N_S} \tag{18.12}$$

According to the previous equation, the probability that the whole group of columns that are subordinate to the same MBL is working, can be expressed as:

$$p_{gr} = p_{gr}(N_{CGR}) = (1 - p_0^{'})^{N_{CGR}} \tag{18.13}$$

The probability of having *k* groups of N_{CGR} working redundancy columns where $0 \le k \le N_{GR}$, is therefore:

$$p_{kgr}(k, p_0^{'}) = \binom{N_{GR}}{k} \cdot p_{gr}^{k} (1 - p_{gr})^{N_{GR}-k} \tag{18.14}$$

The matrix is composed of a number of N_{GM} main bit lines equal to

$$N_{GM} = \frac{N_C}{N_{CGR}} \tag{18.15}$$

Therefore the probability of having N_S working sectors is:

$$p'_{tot} = \sum_{m=0}^{N_{GR}} \left[\binom{N_{GM}}{m} \cdot (1 - p_{gr})^m \cdot p_{gr}^{N_{GM}-m} \right] \cdot \sum_{j=m}^{N_{GR}} p_{kgr}(j, p'_0) \tag{18.16}$$

The yield for the matrix is therefore equal to

$$Y_m = (p'_{tot})^{N_B / N_S} \tag{18.17}$$

In order to better evaluate the theoretical yield based on how many redundancy resources are used, it is necessary to take into account the redundancy rules that depend on the architecture of the device. For instance, in case only one sense amplifier is dedicated to redundancy, it is not possible to apply redundancy on more than one output at a time. In other words, the memory must be rejected if there are two failing columns belonging to the same byte or word, as shown in Fig. 18.5.

Problem 18.2: Calculate the yield Y_m taking into account the fact that the device will be rejected if there are at least two failures involving the same byte or word.

Increasing the number of redundancy elements above a certain value produces limited improvements, especially if the initial yield is very high. It is also important to recall that the area occupation of the device increases as the number of redundancy rows and columns increases, thus reducing the number of parts per wafer. The geometry of the silicon wafer where the devices are diffused is circular, while the device is a rectangle: the coverage factor, i.e. the ratio between the area occupied by the devices and the area of the wafer is, therefore, always smaller than one.

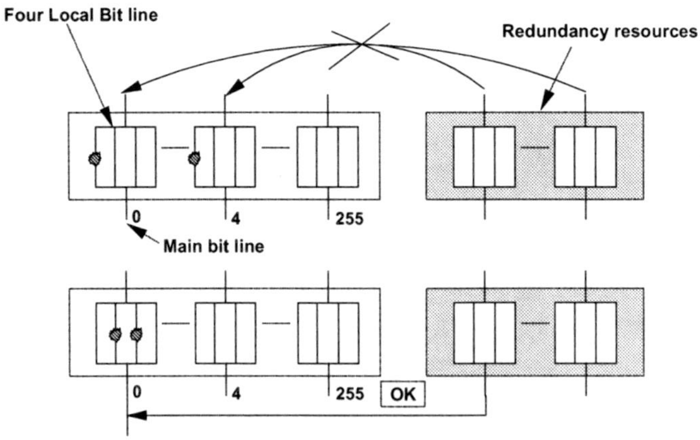

Fig. 18.5. Examples of admissible and non-admissible redundancy

An approximate relation to calculate the number of parts that can be diffused on a wafer whose usable radius[1] is R_0, assuming that the area of the device is A_0, and its form factor (i.e. the ratio between the X and the Y size) is λ, is the following:

$$N(A_0,\lambda)=\frac{\pi R_0^2}{A_0}-\frac{2\pi R_0\sqrt{\lambda}}{\sqrt{A_0(\lambda^2+1)}} \tag{18.18}$$

The number of devices that are operational without the need of redundancy is:

$$N_{SR}=Y_0\cdot N(A_0,\lambda) \tag{18.19}$$

In case column redundancy is used, the number of good memories becomes:

$$N_{CR}=Y_{tot}\cdot N(A_0+\delta A_0,\lambda) \tag{18.20}$$

where δA_0 is the area increase due to redundancy and related control circuitry, which contributes in increasing the area of the periphery of the device.

$$A_0=A_m+A_p,\delta A_0=\delta A_m+\delta A_p \tag{18.21}$$

Let's consider, for instance, a memory composed of 16 sectors, each of which has a size of 512 Kbit. Let's assume that each sector is composed of 2,048 local columns: in order to detect the column to be replaced we need 11 bits.

Let's recall that the UPROM store the failing addresses and that they are composed themselves by non-volatile cells. At the beginning, the logical value stored in the UPROM cells is "1", i.e. they are erased. Therefore address FFFFFh would be identified as failing by default. Therefore for each set of UPROM that stores an address, 11 bits in this example, we also need an additional UPROM, also known as guard UPROM, which certifies the fact that the address stored in the UPROM is really relative to a failing location. Therefore it is necessary not only to check whether the address provided by the user matches a failing address stored in one of the UPROM registers, but also that the corresponding guard UPROM has been programmed.

We can evaluate the total number of UPROM, N_U, as follows:

$$N_U=(11+1)\cdot N_{RED}\cdot N_{SECT} \tag{18.22}$$

where N_{SECT} is the number of sectors and N_{RED} is the number of available redundancy columns. Area increase of the periphery can be expressed as

$$\delta A_p=sf\cdot N_U\cdot A_U \tag{18.23}$$

where A_U is the area of a single UPROM (both matrix and related circuitry) and *sf* is the shrink factor, when applicable.

As far as area increase of the matrix is concerned, it is enough to remember that, regardless the chosen redundancy scheme, the redundancy columns that are

[1] The usable radius of the wafer is smaller than the geometrical one since it is necessary to reserve a small external ring to allow mechanical manipulation of the wafers.

present inside the device are identical to the others. If N_C is the number of columns that compose a sector, we can write:

$$\delta A_m = \frac{N_{RED}}{N_C} A_m \cdot N_{SECT} \tag{18.24}$$

At this point it is possible to choose the number N_C that implies the best trade-off between overall redundancy area occupation and related fault coverage. Figure 18.6 shows how yield varies as a relation of the number of redundancy columns used.

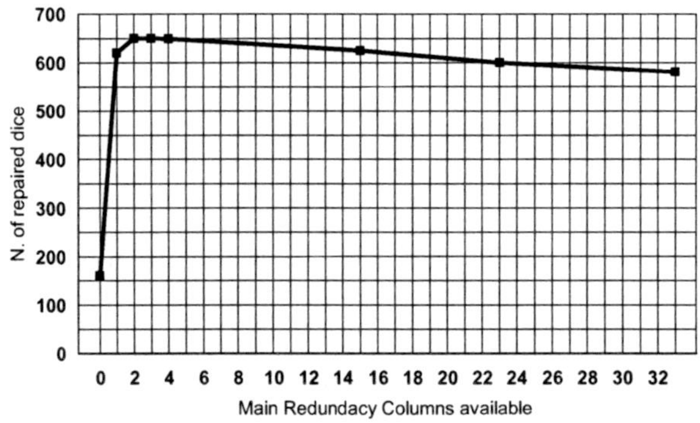

Fig. 18.6. Graph of the yield Y as a function of the number of redundancy columns used

18.4 UPROM Cells

After having analyzed the insertion of redundancy in the overall architecture of the matrix, let's now study the circuit required to read UPROM cells.[2]

There are basically two ways of organizing the non-volatile cells that compose the UPROM: in the former case a real matrix is used; in the latter, they are placed on a single word line.

If we have, for instance, two sectors and two redundancy columns for each sector, we can group the UPROM cells in a 2 x 2 matrix: each row is univocally associated to a given sector, thus avoiding to waste additional UPROM cells just to store the information about which sector has been considered for redundancy. Each time the first sector is addressed, the first row of the matrix is selected and the content of the addressed UPROM cells is read, and it is then compared with the current address.

[2] When we talk about UPROM cell we refer to the Flash cell alone that constitutes the memory element, but many times this term is used to indicate the set of Flash cell together with its circuitry; the context will help clarifying what it is really meant each time.

Matrix organization of the UPROM cells allows saving a lot of space: in fact, all the circuitry required to read, program and test the UPROM cells in independent of the number of rows (i.e. of sectors) that compose the UPROM matrix. The drawback is the time that it takes to understand if the current address matches a failing one, since the UPROM must be read again at every access. Since the UPROM cells are Flash cells, read requires a current to voltage conversion, and this operation is quite long.

An alternate solution is to read all the UPROM cells present on the device in parallel: in order to do it, it is necessary to give up matrix organization, and to introduce for each UPROM cell all the circuitry required for the different operations.

In this way the UPROM cells can be read only once during the VDD ramp-up and their content can be stored in volatile registers that can be directly accessed at each memory access. When the memory is read, it is enough to compare the content of these volatile copies of the UPROM cells against the current address. This solution is very unfavorable from an area perspective, but it is very efficient from a performance standpoint.

In the following sections we will show an example of design of the UPROM circuitry, under the assumption of reading them once at device power-up.

18.4.1 Read Circuitry for the UPROM Cells

The addresses of the failing bits that have been spotted during testing phase are stored in the non-volatile registers composed by the UPROM cells; the comparison between the content of such registers and the current address determines whether to retrieve the data from the matrix or from the redundancy column.

The main issue with the UPROM is that they must be read during the rising ramp of VDD. This consideration also holds true in case of matrix organization for the UPROM cells. In fact in the device there are always other additional UPROM cells used to store the configuration of the internal circuits; for instance, to adapt the voltage regulators to the band-gap voltage on each memory part.

Figure 18.7 shows the electrical scheme of the UPROM cell. Inverters I1 and I2 are latch-connected, whose initial reset is performed by transistors M1 and M2. Inverter I3 acts as the output buffer while M3 is required to bias the drain of the Flash cell F0. At device power-up, POR signal is at logic level "1", thus causing the toggle of the latch where the output of I1 is at "0" and the output of I2 is at "1".

At the end of the POR pulse, voltages V_{UG} and V_{BIAS} are applied and the read of the Flash cell is executed. In case of erased cell, i.e. V_T smaller than 2.5 V, the current flows as soon as both V_{UG} and V_{BIAS} have reached their steady values, 3 V and 1.5 V respectively: in this way the latch toggles again and I1 and I2 are equal to "1" and "0" respectively. The latch remains in its initial state is F0 is written, since no current flows. The size of the inverters of the latch must be carefully chosen keeping into account that:

1. the latch must be able to toggle even in the case of minimum current absorbed by the erased cell

2. in case of written cell, the latch must not toggle because of the current of the parasitic capacitance C1, which starts flowing when the value of V_{BIAS} increases.

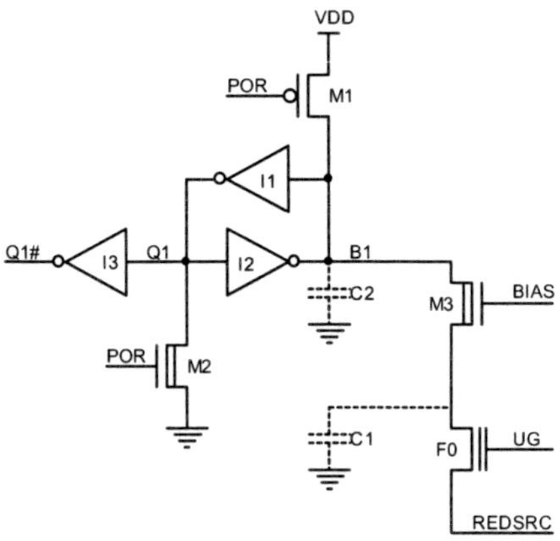

Fig. 18.7. Principle scheme of the circuitry of an UPROM cell

These two requirements are usual contradictory: the former is satisfied if the pull-up of I2 is highly resistive because the current of the cell must be able to lower the voltage of node B1.The latter would require on the other hand a conductive pull-up, so that the charge current for C1 can be provided without perturbing too much the voltage of node B1. As supply voltage decreases, the current in the cell under read, which essentially depends on the gate voltage, decreases as well, and these issues become more critical.

A solution can be to keep the pull-up of I2 resistive, adding a precharge circuit for C1. In this way it is possible to have a read circuitry for the UPROM cells that is more sensitive, and therefore more suitable at low supply voltage, at the cost of a small area increase.

The resulting circuit is shown in Fig. 18.8: the additional part is composed of transistors M4 and M5. P-channel M4 is inserted with the only purpose of protecting the natural transistor M5 from the electrostatic discharges; the signal UPCH, applied at the gate terminal of M4, commands the precharge of the drain node of the cells. To ensure a correct operation at low supply voltage, it is necessary to have the signal V_{UG} to get to 3 V even if VDD is equal to 1.8 V (threshold voltage for the circuit that generates the POR signal), and therefore this node must be boosted.

In order to increase the available current during read, above all at low supply voltage, it is possible to use more Flash cells in parallel for each UPROM cell. In other words, cell F0 of Fig. 18.7 is replaced by the structure shown in Fig. 18.9. The five Flash cells have the poly1 short-circuited, they share the same gate voltage, but one of them has a separate drain contact: four cells are used in read and one is active during program.

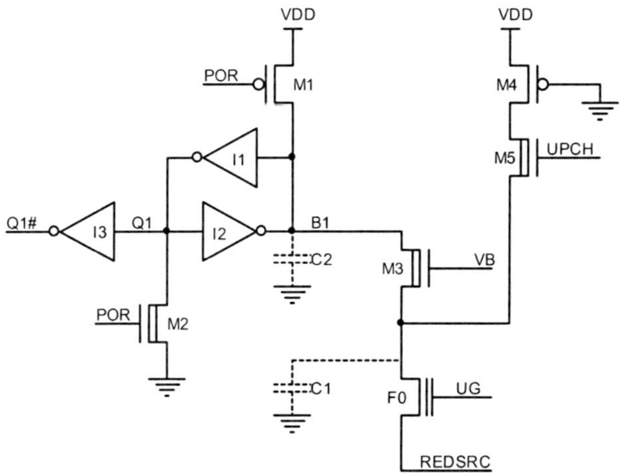

Fig. 18.8. Additional precharge section for the read circuit of the UPROM cell

Programming of a single cell allows reducing consumption; in any case, thanks to the sharing of poly1, the electrons injected in the floating gate gets redistributed and they determine the same threshold voltage step on the other four cells too. In this way the available current during the read phase is amplified by a factor of four.

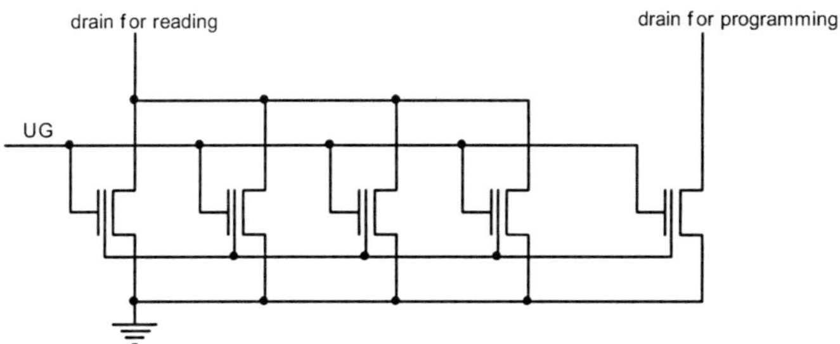

Fig. 18.9. UPROM cell composed of five Flash cells whose poly1 has been short-circuited

18.4.2 Supply Circuitry for the UPROM Cells

Let's now see how to generate the signals required to let the UPROM cell work. In the following, the considerations related to the design of the circuit that generates the UGV signal are presented.

- In order to allow the access to the memory as soon as the supply ramp has finished, it is necessary to synchronize with the POR signal that, in our example, detects the moment when the VDD reaches the value of 1.8V.
- A boost technique is required in order to raise the gate voltage of the UPROM cells up to about 3 V (minimum read voltage) starting from a supply voltage of 1.8V.
- The circuit needs to be on just for the time required to read the UPROM cells, so that the stand-by current is not impacted. The result of the read will be therefore stored in a proper latch.

Problem 18.3: Explain why a latch does not consume in stand-by.

The scheme is shown in Fig. 18.10 where two main parts can be identified: the boost circuit of the V_{UG} and the voltage limiter. When PHI# is high, transistor M1 is turned on and the voltage on C1 is VDD-$V_{T,NAT}$; M2 is turned on too and the voltage difference on C2 is VDD-2$V_{T,NAT}$. As soon as PHI# goes low, the voltage of node F6 is driven to VDD by the output of the inverter I6 and node F9 is driven to 2VDD-$V_{T,NAT}$. This voltage will let M2 turn on, bringing V_{F8} to a full VDD. After a certain delay, given by the inverter chain I1-I5 and by CD1 and CD2, the voltage of node F5 is brought to VDD executing the boost through capacitance C2. At the same time transistors M3 and M4 are driven so that they can transfer the voltage to the output node GV. As we have already seen in Chap. 10, we can calculate the theoretical value of V_{UG} imposing the charge conservation, before (Q_i) and after (Q_f) the boost, on node F8:

$$Q_i = C2 \cdot VDD \tag{18.25}$$

$$Q_f = C_P \cdot V_{UG} + C2 \cdot (V_{UG} - VDD) \tag{18.26}$$

$$V_{UG} = \frac{2 \cdot C2 \cdot VDD}{C_P + C2} \tag{18.27}$$

Assuming a supply voltage of 2 V, a boost capacitance equal to 10 pF and a parasitic load of 2 pF, we get a read voltage of the UPROM equal to 3.3 V.

The four diode-connected p-channel transistors constitute the limitation network of V_{UG}; assuming a voltage drop of 1 V on each of these transistors, it follows that the voltage on node UG cannot go above 4 V. In this way a protection against any over-voltage due to electrostatic discharges is realized, and the boosted voltage is limited whenever the VDD is at its upper specification limit.

As shown in Fig. 18.7, the drain of the UPROM cells is driven by a source-follower n-channel transistor (M3) whose gate is biased at V_{BIAS}. This voltage level must keep the drain voltage around 1 V to avoid soft writing phenomena of the cell. Furthermore it is mandatory that this voltage does not reach undesired levels during electrostatic discharges. The circuit described in the following allows generating this voltage complying with both required conditions.

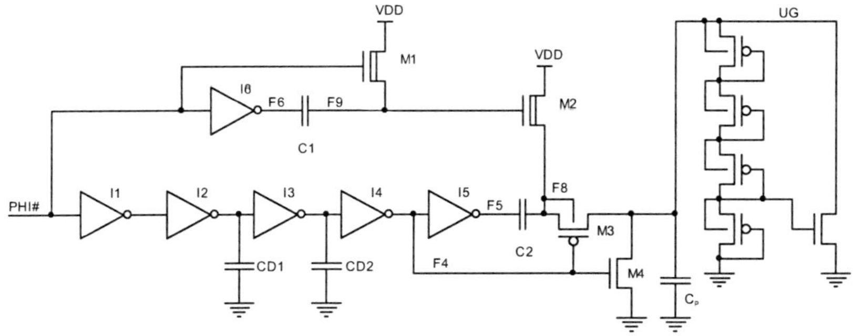

Fig. 18.10. Simplified scheme of V_{UG} generator

The electrical scheme of the circuit is shown in Fig. 18.11. It can be divided into two parts: a timing section and the V_{BIAS} voltage generator. The timing is necessary to apply the drain voltage in a window that is contained in the gate one, in order to limit the stress on the cell. This path is composed of ports I1-I4 and of capacitors C1 and C2. As far as generation of V_{BIAS} is concerned, a scheme based on a simple reaction loop is used, composed of the NOR I5 and transistors M4-M6.

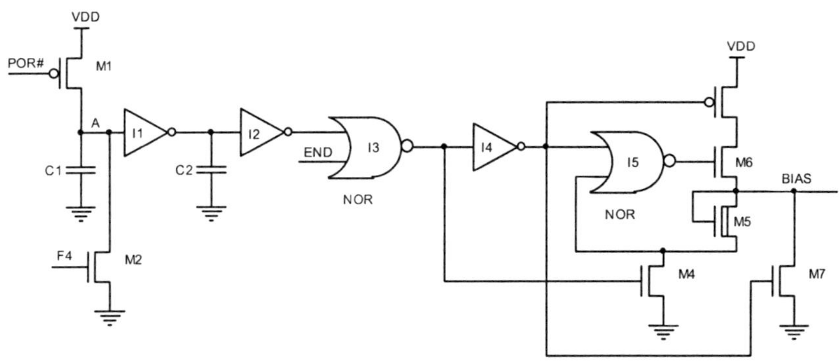

Fig. 18.11. Scheme of V_{BIAS} generator

When the signal F4, which comes from the V_{UG} generator, is low, signal A is kept high by the pull-up M1. In this way the NOR I5 keeps transistor M4 turned off, while M7 is turned on and it keeps node BIAS at ground. During the active phase, signal F4 toggles to high, turning on M2 and causing the output of the NOR I3 to go high. Turning on of M4 allows the current flow through M5 and M6. The regime value of V_{BIAS} is fixed by the trigger threshold of I5 and by the V_{GS} of M5. The presence of the loop composed by I5, M5 and M6 has two big advantages:

– V_{BIAS} voltage is kept at a value equal to the trigger threshold of the NOR port plus the V_{GS} of a natural transistor (about 0.5 V). This is true in case of an electrostatic discharge on VDD too.
– The charge transient of the parasitic capacitance of the BIAS line is very fast thanks to the feedback.

Before the gate voltage is driven to ground, the END signal is activated which, driving the output of the NOR I3 to ground, brings back the circuit in the initial conditions. In this way, as soon as the read cycle of the UPROM cells has finished, the two supply circuits (V_{BIAS} and V_{UG}) are inhibited and the consumption drops to zero. Finally it is really interesting that all the signals are generated starting from the POR signal only; related circuitry is shown in Fig. 18.12.

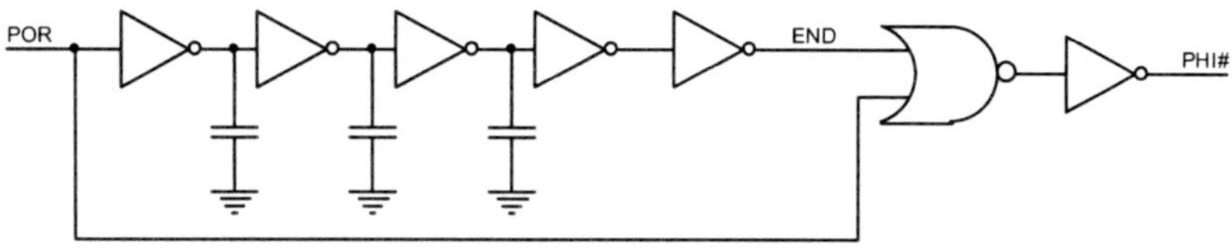

Fig. 18.12. Circuit to generate the control signals of Fig. 18.11

18.5 The First Read After Power On Reset

We have seen in Chap. 5 the generation of the POR signal that, as soon as supply voltage is applied, acts as a reset for all the logic circuits.

Now we want to investigate the relationship between POR, read of the UPROM and the first access to the matrix.

Redundancy circuits modify the internal addressing of the memory so that defective cells are replaced by the working ones, which are present in an additional portion of the matrix. Before executing any operation on the memory it is therefore mandatory to read the UPROM cells that allow a correct detection of the addresses to be replaced.

According to the specifications that are usually followed by non-volatile memories, it is necessary to consider that, once the ramp of the supply voltage VDD has completed, the device must be ready to operate without any further command. In particular, the memory can be read to know the content of a given location. In order to carry out this task, the UPROM cells must have been already read. This read operation is performed exploiting a part of the VDD ramp; therefore the trigger value of the POR, i.e. the signal that starts the read of the UPROM cells, is chosen below VDD. When selecting the trigger threshold of the POR, it is also necessary to consider the noise margin with respect to supply voltage variations: if the threshold is too high, then the device might be reset by the typical noise of VDD.

Let's now analyze with more detail the issue of the first read in the practical case of a memory that requires the word line boost with zero consumption in stand-by. The boost, as explained in Chap. 10, must be the one shot kind. In particular, the boost capacitor must be recharged after every read and all the various phases of the boost must be carefully timed starting from the ATD signal. Let's assume that the device must be able to read between 2.7 V and 3.6 V. The value of the POR, according to the considerations exposed before, is usually set around 1.8 V. The customer can set the desired addresses with CE# low and device not powered and, later on, he can provide the power supply. The ATD signal, which is the base for all the timing chains, is missing; therefore we need an internal signal that starts the read as soon as VDD has reached a proper value. We have to design a circuit that, without consuming in stand-by, can provide such a signal that enables the read of the matrix when the value of VDD is correct.

Figure 18.13 and Fig. 18.14 show the power-up with different speeds for the ramp of the supply voltage. UPHI signal indicates the phase of read of the UPROM and its duration is independent of the speed of the ramp. In case of fast ramp, when the UPHI phase finishes the VDD is already enough to read and, therefore, it is possible to use the falling edge of UPHI to trigger the start of the first read. In case of slow ramp we can see that, when UPHI goes low, VDD is not yet high enough to guarantee a correct read. Therefore we can conclude that the UPHI signal alone is not enough to guarantee a correct operation under all the possible conditions: therefore we must introduce a circuit to generate the signal of first read.

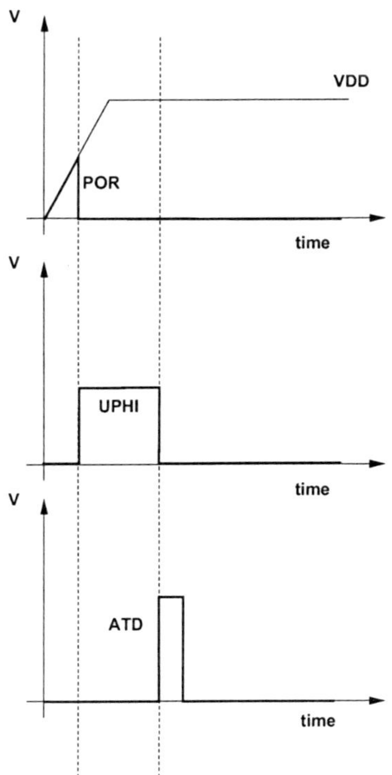

Fig. 18.13. Power-up with a fast ramp on VDD

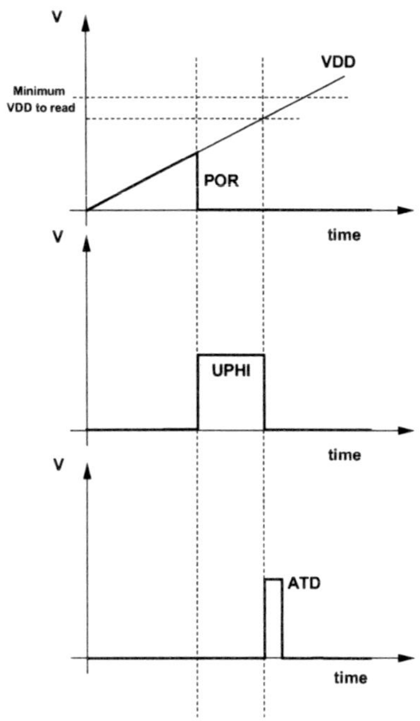

Fig. 18.14. Power-up with a slow ramp on VDD

The operating principle of this circuit is again the same used to generate the POR signal, i.e. the comparison between the VDD ramp and an internal voltage that acts as a reference. The reference scheme is again shown in Fig. 5.49, where the output signal is called READPOR instead of POR. The scheme of the circuit whose task is to generate the reference voltage level is similar to the one shown in Fig. 5.51: the difference is that we use the POR of the device as the enabling signal. Practically the circuit we are dealing with is enabled only when the POR pulse has finished. In this way, the circuit that generated the POR determines the noise margin with respect to VDD and the trigger threshold of the READPOR can be raised up to 2.4 V, which we consider enough to read. The situation shown in Fig. 18.14 for the slow ramps gets modified as shown in Fig. 18.15. Generating an ATD pulse both at the end of UPHI and at the end of READPOR, the first read issue is solved since, by assumption, as soon as READPOR toggles the VDD is enough to read.

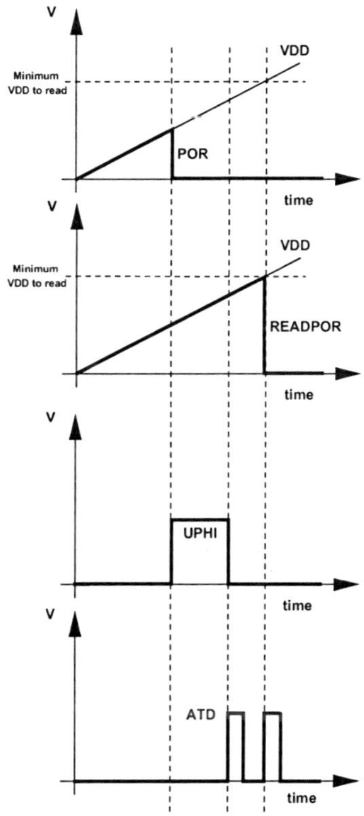

Fig. 18.15. Power-up with a slow ramp on VDD in case a dedicated circuit for the first read is used

18.6 Error Correction Codes

The use of redundancy is limited to defects detected during the factory test: at present, there is no memory device that is able to apply a self-redundancy in case of failure. This is due to the fact that the usage of UPROM cells is complex and, for the time being, the size of the memory device does not justify the implementation of an on-field redundancy. State-of-the-art devices, above all the multi-level ones, rely more and more on error correction codes (ECC) to overcome reliability issues like the charge loss with time.

18.6.1 Elements of Coding Theory

Let's now quickly see the principles of the coding theory, leaving the rigorous mathematical discussion to specialized readings.

Let's define a message as a block of symbols of a finite alphabet; it is usually a sequence of 0 and 1, but it can also be any other number. The message is transmitted over a communication channel that is affected by noise. Because of this noise, the received message might contain one or more errors. Coding theory studies the way redundancy information can be added to the transmitted message so that, at the receiver's side, it is possible to detect or correct a given number of errors. By communication channel we mean any transmission and/or storage process that can be affected by error; for instance, transmission lines for voice and data, write, storage and read process on a CDROM, DVD, HD and solid-state memories.

In block coding the binary sequence is divided into message blocks whose length is fixed. Each block called $\underline{u}$ is composed of k digits of information. There are 2^k different messages in total. Coding transforms, according to certain rules, each input message $\underline{u}$ in a binary n-tuple $\underline{v}$ where $n > k$. This binary n-tuple $\underline{v}$ is the code word of the message $\underline{u}$. In correspondence with the 2^k possible messages there are 2^k code words (among the 2^n possible options). The set of the 2^k code words is the code. The code is significant if there is a bi-univocal correspondence between the message $\underline{u}$ and its code word $\underline{v}$, i.e. the 2^k code words must be all different.

Whenever a vector $\underline{v}$, which might have been altered during transmission, is received, it is necessary to find out which vector $\underline{u}$ had been sent. At this point we need to introduce the concept of distance between the words of a numerical code. The distance $d_H(a,b)$ between two n-tuples of bits is defined as the number of positions where the two vectors differ. A possible representation is shown in Fig. 18.16.

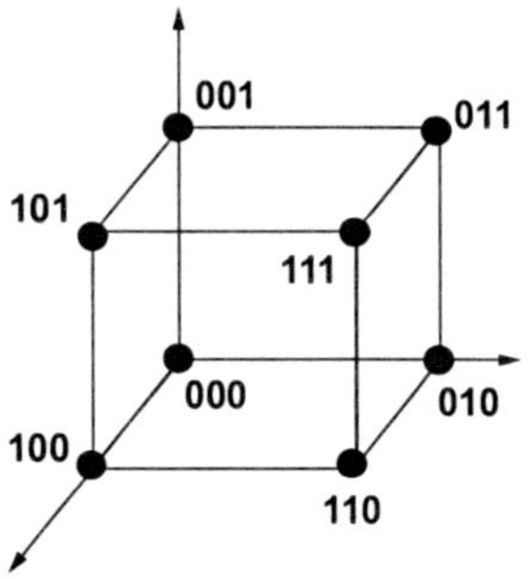

Fig. 18.16. Graphical representation of the distance of the binary words of a code whose length is 3

Let's consider a string of 3 bits and let's place the possible binary terns on the vertices of a cube: it is evident that two consecutive words differ by a unitary distance In this way, if we use all the eight configurations as information, we get a code without redundancy; the minimum distance between two code words is equal to one. An error that alters even just one bit turns a code word into another code word, thus preventing any possible error detection and, consequently, correction.

If we decide to limit the code words to those with an even number of ones (Fig. 18.17), the minimum distance becomes 2. In fact, to go from one code word to another along the only admissible paths, the sides of the cube, we need to do at least two steps. In this case is possible to detect the single error, i.e. the error that "moves" code word to one of the forbidden edges of the cube. Finally, if the code were reduced to the (000) and (111) words only, we could not only detect the single errors, but we could also correct them.

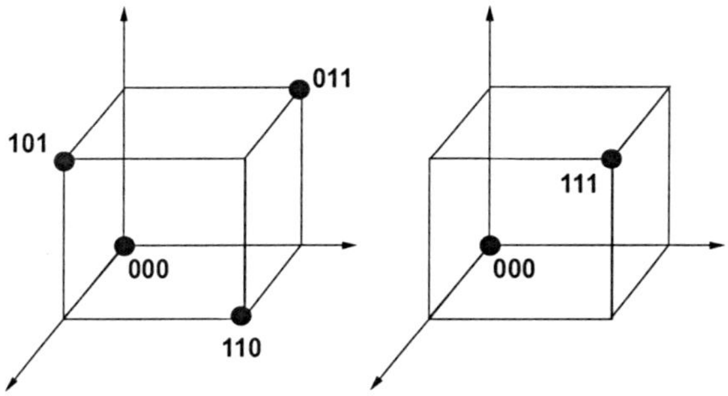

Fig. 18.17. On the left, a code with a minimum distance of 2; on the right, the code reduced to only 2 words allows the correction of a single error

We say that a code has minimum distance d_{min} if there is a code word (different from 0) with distance d_{min} from the all zeros vector, that is to say if there is a word with d_{min} "1".

Indicating with t the number of errors that can be corrected, the following relation holds true

$$t \le \frac{d_{min} - 1}{2} \tag{18.28}$$

Furthermore, Hamming inequality must be true as well

$$\left[\binom{n}{0} + \binom{n}{1} + \ldots + \binom{n}{t}\right] \le 2^{n-k} \tag{18.29}$$

18.6.2 A Memory with ECC

Let's now see the use of a correction code in a memory system. The fundamental concept, as we have seen, is to add a certain number of correction bits, called parity bits, to the data bits.

First of all, the type of failure must be established. Let's assume that we want to detect and correct a bit in a 2-bit data. According to Eq. (18.29), we need three parity bits. In this case the area penalty is very high, but in reality data packets composed of 32, 64 or 128 bits are corrected by a small number of parity bits. Figure 18.18 shows the 5 code bits, two bits of data and three bits of parity together with their decimal value. The fourth column shows, in decimal, the value the code turns into if one bit is modified, either 1 to 0 or 0 to 1.

For instance, let's consider the decimal value 10d, i.e. binary value 01010b. If the first bit changes we have 11010b, i.e. 26d, if the second bit changes we have 00010b, i.e. 2d, 01110b i.e. 14d for the third bit, 01000b i.e. 8d for the fourth bit and, finally, 01011b i.e. 11d for the fifth bit.

Overall value (data + parity)	Data	Parity	Overall value modifying 1 bit	Codes chosen for the different data
O	**OO**	**OOO**	**1, 2, 4, 8, 16**	Code chosen for OO
1	*O1*	*OOO*	*O, 3, 5, 9, 17*	
2	1O	OOO	3, O, 6, 1O, 18	
3	11	OOO	2, 1, 7, 11, 19	
4	**OO**	**OO1**	**5, 6, O, 12, 2O**	
5	*O1*	*OO1*	*4, 7, 1, 13, 21*	
6	1O	OO1	7, 4, 2, 14, 22	
7	11	OO1	6, 5, 3, 15, 23	
8	**OO**	**O1O**	**9, 1O, 12, O, 24**	
9	*O1*	*O1O*	*8, 11, 13, 1, 25*	
1O	1O	O1O	11, 8, 14, 2, 26	
11	11	O1O	1O, 9, 15, 3, 27	
12	**OO**	**O11**	**13, 14, 8, 4, 28**	
13	*O1*	*O11*	*12, 15, 9, 5, 29*	Code chosen for O1
14	1O	O11	15, 12, 1O, 6, 3O	
15	11	O11	14, 13, 11, 7, 31	
16	**OO**	**1OO**	**17,18, 2O, 24, O**	
17	*O1*	*1OO*	*16, 19, 21, 25, 1*	
18	1O	1OO	19, 16, 22, 26, 2	
19	11	1OO	18, 17, 23, 27, 3	
2O	**OO**	**1O1**	**21, 22, 16, 28, 4**	
21	*O1*	*1O1*	*20, 23, 17, 29, 5*	
22	1O	1O1	23, 2O, 18, 3O, 6	
23	11	1O1	22, 21, 19, 31, 7	Code chosen for 11
24	**OO**	**11O**	**25, 26, 28, 16, 8**	
25	*O1*	*11O*	*24, 27, 29, 17, 9*	
26	1O	11O	27, 24, 3O, 18, 1O	Code chosen for 1O
27	11	11O	26, 25, 31, 19, 11	
28	**OO**	**111**	**29, 3O, 24, 2O, 12**	
29	*O1*	*111*	*28, 31, 25, 21, 13*	
3O	1O	111	31, 28, 26, 22, 14	
31	11	111	3O, 29, 27, 23, 15	

Fig. 18.18. An example of correction code for two data; three parity bits are required to detect and correct one bit

If we proceed this way for all the codes, we find out, at the end, that it is possible to find for each data, i.e. 00, 01, 10 and 11, a univocal code. That is to say that the code chosen for data 01 is 13, i.e. 01 of data associated to a parity of 011; modified codes are 12, 15, 9, 5 and 29. These modified codes are unique (i.e. they belong to data 01 and parity 011 only) with respect to the ones chosen to identify the other data. It is important to highlight that we are able to correct an error either in the data or in the parity, i.e. the 5 bits have the same "dignity" for the correction code.

Let's now see what really happens inside the memory.

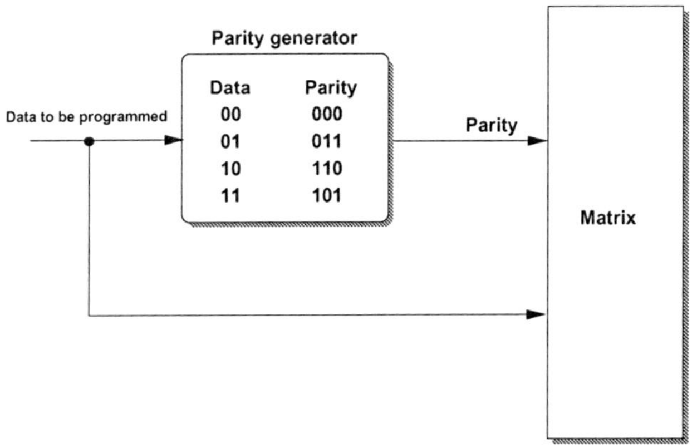

Fig. 18.19. Generation of the parity bits during data program. Parity bits are written in cells of the same type as the one used for the data.

The first step takes places during program, as shown in Fig. 18.19. According to the correspondence shown in Fig. 18.18, the parity is calculated on the input data.

Both data and parity bits are written at the same address in the matrix. When we read from the matrix, both data and parity are retrieved (Fig. 18.20); the data read by the sense amplifier is used to re-calculate the parity and then the two parities, re-calculated and read, are compared. If they match, then the data is correct and the outputs are enabled. Otherwise an error is present.

At this point both data and parity bits are sent to a block that contain the univocal codes that we have obtained in Fig. 18.18. So if we read 18d it means that it comes from the modification of one bit of 26d, i.e. of data 10. The re-computed data is passed through the multiplexer, which this time is enabled to let the modified data pass, towards the outputs. In this way we can recover the correct data. The implementation of the ECC is significantly expensive not only in terms of area occupation inside the matrix because of the additional section containing the parity bits, but also because of the additional logic that heavily impacts the access time.

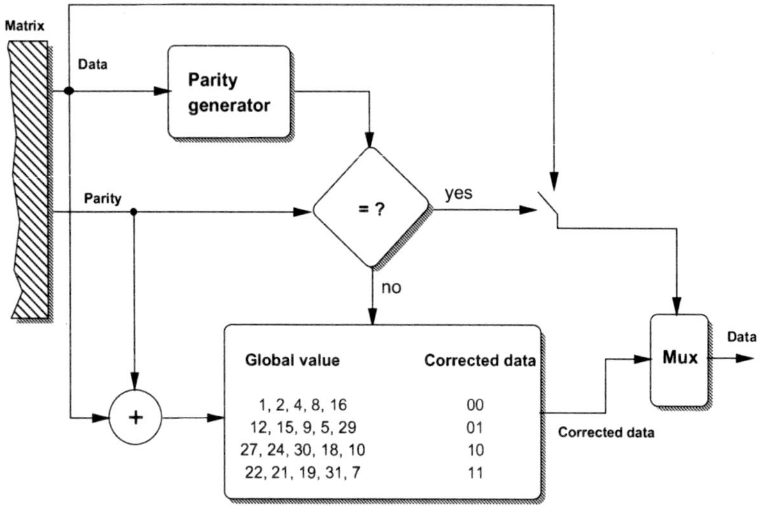

Fig. 18.20. During read, the stored bits are compared with the ones calculated on current read

Bibliography

B. Benjauthrit, L. Coday, and M. Trcka, "An overview of error control codes for data storage," in 1996 IEEE Int. Non-Volatile Memory Technology Conf., pp. 120-126, (1996).

J.M. Berger, "A note on error detection codes for asymmetric channels", Information and Control, no. 4, pp. 68-73, (1961).

R.E. Blahut, Theory and Practice of Error Control Codes. Reading, MA: Addison-Wesley Publishing Company, (1983).

C.L. Chen, "Symbol error correcting codes for computer memory systems," IEEE Trans. Comput., vol. 41, no. 2, pp. 252-256, (Feb. 1992).

C.L. Chen, "Symbol error correcting codes for memory applications," in Proc. Annual Symposium on Fault Tolerant Computing, pp. 200-207, (1996).

C.L. Chen, "Some results on symbol error-correcting codes," in Proc. 2000 IEEE Int. Symp. on Information Theory, p. 475, (2000).

C.L. Chen, M.Y. Hsiao, "Error-correcting codes for semiconductor memory applications: a state-of-the-art review," IBM J. Res. Develop., vol. 28, no. 2, pp. 124-134, (Mar. 1984).

T. Cho, Y.-T. Lee, E.-C. Kim, J.-W. Lee, S. Choi, S. Lee, D.-H. Kim, W.-G. Han, Y.-H. Lim, J.-D. Lee, J.-D. Choi, and K.-D. Suh. "A dual-mode NAND flash memory: 1-Gb multilevel and high-performance 512-Mb single-level modes," IEEE J. Solid-State Circuits, vol. 26, no. 11, pp. 1700-1706, (Nov. 2001).

D.J. Costello, Jr., J. Hagenauer, H. Imai, and S. B. Wicker, "Applications of error-control coding," IEEE Trans. Inform. Theory, vol. 44, no. 6, pp. 2531-2560, (Oct. 1998).

J. Cunningham, "The use and evaluation of yield models in integrated circuit manufacturing", IEEE J. Solid-State Circuits, vol. 3, pp. 60-71, (May. 1990).

J.A. Fifield and C.H. Stapper, "High-speed on-chip ECC for synergistic fault-tolerant memory chips," IEEE J. Solid-State Circuits, vol. 26, no. 10, pp. 1449-1452, (Oct. 1991).

H.L. Davis, "A 70-ns word-wide 1-Mbit ROM with on-chip error-correction circuits," IEEE J. Solid-State Circuits, vol. 30, no. 5, pp. 958-963, (Oct. 1985).

T. Fuja, C. Heegard, and R. Goodman, "Linear sum codes for random access memories," IEEE Trans. Comput., vol. 37, no. 9, pp. 1030-1042, (Sept. 1988).

K. Furutani, K. Arimoto, H. Miyamoto, T. Kobayashi, K. Yasuda, and K. Mashiko, "A built-in Hamming code ECC circuit for DRAM's," IEEE J. Solid-State Circuits, vol. 24, no. 1, pp. 50-56, (Feb. 1989).

R.M. Goodman, "On-chip ECC for multi-level random access memories," in Proc. 1989 IEEE/CAM Information Theory Workshop, pp. 7-4, (1989).

R.M. Goodman, M. Sayano, "The reliability of semiconductor RAM memories with on-chip error-correction coding," IEEE Trans. Inform. Theory, vol. 37, no. 3, pp. 884-896, (May 1991).

S. Gregori, P. Ferrari, R. Micheloni, and G. Torelli, "Construction of polyvalent error control codes for multilevel memories", in Proc. 7th IEEE International Conference on Electronics, Circuits, and Systems, pp. 751-754, (Dec. 2000).

S. Gregori, O. Khouri, R. Micheloni, and G. Torelli, "An error control code scheme for multilevel Flash memories," in Records 2001 IEEE International Workshop on Memory Technology, Design and Testing, pp. 45-49, (Aug. 2001).

S. Gregori, et al., "On-Chip error correcting technique for new generation Flash memories", IEEE Proceeding of the, Vol. 91, No. 4, pp. 602-616, (April 2003).

R.W. Hamming, "Error detecting and error correcting codes," Bell Syst. Tech. J., vol. 26, pp. 147-150, (1950).

A. Hocquenghem, "Error corrector codes" (Codes correcteurs d'erreurs), Chiffres, no. 2, pp. 147-156, (1959).

H.L. Kalter, C.H. Stapper, J.E. Barth, Jr., J. DiLorenzo, C.E. Drake, J.A. Fifield, G.A. Kelley, Jr., S.C. Lewis, W.B. van der Hoeven, and J.A. Jankoski, "A 50-ns 16-Mb DRAM with a 10-ns data rate and on-chip ECC," IEEE J. Solid-State Circuits, vol. 25, no. 5, pp. 1118-1128, (Oct. 1990).

M. Kubo and S. Chou, "Fault tolerant techniques for memory components," in 1985 IEEE Int. Solid-State Circuits Conf. Dig. Tech. Pap., pp. 230-231, (Feb. 1985).

P. Mazumder, "An on-chip ECC circuit for correcting soft errors in DRAM's with trench capacitors," IEEE J. Solid-State Circuits, vol. 27, no. 11, pp. 1623-1633, (Nov. 1992).

T. Michalka et al., "A discussion of yield modeling with defect clustering,circuit repair and circuit redundancy", IEEE J. Solid-State Circuits, vol. 3,pp. 116-127, (Aug. 1990).

T. Nakayama, Y. Miyawaki, K. Kobayashi, Y. Terada, H. Arima, T. Matsukawa, and T. Yoshihara, "A 5-V-only one-transistor 256K EEPROM with page-mode erase," IEEE J. Solid-State Circuits, vol. 24, no. 4, pp. 911-915, (Aug. 1989).

W.W. Peterson and E. J. Weldon, Jr., Error Correcting Codes, 2nd ed. Cambridge, MA: M.I.T. Press, (1972).

F.I. Osman, "Error-correction technique for random-access memories," IEEE J. Solid-State Circuits, vol. 17, no. 5, pp. 877-882, (Oct. 1982).

B. Polianskikh and Z. Zilic, "Design and implementation of error detection and correction circuitry for multilevel memory protection," in Proc. 32nd IEEE Int. Symp. on Multiple-Valued Logic, pp. 89-95, (2002).

T.R.N. Raho and E. Fujiwara, Error Control Coding for Computer Systems. Englewood Cliffs, NJ: Prentice Hall, (1989).

P. Ramanathan, K. K. Saluja, and M. Franklin, "Testing check bits at no cost in RAMs with on-chip ECC," IEE Proceedings-E, vol. 140, no. 6, pp. 304-312, (Nov. 1993).

D. Rossi, C. Metra, and B. Riccò, "Fast and compact error correcting scheme for reliable multilevel Flash memory," in Proc. 2002 IEEE Int. Workshop on Memory Technology, Design and Testing, pp. 27-31, (2002).

C.V. Srinivasan, "Codes for error correction in high-speed memory systems – part I: correction of cell defects in integrated memories," IEEE Trans. Comput., vol. 20, no. 8, pp. 882-888, Aug. (1971).

C.H. Stapper and H.-S. Lee, "Synergistic fault-tolerance for memory chips," IEEE Trans. Comput., vol. 41, no. 9, pp. 1078-1087, (Sept. 1992).

T. Tanzawa, T. Tanaka, K. Takekuchi, R. Shirota, S. Aritome, H. Watanabe, G. Hemink, K. Shimizu, S. Sato, Y. Takekuchi, and K. Ohuchi, "A compact on-chip ECC for low cost Flash memories," IEEE J. Solid-State Circuits, vol. 32, no. 5, pp. 662-669, (May 1997).

T. Toyabe, T. Shinoda, M. Aoki, H. Kwamoto, K. Mitsusada, T. Masuhara, and S. Asai, "A soft error rate model for MOS dynamic RAM's," IEEE J. Solid-State Circuits, vol. 17, no. 2, pp. 362-367, (Apr. 1982).

R. Vancu, L. Chen, R. L. Wan, T. Nguyen, C.-Y. Yang, W.-P. Lai, K.-F. Tang, A. Mihnea, A. Renninger, and G. Smarandoiu, "A 35ns 256k CMOS EEPROM with error correcting circuitry," in 1990 IEEE Int. Solid-State Circuits Conf. Dig. Tech. Pap., pp. 64-65, (Feb. 1990).

J. Yamada, "Selector-line merged built-in ECC technique for DRAM's," IEEE J. Solid-State Circuits, vol. 22, no. 5, pp. 868-873, (Oct. 1987).

T. Yamada et al., "A 4-Mbit DRAM with 16-bit concurrent ECC", IEEE J. Solid-State Circuits, vol. SC-23, pp. 20 - 25, (Feb. 1988).

G.-C. Yang, "Reliability of semiconductor RAMs with soft-error scrubbing techniques," IEE Proc. – Comput. Digit. Tech., vol. 142, no. 5, pp. 337-344, (Sept. 1995).

19 The Output Buffer

19.1 Introduction

The role of the output buffer is to provide the data acquired during a read operation to the external world. The output load is usually represented by a capacitor C_{LOAD} whose value may vary from 30 pF to 100 pF. An inverter can represent the simplified structure of the output buffer: the pull-up charges the output if an erased cell is read, while the pull-down discharges the output if the cell is written.

The budget for the access time can be partitioned in four basic parts:

1. time for input buffer and ATD: 10%
2. time for row and column decoding & voltage setting: 30%
3. time for read, sense amplifier commutation: 40%
4. time for output buffer commutation: 20%

The output buffer is therefore one of the key elements for the read path, whose performance heavily influences the overall access time. The use of lower and lower power supply voltage, to reduce the power dissipated by the devices[1], causes a slowdown in signal propagation. Representing the output buffer as a simple inverter, commutation time is defined as the time required to bring the output to VDD / 2 (CMOS level) starting from the moment when the data arrives from the sense amplifier.

Let's determine the relationship between commutation time and supply voltage. Under the hypothesis that the "final" MOS always work in the saturation region (i.e. providing a constant current), we have:

$$I = C_{LOAD} \frac{\Delta V}{\Delta t} \tag{19.1}$$

$$I = \frac{1}{2} \mu C_{ox} \frac{W}{L} (V_{GS} - V_T)^2 \tag{19.2}$$

from which the T_{switch} required to reach half the supply voltage can be calculated:

$$T_{switch} = C_{LOAD} \frac{VDD}{2} \frac{1}{\frac{1}{2} \mu C_{ox} \frac{W}{L} (V_{GS} - V_T)^2} \tag{19.3}$$

The relationship between size of the transistors and VDD to get a certain commutation time can be determined from the previous equation:

---

[1] Dissipated power is quadratically related to the value of the supply voltage.

$$\frac{W}{L} = C_{LOAD} \frac{VDD}{2} \frac{1}{\frac{1}{2}\mu C_{ox}(VDD - V_T)^2} \frac{1}{T_{switch}} \tag{19.4}$$

V_T is the absolute value of either p-channel or n-channel threshold voltage (depending on which output transition is considered). The aim is to charge and discharge the output capacitance as fast as possible, in order to reduce commutation time; on the other hand, this task requires a great current and therefore large output transistors and the related area occupation issues. The load for the 5 V devices is 100 pF; the reduction of the VDD has brought the use of a smaller load, down to 50 pF or even 30 pF. Let's consider the case of charging or discharging a capacitance of 100 pF in 10 ns, with a voltage swing of 3 V; from Eq. (19.1), a current of 30 mA is required. This calculation assumes an ideal behavior of the transistor as a current generator. In reality, the sizing of the transistors is done in such a way that during the first nanoseconds the current is 5–10 times higher than the calculated one. In the case of 16 simultaneously switching buffers, a peak current of 1 A or more is possible, and all the circuits of the device can be negatively affected by such a huge current.[2]

Problem 19.1: Assuming a device where memory locations are consecutively read with a period of 100 ns, calculate the mean current consumption for 16 output buffers and consequently calculate the temperature that the device internally reaches, considering a supply voltage of 5 V.

From the previous consideration, it is easy to imagine that the commutation of the output buffers can influence, by transmission through VDD and GND, the MAT and REF nodes that constitute the inputs of the differential stage of the sense amplifier. To avoid noise issues, it is better to latch the input data of the output buffer. The ENDREAD signal that we have described in the chapter on the synchronization signals can be used. Assuming an active high enable signal, the data can be stored by either a D-type flip-flop active on the rising edge of the ENDREAD or by a latch enabled by the ENDREAD itself. Let's suppose that a spurious commutation takes place in the sense amplifier caused by the noise induced by the commutation of the output buffers. The output buffer comprising a D-type flip-flop is not affected, since the data is already stored. If a latch is used and the ENDREAD signal is still high, then the input transition is detected and therefore a wrong data state is sent out. For this reason, the flip-flop solution is preferred, as shown in Fig. 19.1.

The pull-up and the pull-down are not driven by the same signal in order to prevent the simultaneous conduction and consequently the waste of current (known as crowbar current) flowing between VDD and GND. The function of the block called CONTROL LOGIC, shown in Fig. 19.1, is designed to deliberately prevent a time interval when both transistors of the final stage are turned on. The circuit shown in Fig. 19.2 is triggered by a commutation of the input turning off the currently active transistor and, only afterwards, turning on the other.

[2] Just to give an example, a normal flat-iron for clothes sinks a DC current of some Ampere (4 ÷ 5 A), while a domestic differential switch can sink up to 25 A.

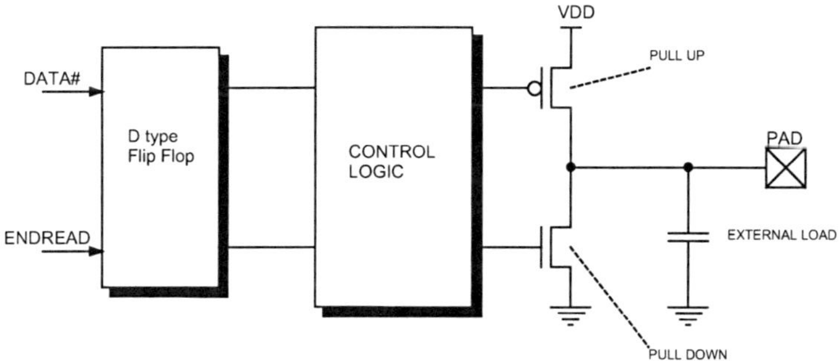

Fig. 19.1. Schematic of the output buffers where the pull-up, the pull-down and the flip-flop for storing the data coming from the sense amplifier are shown

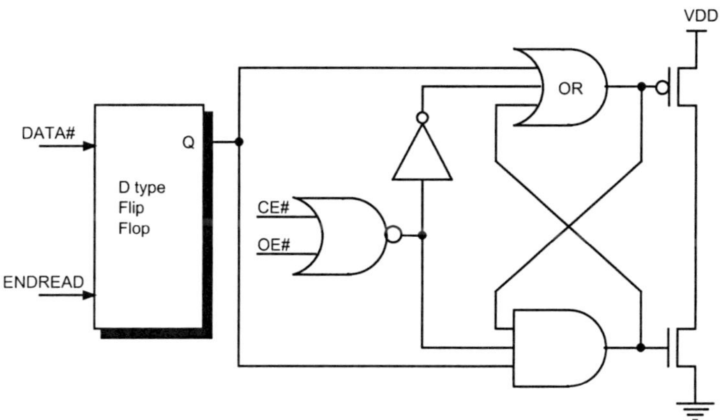

Fig. 19.2. Logic network to avoid crowbar current

Let's consider the case when the sense amplifier consecutively reads an erased cell (DATA# = 0) and then a written cell (DATA# = 1). Both CE# and OE# signals are low, since we are in read mode. After the first memory access, the pull-up is on and the pull-down is off, so both the gates are at ground. When the new data comes, the output of the OR cell goes to "1"switching off the pull-up, while the output of the AND cell can toggle and turn on the pull-down only when all the addresses are high, and therefore the pull-up is already off. The sequence is obviously inverted (the pull-down is turned off first, then the pull-up is turned on) for the opposite transition (DATA# goes from "1" to "0").

In the following sections, different types of output buffers are shown.

19.2 NMOS Output Buffer

Figure 19.3 shows an output buffer in NMOS technology. The task of the OE# signal is to turn off both the pull-up and the pull-down; when OE# is high, the output node PAD is no longer driven by the output buffer, but it is in a high impedance (tristate) condition.

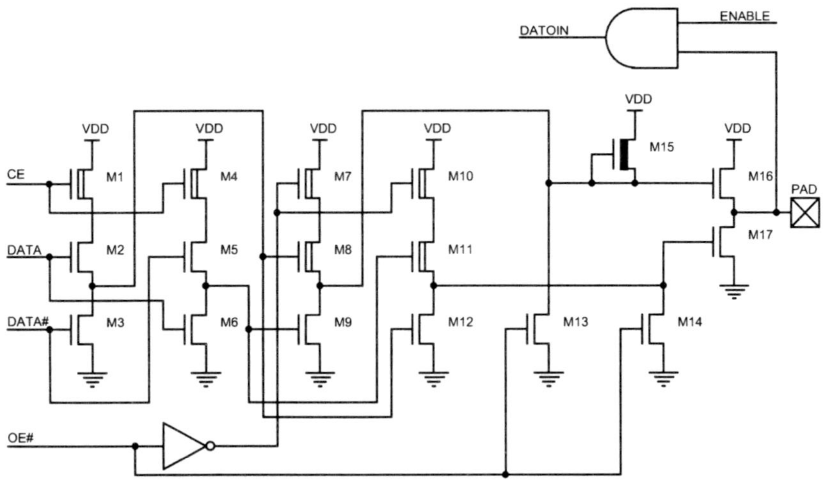

Fig. 19.3. Output buffer in NMOS technology

Problem 19.2: Design a lab experience to find out if a node under test is in high impedance (tristate) condition.

In this condition, both M7 and M10 are off, so that the gates of both M16 and M17 are driven low. The inputs to the buffer are both the DATA signal (from the sense amplifier) and its complement, because in a NMOS inverter it is not possible to turn off a transistor and to turn on the other using the same signal. The two complementary signals are applied to two symmetrical structures. The need of using this circuitry is related to the use of the NMOS technology. The output of the differential stages of the sense amplifier does not have a dynamic swing equal to the supply voltage and, therefore, differential architectures must be implemented to speed up commutations. M2 and M5 are LVS (enhancement), while in the following structure M8 and M11 are natural, in order to apply a sort of initial filter to hinder the commutation, preventing spurious variations from making the output erroneously toggle.

After the first filter is overcome, M8 and M11 are natural so that the commutation is very fast, since their threshold is much lower than the LVS threshold voltage, and therefore the turn on time is less. Finally the task of M15 is to bring and keep the gate of M16 to VDD as soon as the transient is over, so that the pull-up is brought to the maximum available V_{GS}.

It is clear that the value of the output voltage cannot reach the value of the supply, but rather VDD-V_T. In this case the outputs are TTL compatible and reaching VDD is not required.

Problem 19.3: Is it possible to design an output buffer where a depletion transistor is used instead of M16, even if differently connected?

The output buffer of the Flash memory also contains an input buffer, which was not described in the chapter about the input buffer. The communication of the device with the external world is limited to writing for the EPROM, while the Flash device requires a more complex protocol so that the progress of the write operation can be monitored. Suspending the normal operation of the output buffers and enabling the input buffer that is embedded in the output buffer achieve this task. The driver composed of M16 and M17 is set to high impedance and the circuit, represented by the AND logic port, is enabled to propagate the data as input to the device. The data might be either data to be written (the corresponding address is provided on the input pads) or commands to be executed by the device. The specification for this input buffer is more relaxed than the address buffers and other common inputs, especially in terms of speed, therefore its design is actually implemented with the logic shown, together with logic ports that allow turn on, turn off and power consumption control of the buffer itself.

19.3 A CMOS Super Output Buffer

In order to reduce commutation time for the output buffers, the form factor of both the pull-up and the pull-down must be carefully designed to achieve the minimum delay. While a short channel length provides the best performance, the minimum channel length is set by the rules applied to implement Electron Static Discharge protection.[3] The gates of both the pull-up and the pull down are usually biased at GND and VDD, i.e. the V_{GS} is driven to full VDD. As the supply voltage of the devices decreases, it is necessary to increase the size of the final stage to maintain the performances to be equal to the previous devices, which used higher VDD. A possible solution is to drive the two final transistors with a V_{GS}, in modulo, greater than VDD. To achieve this result, a boost operation is carried out. This operation is composed of two phases: the charging of the C_{BOOST} capacitance and then the boost of the desired node. In our case, the gate of the pull-up must go below ground, while the gate of the pull-down must go above VDD. Two boost capacitances can be used, one for the final p-channel and one for the n-channel. A more area efficient method is to use a clever single capacitance, to generate the boost for both transistors without wasting too much space, as shown in Fig. 19.4.

[3] This topic will be discussed in Chap. 21.

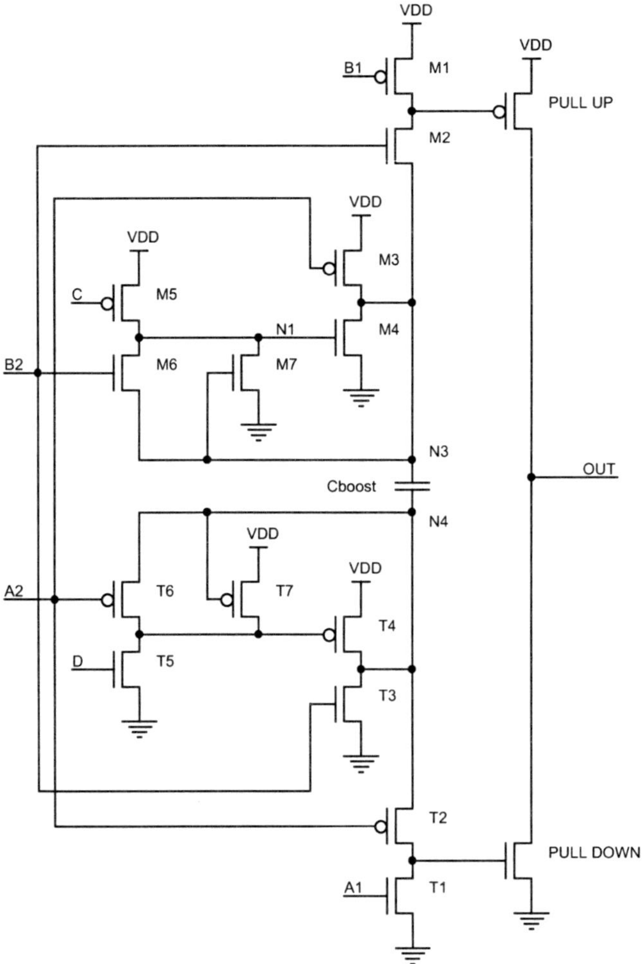

Fig. 19.4. The output buffer and its boost capacitor that allows boosting both the pull-up and the pull-down

At the beginning, C_{BOOST} is precharged, through M4 and T4, so that the voltages of the nodes N3 and N4 are ground and VDD respectively. M2 and T2 transistors isolate the capacitance from the final transistors. Let's consider that the data coming from the sense amplifier requires that a "0" be driven at the output; M3 is turned on, so that V_{N3} is set to VDD and, thanks to the boost, V_{N4} follows it up to 2 VDD, taking with it the gate of the pull-down. Of course, T3, T4 and T6 must be turned off not to leak away the charge stored on the boost capacitor. On the other hand, if the output must be charged to "1", the voltage of node N4 is forced to ground by T3; V_{N3} and the gate of the pull-up are therefore set to minus VDD.

B1 and B2 signals, as well as A1 and A2, are in a temporal sequence; the delay (achieved by a series of two inverters properly dimensioned) must guarantee that the charge stored on C_{BOOST} does not leak towards either VDD or GND through M1 and T1.

An important part of the whole structure is the circuit that turns M4 and T4 off that, during the boost phase, brings back on their gates the voltage present on the drain. This operation is necessary to avoid undesired charge leakages and not to limit the voltage value that can be reached by the gate of the pull-up and pull-down thanks to the boost. Let's consider for instance T4 (similar considerations apply on M4). If a voltage equal to VDD were imposed on the gate, when the drain reaches a voltage equal to VDD + $V_{T,P}$, T4 would turn on and the boost voltage on the pull-down would be clamped.

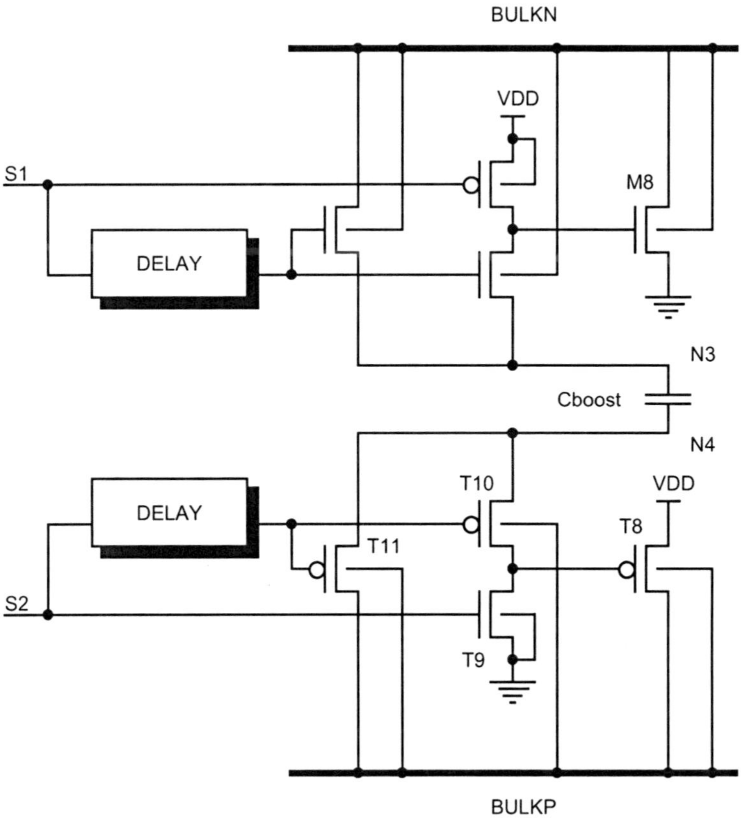

Fig. 19.5. The control network for the well that must be added to the output buffer of Fig. 19.4

As we have seen previously, N3 and N4 nodes work in a voltage range higher than power supply: therefore it is mandatory to verify that none of the junctions of the transistors connected to these terminals can be forward-biased. The usual connection (bulk of the PMOS to VDD and bulk of the NMOS to GND) does not necessarily guarantee reverse biasing of the junctions during the boost: all the bulks of the PMOS connected to N4, during the boost of the pull-down, in case the voltage is higher than VDD + $V_{T,pn}$, have a voltage high enough to forward bias the drain / n-well junctions.

Therefore the circuit shown in Fig. 19.5 is added to the network of precharge and boost: this circuit controls the body of both the p-channel and the triple-well n-channel transistors shown in Fig. 19.4.

Problem 19.4: Find out which transistors in Fig. 19.4 require a connection to the circuit shown in Fig. 19.5.

During the precharge of C_{BOOST}, the node BULKN, to which all the ip-wells of the triple-well transistors are connected, is kept to ground by M8 (S1 = 0). The node BULKP, to which the n-well of the PMOS connected to N4 are attached, is at VDD thanks to T8 (S2 = 1). Let's see the case of boost on the pull-down. Signal S2 is set low, T9 turns off and, after a certain delay, T10 and T11 turn on. Node BULKP is brought by T11 to the boosted voltage V_{N4}. T10 blocks the path toward VDD by biasing the gate of T8 to the same voltage as its drain. The DELAY block is required to guarantee the complete turn-off of T9 before the boosted voltage is transferred. Similar considerations apply for BULKN.

19.4 The "High Voltage Tolerance" Issue

Integrated circuits assembled on a board (microprocessors, memories, converters, etc.) are usually fabricated with different technological processes. There is no point indeed in using a non-volatile memory process, with all its complications, to realize a purely logic component. Therefore it may happen that devices operating at different supply voltages have to communicate with each other on the same board. One of the issues is the interfacing of the different voltages. Some devices offer the option of supplying the core at VDD and the I/O buffers at VDDQ. Nowadays, several commercial memories have a core working with a VDD equal to 3 V and a VDDQ equal to 1.8 V, so that they can be interfaced with logic chips usually working at 1.8 V only.

The specification known as *high voltage tolerance* is mandatory for the devices where core and I/O power supplies are not separated; in this case, a chip supplied with 3 V only must share the *DATA BUS* with other chips operating at higher supply voltages, for instance 5 V (Fig. 19.6).

When a device has to use the common DATA BUS, the microprocessor, if present, acts on either OE# or CE# signal, to switch off the other devices: the corresponding output buffers are put in tristate, switching off both the pull-up and the pull-down. In Fig. 19.6, biasing for the final stages are shown in the case of a device supplied with 5 V that drives the DATA BUS. Both the bulk and the gate of the turned off pull-up transistors are connected to their supply voltage (3 V), causing a forward biasing of the drain-body junction (Fig. 19.7).

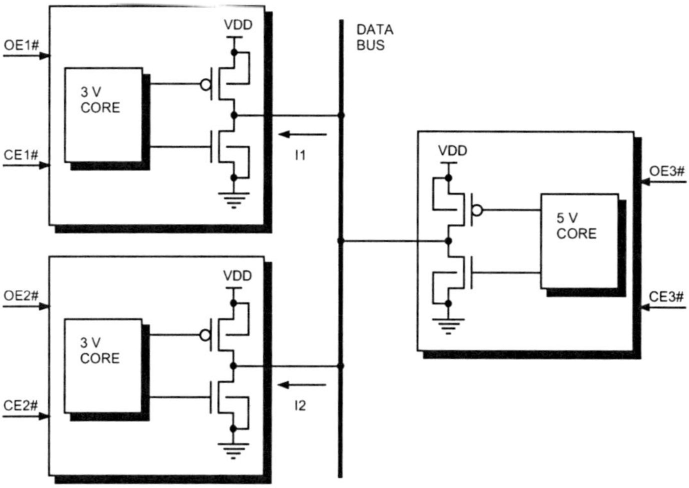

Fig. 19.6. Sharing of the DATA BUS by devices supplied at different voltages

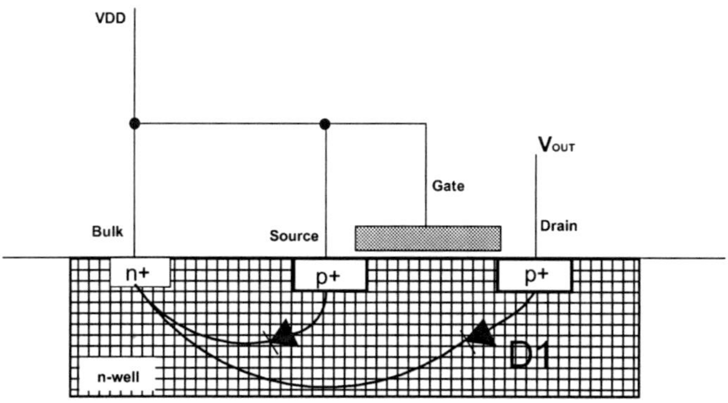

Fig. 19.7. When V_{OUT} > VDD + $V_{T,pn}$ drain-body junction (D1) is forward-biased

A device with a forward-biased junction is always at risk, since the current doesn't flow in the intended paths, but it passes through the substrate, usually causing some undesired parasitic effects. Furthermore, the forward biasing of the p-channel transistors in the 3 V circuits by 5 V signals can induce a consumption of the device at 5 V, because of the current flowing from the 5 V into the 3 V supply.

Several manufacturers solve the issue by stating in the specification that a diode must be placed on the board in parallel to the pull-up: this diode must have a lower threshold so that it absorbs the current. Another way is to accept power consumption for the device at 3 V. Let's see if it is possible to adequately solve both the problems.

We have to build a circuit, let's call it HVT, that can compare the voltage present on the output pad with VDD and provide, as an output, the highest available voltage; we can also use this voltage to bias the body of the pull-up. The simplest solution would be to use a voltage comparator; due to the severe specifications of consumption in stand-by, more complex structures must be used. Since the trigger for the forward biasing of the pull-up depends on the threshold of the drain / body junction of the p-channel, triggering voltage for the HVT circuit must be lower than this value; for instance, we can use the threshold of a natural transistor $V_{T,NAT}$ (~ 300 mV), which is lower than the turn on of the *pn* junction (~ 700 mV). The transfer characteristic for the HVT circuit is drawn in Fig. 19.8. Of course, the circuit that we are designing is active only when the buffer is in tristate, the time when another device is enabled to set the outputs through the common bus.

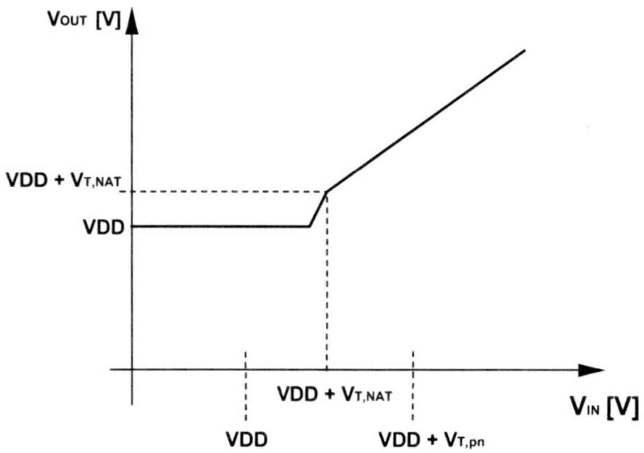

Fig. 19.8. Transfer function for the HVT (High Voltage Tolerance) circuit

Figure 19.9 shows the position of HVT inside the circuitry of the output buffer. During the read phase, the buffer is active; OE# signal is low and M2 and M3 transistors and P2 transfer gate are turned off.

M1 transistor and P1 transmission gate are turned on and the HVT circuit is disabled (Fig. 19.10a). By setting OE# high, the buffer goes into tristate, the previously turned off transistors are turned on, connecting the output of the HVT circuit to both the gate and the bulk of the pull-up (Fig. 19.10b).

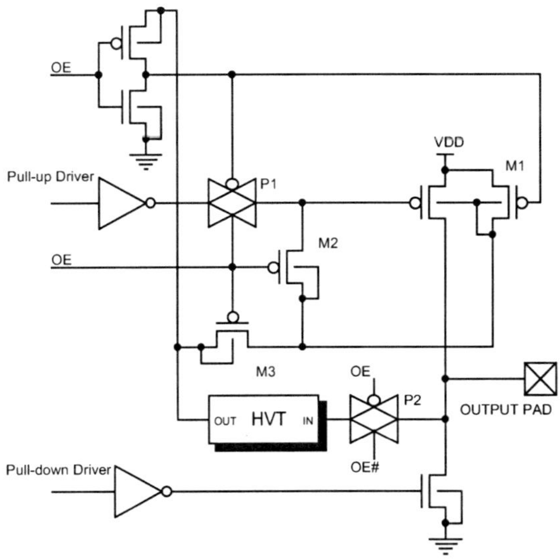

Fig. 19.9. Schematic of the output buffer with its HVT circuit to comply with high voltage tolerance specification

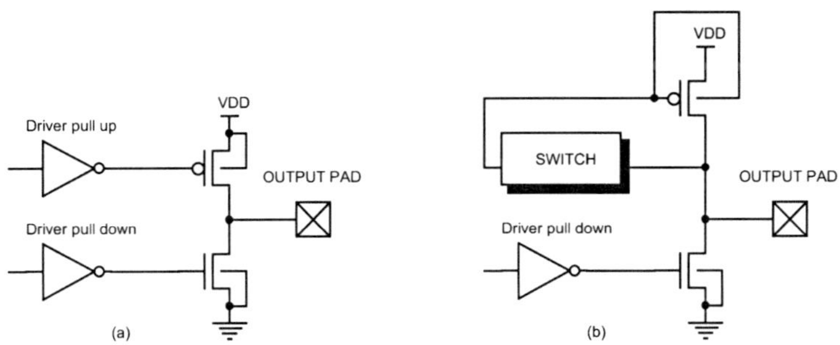

Fig. 19.10. Circuit of Fig.19.9, in case of OE = 1 (**a**) and OE = 0 (**b**)

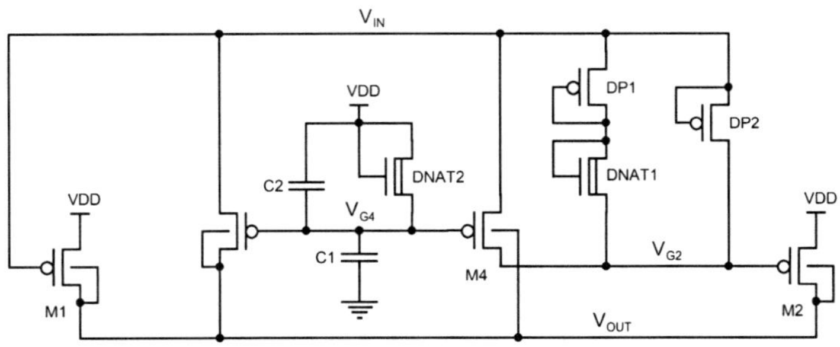

Fig. 19.11. Electrical scheme of the HVT circuit

HVT circuit is drawn in Fig. 19.11. M1 keeps the output at VDD when

$$V_{IN} \leq VDD - |V_{T,P}| \tag{19.5}$$

Diode-connected transistors DP1 e DNAT1 cause a voltage drop such that

$$V_{G2} = V_{IN} - \left(|V_{T,P}| + V_{T,NAT}\right) \tag{19.6}$$

and therefore M2 transistor is on until V_{IN} reaches (VDD + $V_{T,NAT}$), keeping the output at VDD.

M3 and M4 transistors turn on when their drain (V_{IN}) reaches a voltage value of

$$V_{IN} = VDD - V_{T,NAT} + |V_{T,P}| \tag{19.7}$$

bringing V_{IN} voltage to the output and on the gate of M2. This causes the simultaneous turn on of M3 and turn off of M2. Finally the diode DP2 allows the gate of M2 to go low when V_{IN} falls.

The circuit described above correctly works when the supply voltage varies if the node that drives the gates of both M3 and M4 can correctly follows the variations of VDD. C1, C2 and DNAT components, shown in Fig. 19.12, are used to achieve this result.

When V_{IN} (= VDD) increases, C1 capacitor is charged by the DNAT2 diode and therefore V_{G4} is equal to (VDD – $V_{T,NAT}$); when V_{IN} decreases, the diode is inverse-biased, therefore the discharge takes place through C2. C1 and C2 capacitors must be dimensioned in such a way that the value reached by V_{G4} is lower than (VDD – $V_{T,NAT}$).

Considering the capacitive division, the following relation must hold:

$$\frac{C2}{C1 + C2} \cdot VDD < VDD - V_{T,NAT} \tag{19.8}$$

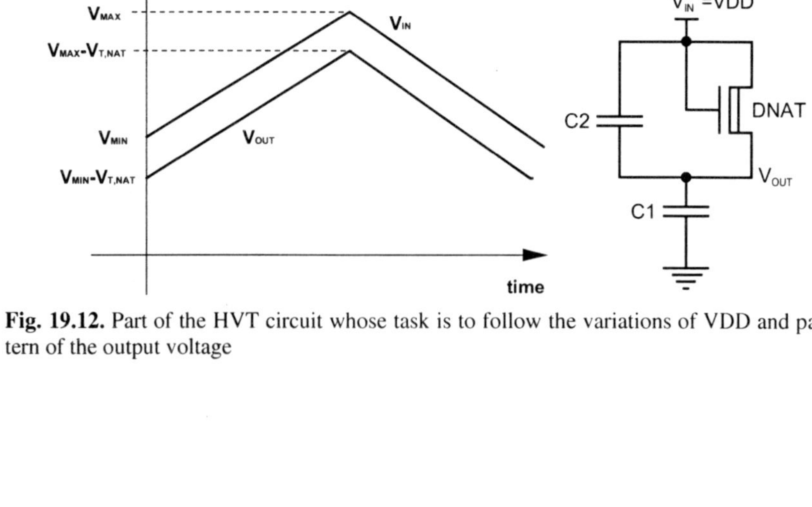

Fig. 19.12. Part of the HVT circuit whose task is to follow the variations of VDD and pattern of the output voltage

19.5 Noise Induced on the Signal Circuitry by Commutation of the Output Buffers

Device pads are connected to the external terminals of the package, called pins, by means of bonding wires that can be modeled as an inductance L. Problems arise during the commutation of the output buffers that, as we have seen, have to provide the data sensed during read to the external world. During the turn on of the pull-up and the pull-down, a voltage drop occurs on the inductance of the bonding wires of VDD and ground, so that the value of the internal VDD node decreases while the value of the internal ground node increases. The amount of the voltage drop depends on both the value of the inductance and the current flowing through it according to the equation

$$v = L\frac{di}{dt} \tag{19.9}$$

If all the N buffers, where N is equal to 8 for a byte and 16 for a word, have a simultaneous switching, the voltage drop is proportional to N.

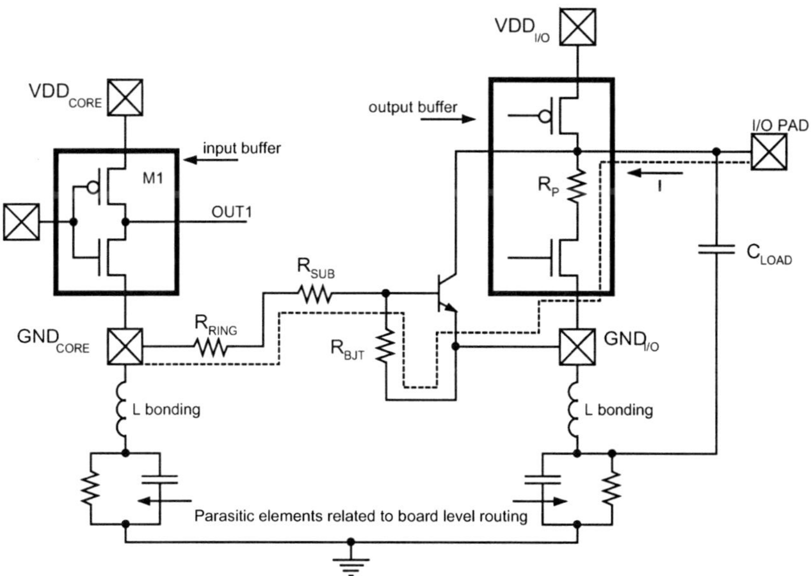

Fig. 19.13. The discharge current of the output buffer returns to the inputs through the dashed path

Figure 19.13 shows the parasitic connections present on the device that are involved in the discharge commutations of the output capacitances. Ground and VDD pads have been separated, those for the outputs ($GND_{I/O}$ and $VDD_{I/O}$) and those for all the rest of the device (GND_{CORE} and VDD_{CORE}), in order to increase the noise immunity. Resistance R_p, inserted between the external pad and the drain of the pull-down is used, together with the npn lateral transistor, to protect the output buffer from the electrostatic discharges between the I/O pad and GND. The discharge current of the pull-down causes an increase of the value of the $GND_{I/O}$ node that, in turn, is reflected on the GND_{CORE} pad through the resistive dashed path in Fig. 19.13. R_{BJT} is the resistance associated to the biasing of the ESD protection bipolar, R_{SUB} is the resistance of the substrate (that is common to all the device) and R_{RING} is the resistance of one of the several biasing rings for the n-channel transistors near the input buffers.

Let's suppose that a TTL-level signal is presented to the input. As a result of the commutation of the pull-down of the output buffer, the value of the ground increases, V_{GS} of the input pull-down decreases and therefore the pull-up, PMOS M1, might be able to bring the node OUT1 high, e.g. causing the generation of a new ATD and, subsequently, an undesired read phase.

The usually implemented solutions are to separate as much as possible the different ground networks and to realize a greater resistive path between the two by interposing biasing shields and by confining the anti-latch up rings of the different grounds in distinct zones. There is also the possibility of isolating the ground of the outputs from that of the circuitry. If triple well technology is available, the n-channel of each output buffer can be placed in a separate triple-well tub.

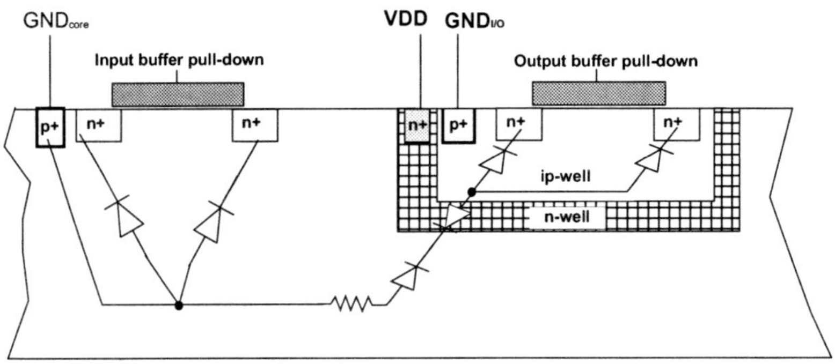

Fig. 19.14. Placing the pull-down of the output buffer in triple well eliminates the resistive path between the output and the input

Figure 19.14 represent the cross section of the process for the two n-channel transistors, the one of the input buffer connected to GND_{CORE} and the one for the output buffer placed in the triple well.

In order to define the global network of supplies and grounds of a device, it is also mandatory to take into account the noise induced on the supplies by the commutation of the output buffers that occur when they charge the output capacitances.

Figure 19.15 shows the traditional schematic of the connection to both ground and power supply of the output buffers and of the signal circuitry for a non-volatile memory. The equivalent circuit is shown in Fig. 19.16. The values of the load capacitances, external to the pads, of the inductances for the bonding wires and of the resistance of the traces have been estimated to be 100 pF, 10 nH and 0.5 Ω respectively; the values of the capacitances seen between the internal pads of VDD and GND are typically around 20 pF and 1 nF, respectively, for the whole 16 output buffers and for the signal circuitry. From the previous analysis, we already know that the input circuitry is not immune to the fluctuations of the $GND_{I/O}$, node, because of the substrate resistance whose value is very low, i.e. some Ohm.

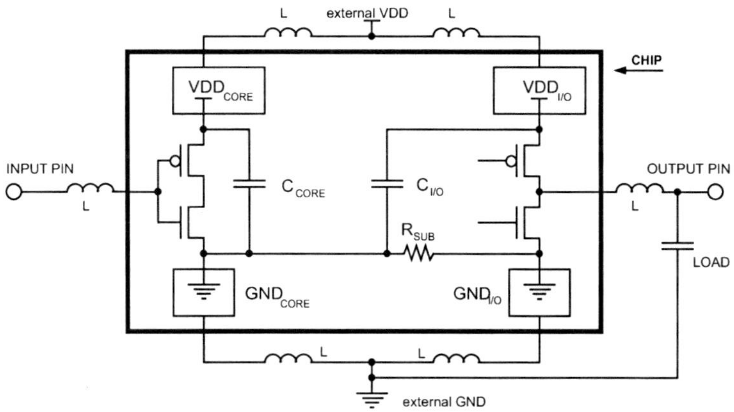

Fig. 19.15. Schematic of the connections of power and ground for a non-volatile memory without triple well

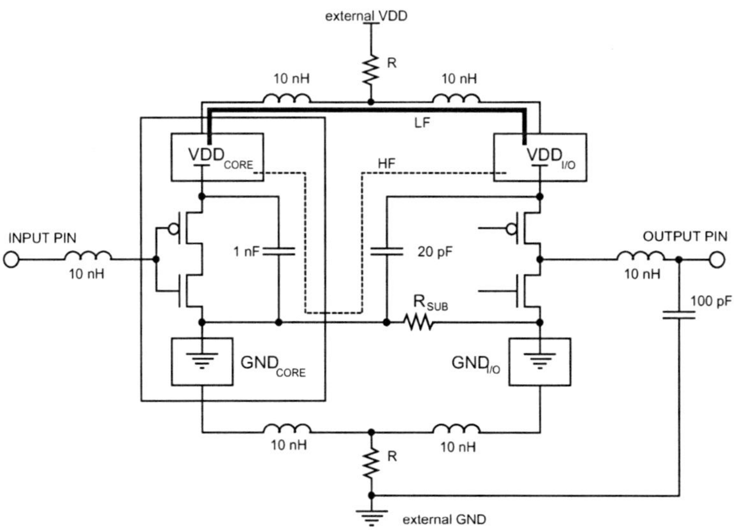

Fig. 19.16. Equivalent circuit for the classic connection of the output buffers. In the frame on the left an input buffer is shown. R = 0.5 Ω, R_{SUB} = 1

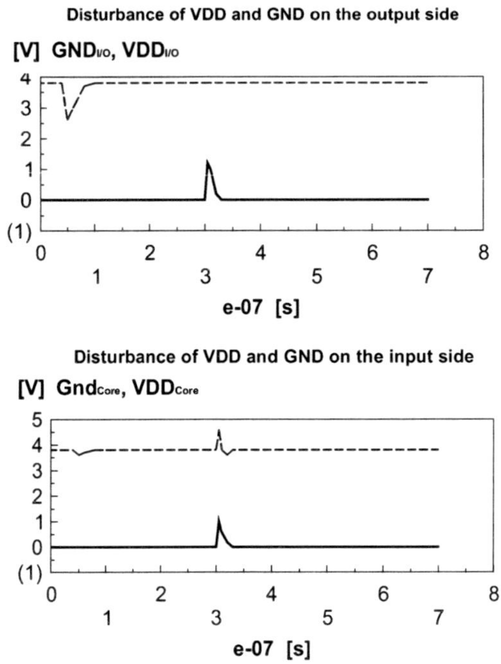

Fig. 19.17. Simulation for a noise present on the $VDD_{I/O}$ node

Let's now examine the effect on the signal circuitry of noise on the $VDD_{I/O}$ node. Figure 19.17 shows the timing diagram of the VDD_{CORE} voltage of the signal circuitry perturbed by a simultaneous switching of 16 output buffers. VDD is set to 3.6 V and furthermore the signals applied to the pull-up and the pull-down transistors of the output buffers do not toggle at the same time to avoid a crowbar current flowing through them. It is clear that the fluctuation of the VDD_{CORE} is significant both in absolute value and duration, and the fluctuation can induce spurious commutations of the input buffers. The reason for this sensitivity is due to the two paths through which a fluctuation on the $VDD_{I/O}$ node can propagate to the VDD_{CORE} node.

The former (LF) is shown in bold in Fig. 19.16 and it turns out to be dominant at low frequencies when the inductances are equivalent to short circuits and the capacitances to open circuits; the latter (HF), shown as a dashed line, is dominant for frequencies higher than 100 MHz, corresponding right to the commutation timings of the input signals of the output buffers. In this frequency range, the capacitances completely transfer the fluctuation of the $VDD_{I/O}$ node to the one of VDD_{CORE}. In conclusion, it is mandatory to introduce latches for data in non-volatile devices and that the noise margin of the input buffers of the addresses must be increased in order to reduce the sensitivity of the signal circuitry to this phenomenon.

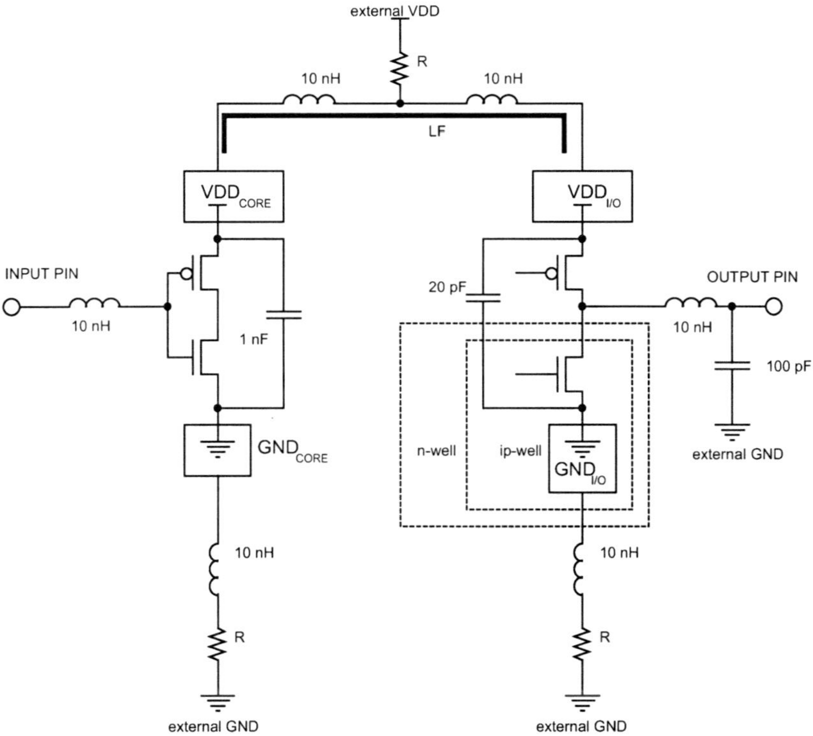

Fig. 19.18. Scheme of the connection of the output buffers in triple well. R = 0.5 Ω

Placing the NMOS transistors of the output buffer in a triple well structure is intended to eliminate the HF path shown in Fig. 19.16 for both the fluctuation of $GND_{I/O}$ and of $VDD_{I/O}$ originating in the output buffers; the resulting scheme is shown in Fig. 19.18. The transfer of noise due to the buffers towards the signal circuitry should only take place then through the low-pass network that is pertinent to VDD supply, and consequently the amount of noise of VDD_{CORE} and its frequency content, as shown in Fig. 19.19. The implementation of triple well structures will lead to increased noise immunity with respect to the conventional solution, at least at high frequency, for which the equivalent circuit of the network is shown in Fig. 19.18.

Unfortunately the use of a triple well structure introduces two additional parasitic capacitances, the former between the ip-well and the n-well, the latter between the n-well and the substrate, which re-establish a connection between the nodes $VDD_{I/O}$, $GND_{I/O}$ and GND_{CORE}, as shown in the equivalent circuit of Fig. 19.20. The value of these capacitances is determined by the area occupied not only by the NMOS of the buffers, but also of the ESD protection bipolars placed in the triple well.

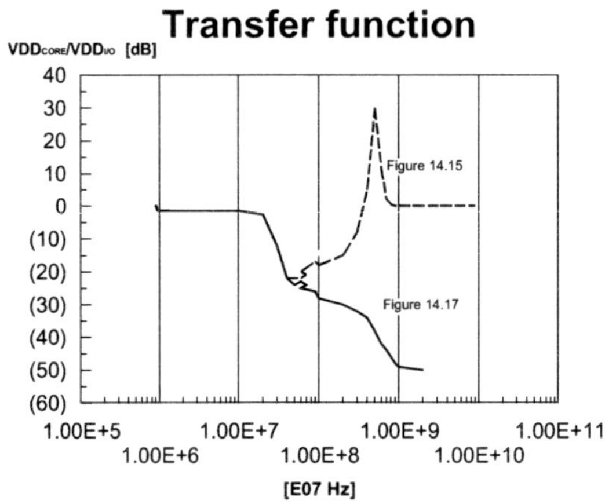

Fig. 19.19. Reduction of the noise achieved by placing the buffer in triple well. The transfer characteristics VDD_{CORE} / $VDD_{I/O}$ pertinent to Fig. 19.16 and 19.18 are shown

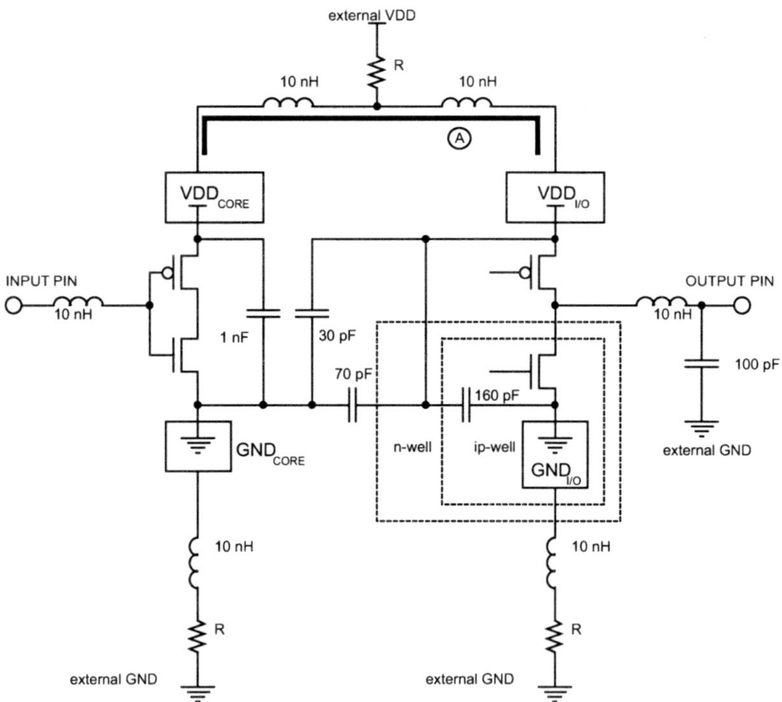

Fig. 19.20. Scheme of the connection of the output buffers in triple well and related parasitic capacitances

The capacitance between the ip-well and the n-well is about 10 pF, while the one between the n-well and the substrate is about 5 pF for each buffer for a typical 0.6 μm Flash process. The simulation of the simultaneous switching of the output buffers in triple well, performed under the same conditions as with the conventional structure, unfortunately highlights a greater coupling between the nodes $VDD_{I/O}$ and $GND_{I/O}$ precisely because of the presence of the capacitances between the wells; anyway, these capacitances realize an improved filtering of the noise towards the signal circuitry with respect to the scheme of Fig. 19.16, and in fact the noise on VDD_{CORE} and GND_{CORE} is halved, at least as peak value. Of course the entity of the peak is influenced by the value of the additional parasitic capacitances. Therefore, placing the output buffer in an isolated well is desirable to reduce the sensitivity of the input circuitry to the switching noise of the outputs.

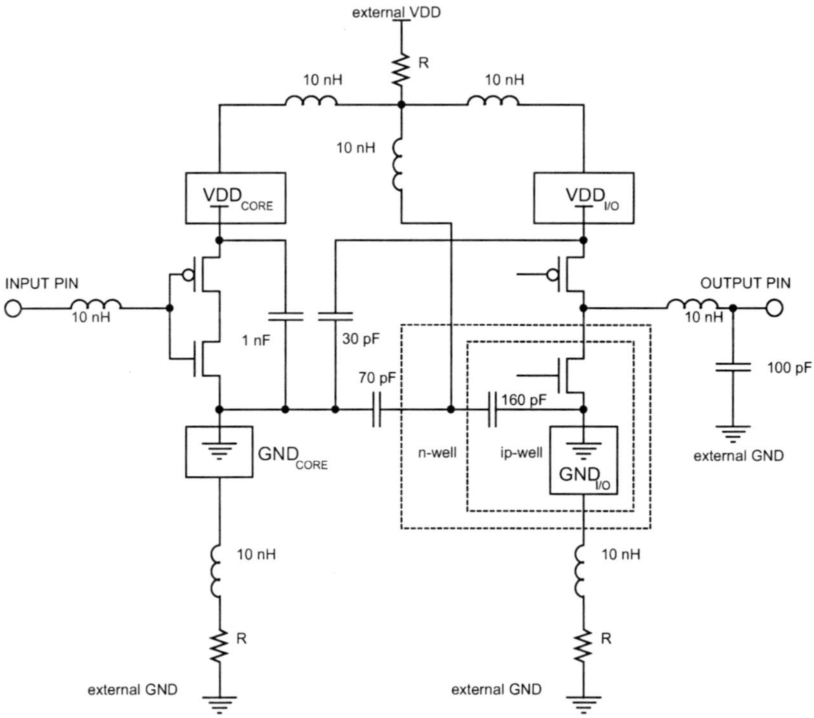

Fig. 19.21. Scheme of the connection of the output buffers in triple well in the case of dedicated connection to VDD for the n-well

Figure 19.21 shows an alternate solution for the scheme of Fig. 19.20, which requires a separate bonding to VDD for the n-well of the output transistors placed in triple well. In this way the coupling between $VDD_{I/O}$ and $GND_{I/O}$ is reduced; consequently, for the commutation that involves the pull-up transistors of the output buffers, the noise transferred to the signal circuitry decreases, while, for the commutation that involves the pull-down transistors, the situation is the same as in the case shown in Fig. 19.20. On the other hand, an AC analysis of the network of Fig. 19.21 proves that, in the transfer of noise between $GND_{I/O}$ and $GND_{I/O}$, the dominant impedance is the series of the capacitances of the wells, which has not changed with respect to the scheme of Fig. 19.20; therefore the noise immunity of $GND_{I/O}$ has not improved.

In order to complete the analysis, let's consider the case where the input buffers are placed in an isolated well also. The equivalent circuit for this configuration is shown in Fig. 19.22 and, with respect to Fig. 19.20, it includes the capacitance between the ip-well and the n-well of the input buffer, and the network for the connection of the input buffer itself to ground. Simulations prove that the noise induced on the ground of the input buffer is reduced if compared to the previous cases. On the other hand it is necessary to bear in mind that the implementation of the scheme of Fig. 19.22 greatly increases the area occupation, because of the ESD protections that must be added inside and outside the triple well. Anyway, depending on the chosen technology and specific device layout, placement of the ESD protection bipolar inside the triple well can be avoided.

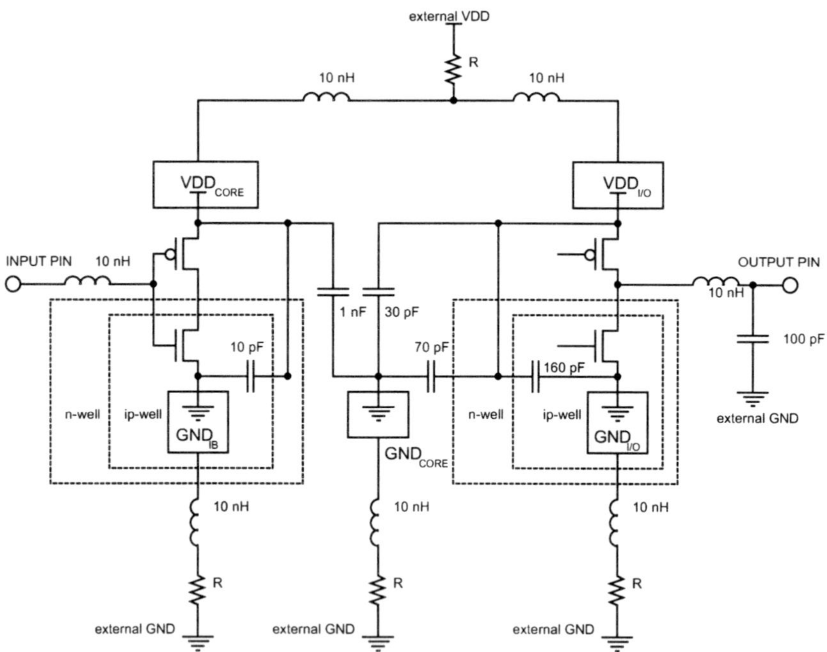

Fig. 19.22. Scheme of the connection of both the output and the input buffers in triple well

To sum up, the design can be realized by implementing four ground pads and three supply pads. Two ground pads can be dedicated only to the ground of the pull-down in triple well of the output buffers, $GND_{I/O1}$ for eight buffers, $GND_{I/O2}$ for the remaining eight, and the two pads are then connected to the same package pin. GND_{IB} is used for the ground of the NMOS of the input buffers only, and GND_{CORE} for all the remaining circuits inside the device. Partitioning of the ground networks aims at dividing and isolating the current paths as much as possible.

Three pads are used for the supplies: the first dedicated only to the source of the final p-channel transistors of the output buffers, the second used to bias the n-well tubs that constitute both the bulk of the output p-channel transistors and the n-well that constitute the "surrounding" of the n-channel in triple well of the output buffer, and finally the third for the entire signal circuitry.

Bibliography

R. Achar, M. Nakhla, "Simulation of high-speed interconnects", Proc. IEEE, Vol. 89, pp. 693-728, (May 2001).

H.B. Bakoglu, Circuits interconnections and packaging for VLSI, Addison Wesley Pub. (1990).

B. Dipert and M. Levy, Designing with Flash Memory, ANNABOOKS, San Diego, (1994).

G. Campardo, M. Zammattio, S. Ghezzi, "Integrated device with pads", USA patent No. 5,923,076, (July 13, 1999).

G. Campardo, S. Zanardi, M. Branchetti, "Low noise output buffer for semiconductor electronic circuits", USA patent No. 6,060,753, (May 9, 2000).

G. Campardo, S. Zanardi, A. Ghilardelli, "Output stage for a memory device and for low voltage applications", USA patent No. 6,215,329, (April 10, 2001).

G. Campardo et al., "High voltage tolerance output stage", USA patent No. 6,150,844, (November 21, 2000).

C. Lam, S. Ali, et al., "Three dimensional modeling of multichip module interconnects", IEEE Trans. Comp. Hybrids, Manuf. Techn., Vol. 16, pp. 699-704, (November 1993).

Packaging Handbook, Intel, (1993).

20 Test Modes

In addition to User Mode, there is also a "hidden" method for customer access to both the matrix and circuitry to analyze the operating behavior of the device.

Access to this kind of investigation is not explained in the manual because the chip manufacturer, only, should know the enabling code. This chapter will describe the main available test modes, without showing the access procedure; just recalling that it is usually composed by a sequence of instructions issued together with some third level on specific control pins.

20.1 Introduction

Test modes will be defined as all operation of circuitry performances unknown to the user. There are mainly two reasons to not implement such operations: (1) many of these tests must be performed exactly knowing the cell physics. Since all voltages are applied to the matrix directly from the exterior, the operator must mimic the correct timing sequence, usually carried out by the circuitry; (2) the inferred information can be quite sensitive, and the customer is not always able to correctly analyze it. Therefore, test usage only will be presented.[1]

20.2 An Overview on Test Modes

In this section, some test modes are provided with a short description; usage is not extensively presented because it can be easily inferred. Other test modes are explained in greater detail in following sections.

Measurements on UPROM

All the operations performed on these non-volatile registers are applied in test mode only: Therefore, UPROM can be programmed, erased and verified as any other matrix cell. Furthermore, UPROM can be disabled to separately analyze both the normal and the redundant path. This method applies both to redundancy and configuration UPROM.

---

[1] Besides that, commands can vary among different manufacturers.

OTP Rows

OTP is usually divided into two sets: one is reserved for the user, the other for the chipmaker. Access to these rows is not defined in the datasheet. Therefore, only key customers are permitted to use them, – i.e. to implement a hardware password to prevent device cloning. A keyword written in the OTP and periodically read by the code written in the Flash addressable space prevents illegal program reproduction. Since the OTP write operation is unknown, it is not possible to precisely copy the code, and therefore hackers cannot clone it.

Redundancy Matrix

Both redundancy rows and columns can be accessed in order to verify their functionality before usage. Therefore it is possible to see the entire matrix, which is composed of addressable cells (accessed by external addresses) and redundancy cells (accessed in test mode only). These cells can be independently read and written, while they are erased together with the sector they belong to (by sharing the source node).

Reference Matrix

Reference cells can be completely accessed to be set to the desired threshold voltage. Read, Program and Verify are usually independent for each cell, while Erase is common, for sake of simplicity.
For uniformity, Erase of all cells (of the Matrix, the UPROM and the Reference) is performed following, as much as possible, the same procedure.

Sector Erase

Since the Erase time of 1 Mbit Flash is around one second, it is evident that the possibility of erasing more sector altogether is a great benefit to decrease testing time. Some devices also features Full Chip Erase, i.e. the simultaneous erase of the whole matrix.

Parallel programming

This test mode allows parallel programming of more cells with respect to user mode, in order to reduce writing time of the whole matrix during EWS.

EPROM-like test modes

These test modes mimic device behavior into the EPROM that do not have an internal, dedicated state machine. EPROM-like test modes require an external "intelligence" to properly time the application of the various voltages. Therefore it is possible to provide both VPCX (gate voltage) and VPD (drain voltage) from the outside during Program, and to verify the behavior of the matrix independently of the charge pumps. Negative voltages are usually generated on chip. The problem of forcing a negative voltage from outside is related to the ESD protections connected to the pad. In this case, protection is further complicated by the fact that any n-type junction directly connected to the substrate must be avoided.

Problem 20.1: Design an ESD protection structure for negative voltages using a triple well process.

Number of attempts

User mode algorithms are able to repeat both erase and program pulses before considering a memory location as fail. The number of attempts can be reduced in order to examine (and stress) the behavior of the matrix in a situation worse than the real usage.

Sync signals

Voltage values of the integrated circuit nodes can be probed by special instrumentation that is capable of displaying them on a screen as if it were an oscilloscope. The issue is that a trigger, i.e. a synchronization signal, is required. Slow operations, like erase (which lasts some tens of milliseconds) do not provide any trigger. To overcome this limitation, a test mode is available: this generates a sync signal that is available on a pad.

The state machine

Present Flash memories embed very complex algorithms. In test mode, it is possible to execute only some portions of both the program and erase algorithms.

For example, program all zero, i.e. a complete program and verification of the matrix, can be performed alone. The same concept applies for both electrical erase and soft program.

It is also possible to program repetitive patterns that are used in EWS to verify the correct functionality of matrix cells. A typical example is the *checkerboard* program: the first row is written as 1010101010..., the second as 0101010101 and so on.

Analysis of internal nodes

There are specific test modes to measure the internally generated and regulated voltages in device pads, such as VPD and VPCX. Several logic nodes can be studied externally by storing them on a register bank whose content can be read similar to a normal matrix location.

20.3 DMA Test

This section describes the "prince" of the test modes, the DMA[2] test whose main purpose is to connect the cell terminals directly to the external pads. The significance of this test is that device characterization is always difficult and time con-

[2] DMA is the acronym of Direct Memory Access.

suming. It is difficult to "filter out" the interference of the matrix circuitry, so that the behavior of the cells is "disrupted". Many incorrect performances can result from in the circuitry problems, like voltages applied in the wrong way, voltage spikes etc. The possibility of analyzing single cells is therefore a valuable contribution. DMA allows bypassing both the sense amplifier and the output buffer by connecting the drain of the cell directly to the output pad, as shown in Fig. 20.1.

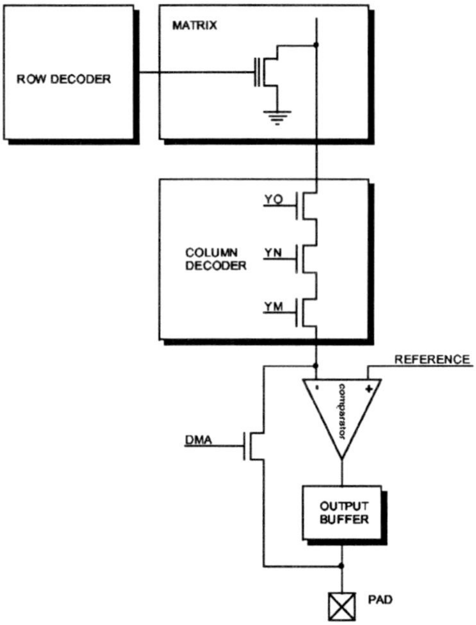

Fig. 20.1. DMA test directly connects cell drain to the pad, bypassing both the sense amplifier and the output buffer

Decoders work correctly, but the sense amplifier is switched off, while the output buffer is in high impedance. On the external pad, a voltage equivalent to the one forced by the sense amplifier in read (~ 1 V) is applied. Measuring the current supplied by the generator enables the calculation of both the threshold voltage and gain of the cell. Cell characteristic can be obtained by implementing, besides DMA, another test mode that allows controlling VPCX voltage, i.e. row biasing, through another external pad. By varying the VPCX, and measuring cell current in DMA, it is possible to draw its characteristic. Cells distributions, that have been deeply discussed and of crucial importance in Flash memory design, are calculated using this test. Gain can be computed by means of two DMA measures at different VPCX values.

All the cells inside the device, matrix, redundancy, reference "mini" matrix, UPROM, even the mirror transistor of the reference, have a parallel path to bring their drain on output pads to tap their current.

20.4 Fast DMA

It is important to know that in DMA, measuring the current absorbed by the cell is a very slow procedure, milliseconds, which poses a considerable hindrance when the test is applied on a multi-million cells matrix[3]. Another test is used to expedite the task of acquiring cell distributions. A fixed current (that can be both controlled and observed from the outside) is provided to the reference branch of the sense amplifier and is compared with the one flowing through the addressed matrix cell. The advantage is that read operation is very fast, since current to the voltage conversion is performed by the sense amplifier. If, for example, the reference current is 10 μA, it is possible to distinguish all the cells that absorb a higher amount of current. By varying the current in small increments, or changing VPCX value, it is possible to draw an histogram that is similar to cell distribution graph, but it is much faster to obtain.

20.5 Oxide Integrity Test

Drain stress and gate stress tests can be used to judge the oxide quality. The former is used to stress all the cells belonging to the same column where the cell to be programmed is located. These cells have the gate connected to ground and the drain biased at the programming voltage. If the thin oxide quality is not at its peak, the electrons stored in the floating gate may "escape" attracted to the drain contact by the positive voltage (Fig. 20.2).

Erased cells might suffer from a drain stress as well if they are written through the interpoly oxide, even if this effect is usually less relevant. Finally, erased cells might become over-erased thus being depleted. To verify the quality of tunnel oxide during EWS, written cells undergo a collective drain stress. This is followed by a verify to control the permanence of the written threshold voltage. This test mode allows applying programming voltage to all the columns of the matrix, sector after sector, while the gate is biased at ground.

On the other hand, gate stress testing provides information on the state of a cell whose gate is shared with the cell gate under program while its drain is floating (Fig. 20.3). In this case, an erased cell might appear as written because the electric charge may be attracted from the drain junction onto the floating gate towards gate voltage. This might occur because of either tunnel oxide features resistive paths, or a trapped charge that causes a potential barrier represented by the oxide to decrease.

During EWS, gate stress is applied to further check the quality of the oxides. It is also applied to reproduce stress condition the cells undergo due to matrix organization; where many cells share the same gate, while others share the same drain.

---

[3] In case of an 8 Mbit Flash, if the acquisition time is about 3 ms the overall time is $3\ \text{ms} \cdot 8 \cdot 1{,}024 \cdot 1{,}024 = 25{,}165.8\ \text{s} = 7\ \text{h}$.

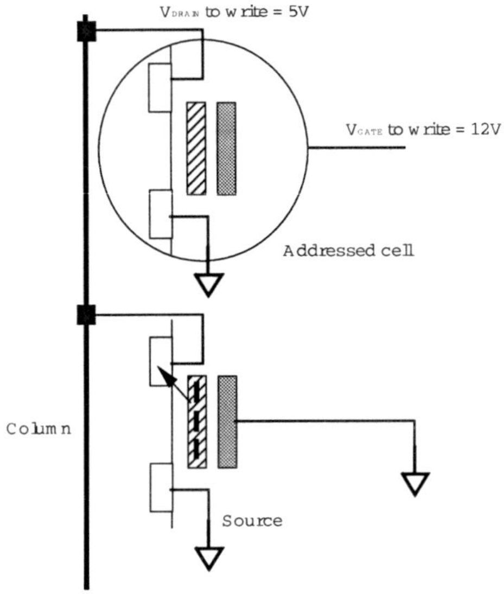

Fig. 20.2. Drain stress effect on the cell belonging to the same column where the cell under program is located

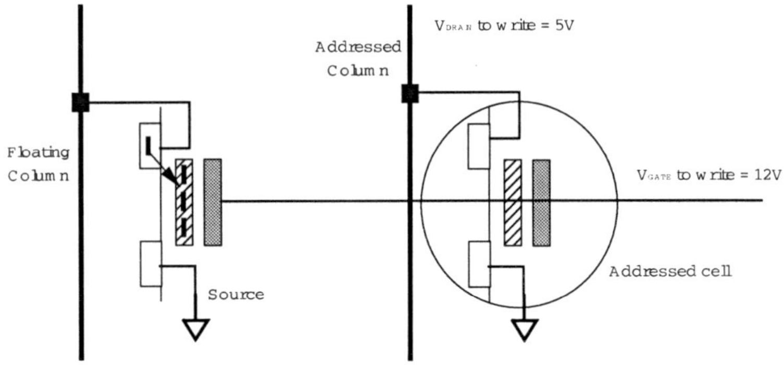

Fig. 20.3. Effect of gate stress on the cell belonging to the same row where the cell under program is located

Finalization of test modes is usually the last step of device design. The design is completed and ready to be manufactured. While the chip is under development, both designers and product engineers work together to define EWS, Final Test (i.e. the test performed on packaged parts), characterization programs, and related hardware, to be prepared when the first wafer is manufactured.

Bibliography

G. Campardo et al., "Data protection method for a semiconductor memory and corresponding protected memory device", USA patent No. 6,286,086, (September 4, 2001).

C. Casagrande, "Flash memory testing", in Flash memory, P. Cappelletti et al., Ed Norwell, Ma: Kluwer, (1999).

21 ESD & Latch-Up

After circuit design and layout, there are additional issues to be covered before releasing the design for fabrication. Two of the more difficult issues to address are Electro Static Discharge (ESD) protection and "latch-up". Over the years, ESD protection has grown in importance as semiconductor technology processes have become smaller and more complex. "Latch-up" is the undesired enabling of parasitic bipolar transistors which may cause irreversible device damage.

21.1 Notes on Bipolar Transistors

Today, most VLSI designers do not use bipolar transistors. Designers primarily focus on designs using standard CMOS circuits. However, it is worthwhile to discuss bipolar junction transistors (BJTs). They appear in any solid state CMOS circuit as parasitic devices. In order to understand the operation of a bipolar transistor, it is necessary to recall how an ordinary *p-n* junction works.

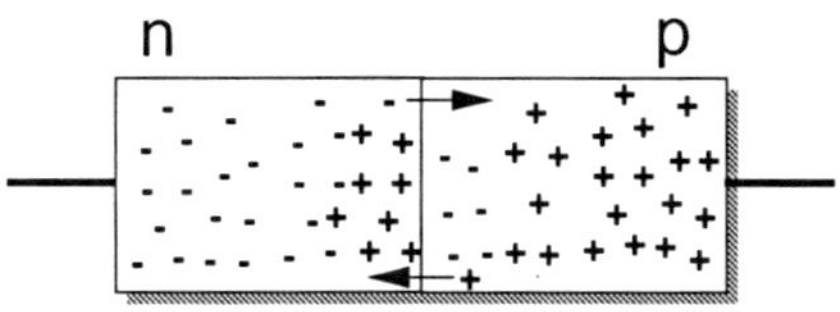

Fig. 21.1. Diffusion current due to the different charge concentration in the two regions, *p* and *n*

A junction (see Fig. 21.1) is formed by the joining of two complementary doped semiconductor materials. In the case where no external electrical field is applied, a transient diffusion current occurs. This causes a charge migration at the interface between the *p* and *n* regions. Positive charges are excessively present in the *p* region; these charges are defined as majority carriers in this region, while the negative charges are defined as minority carriers.

Majority carriers of the *p* region flow into the *n* region. Conversely, we see the opposite effect in the *n* region: majority carriers of the *n* region (negative charges) flow into the *p* region. This flow is caused by the different concentration of charges in the two zones. Charges that cross the junction recombine as soon as they reach the complementary zone. As a result, doped atoms placed near the interface have no free charges (see Fig. 21.2). A double layer of fixed charges, called the depletion region, is generated. The depletion region's electrical field hinders further diffusion of majority carriers from one layer to the other. It favors the drift of the minority carriers in the opposite direction.

When equilibrium is reached, the electrical field in the depleted zone balances the diffusion of majority carriers from each layer. However, a drift of minority carriers still exists in the opposite direction. Overall current through the junction is equal to zero for both electrons and holes since the junction is not biased.

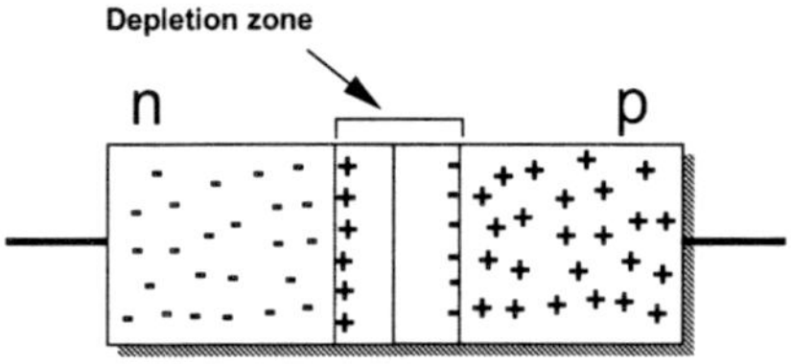

Fig. 21.2. Equilibrium is reached as soon as depletion zone is formed

This phenomenon occurs in doped semiconductors since current is composed of two kinds of carriers. Their movement is due not only to the presence of an electrical field (drift), but also to the different concentrations of holes and electrons (diffusion). This effect is similar to salt dissolving in water. The "depth" of the depletion region is inversely proportional to doping (see Fig. 21.3). If the *n* region is less doped than the *p* region, the depletion region will be wider in the *n* region. An explanation can be given based on electrostatic considerations: the more heavily doped region opposes the penetration of carriers with an opposite charge.

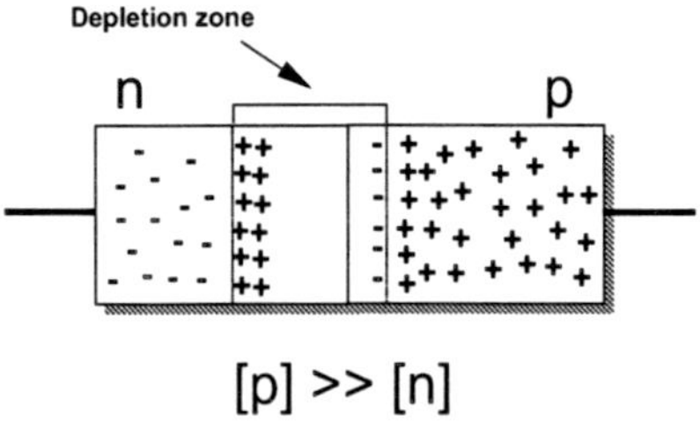

Fig. 21.3. Depletion region is wider in the less doped zone

An additional feature of the depletion region is its behavior towards minority and majority carriers. In the *p* region, majority carriers are the holes, and minority carriers are the electrons. Majority carriers cannot cross the depletion region since the majority carriers view the depletion region as an open circuit. However, the minority carriers see an electrical field that is generated by the forming of the depletion region. This allows minority carriers to cross the depletion region. The minority carriers view the depletion region as a short circuit. Similar rules apply for carriers in the *n* region.

The depletion region behaves differently when an external voltage is applied. The depletion region behaves like an electrostatic barrier onto which the field induced by the external biasing is superimposed. The overall effect is that the equilibrium between diffusion and drift goes out of balance. To preserve the overall voltage balance, external biasing is compensated by a resistive voltage drop and by the voltage in the depletion region.

When the *n* region is positive with respect to *p* region, the junction is defined as "reverse-biased". After the depletion region is formed, the minority carriers cross the depletion region and arrive at the other side. In this case they move by drift since they appear to be majority carriers to the external battery (see Fig. 21.4). The battery cathode "extracts" negative charge out of the *n* region, thus decreasing charge concentration and therefore enlarging the depletion region. The same happens in the *p* region. The double layer grows until its voltage drop balances battery voltage, stopping the extraction of majority carriers. Widening of the depletion region prevents majority current flow. Only a small minority current can flow (i.e. the reverse current of the diode).

If the external voltage is too high, the electrical field inside the barrier can be so intense such that charge accelerates across the junction with enough energy to pull electrons out of the lattice. This phenomenon begins an effect of electron-hole multiplication. There now exists a corresponding increase of both unbound charges and current. This effect is known as "junction breakdown". This condition can be reversible if thermal dissipation does not exceed certain limits.

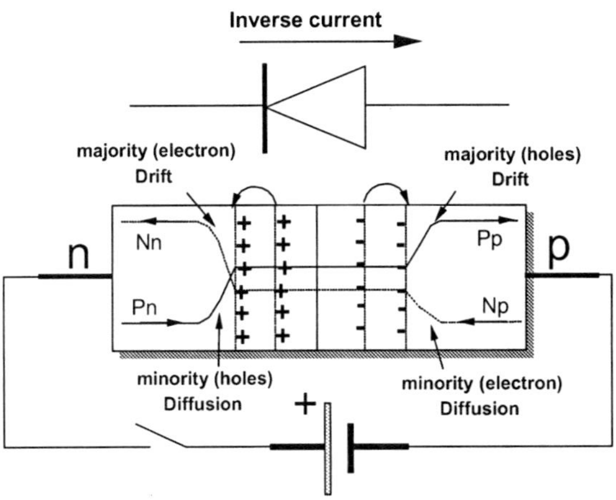

Fig. 21.4. A reverse-biased *pn* junction. Application of an external voltage widens the depletion region, letting a minority current flow through the junction.

Figure 21.4 also shows drift and diffusion currents for majority and minority currents inside both regions. Minority carriers of the *n* region (Pn) move, by diffusion, towards the depleted region. Next, they pass through the depleted region and arrive in the *p* region where they are now majority carriers. When in the *p* region, the majority carriers move by drift towards the battery contacts. Since current direction is, by convention, one of the positive charges, direction of the current in the reverse-biased diode can be derived.

If we reverse the battery contacts as shown in Fig. 21.5, the opposite effect is achieved: the depletion region becomes narrower and the diffusion current (the current due to the majority carriers) is present. Narrowing of the depletion region is due to the external voltage that permits majority carriers to drift towards the depletion region. The barrier becomes narrower and diffusion current inside the junction prevails over drift current since the electrical field of the double layer is reduced. As soon as the depletion zone is passed, charges become minority carriers and move by diffusion towards external contacts. For example, electrons move by diffusion from the *n* to the *p* region where they become minority carriers. In order to maintain electrical neutrality, the charges with the opposite sign, the majority carriers, are pulled from the battery contacts.

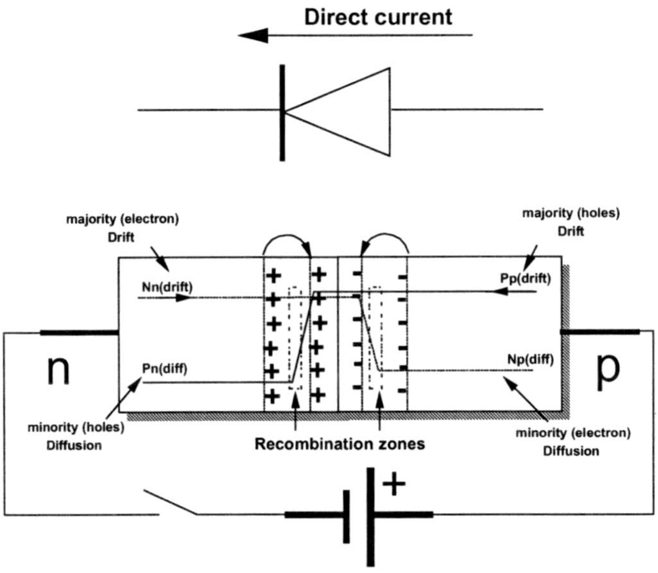

Fig. 21.5. Forward-biased *pn* junction. By applying an external potential, depletion region width decreases, and the flow of the majority carrier through the junction is allowed in the direction shown.

There is an initial time period during which the depletion region is formed and minority carriers that are crossing the barrier violate electrical neutrality. Majority carriers, supplied by the battery, form a charge distribution that corresponds to one of the diffusing minority carriers. This way, the majority carriers compensate for minority carriers and the carriers do not repel each other. The diffusion current can sustain itself.

When the initial time period is over, the depletion region is electrically neutral. This neutrality is violated when minority carriers recombine. The recombination that occurs in the space behind the depletion zone pulls more majority carriers from the battery contacts. At this point, there are both diffusion and drift currents from complementary charges in the two regions. Their values can be controlled by modifying the barrier voltage in the double layer.

Now that the typical *pn* junction has been reviewed, operation of the bipolar transistor can be examined. A bipolar transistor is obtained by placing two junctions side by side. The purpose of a bipolar transistor is to realize a current-controlled, current amplifier (an amplifier is a device capable of modulating the current flowing through the load by means of a smaller signal current).

After examining the *pn* junction, it is apparent that varying the current inside a junction can be accomplished by modifying the bias at its ends. Moreover, we can have either drift or diffusion current depending on the kind of biasing. Two concepts must be kept in mind:

- A reverse-biased junction behaves as a short circuit for the minority carriers and as an open circuit for the majority ones.
- In a forward-biased junction, both drift and diffusion currents coexist.

Using a forward-biased junction, a flow of both majority and minority carriers is generated. If we "attach" a reverse-biased junction to the forward-biased junction, the former separates the minority carriers and prevents the majority carriers from passing, thus producing a charge flow.

Enabling a bipolar transistor is done by forward-biasing the base-emitter junction and reverse-biasing the base-collector junction. In a forward-biased junction, both diffusion and drift currents coexist. Therefore the majority carriers in the *n* region (which we call the emitter) diffuse in the *p* region, where they are minority carriers. If the base-collector junction is reverse-biased, former minority carriers are dragged from the *p* region to the *n* region, since they view the depletion region as a short circuit. However, the drift current of the majority carriers (holes) in the base flows through the emitter towards the external battery (Fig. 21.6).

By doping the emitter significantly more than the base, it is possible to have a collector current much higher than the base current. Using this property, the device can potentially amplify the input signal. However, the critical manufacturing focus is the proximity of the direct-biased junction to the reverse-biased junction. The diffusion and drift currents must coexist. If the base-collector junction is too far apart, the diffused charge in the base recombine before reaching it. For this reason it is not possible to make a transistor by simply coupling two diodes side by side.

Problem 21.1: In Fig. 21.7a, what does Rb represent? In Fig. 21.7b, what are the limits of V_{IN}? What happens if, while a bipolar is working as an amplifier, collector current is suddenly interrupted?

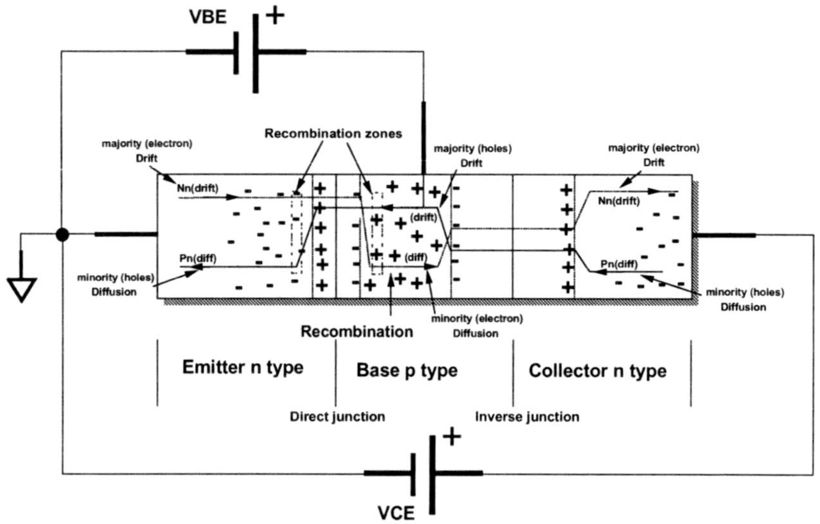

Fig. 21.6. *npn* bipolar transistor. Base-emitter junction is forward-biased, collector-base junction is reverse-biased

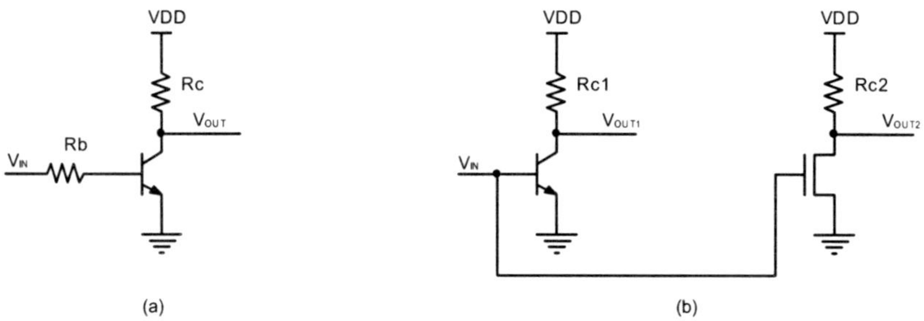

Fig. 21.7. Schematics pertinent to problem 21.1

21.2 Latch-Up

When dealing with a CMOS process, it is important to understand the latch-up phenomenon in order to avoid it. Figure 21.8 shows the cross section of two MOS transistors, an *n* type and a *p* type. Parasitic bipolar transistors exist in two regions with different polarity. They must be turned off during device operation to avoid undesired behavior. Among all of them, the most troublesome are those shown in Fig. 21.8. Figure 21.9 depicts the bipolar transistors only. The resulting network is called a Silicon Controlled Rectifier (SCR). An SCR is a positive reaction circuit that virtually short-circuits the two supplies causing both high power consumption and, possibly, permanent damage.

Both n^+ biasing in the n-well and p^+ biasing in the substrate are required to bring the two regions to a known potential. VDD is used for the n-well and ground is used for the p-well. In case of charge injection into the *p* substrate, a voltage drop occurs on R4 and transistor B2 can turn on. This causes a voltage drop on both R2 and R1 and also causes B1 to turn on as well.

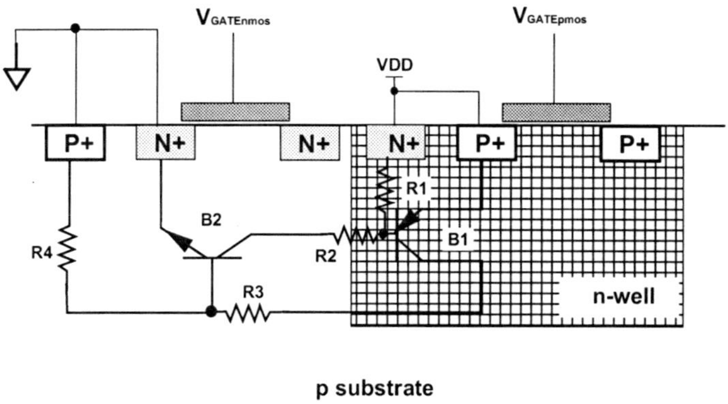

Fig. 21.8. Main bipolar transistors involved in latch-up configuration

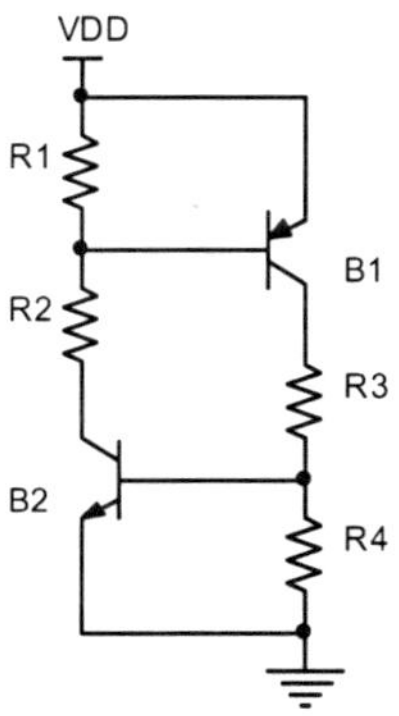

Fig. 21.9. Electrical scheme of SCR

At this point B1 raises the potential of the base of B2. This causes B2 to conduct more. Eventually, a race condition exists between the two transistors and the device is destroyed. This is due to the high current causing substantial heating. The devices are no longer able to dissipate this heat.

There are several methods to remedy this issue. One of these methods is placing rings (biasing contacts, as those shown in Fig. 21.8) that completely surround the transistors and collect the charge that could potentially enable the parasitic bipolar transistors. It should be noted that the charge injected into the *p* substrate are of positive type (i.e. holes). Also note that the charge is potentially responsible for turning on B2. Therefore the rings to collect this charge are of type *p* connected to ground. Conversely, in the n-well, the rings are of *n* type and connected to VDD.

Emphasis on biasing rings is taken into account during the layout phase. In non-volatile memories, this issue is much more important since internal nodes are boosted to high voltages during program and erase, considering also that high currents are present.

Another important item to look for are undesired floating nodes. Floating nodes may cause spurious biasing and charge injection into the substrate. This could subsequently trigger a latch-up condition.

Normally, robust anti-latch-up rings are placed around high voltage sections. Epitaxial substrate allows a reduction of the values of R1 and R4 in Fig. 21.9. Epitaxy is the process in which a doped layer with the same lattice structure is deposed on the silicon substrate. Using this process, it is possible to limit the number of contacts to the substrate since latch-up occurrence is limited by the low resistivity of the substrate itself.

Once the device has been fabricated, its robustness against latch-up is tested. On all of the chip I/O pins, currents on the order of hundreds of milliamps are both injected and drawn. A latch-up condition is determined by observing the current consumed from the power supply (i.e. the VDD supply). Quality of the device varies with respect to the amount of current required to trigger a latch-up condition.

Problem 16.2: On the plane (VDD, I_{DD}) draw consumption figure for a device where latch-up occurs.

21.3 Bipolar Transistors Used in Flash Memories

Besides parasitic bipolar transistors, there are some bipolar transistors explicitly used in flash memories. For example, the vertical BJT is widely used in band-gap circuitry. The lateral BJT is widely used as protection against Electro Static Discharge (ESD). The base of the lateral BJT is the *p* substrate, while the collector and emitter are realized by two junctions in n^+ active area. Figure 21.10 shows an example of a lateral bipolar transistor. Both the layout and a cross-section of the device are shown. The depicted resistor corresponds to the base resistance which is given by the nearest biasing contact to the substrate.

Enabling this bipolar is achieved by raising the collector with respect to the ground node. The collector itself attracts negative charges and injects positive charges into the substrate, which constitutes a current. This current, collected by the ground node, raises the potential at the terminals of the diffused resistor until the bipolar turns on. This effect causes a current to flow between collector and emitter. This style of transistor is used in the ESD protection circuit because its triggering voltage is so high. The component is turned off during normal operation of the device and only becomes active during electrostatic discharge.

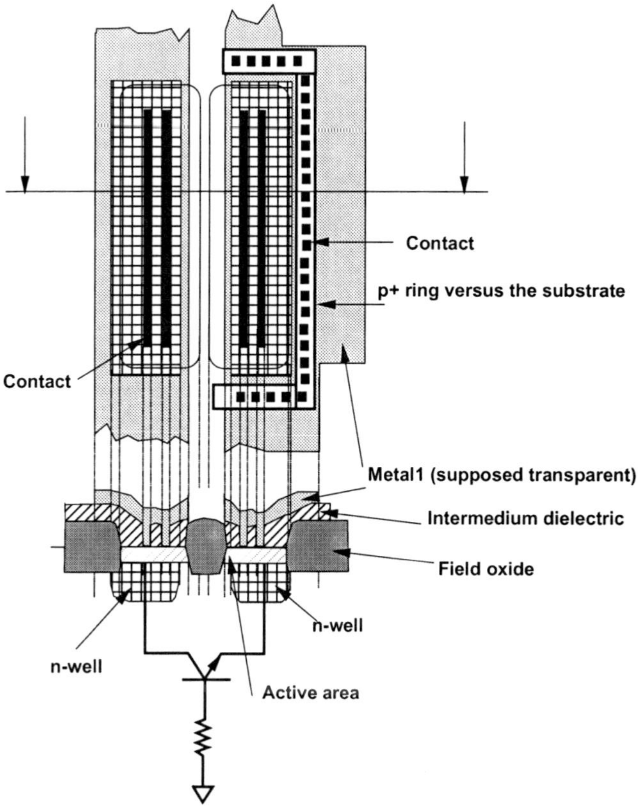

Fig. 21.10. Lateral bipolar transistor used for ESD protection: layout, section and electrical symbol

The structure of the vertical bipolar transistor, typically used in band-gap circuitry, is quite different. It is similar to transistors found in a bipolar process. An example of a layout of a vertical BJT is shown in Fig 21.11. It is a *pnp* bipolar connected in a diode configuration (i.e. the collector is short-circuited to the base).

The shape of the layout of a vertical bipolar transistor is circular since a large amount of current is required to flow through the device. Sharp corners are avoided since corners typically lead to current crowding that might cause breakdown. In the event high current consumption is not of concern, the layout can be made using more conventional methods featuring rectangular regions that make the mask generation task easier.

Problem 16.3: Draw the electrical scheme relative to the layout of Fig. 21.11.

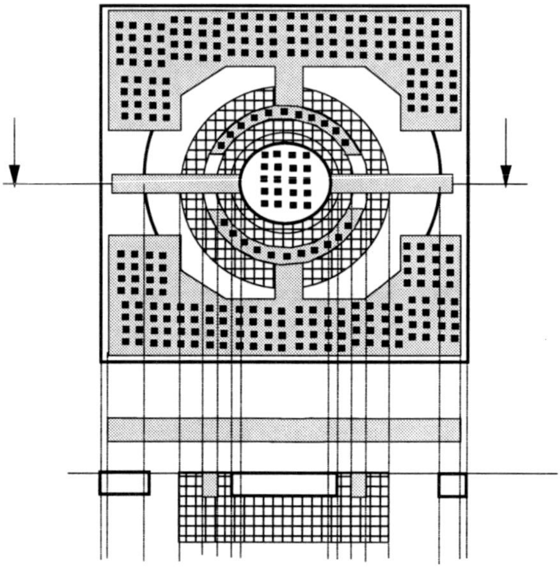

Fig. 21.11. Vertical Bipolar Transistor layout and section

21.4 Distribution of Power Supplies and ESD Protection Network

Planning for power supply distribution is carried out during initial device floor-planning and is generally considered the "backbone" of a robust device. It must be taken into account that both ground and power paths cannot be considered equipotential throughout the device. Furthermore, power supply metal must be properly sized so that they comply with electro-migration criteria.[1]

In an effort to minimize the noise induced by output buffer switching, power supplies are typically split. Moreover, using multiple pads for power supplies improves the quality of the power supply network. It is better to divide the supplies internally even if the pad used is common with another supply. The double wire bonding will act as a filter to reduce noise propagation.

Figure 21.12 shows a device where 16 outputs are divided into two equal banks. The charge pumps are located on the side next to the inputs. A possible supply network is shown where the metal of output buffer grounds, $GND_{I/O1}$ and $GND_{I/O2}$, are separated such that switching current flows on two impedances thus

[1] Current flow heats the conductors (Joule effect) causing a movement of the "grains" that compose the metal layer. It is therefore very important to size the wires bearing in mind the dissipation they undergo. A rule of the thumb says that for each micron of width a current of about 1 mA is allowed.

generating less noisy transients. On the right there are two additional ground pads. One pad is for the input buffers. The other pad is used by the rest of the circuitry. The three VDD pads previously described are shown as well.

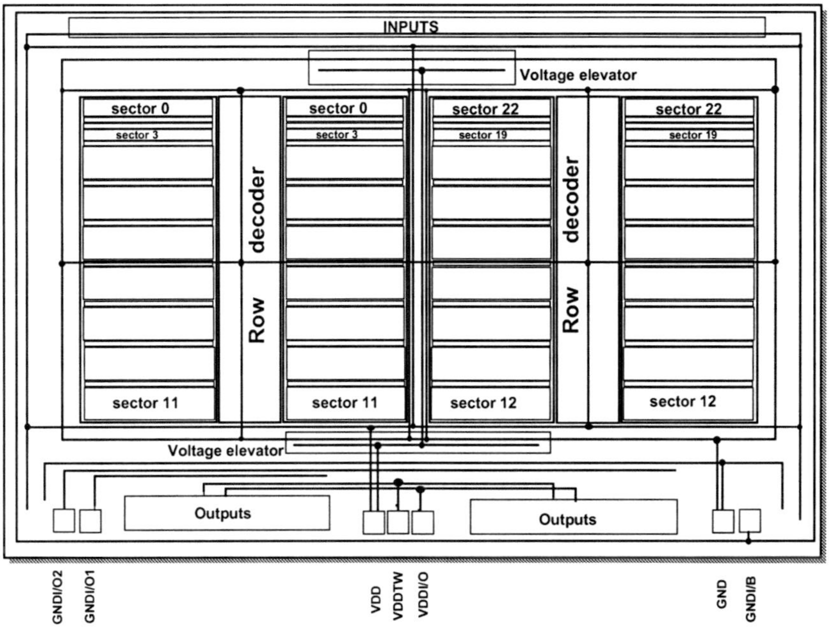

Fig. 21.12. An example of distribution of the main supply lines

Splitting ground and VDD is an empirical process driven solely by experience. It is difficult to build an equivalent electrical network that can explicitly demonstrate the quality of potential solutions through simulation. Modeling is much easier at output buffer level since the problem is much more manageable in a small, fixed area.

The power supply network should not only have electrical connections that make the device work properly, but also satisfy the requirements for Electro Static Discharge protection. It is a common experience to get a "shock" touching the door handle after walking on a carpet or touching the car after driving on a windy spring day. This phenomenon is caused by the accumulation of "static electricity". It can also happen with electronic devices when they are manipulated during board assembly and soldering.[2]

[2] When using the solder for electrical circuit board at home, how should the solder-to-network connection be done to be sure not to damage the circuit that is being soldered?

The voltages involved are on the order of some thousands of volts. Their duration is as little as a fraction of a nanosecond. The damage that ESD can cause to an integrated circuit range from oxide breakdown to junction breakage. Therefore the communication channels between the device and the external world (i.e. the pads) must have proper protection to stop these accidental discharges. In the previous examples, a lateral bipolar transistor is used that is normally off and turns on only when the discharge occurs. The threshold of this component is very high so that its intended use is guaranteed under normal operating conditions.

The main task of this element is to absorb the charge by forming a low impedance path, while, at the same time, preventing undesired propagation of this charge to the internal elements of the device. The study of the protection network for a complex device deserves its own dedicated book! What follows provides a simple functional description of the way the discharge propagates.

A protection network is the set of all possible discharge paths while taking into account that discharge can happen between any two pads. Besides providing the discharge with a propagation path to "close the loop" between the positive and the negative terminal, it is also important to protect both the internal circuit and the memory matrix. It is not uncommon that, after a discharge, some of the cells inside the memory matrix have been altered (either erased or programmed).

For instance, the discharge path between an input and VDD (Fig. 21.13) turns on the bipolar placed between the input pad and ground. A high voltage on the collector injects holes into the substrate via avalanche breakdown. A voltage drop occurs on the diffused base resistance. The discharge flows to VDD through the substrate resistance and the diode between the *p* substrate and the n$^+$ type biasing junction connected to VDD present in the nearer n-well. Therefore, for each kind of discharge, either positive or negative, there must exist propagation paths composed of bipolar transistors, diodes, substrate resistors and connection metals that let the discharge flow instead of letting it destroy either the gate oxide or a junction.

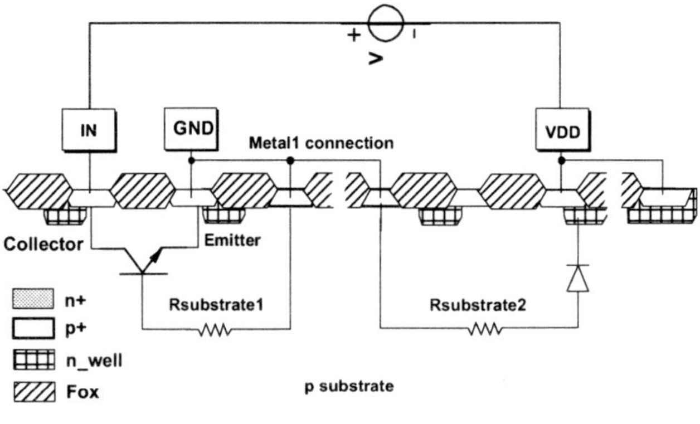

Fig. 21.13. An example of a ESD path discharge

One of the primary rules to follow is to not connect n-channel transistors directly to supplies. During a discharge, when the chip is not powered, an NMOS transistor has a gate to ground (it is turned off). Any discharge voltage is totally across its drain and its gate. A PMOS is generally turned on, and, therefore, it doesn't undergo such high electrical fields. Suggestion is to always interpose a PMOS between supply and the drain of the NMOS that are connected to the supply if need be, in order to mimic an inverter-like schematic as much as possible.

The study of the ESD network is a fundamental topic of modern integrated circuit design. Today, devices are designed and specified to withstand a discharge of up to 2 kV. Most internal tests show that devices can withstand up to 4 kV. ESD protection related bipolar transistors should be present on each input and output of an integrated circuit. An entire network is designed for the power supplies, using, in the case of VDD, at least four bipolar transistors placed in the corners of the layout, symmetrically, to protect the entire device against the internal propagation of static discharge.

Bibliography

B.T. Ahlport et al., "CMOS?SOS LSI input/output protection networks", IEEE Transactions on Electron Devices, Vol. ED-25, No. 8, pp. 933-938, (August 1978).

S. H. Cohen, G. Caswell, "An improved input protection circuit for CMOS/SOS ARRAY", IEEE Transactions on Electron Devices, Vol. ED-25, No. 8, pp. 926-932, (August 1978).

C. Duvvury, et al., "A Synthesis of ESD input protection scheme", EOS/ESD Symposium Proccedings, pp. 88-98, (1991).

C. Duvvury, et al., "Dymanic gate coupling of NMOS for efficient output ESD protection", IEEE, IRPS, pp. 141-150, (1992).

W.S. Feng, et al., "MOSFET drain breakdown voltage", IEEE Electron Device Letters, Vol. EDL-7, No. 7, pp. 449-450, (July 1986).

E. Fujishi, et al., "Optimized ESD protection circuits for high speed CMOS/VLSI", Custom Integrated Circuits Conference, pp. 569-573, (1984).

F.C. Hsu, R.S. Muller, C. Hu, "A simplified Model of short channel MOSFET characteristics in the breakdown mode", IEEE Transactions on Electrical Devices, Vol. ED-30, No. 6, pp. 571-576, (June 1983).

G.J. Hu, "A better understanding of CMOS latch-up", IEEE Transaction on Electron Devices, Vol. ED-31, No. 1, pp. 62-67, (January 1984).

G. Krieger, "Nonuniform ESD current distribution due to improper metal routing", EOS/ESD Symposium Proccedings, pp. 104-109, (1991).

I.M. Mackintosh, "The electrical characteristichs of silicon P-N-P-N Triodes", Proceeding of the IRE, pp. 1229-1235, (June 1958).

T.J. Maloney, "Designing MOS inputs and outputs to avoid oxide failure in the charged device model", EOS/ESD Symposium Proccedings, pp. 220-227, (1988).

P.S. Neelakan Taswany,"MOS scaling effects on ESD-based failure", IEEE Custom Integrated Circuits Conference, pp. 400-403, (1986).

Troutman, Latch-up in CMOS Technology The problem and Its Cure, Kluwer Academic Publishers, (1986).

22 From Specification Analysis to Floorplan Definition

In this chapter we are going to carry out a hypothetical feasibility study, in order to analyze the design choices that allow us to satisfy the specifications, written after having performed a market research involving the main customers for Flash memories. All the main blocks are investigated, even if the circuital level of detail is not reached, using the notions learned in previous chapters. The analysis will result in floorplan preparation, which will allow us to estimate the area occupation of the device[1].

22.1 Introduction

Aim of this chapter is to get to device architecture through its feasibility study. A hierarchical approach is used, starting from the row decoders towards both input and output circuitry.

The first decision to make is how to shape the memory matrix, i.e. the number of rows and columns. Our memory is going to be a low-voltage, bi-level 8 Mbit, divided according to the specification as follows:

- 15 sectors, 64 Kbyte each
- 1 sectors, 32 Kbyte each
- 2 sectors, 8 Kbyte each
- 1 sectors, 16 Kbyte each

ordered according to the memory map shown in Fig. 22.1.

22.2 Matrix Organization

In order to define the shape of the matrix, specifications must be taken into account. The number of cycles required is equal to 10^5 for each sector; therefore it is necessary to isolate them each other, in such a way that no spurious influences oc-

[1] The sizes of the various blocks are related to a hypothetical process, where the channel length of the transistors is 0.6 μm. What we are interested is the relationship of the sizes of different blocks, rather than absolute size. In fact, the real device can be realized in such a way that planar dimensions are reduced with respect to the original process, in order to decrease overall chip area.

cur between cells lying in different sectors during bth write and erase operations. In fact, as we know by now, owing to the matrix organization the cells that are on the same column share the drain contact, while those on the same row share the gate contact.

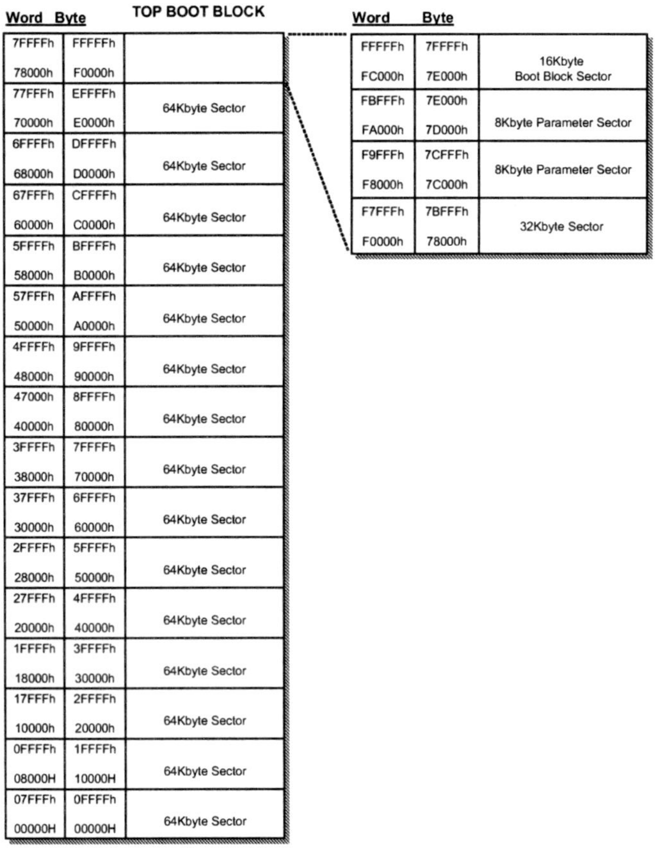

Fig. 22.1. Memory map of the device, which is divided into 19 sectors: 15 equal sectors (64 Kbyte), boot block (16 Kbyte), two parameters (8 Kbyte each) and finally a small sector of 32 Kbyte

Let's recall Fig. 3.12: if we want to program the cell shown in the figure we must bias its gate at about 12 V and its drain to 5 V. Of course, all the cells on the same row have the gate at 12 V, while the cells on the same column have the drain at 5 V. Therefore a cell whose gate is at ground but belonging to the same column where a cell under program is found, has its drain at 5 V, thus suffering from the drain stress effect, that tends to erase it. We have already seen in Fig. 3.13 the effect of this stress on the written cells. Every technological process has its own drain stress chart. Latest generation's cells are much more sensitive to this parasitic effects, owing to their reduced dimensions.

If the column is composed of 1,024 cells and we want to program them all, in a worst case scenario we will use the maximum allowed programming time for each cell, say 250 μs, so that the time of drain stress applied to the first cell of the column is

$$250\mu s \cdot 1{,}024 = 0.256 s \tag{22.1}$$

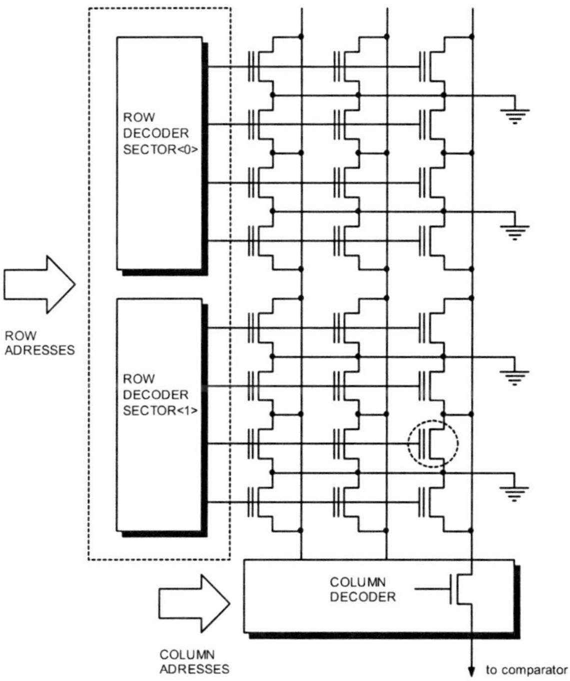

Fig. 22.2. In the sectors with the drain in common, the cells on the same column belonging to different sectors suffer from electrical stress, and previously programmed cells are undesirably erased

Let's assume that the sectors are realized by row, as shown in Fig. 22.2: we program the cells of sector 1 once, and we will never modify them throughout the entire device's life.

Let's now cycle 100,000 times the sector 0. If, this time, we assume that the cells are always programmed using minimum pulses of 10 μs, then the drain stress on the cells belonging to sector 1 is applied for a time equal to

$$10\mu s \cdot 1024 \cdot 10^5 \tag{22.2}$$

At the end of the cycles on sector 0, all the cells in sector 1 turn out to be erased because of drain stress. Realizing the sectors by column might solve this issue. Different sectors would be then subject to gate stress[2] whose influence is usually less severe than drain stress because of the different capacitive ratios. This kind of organization, on the other hand, calls for a source erase, which is difficult to implement when a voltage supply of 3 V is required. Therefore the sectors must be organized by row, and negative gate erase must be used. Thus we must design a hierarchical column decoding to separate the drains of adjacent sectors by isolating them during the different operating modes, as shown in Fig. 22.3.

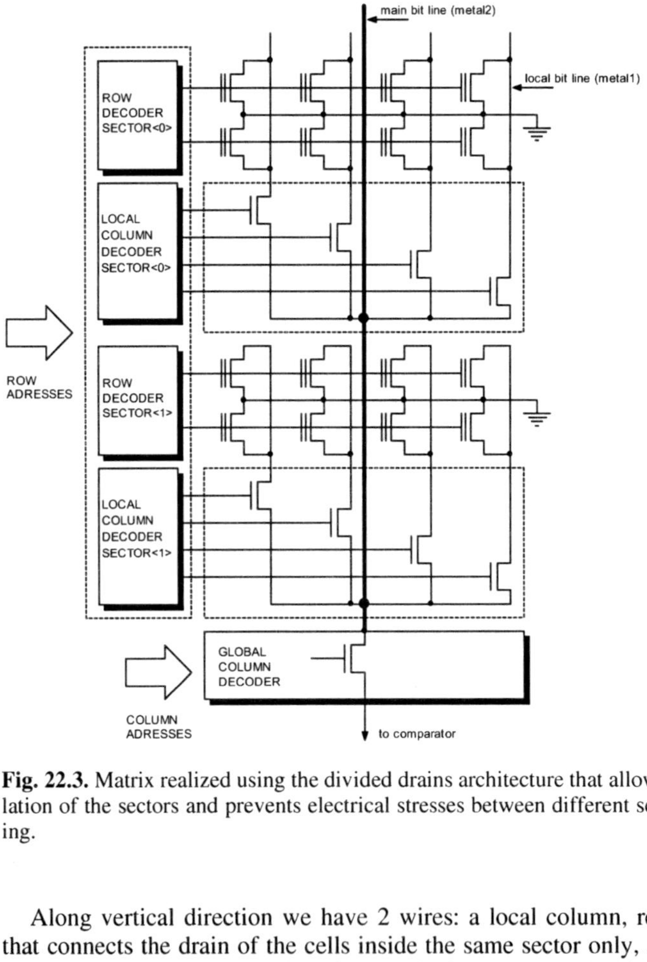

Fig. 22.3. Matrix realized using the divided drains architecture that allows the complete isolation of the sectors and prevents electrical stresses between different sectors from happening.

Along vertical direction we have 2 wires: a local column, realized in metal1, that connects the drain of the cells inside the same sector only, and another vertical wire, global column, realized in metal2, that connects the hierarchical decoders

[2] It is the stress caused to a cell that shares the same row as the cell that we are programming.

to the overall column decoder. The choice of using a global column every four local columns is dictated by the size of the cell, which finally constrains the size of the column pass transistor. Therefore the global column decoder addresses four local columns and aim of the local decoder, also known as YO decoder, is to select a local column between the four local ones. Every operation on the cells activates YO decoding of a single sector only, in order to isolate its cells from those of the other sectors. The structure is usually more complex than shown in Fig. 22.3. In fact, the cell pitch along X direction is usually smaller than the pitch of the transistors of the YO decoder, and therefore we cannot place four selection transistors in the pitch of four cells.

Then two local decoders for each sector are used: the former is above the sector and it is able to decode the external columns in the group of four, the latter, below the sector, for the internal columns. At the end, as shown in Fig. 22.4, YO decoder has been divided into two half-decoders to overcome the issue of space.

Summing up, our sectors are realized by row, the number of rows defining the size of the sector. Negative gate erase is used, and the stress of the cells is mitigated by implementing separated-drains architecture. YO decoders of each sector are driven by the row addresses that select the sector inside the matrix.

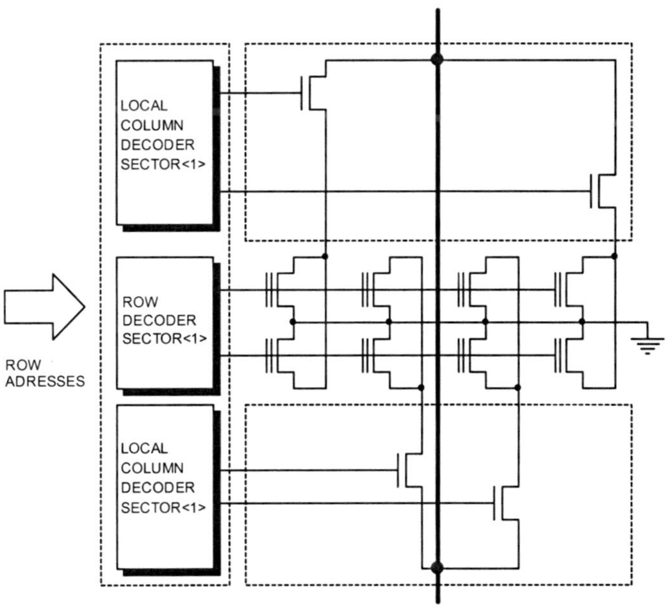

Fig. 22.4. Real arrangement of the local column decoder that is divided into two parts owing to the fact that the bit line selection transistor cannot be placed in the pitch of the matrix cell

22.3 Matrix Row Dimensioning

Now we need to decide the physical structure of the matrix. Guiding principles for this choice are the following:

1. Area occupation must be as small as possible, because it means more devices on a wafer and therefore a higher profit.
2. Memory access time, i.e. the time that the device takes to deliver the content of the addressed location on the outputs, must be, according to our hypothetical specification, less than 100 ns.

Every memory, either volatile or not, has different access modes to the matrix. Figure 22.5 shows the address access time T_{ADD} and Chip Enable access time $T_{CE\#}$.[3]

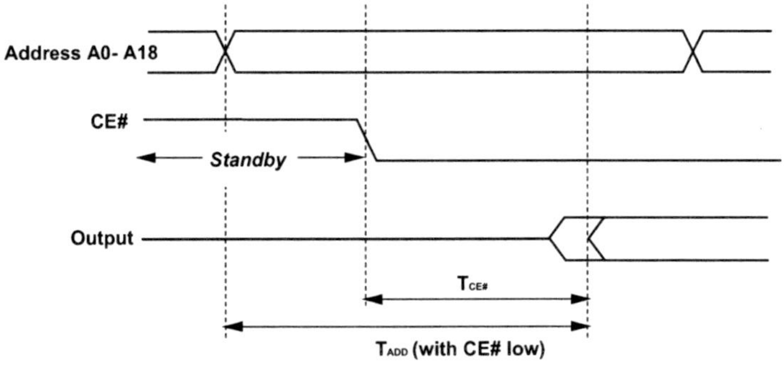

Fig. 22.5. Specification for both Chip Enable and address access time. Device is in standby when CE# pin is high.

T_{ADD} assumes that the device is powered up and all the service circuitry is turned on: it is the time between the address valid and the delivery of the data on the output pins. The limit of 100 ns must be guaranteed for Chip Enable access too. The main difference between these two read modes is that CE# mode assumes that the device is initially in a standby condition (CE# high), i.e. powered but "asleep" in all its functions. According to the specification, current consumption in standby must be, as a rule, at least three orders of magnitude lower than consumption in read mode, i.e. around 5 μA! It means that the non-CMOS circuitry, like the input buffers, must be turned off to avoid consumption. Access from CE# therefore implies the turning on of some parts of the device. $T_{CE\#}$ specification is the most difficult to comply with, and can be taken as a reference point for the read. Our aim is to design a memory whose $T_{CE\#}$ access time is equal to 100 ns under worst case conditions, i.e. a temperature of 125°C, TTL input levels and 100 pF of output load.

Let's now get back to the considerations on the matrix and on sector partitioning. Access time is therefore one of the main parameters to define the quality of

[3] In the specification CE bar pin is often referred to as E#. Symbol # is used to indicate complement.

the memory: the analysis that we perform on the delay introduced by the row will guide our following steps. We have seen that the cells are organized as a matrix. Local column is realized in metal and it connects the drains of the cells. In the same way, the cells of the same row are connected by a polysilicon wire, which also acts as their gate.
Figure 22.6 shows the section of a word line in the middle of the polysilicon line[4]. Assuming that coupling can be represented as plain parallel plates capacitors and that the thickness of the two oxides is 200 Å e 120 Å respectively, it is easy to calculate the capacitance of a cell, i.e. 0.4 fF. Let the resistivity per square of poly2, the layer that constitutes the row, be 6 Ω/□. A layer of silicide (a metal compound that lowers poly2 resistivity, whose normal value is 50 Ω/□) is usually deposed over the polysilicon.

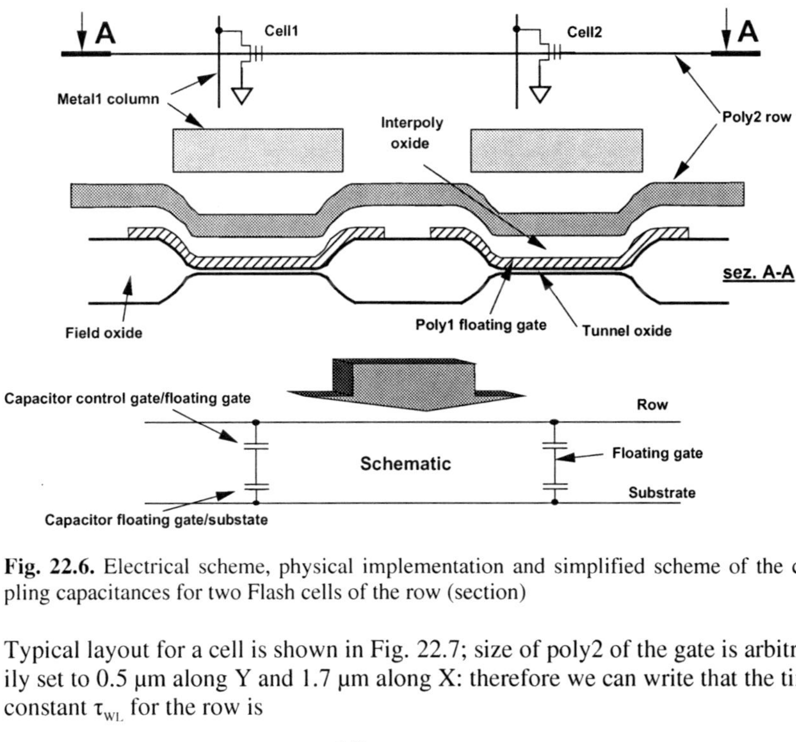

Fig. 22.6. Electrical scheme, physical implementation and simplified scheme of the coupling capacitances for two Flash cells of the row (section)

Typical layout for a cell is shown in Fig. 22.7; size of poly2 of the gate is arbitrarily set to 0.5 μm along Y and 1.7 μm along X: therefore we can write that the time constant τ_{WL} for the row is

$$\tau_{WL} = (RC)_{WL} = \frac{1.7\,\mu m}{0.5\,\mu m} \cdot 6\Omega \cdot 0.4\,fF \cdot N_{cell} \tag{22.3}$$

N_{cell} being the number of cells that compose the word line.

[4] The polysilicon of the control gate (poly2) is doped in a different way than the polysilicon used for the floating gate (poly1).

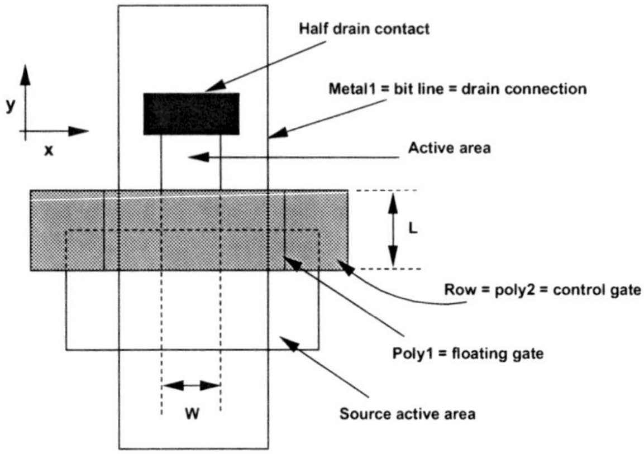

Fig. 22.7. Plan view of a Flash cell where its main layers are shown

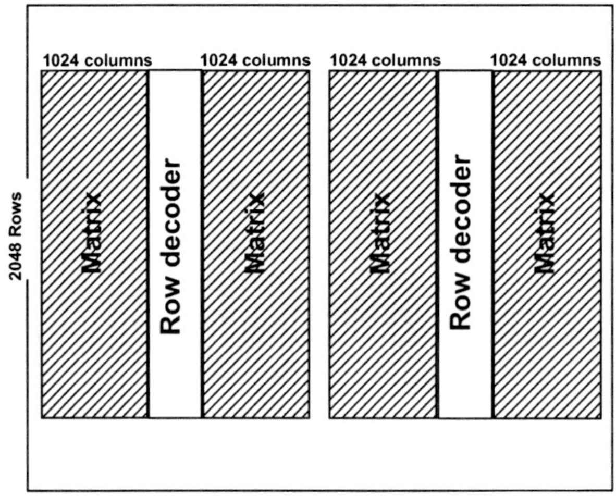

Fig. 22.8. Organization of the device with four semi-matrixes and two row decoders

By imposing that the rise time of the row be less than 10% of the whole access time, the value obtained for the time constant is 10 ns: therefore the number of cells in the row is 1,107. Rounding this result to the nearest power of two, it follows that the number of cells in our row must be 1,024.

Anyway we cannot organize the memory with a single row decoder placed between two matrixes of 1,024 columns each, because the resulting device would be too long, and the ratio between X and Y dimension would be unacceptable. Furthermore, in order to divide the parasitic loads, we can think of a device with two row decoders and therefore a matrix organized as 4,096 columns by 2,048 rows, as shown in Fig. 22.8.

22.4 Dimensioning the Sectors

Let's now wonder how we should realize the sectors inside the matrix. The device should be organized in such a way that it can be read using a parallelism for the output data either by 8 or by 16.

The pin called BYTE# allows us to decide the width of the output. When BYTE# is low, by-8 organization is selected: addresses from A<0> to A<18> are used, while the Input/Output pin DQ<15> acts as an address that selects either the upper or the lower byte of the memory word. The pads of the unselected data remain at High-Impedance. On the contrary when BYTE# is high the memory is organized by 16; A<19:0> act as addresses while all the 16 output pins are active.

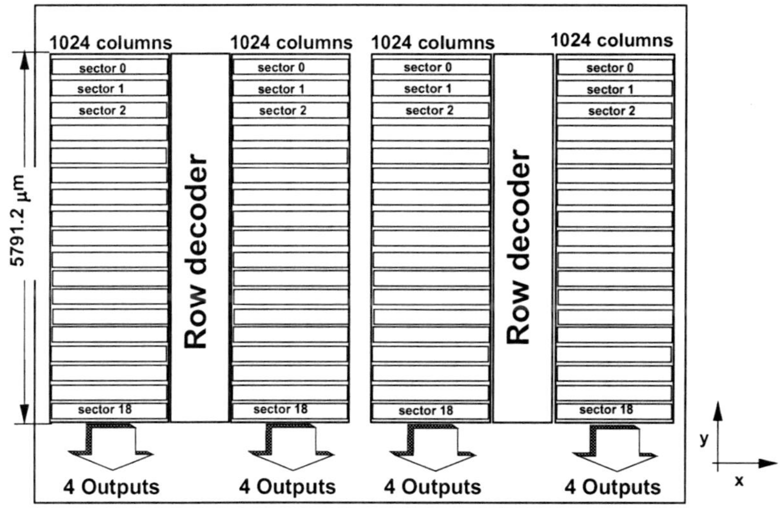

Fig. 22.9. First attempt of sector partitioning. Each sector is divided among all the four semi-matrixes that produce four outputs each

A first example of a possible architecture is shown in Fig. 22.9. Each semi-matrix provides 4 output bits, giving a total number of 16, and every sector is divided among all the four semi-matrixes. The issue with this organization lies in the size of the local column decoders, which in turn heavily affect the overall size of the device. Let's assume that the Y pitch of the cell is 1.9 μm. The 64 Kbyte sectors, realized according to Fig. 22.9 using 1,024 cells for each row, have a dimension along Y direction equal to

$$\frac{64\cdot 1{,}024\cdot 8}{4\cdot 1{,}024}\cdot 1.9\mu m = 243.2\mu m \tag{22.4}$$

To this size, the space occupied by the local decoders Y0 must be added, which can be estimated to be about 100 μm. Each 64 Kbyte sector has a Y dimension of 343.2 μm, while the 8 Kbyte sector has a Y dimension of 130.4 μm. It is clear that local decoders have a bigger impact on smaller sectors.

Overall, the Y dimension, calculating only the occupation of the sectors and of local decoders, neglecting the space required for the routing of the signals, is equal to

$$Y = 343.2\mu m \cdot 15 + 221.6\mu m + 160.8\mu m + 130.4\mu m \cdot 2 = 5{,}791.2\mu m \tag{22.5}$$

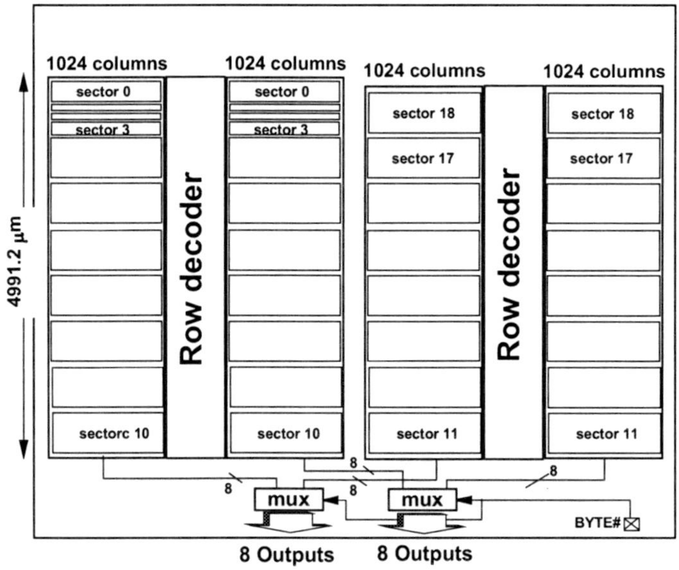

Fig. 22.10. Each sector is divided between two semi-matrixes. Multiplexers are used to handle by-8 and by-16 reads, and they are driven by BYTE# pin.

Another possible organization of the matrix is shown in Fig. 22.10. In this case the sectors are divided between two semi-matrixes. Let's redo the math to see how Y dimension is modified.

For the 64 Kbyte sectors, we have

$$\frac{64 \cdot 1{,}024 \cdot 8}{2 \cdot 1{,}024} \cdot 1.9\mu m = 486.4\mu m \tag{22.6}$$

to which we must add the 100 μm of YO decoder. Overall, Y dimension is equal to

$$Y = 586.4\mu m \cdot 7 + 343.2\mu m + 221.6\mu m + 160.8\mu m \cdot 2 = 4{,}991.2\mu m \qquad (22.7)$$

The difference between solutions of Fig. 22.9 and Fig. 22.10 is remarkably 800 μm, therefore the latter is preferable.

22.5 Memory Configurations

The sectors have different sizes because some of them are used by the customer as data bank, while others are used to store the code. It means that the two 4 Mbit matrixes cannot have the same size. Looking at Fig. 22.10, we can see that the left block is higher than the one on the right.

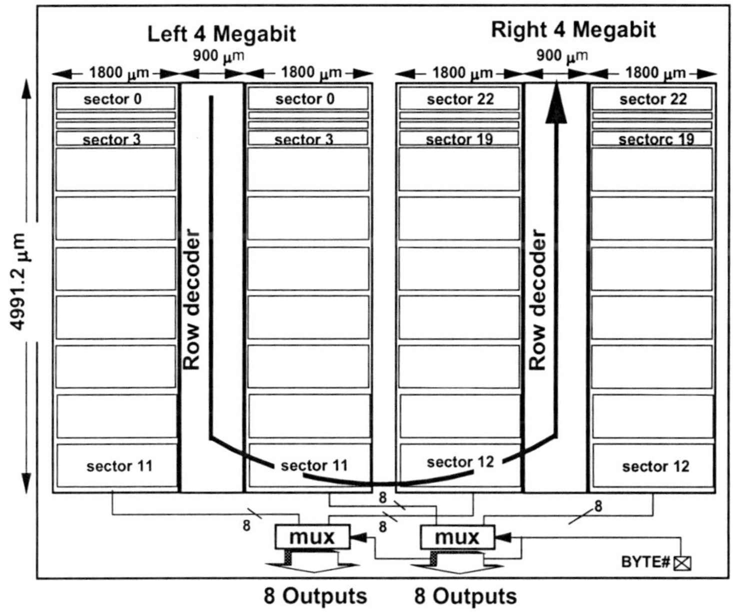

Fig. 22.11. Matrix floorplan with sector partitioning and sequential indication of the addresses. The 4 Mbit on the right is, for the memory map, an extension of the 4 Mbit on the left. The arrow shows the direction of the address increment

An additional feature can be provided by making these two blocks symmetrical as shown in Fig. 22.11. The Boot Block is the sector that at device power up is automatically queried by the external microprocessor. By incrementing row addresses, the matrix is scanned as shown by the arrow in Fig. 22.11. Options are to have the boot either at address 00000h (TOP configuration) or at FFFFFh (BOTTOM configuration).

Our device can be therefore configured in four different ways:

- 16 equal sectors (64 Kbyte)
- TOP
- BOTTOM
- 14 sectors of 64 Kbyte + 2 sectors of 32 Kbyte + 2 sectors of 16 Kbyte + 4 sectors of 8 Kbyte.

Desired configuration is chosen by programming some non-volatile registers internal to the device: this operation is perform during factory test. For instance if TOP configuration is required, the device sees the four small sectors on the left as separated, while the four sectors on the right are virtually united. At this point, estimating an occupation for the row decoder equal to about 900 μm, we obtain that the occupation along X direction is:

$$X = 1.7\mu m \cdot 1{,}024 \cdot 4 + 900\mu m \cdot 2 \cong 8{,}763.2\mu m \qquad (22.8)$$

Overall area occupation for the matrix, the local column decoders and row decoder is therefore equal to

$$A = 8{,}763.2\mu m \cdot 4{,}991.2\mu m \cong 44mm^2 \qquad (22.9)$$

22.6 Organization of Column Decoding

Let's get back to Fig. 22.11. We have chosen to organize the device using two row decoders and four semi-matrixes. Each sector is then divided between two semi-matrixes that are adjacent to a row decoder. Since the device is organized by-16, i.e. a data is composed of 16 bits, each semi-matrix will provide 8 output data, and each decoder will simultaneously activate the two semi-matrixes that are placed on its sides.

Figure 22.12 shows the disposition of the read circuitry, i.e. the sense amplifiers, placed below global column decoders. Totally we will have 32 sense amplifiers, whose outputs will be selected by two subsequent multiplexers to be able to converge on the 16 output pins in case of read by-16 or to choose the 8 most significant bits or the 8 least significant bits in case of read by-8. It is reasonable that the 8 sense amplifiers placed under the single semi-matrixes are designed in such a way that they occupy, along X direction, the same length as the semi-matrix. Let's estimate the sense amplifier strip to be 130 μm along Y direction.

Lets' recall that inside the sector we have a global column (MBL) every four local columns (LBL). Each sector has 1,024 columns and, therefore, on the sense amplifiers 256 column converge, which must be properly decoded to be able to select, at the end, the eight local columns that compose the byte.

The simplest, and less efficient, way to decode 256 different paths, as is in our case, is to generate 256 signals, i.e. 256 driving circuits and 256 levels of metal to bring around.

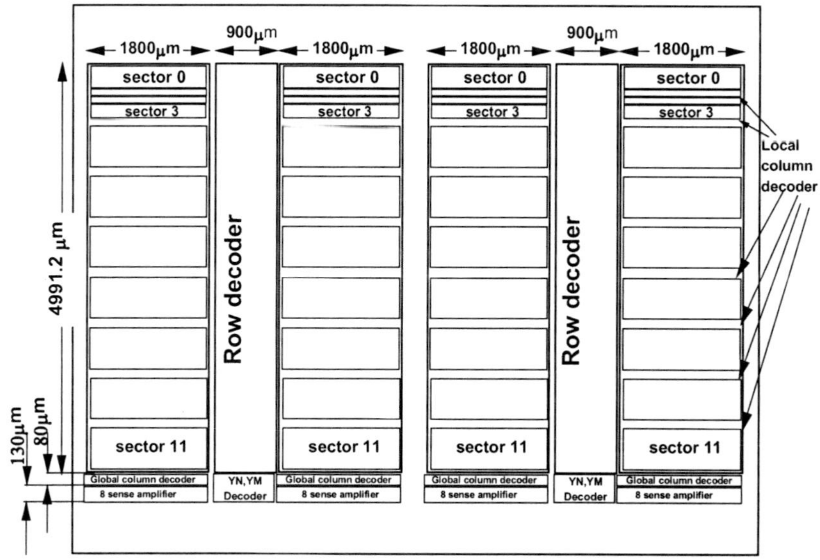

Fig. 22.12. Complete floorplan of the sense amplifiers and of column decoders

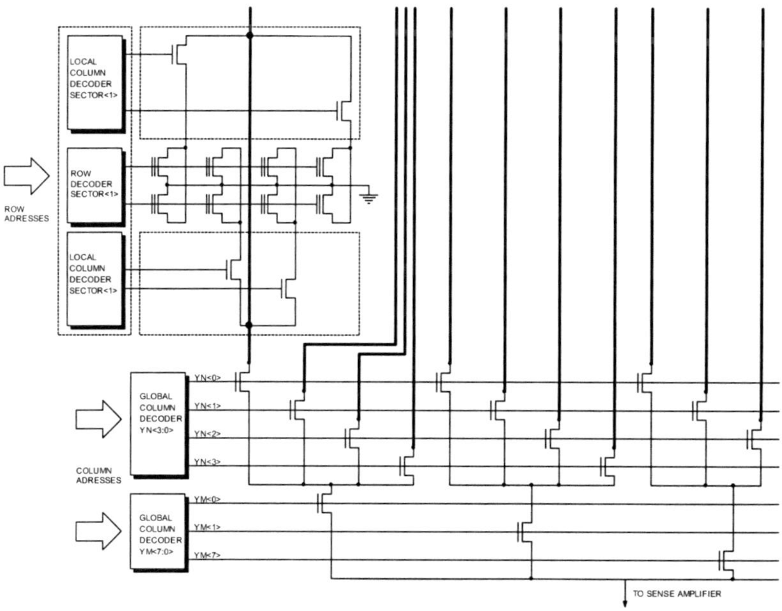

Fig. 22.13. Complete scheme of the column decoding, both local and global, for an output. MBLs are grouped by four and decoded by the YN signals. Then YM signals select one among the 8 packets of 4 MBLs to be able to converge to the sense amplifier

In reality, global column decoders are organized hierarchically, i.e. using two subsequent decoders. Since 8 columns are read at the same time, we must select one global column every 32 (256/8), i.e. we need 5 address bits. Let's divide these bits into two groups: two bits generate YN<3:0> signals, and the left three bits generate the eight YM signals. Once the MBL has been selected, the four YO signals select the local bit line. The scheme shown in Fig. 22.13 summarizes the structure of the column decoders for a semi-matrix.

As far as calculation of area occupation is concerned, we can assume that global decoders occupy a space equal to 80 μm along Y direction.

22.7 Redundancy

As we have seen in Chap. 18, the complexity of the fabrication process heavily penalizes the yield of the device. In order to overcome these issues, redundancy techniques are used, i.e. the introduction of additional memory cells, which are able to replace the bad ones, transparently to the user. In present devices, this operation is performed in factory only, to repair just the damages caused during fabrication. For our feasibility study, where a fundamental role is played by the access time specification, we decide to use the architecture of column redundancy only, by adding a sense amplifier dedicated to read from the redundancy matrix.

Figure 22.14 shows the schematic representation of the device; the external semi-matrixes occupy along the X a bigger space than the internal semi-matrixes because the redundancy columns have been placed on the sides only.

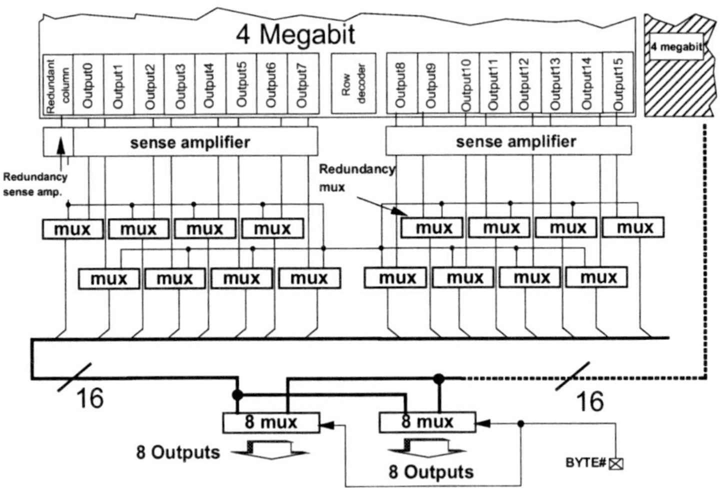

Fig. 22.14. Scheme of the read organization where the two levels of multiplexing, one for redundancy and one for x8 and x16 read, are shown

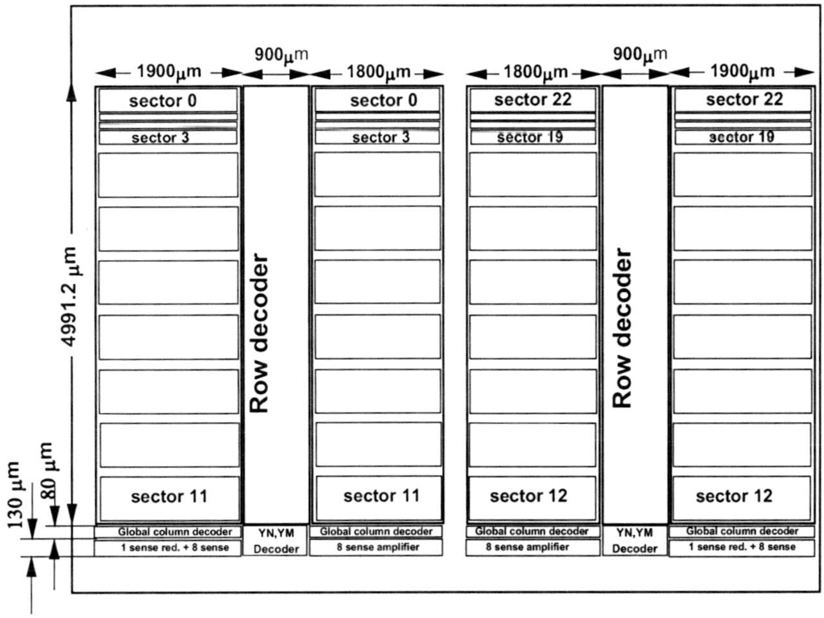

Fig. 22.15. Floorplan including redundancy matrix and its sense amplifier

Figure 22.15 shows the new device floorplan, where the redundancy columns and related circuits have been added.

From the analysis carried out in the chapter on redundancy, we know that different solutions can be applied to our current example. In this case it is necessary to take into account on one hand the additional area occupation and on the other hand the failure recovery capability. Let's assume to use 48 redundancy columns, eight of which can be individually addressed, i.e. they can be associated to a single local bit line. The remaining 40 columns can be used in packets of four, in order to replace whole main bit lines.

This solution allows recovering up to nine single defects every 4 Mbit. The UPROM that are needed to handle the redundancy columns are placed above the two external semi-matrixes. X dimension is equal to the length of the word line, i.e. 1,900 μm, while Y dimension can be estimated to be 550 μm.

Figure 22.16 shows the block diagram of the device, as complete as possible considering the level of this discussion, where the architecture of the UPROM allows reading at power up of all the cells in parallel, in order to avoid any delay on access time.

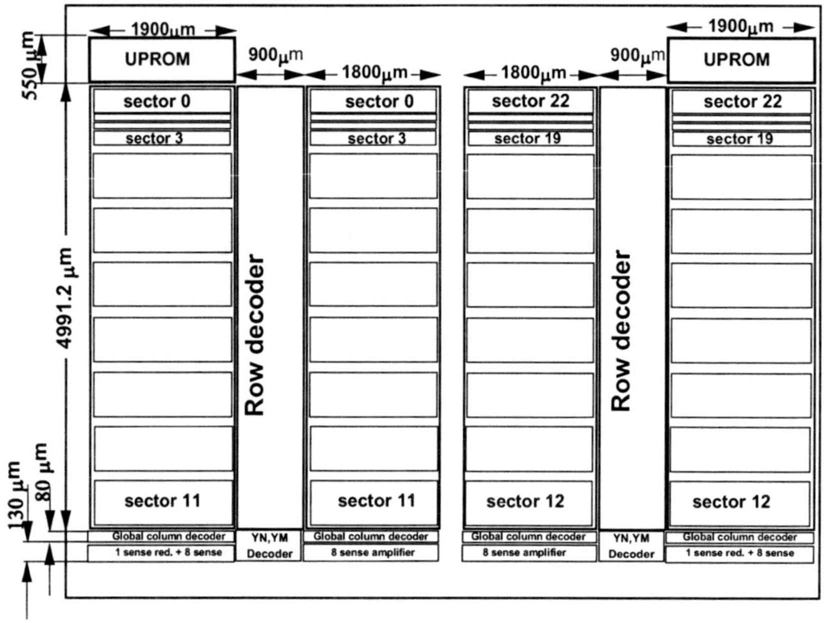

Fig. 22.16. Floorplan. UPROM blocks contain real non-volatile cells and the logic required to write, erase and read them.

22.8 First Considerations on Read Mode

Considering the tight limits of both power consumption and access time from standby that we have settled, we decide to use local boost, in accordance with all the considerations exposed in Chap. 10. Now we want to estimate the area occupation of the boost circuitry, in order to define the architecture of the device.

As described in Chap. 10, a feature of the distributed boost architecture is that the boost is applied only to the row decoder of the addressed sector. Now we need to estimate the size of the parasitic capacitor C_{LOAD} corresponding to a sector of 256 rows.

Assuming for the row driver a p-channel whose width is 30 μm and with minimum length (0.9 μm), we can estimate 40 pF for the parasitic capacitance associated to the source junctions of the transistors. On top of that, the value of the capacitance of the n-well tub that contains the PMOS transistors must be added, the total becoming 90 pF. Finally there is the capacitance due to interconnection lines. All in all, it is reasonable to consider a total parasitic capacitance of 100 pF.

By analyzing distributions, we can say that it is sufficient to raise the word line voltage of 1 V above power supply in worst case (i.e. when it is equal to 2.5 V). According to Eq. (10.3) we can write that boost capacitance must be equal to

$$C_{BOOST} = \frac{2}{3} C_{LOAD} \tag{22.10}$$

Area of C_{BOOST} can be calculated under the hypothesis of plain & parallel plates capacitor using the following equation

$$C = \varepsilon_0 \varepsilon_{ox} \frac{A}{t_{ox}} \tag{22.11}$$

Given an oxide thickness of 250 Å, we get an area of

$$A = \frac{100 \cdot 10^{-12} F \cdot 250 \cdot 10^{-8} cm}{11.7 \cdot 8.854 \cdot 10^{-14} F \cdot cm^{-1}} \cdot \frac{2}{3} = 0.16 \cdot 10^{-3} cm^2 \tag{22.12}$$

If we assume that the capacitor is as long as a semi-matrix, i.e. almost 2 mm, we get an increment along Y of about 50 μm for each sector. On top of that, control and charge circuits for the boost capacitors must be added: for this, we can estimate an area occupation equal to half the space occupied by the capacitors themselves. Summing up, we estimate that every 512 Kbyte increases its height of 100 μm. The first and the last four sectors have a total parasitic capacitance C_{LOAD} equal to the capacitance of the other sectors. Figure 22.17 shows an updated floorplan.

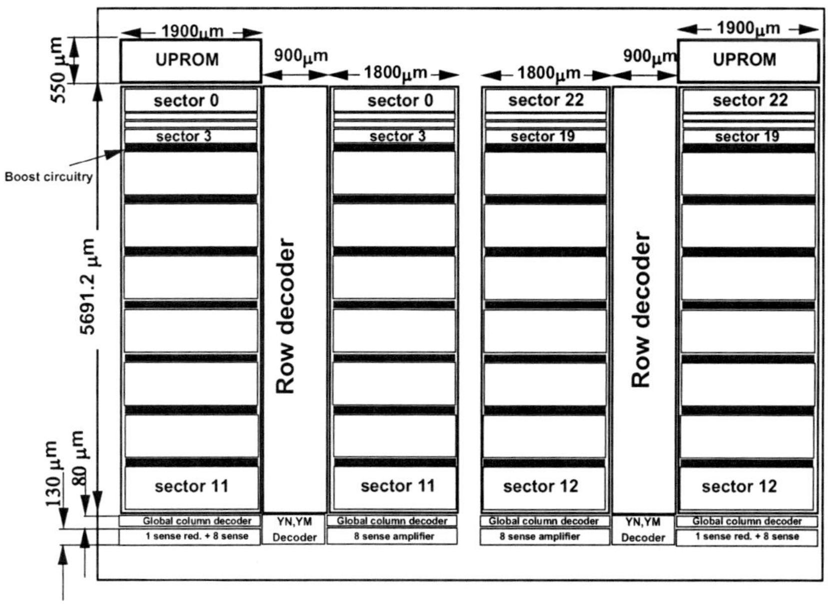

Fig. 22.17. Floorplan including boost circuitry

22.9 Architecture of the Reference

Reading a Flash cell, i.e. determining its logic state, mainly depends on the correct operation of the reference cell. In fact it is by comparing the current of the addressed cell with the current of the reference that the sense amplifier decides if the stored data is either a “0” or a “1”. We know that in case of EPROM cells, which are non-volatile but that can be erased by means of UV radiation only, placing the reference and the data cell as close to each other as possible solves the issue related to the reference. As shown in Fig. 22.18, a matrix column is used, one for each output, in order to realize the reference in read.

There are several advantages that come with this approach: the spread of the values of the reference cells with respect to the matrix ones is as small as possible, reference columns are realized inside the matrix and therefore are equal to the matrix ones. The differentiation occurs in the separate column decoding while row decoding is common: to be more precise, the word line is common to both the matrix and the reference cells.

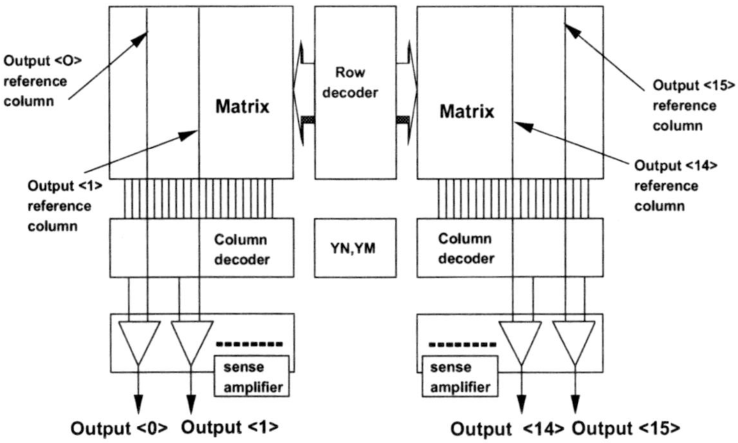

Fig. 22.18. Architecture where the reference is inserted inside the matrix. A column of reference cells for each output is present. The matrix row is the gate for both the matrix cells and the reference ones

In this way the reference is turned on together with the cell to read, thus eliminating the timing issues, and the loads on the branches of the sense amplifier are naturally balanced. This kind of organization is not suitable for a Flash cell, as discussed in Chap. 12. The reference is therefore placed outside the matrix.

Small matrixes of cells are realized, whose size is for example 10 by 10 cells, and the innermost cells are used to avoid border effects[5]. In this case, it is possible to write and/or erase them during testing phase in order to get the best possible reference. For sake of simplicity, a single small matrix is used, whose reference cells are shared by all the sense amplifiers.

The small matrix is placed symmetrical with respect to the sense amplifiers in order to balance the parasitic loads, i.e. not penalizing a sense amplifier with respect to the others.

22.10 Read Problems for a Non-Static Memory

We define *dynamic* those circuits, and therefore those devices, whose operation is related to the phases of a clock, often provided externally, that allows for the use of precharged nodes, i.e. initially biased to a given voltage value. In static circuits, on the other hand, voltage levels are always power supply voltages, apart from transients. The device that we are going to realize, given the required features, does not have an external clock, but nevertheless it falls into the dynamic category; to be more precise, we should say semi-dynamic (or semi-static), because it is a sort of hybrid.

We have seen that a boosted voltage is a voltage generated on a floating node using a capacitor. In other words, the over-voltage is not sustained by a voltage generator, but by the charge stored on the plates of a capacitor. A node biased like that does not remain indefinitely at the achieved voltage value, but it tends to get back to its natural state, i.e. either ground or VDD, losing its charge by leakage. Therefore either the sense amplifier reads at the right time or it does not read anymore, because the voltage of the word line does not remain at its boosted value forever. So we need a signal that gives a correct timing to the sense amplifier.

In order to solve the issue, we realize two "dummy" read paths, that work in parallel with the real one: the former dummy path reads a written cell, the latter reads an erased cell. If the read time of the dummy sense amplifiers is the same, with some margin, as the real sense amplifiers, we can use them to understand when the sense amplifier that reads from the matrix has carried out its task. It means that we must add two sense amplifiers that we place, as is with the small reference matrix, in a central position. The number of sense amplifiers grows to 36. Figure 22.19 shows the updated floorplan for the device.

[5] The rows and the columns that constitute the border of the matrix are different from the others because of "border effects", due to the abrupt variation of the geometries between the internal and the external of the matrix. This issue is solved by introducing unused rows and columns that are used just as a frame for the whole matrix.

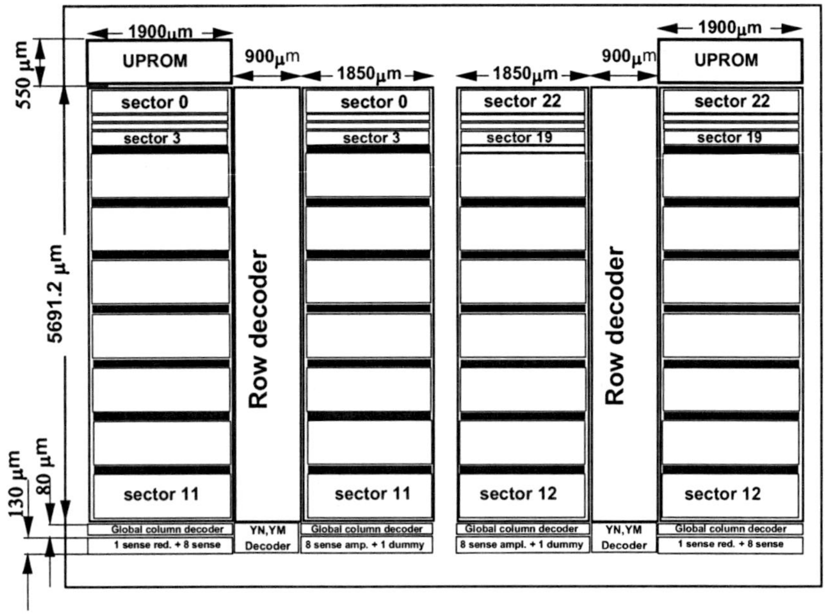

Fig. 22.19. The two dummy sense amplifiers allows, in read, the generation of the end boost signal

22.11 Erase and Program Circuits

Between sense amplifiers and output buffers it is necessary to place a logic that allows programming, verification and multiplexing for by-8 and by-16 read. Furthermore this logic must handle a part of the algorithm that allows deciding how many bits must be simultaneously programmed.

The issue with programming is mainly related to the current required that we can estimate to be 300 μA per cell. Voltages required to bias the gate and the drain of the cells are produced by charge pumps starting from VDD power supply, whose minimum value is 2.5 V. The value for the drain voltage is about 4.5 V; in case of parallel programming of 16 cells, a current around 5 mA must be provided. A charge pump featuring such characteristics requires very large capacitors whose turn on time is very long. As a consequence, it is preferable to realize several smaller charge pumps, in order to place them more easily in the device floorplan.

Write algorithm verifies, after every pulse on the drain, the result of the operation and the subsequent pulse is given only to the bits that haven't pass the verification, thus avoiding a useless stress to those cells that have reached the target threshold voltage after a single pulse. In order to limit the required current, thus reducing area occupation of drain pumps, we decide to program in parallel no more than eight cells, even if the device is configured by 16. Since write time is one the key parameters when evaluating memory quality, it is necessary to use any trick that can reduce it. Control logic circuitry, for instance, normally handles programming byte by byte, but it is able to work by word, in case a maximum number of eight bits must be simultaneously programmed. Specification calls for a programming time equal to 10 μs/byte, and the maximum time for a sector is 30 s.

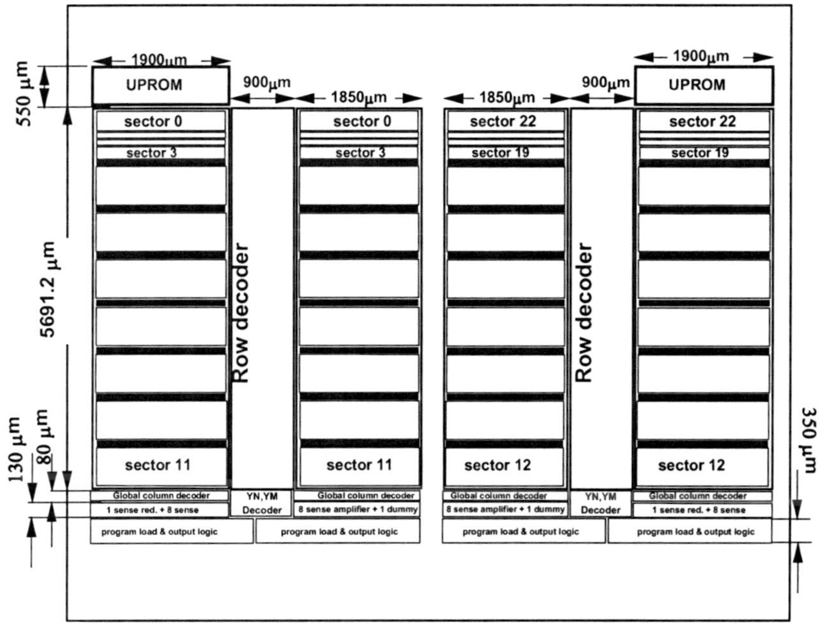

Fig. 22.20. Floorplan including the output logic and the program loads

When we consider area occupation, we should take into account the transistor, called program load, which is in charge of delivering the regulated voltage to the drain of the cells. This transistor, in order to work in every corner of the specification, must be at least 100 μm wide[6]. Given the chosen structure, column decoding is the same in program as in read; therefore we must account for a program load for each sense amplifier.

[6] The current that a PMOS transistor can provide is about 150μA/square, when both V_{GS} and V_{DS} are at their maximum value. Therefore we can say that in order to get the typical value of 300 μA, we must be able to provide at least 500 μA and, in case the transistor is not working in saturation, we need at least 30 times more, i.e. 90 squares.

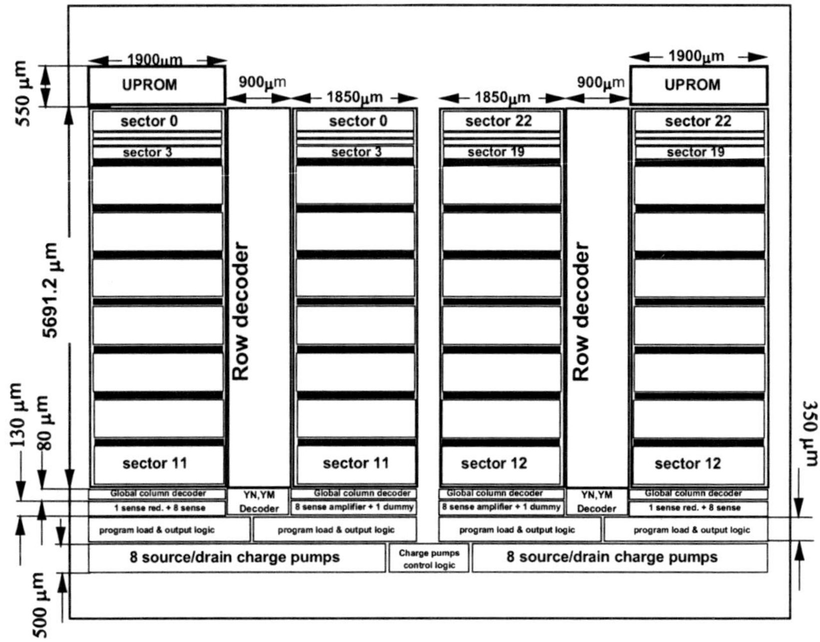

Fig. 22.21. Source and drain pumps occupy the whole X dimension of the device and 500 μm along Y

Summing up we can estimate an height of 350 μm for both the logic and the program loads, and the shape of the device is as shown in Fig. 22.20. The charge pumps we described above will be placed in a strip between the logic and the output buffers. Their dimension can be estimated starting from the size of the capacitors that constitute the pump. Let's assume a current requirement of 300 μA for each cell, having 5 V at the drain: we must then pump the drain node of a quantity equal to VDD, in the worst case of power supply of 2.5 V. The charge pump is based on the continuous boost principle. An oscillator continuously charges a capacitor, by pumping the output node of the parasitic capacitor. The current that the capacitor can provide is the one given by its discharge during the stable intervals of the clock. If we assume a specific capacitance for thick oxide capacitors equal to 1 fF/μm^2 and that 300 μA must be provided in 100 ns causing a variation of 100 mV at the terminals of the capacitor, we get a capacitance equal to

$$C = \frac{300\,\mu A \cdot 100\,ns}{100\,mV} = 300\,pF \tag{22.13}$$

And such a capacitance requires an area occupation of at least 300,000 μm^2. Programming eight cells in parallel, the area reserved for the capacitors accounts for about 2.4mm^2. Assuming that the area required by the circuitry of the pumps is equal to the area required by the capacitors, we get at the end to approx 5mm^2, i.e. a strip as long as the device and 500 μm tall, as shown in Fig. 22.21.

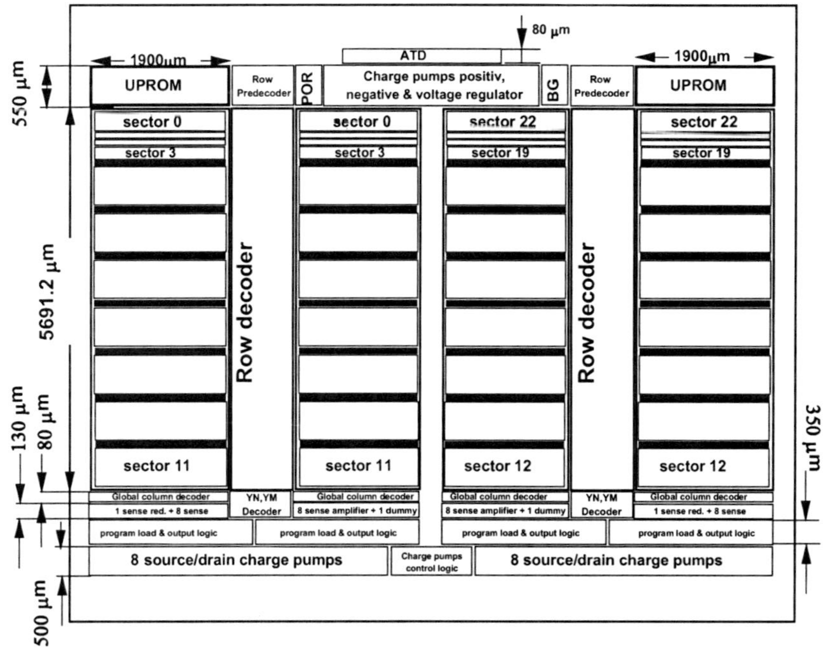

Fig. 22.22. Floorplan inclusive of gate pumps and regulators and circuits to generate ATD and POR

The pumps used during program are also used to bias the source of the cells during erase.

In the space between the two UPROM banks, pumps and regulators required to generate word line voltages must be allocated. It is necessary to have a positive pump to be used during program, and a negative one for erase.

Updated floorplan is shown in Fig. 22.22, where the circuits to generate POR (see Chap. 5) and ATD signals have been added.

In order to complete the preliminary analysis of the device size we need to take into account the pads and the control logic that implements the algorithms.

22.12 Pad[7] Placement

Let's assume that the market survey has pointed out that, owing to the different customer requirements, our device must be mounted into three different packages:

---

[7] PAD indicates the places where bonding wires are soldered; such wires connect the integrated circuit to the external pins of the package. The name derives from the fact that these are very large structures, in the order of 100 μm x 100 μm, that are visible to the naked eye.

TSOP48, TSOP40 and SO44. These are all surface mount packages, where the number of pins is different.

In order to accommodate this requirement, it is often necessary to add some pads to the device, which can be selected by means of non-volatile registers. The analysis of all the configurations leads to the definition of the required pad placement. Let's assume that the most problematic package is TSOP40 owing to the fact that the cavity, i.e. the space reserved to the chip, allows for a maximum occupation along X of 10 mm. The space that must be left between the edge of the cavity and the chip is at least 0.5 mm on each side. Consequently, size along X of our device cannot be greater than 9 mm. If we look at the floorplan, we find out that X dimension of the device is already at the limit. In order to completely define the dimension along X, we must take into account the signals that, starting from the top of the device, go along Y down to the bottom and vice versa; among these signals we can find supplies, whose width is quite big (about 50 μm), that we will define later. In practice, we decide that the device must not have an X dimension greater than 10 mm.

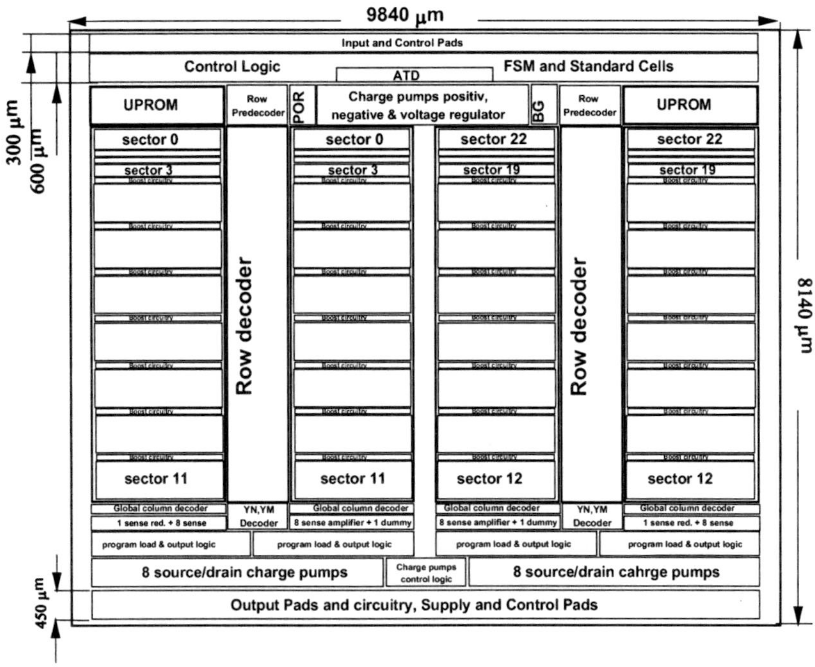

Fig. 22.23. Complete floorplan

We can estimate the area occupation of the pads recalling that the size of one pad is approx 150 μm x 150 μm, plus the space for the circuitry and the Electro-Static Discharge protections. At the end, we can estimate along the Y a size of 450 μm for each pad.

Finally we must allocate enough space in the top portion of the device for control logic, which is composed of PLA and glue logic, realized using standard cells methodology.

Cost analysis requires that the maximum area of our memory must be 62 mm^2, considering that a 15% linear shrink will be applied; above this value, the device cannot be produced because of the price that the market is willing to pay for memories of such a size. Since we fixed X dimension to be 10 mm (before shrink), Y dimension must be approx 8.65 mm. The space left for the logic is therefore equal, along Y, to approx 600 μm. Complete floorplan is shown in Fig. 22.23.

22.13 Control Logic and Related Circuitry

We are going to realize the control logic for program and erase using two different finite state machines, implemented as PLA. Since the algorithms need several count operations, it is necessary to introduce dedicated counters.

Time Counter checks, during program phase, the duration of the pulse itself and the sequencing of the different operations. In this case, the pulse lasts some microsecond. During erase phase, the pulse lasts ten milliseconds. Both pulses are repeated a given number of times until the desired threshold voltage of the cell is obtained. End-count checks are implemented too, so that the user can be warned in case of failure of the operation, i.e. in case the cell has not reached the desired threshold voltage after the maximum number of attempts.

Sector, row and column counters are used in the programming phase, called preconditioning, which occurs before erase. In the sector under erase, all the bytes are sequentially scanned, programmed and verified, before starting the real erase operation.

At the end of the erase, again thanks to the scanning performed by the counters, the over-erased cells, if any, are recovered by the soft-program algorithm.

The two PLA are interfaced by a logic and all the blocks are connected each other through the GLUE LOGIC block, which has been represented as a unique block, but in reality it is distributed across all the available space. This block is also responsible for handling the test modes (Chap. 20) that are used to analyze the behavior of the chip. Finally we have the CUI, Command User Interface, whose task is to receive the commands from the external, decode them and activate the state machines as required. The block diagram of the control logic required to operate a Flash memory is shown in Fig. 22.24.

In this chapter we have imagined the feasibility study that is carried out before starting a project, in order to understand the validity and industrial advantage that derives from its realization. We have rapidly gone through all the main building blocks, trying to understand the behavior and to justify their use. In this way, the device is born. We can say that now it is just like a jigsaw puzzle where we have assembled the frame and some parts, so that we know the size, but the contour of the resulting picture is not well defined. Our task is to focus the picture, defining it using a lot of skill and a good amount of fantasy. At the end of the job, we will compare the forecast done during feasibility study with the final result in order to understand our mistakes … if any!

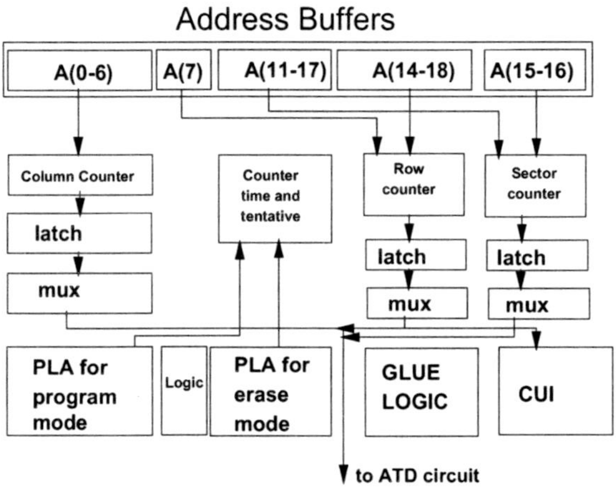

Fig. 22.24. Block diagram of the control logic

Bibliography

B. Venkatesh et al., "A 55ns 0.35um 5V-Only 16M Flash Memory with Deep-Power-Down", ISSCC96.

R. Micheloni, M. Zammattio, G. Campardo, O. Khouri, and G. Torelli, "Hierarchical sector biasing organization for Flash memories", in Records 2000 IEEE Int. Workshop on Memory Technology, Design and Testing, pp. 29-33, (Aug. 2000).

23 Photo Album

23.1 Introduction

In this chapter, a collection of photos realized by means of an electronic microscope will be presented. Many of the structures discussed in the previous chapters will be shown. As a suggestion, the reader should try to recognize all the elements we have dealt with.

An index of contents prefaces the photos, while comments are present in both pictures and captions. Figures from 23.26 to 23.34 have been repeated for sake of clarity; the first figure shows the layout whereas, in the second picture, the indications of the main circuit blocks is overlaid to the layout itself. The authors would like to thank STMicroelectronics Srl, Agrate Brianza, Milan, Italy, for the kind permission.

23.2 Figures Index

23.3 The Photos

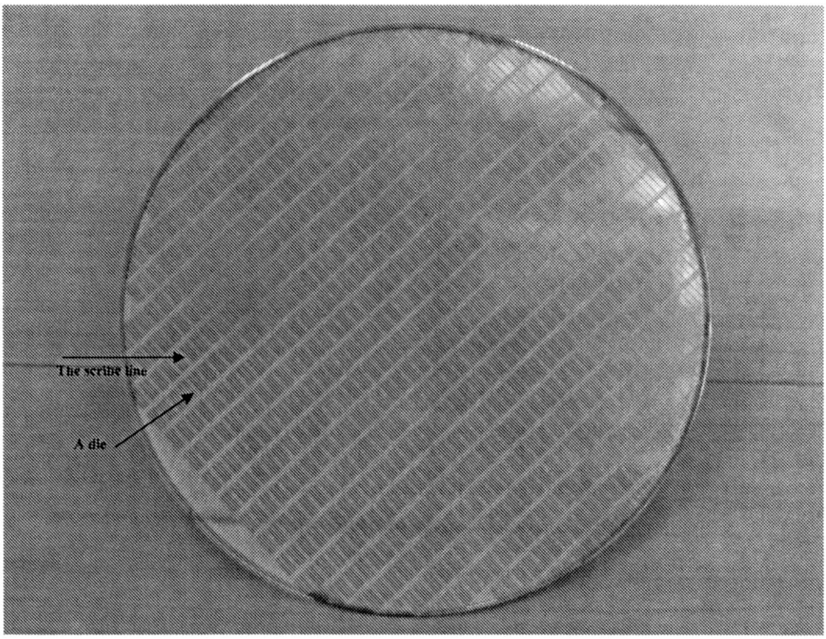

Fig. 23.1. Eight-inch wafer. An eight-inch wafer containing Flash memory devices. In the photo, a single die and a scribe line, i.e. the space left between two contiguous devices to allow separating the dice and to assemble them into the package, is shown in foreground

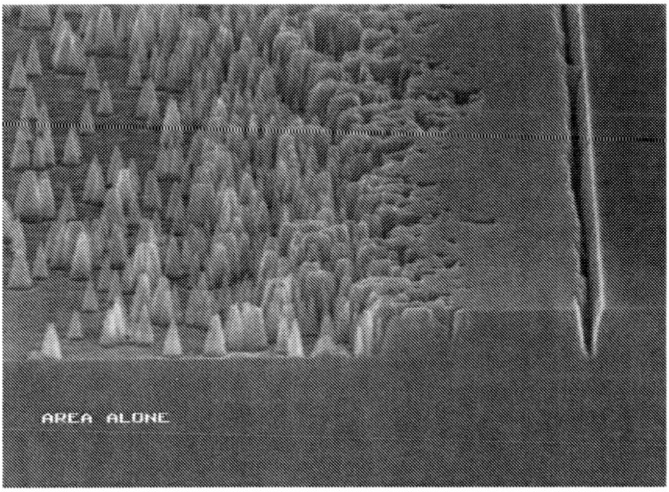

Fig. 23.2. Over etch. Low selectivity etching produces a non-planar silicon surface. The formations that can be distinguished on the left are due to the anisotropic etching

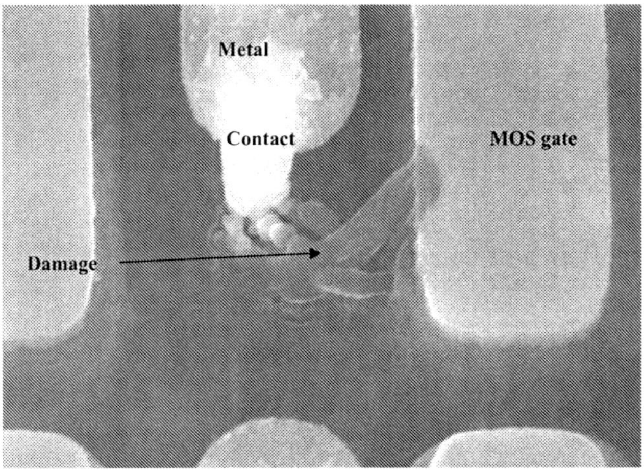

Fig. 23.3. Snap-back damage. The snap-back effect provokes an anomalous current flow with the consequent transistor self-heating. In this case, the metal of the contact smelts, short-circuiting the transistor gate as a result

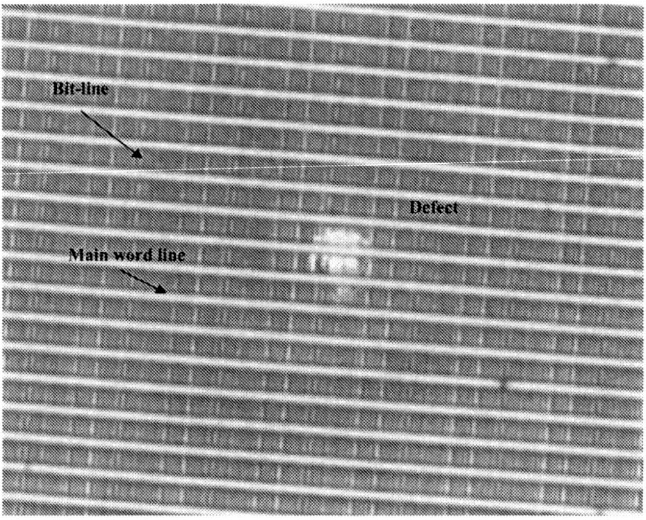

Fig. 23.4. Particle defect. A portion of the memory array where the main word line and the bit line are highlighted. The defect is represented by the bright spot in the middle, probably a particle of intermetal dielectric that has damaged the entire structure. Dust particles with diameter around one micron are comparable with the size of the structures

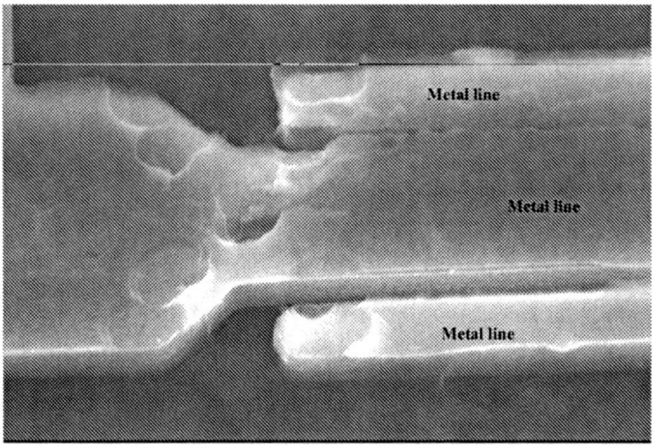

Fig. 23.5. Electromigration defect. The current flow, higher than the designed one, produces an excessive heating of the structure, with the consequent metal smelting and irreversible damage

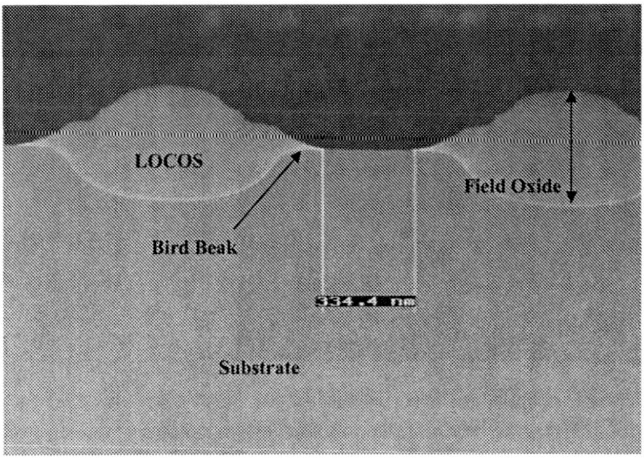

Fig. 23.6. LOCOS insulation. We have seen that the insulation allows separating the different areas that form the transistor to guarantee that no undesired paths or parasitic components are involved in the functioning. The LOCOS insulation is obtained by growing the oxide. This provokes the so-called beak, which is a limitation for the reduction of the device size

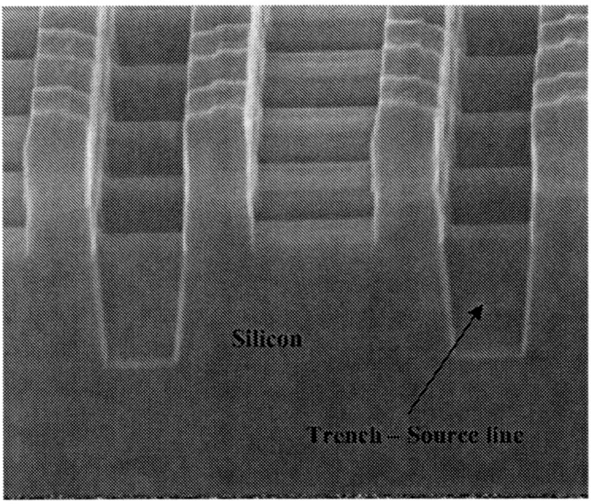

Fig. 23.7. Realization of the trench insulation obtained by means of an excavation in the silicon surface. Subsequently, the hole is filled with oxide so as to prevent the formation of the beak

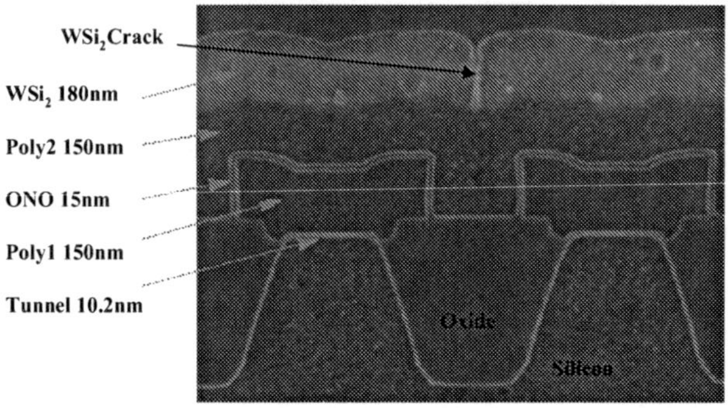

Fig. 23.8. Trench Flash cell architecture. The structure of the Flash cell is realized by a trench insulation process. The typical values of the different layers composing the structure are reported. In the photo, two contiguous Flash cells are shown. It is possible to distinguish the tunnel oxide, the poly1 floating-gate, the interpoly insulator, the ONO layer, the poly2 control gate, which is continuous since it forms the word line and, finally, silicide deposed on the row to lower the resistivity. The silicide deposition is defective due to the cracking highlighted

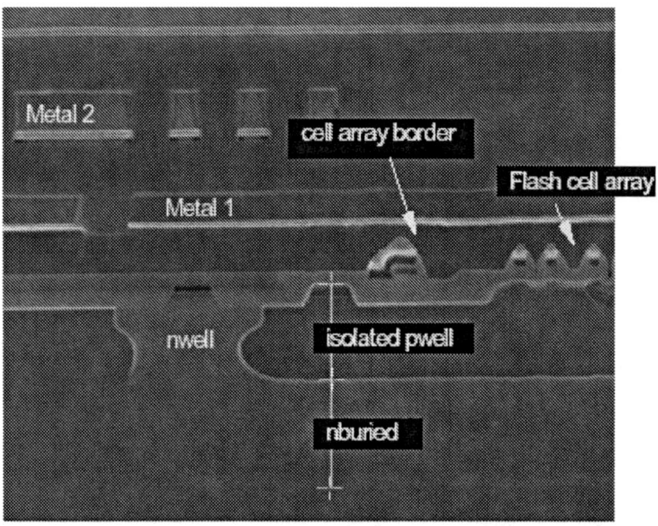

Fig. 23.9. Triple-well structure. A triple-well structure realized beneath the array is shown. The n-buried is connected to the shallow n-well to form the insulation tub for the insulated p-well. The array cells and the metal1 that forms the bit line can be distinguished on the right

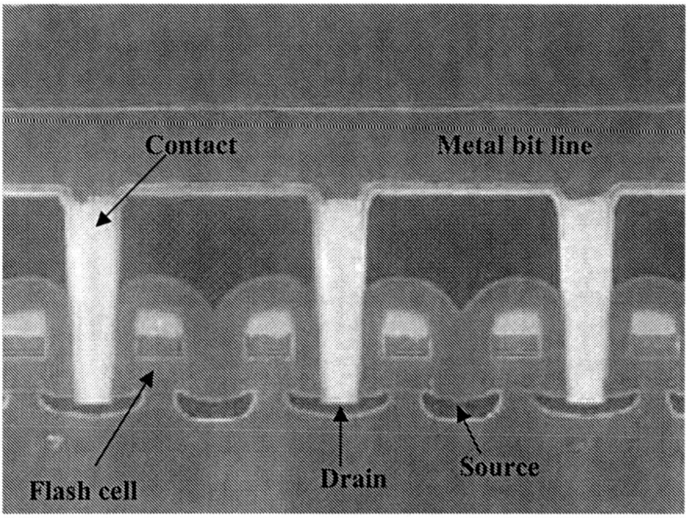

Fig. 23.10. Matrix bit line selection. This section shows six Flash cells along a bit line. The drain and the source are shared between two contiguous cells, the bit line metal contacts the drain junctions. Notice the vertical size of the contacts with respect to the cell height; this highlights the difficulty in realizing the contacts without damaging the cell structure of the Flash memory

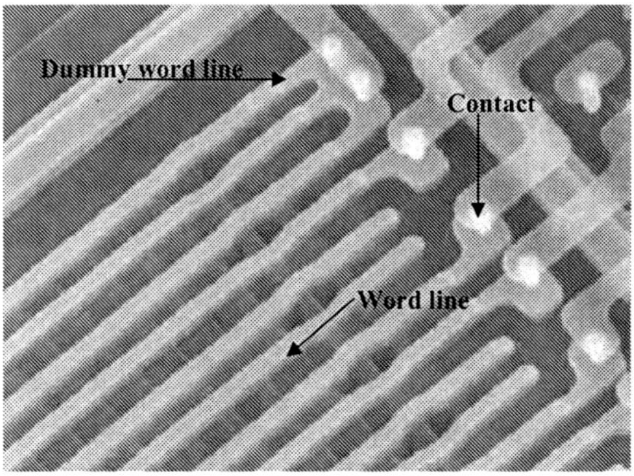

Fig. 23.11. Matrix dummy word lines. The dummy rows at the end of the sector are contacted by means of metal. The active word lines of the array are also visible

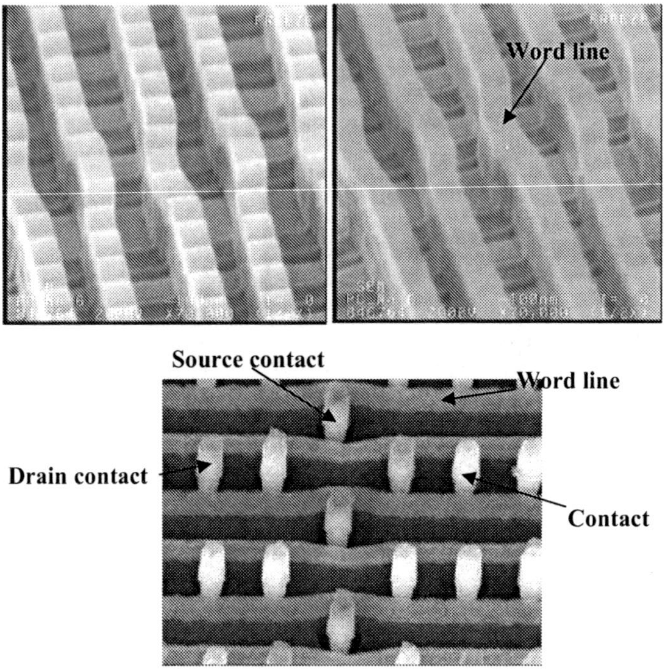

Fig. 23.12. Matrix word lines. The figures above show the rows of the array in perspective, and those that still have the contact plugs after the oxide removal. Notice the drain and source contacts placed on opposite sides

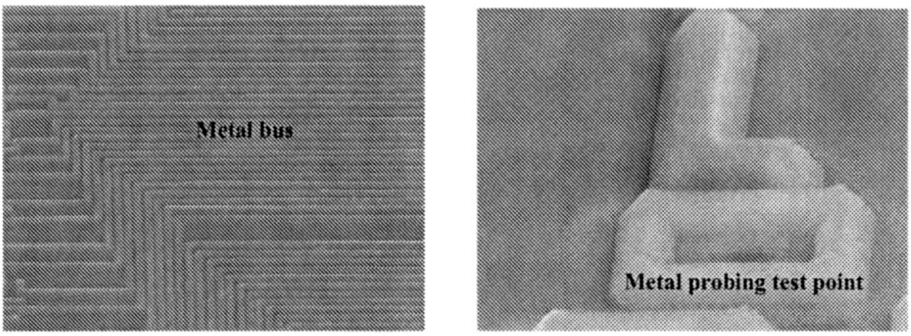

Fig. 23.13. Metal bus. A signal bus realized of metal can be recognized on the left, note that the curvature of the angles is the same for all the metals. On the left a test pad, realized by means of a metal ring so as to easily insert the microprobes used during laboratory analyses

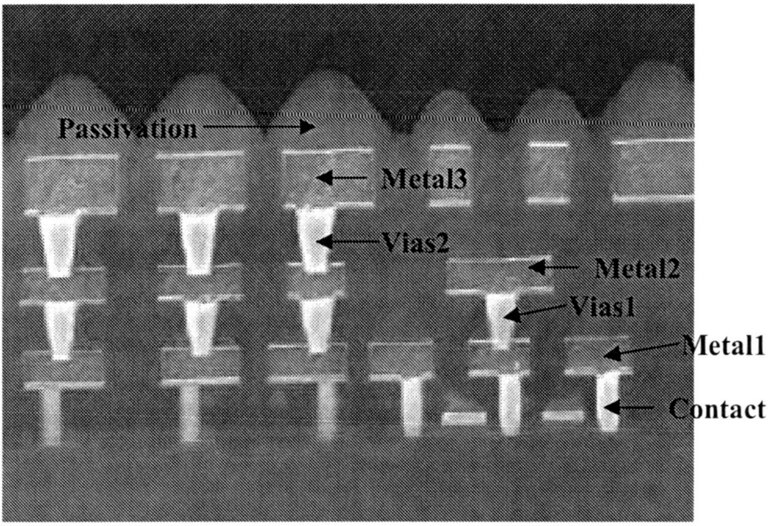

Fig. 23.14. Stacked contact. In the photo, three metalizations and their connections can be recognized. It is important to observe that the connections are located the one over the other, thus minimizing the space occupation

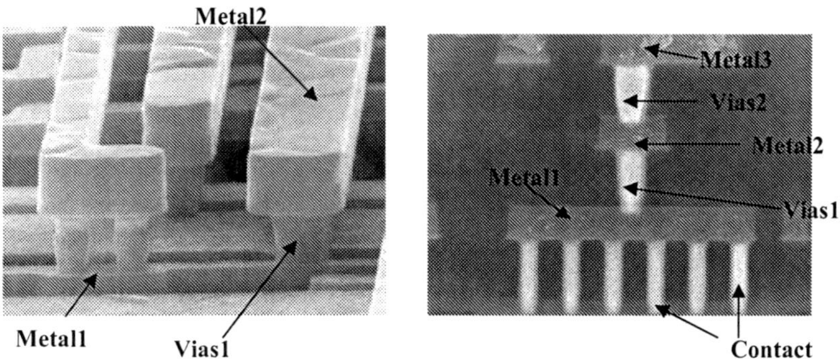

Fig. 23.15. Metal connections by vias (1). On the left, the intermediate oxide has been removed to make the via1 connections visible, while a stacked connection is shown on the right, where the multiple contacts that reduce the resistivity between metal1 and the underlying layer (active area or poly2), realized by means of six contacts, can be noted

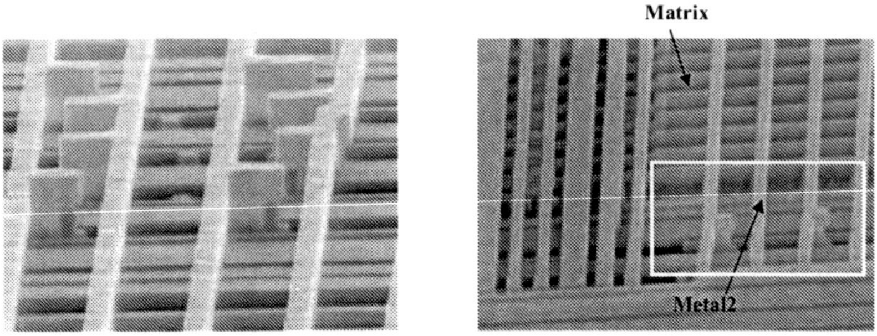

Fig. 23.16. Metal connections by vias (2). On the right, a portion of array sector with metal2 stripes of the circuits all around. On the left, magnification of the portion shown on the right

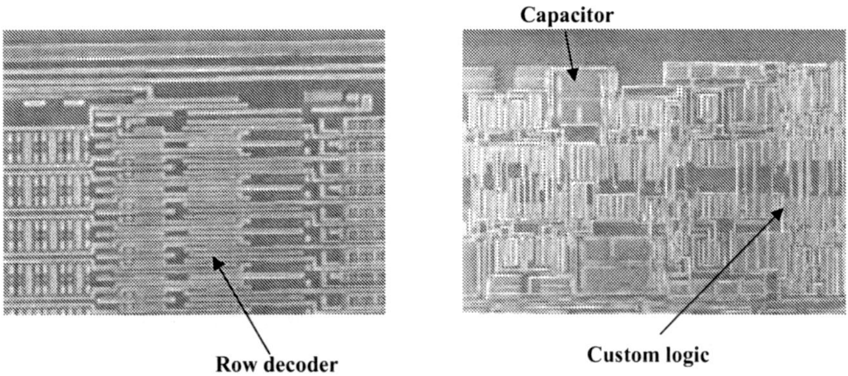

Fig. 23.17. Metal circuitry. On the left, an image of the row decoder where the regularity of the structure, derived from the regularity of the array, is evident. On the right, the lack of regularity of the custom circuitry can be observed. There is order only in the direction of the connection lines that, in this case, go in the top-down direction. Generally, the different connection layers (metallizations) are realized in a given direction to improve connectivity. Thus, as in the figure metal2 is shown, metal3 will be laid out mainly in the horizontal direction

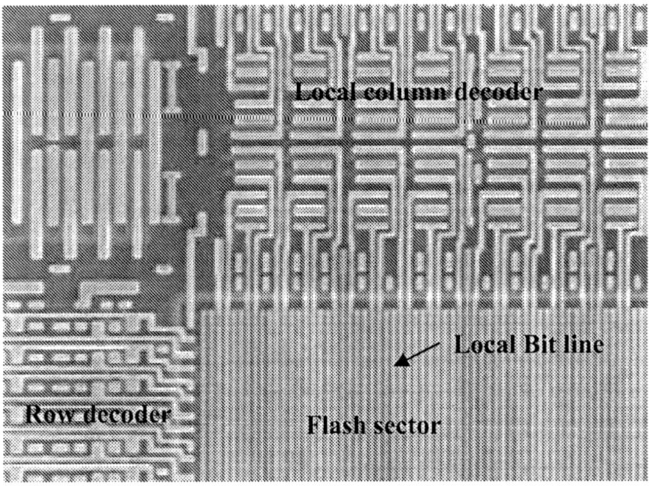

Fig. 23.18. Local decoder (1). This photo reports the local column decoder, a portion of sector and the local bit lines. On the left, the row decoding. The layer is always a metallization, metal1 in this case. The bit lines can also be recognized

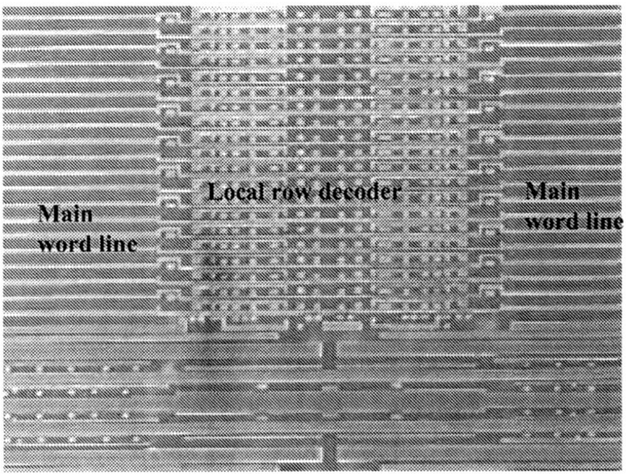

Fig. 23.19. Local decoder (2). Metal3 with the local row decoding between two sectors and the related main word lines

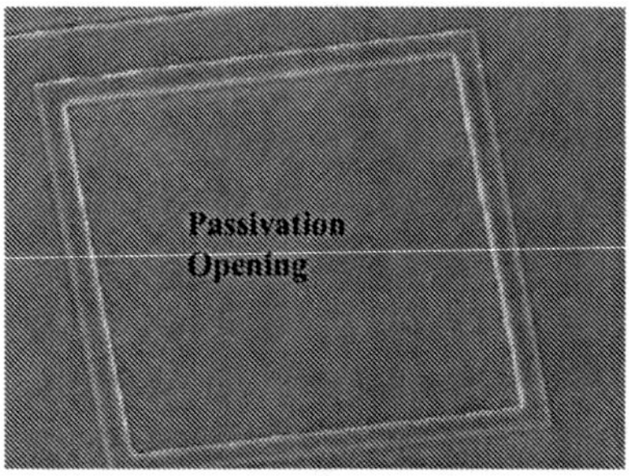

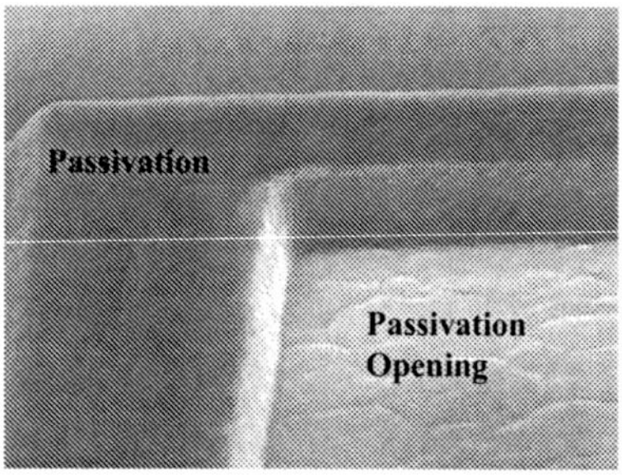

Fig. 23.20. Pads. The passivization openings, covered by metallization islands, used to solder the bondage wires that connect the device with the package pins

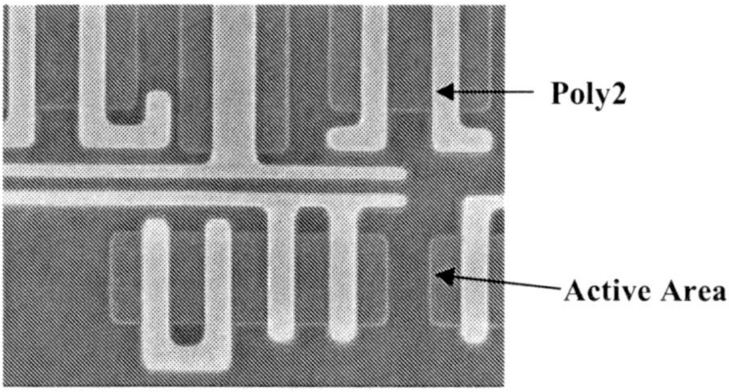

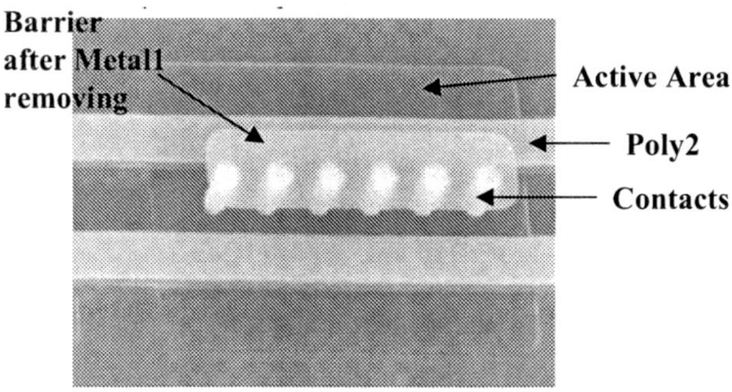

Fig. 23.21. MOS transistor (1). The first photo shows the layout of some Mos's. The shape of the active area overlaid with the poly2 gates is evident. The picture below shows a MOS in perspective, the contacts are used to connect metal1 with the underlying active area

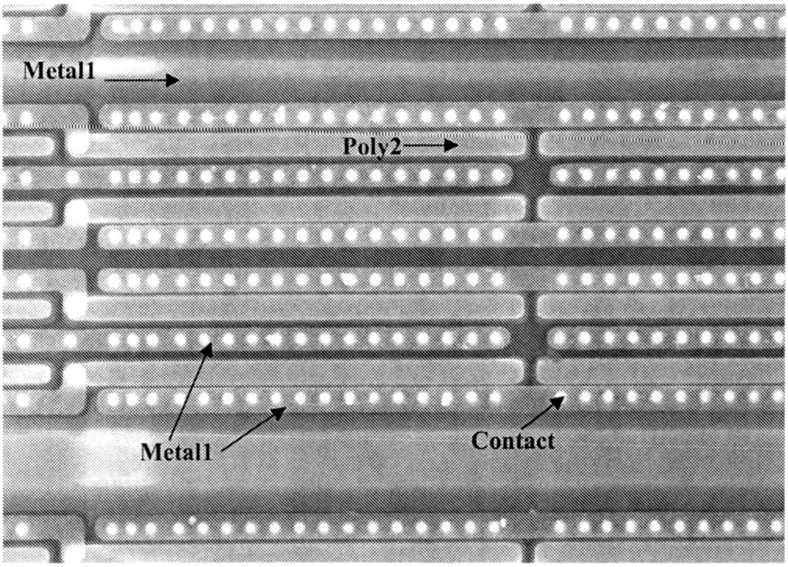

Fig. 23.22. MOS transistor (2). Poly2 gate that creates the MOS transistors and the metal1 that contact the underlying active area

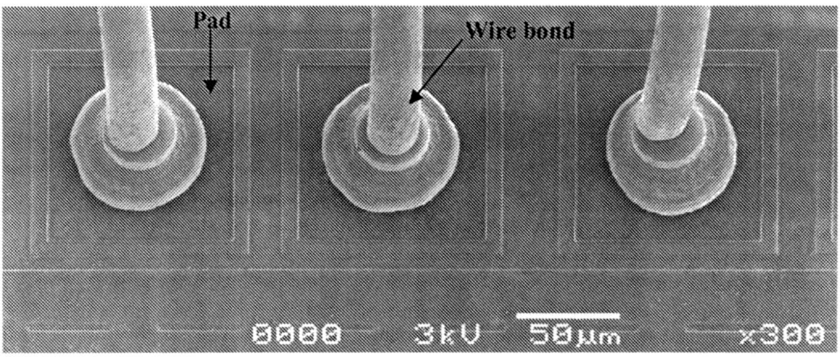

Fig. 23.23. Pad bond wire (1). Three pads of a device and the connections obtained by fabricating a ball soldered on the pad and gold wires that connects the balls with the external pins of the package. Note that the wire on the right is almost detached, which pinpoints a problem of packaging

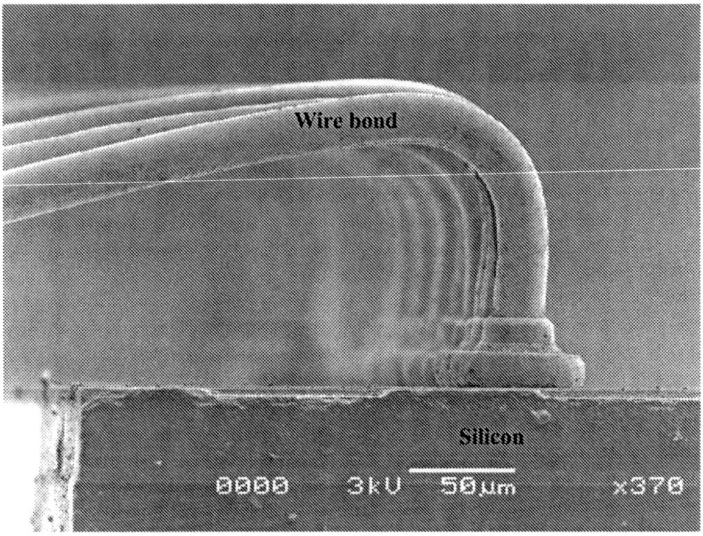

Fig. 23.24. Pad bond wire (2). A second image of a wiring bonding between silicon and pins (not shown in the photo). Note the wire curvature, designed to obtain the best connection and mechanical resistance

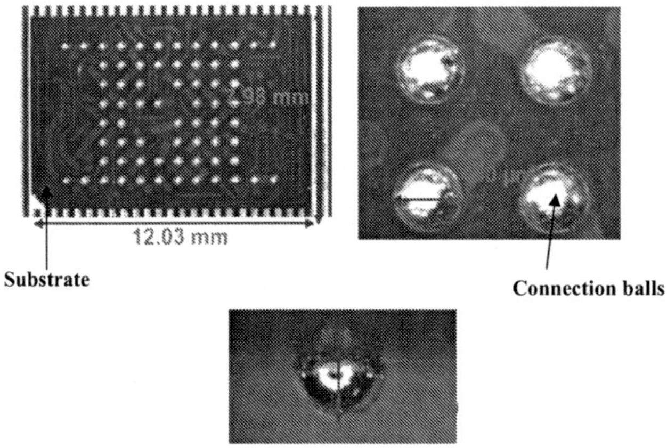

Fig. 23.25. BGA package. The layout of a kind of package widely used today, especially in wireless applications, is shown in the right top corner. The chip is soldered on the opposite side and is connected to the substrate, which is a real board with metal stripes that allows connecting the pads to the balls, by means of the wire bonding. The balls are then soldered to the board and the final connections are fabricated. The advantage of this package is the miniaturization and light weight

Fig. 23.26. EPROM chip memory. Complete layout of a 512kbit EPROM device in NMOS technology

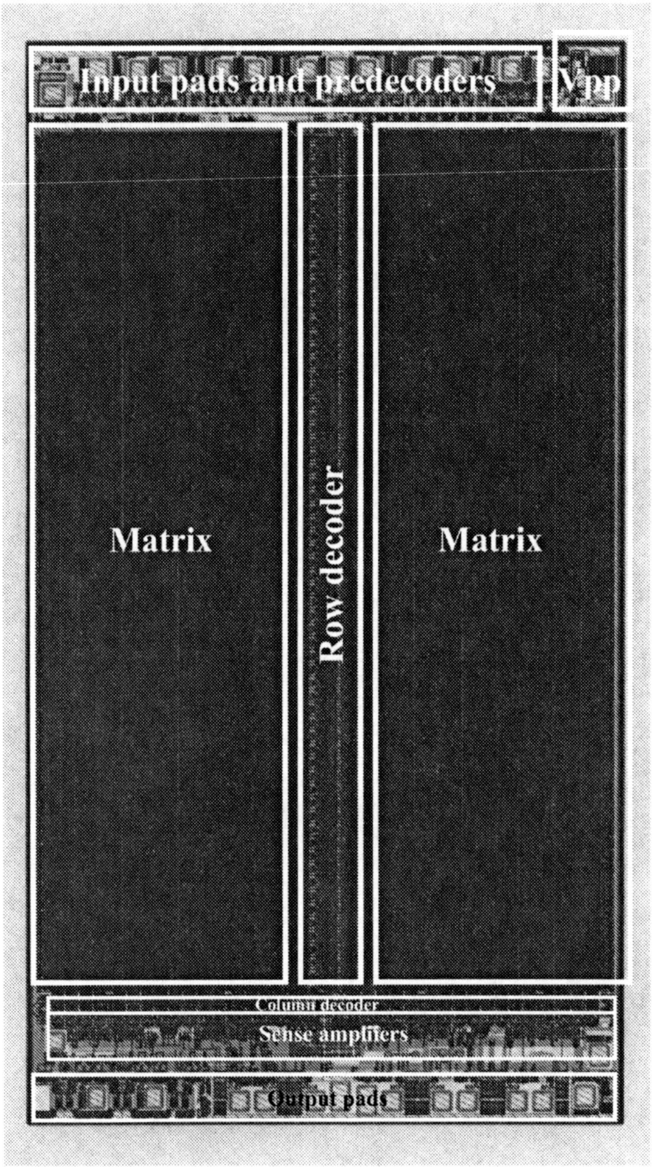

Fig. 23.27. The array, divided into two semi-arrays is highlighted, with the row decoder in between. The column decoder is located near the sense amplifiers. In the top part are the input pads and both row and column predecoders. Finally, the VPP pads with the related circuitry is shown

Fig. 23.28. ASM memory chip. ASM is the acronym of Application Specific Memory. In this case, an EPROM memory (realized in CMOS technology) is used to realize a FIR filter

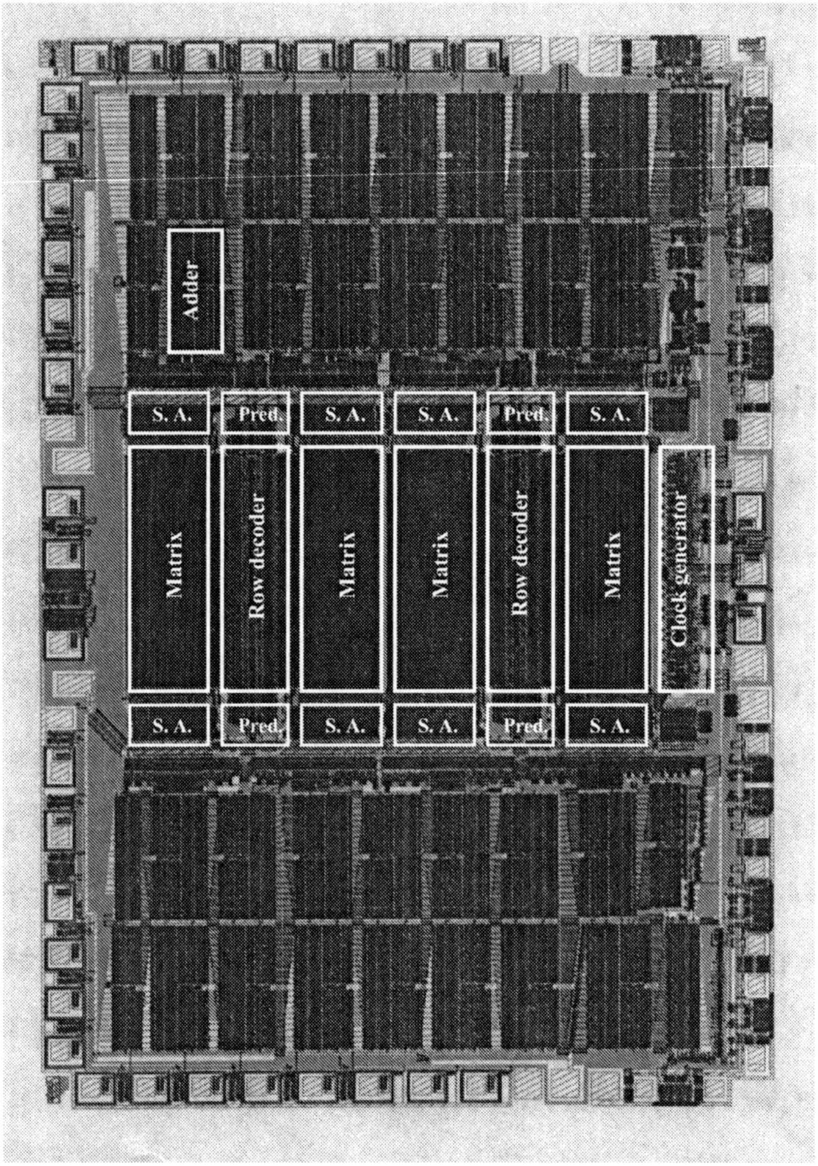

Fig. 23.29. This device realizes a FIR filter using an EPROM memory, divided into four semi-arrays to diminish the row lenght and speed up read. The content of the memory is a look-up table whose outputs are added by a fast adder to produce the digital processing that realizes the required filter. The "S.A." acronym stands for sense amplifier, and "pred." stands for row predecoder. In this case, the column decoder is not present to speed up read

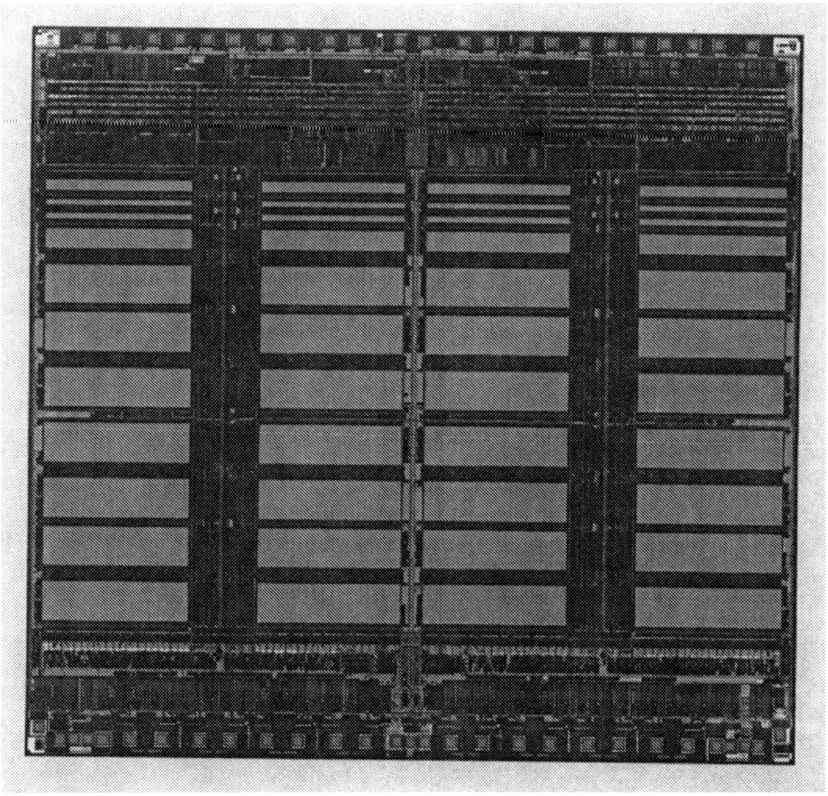

Fig. 23.30. Flash chip memory. Complete layout of an 8 Mbit Flash memory in CMOS technology. This is the device described in Chap. 22. The different blocks are described in detail in Fig. 22.23 to which the reader should refer for the analysis of the circuit blocks

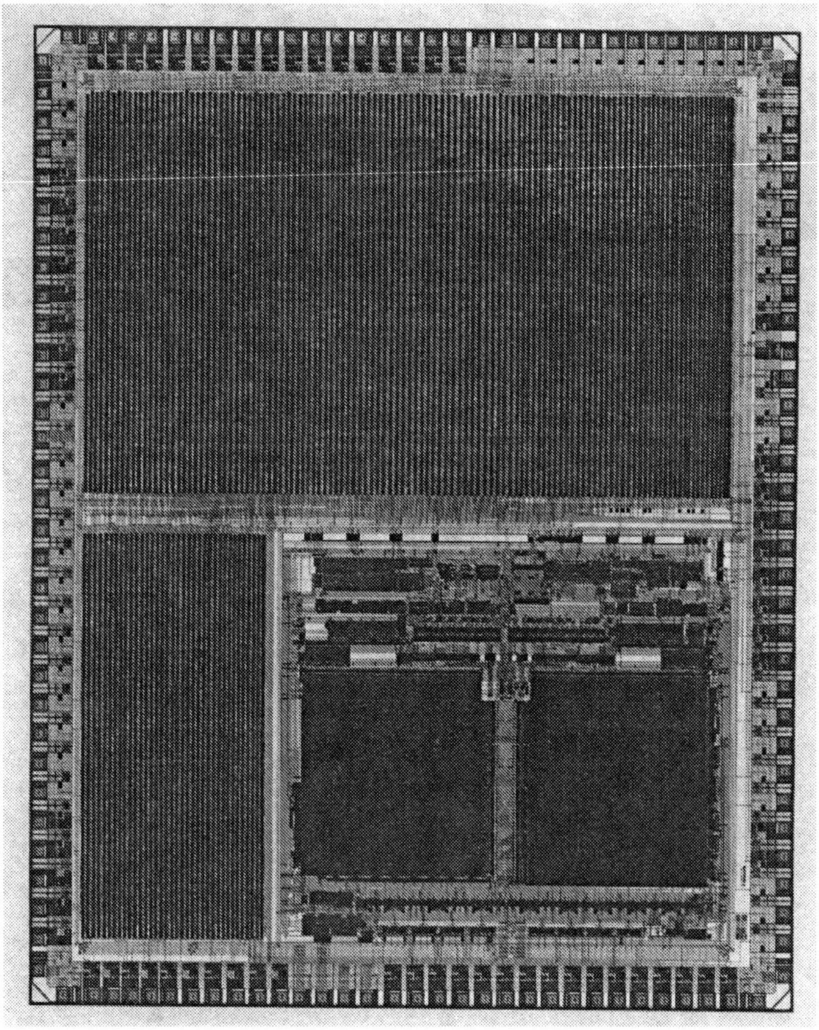

Fig. 23.31. ASM memory chip. Complete layout of a 2 Mbit Flash device with an ASIC in CMOS technology for automotive applications

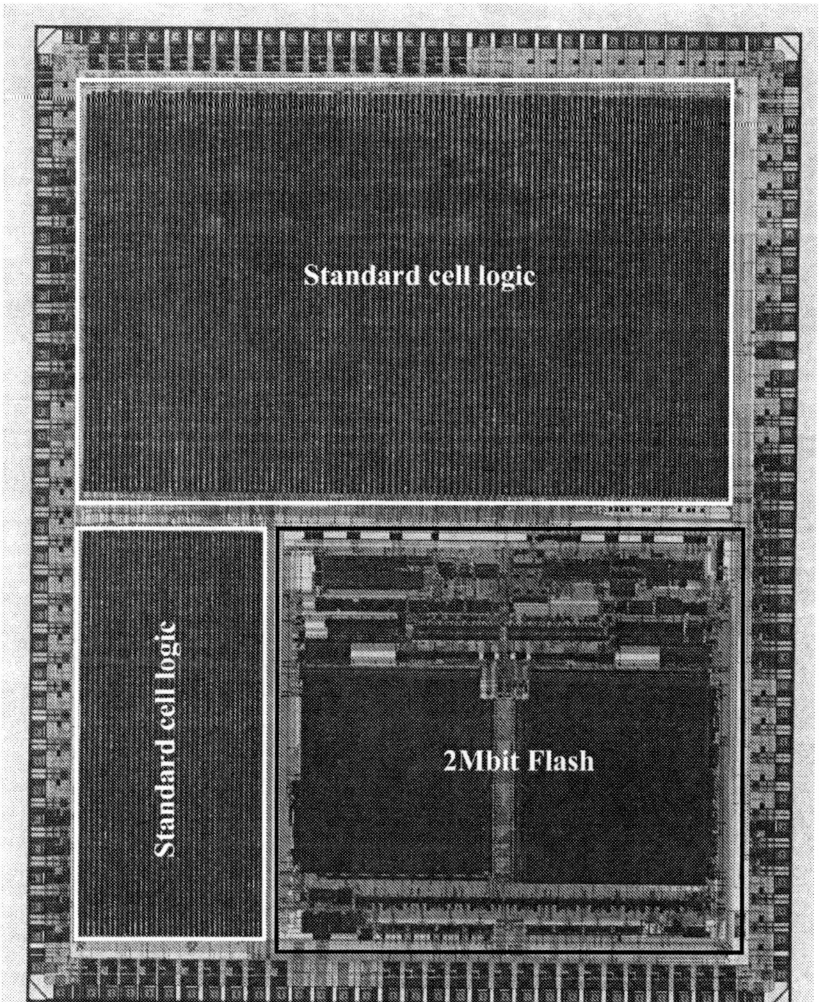

Fig. 23.32. ASM memory chip. The 2 Mbit Flash memory is located inside a logic device, pad limited, designed with the standard cells technique. Note the regularity of the logic cells compared with the part of memory realized by designing custom transistors almost one by one

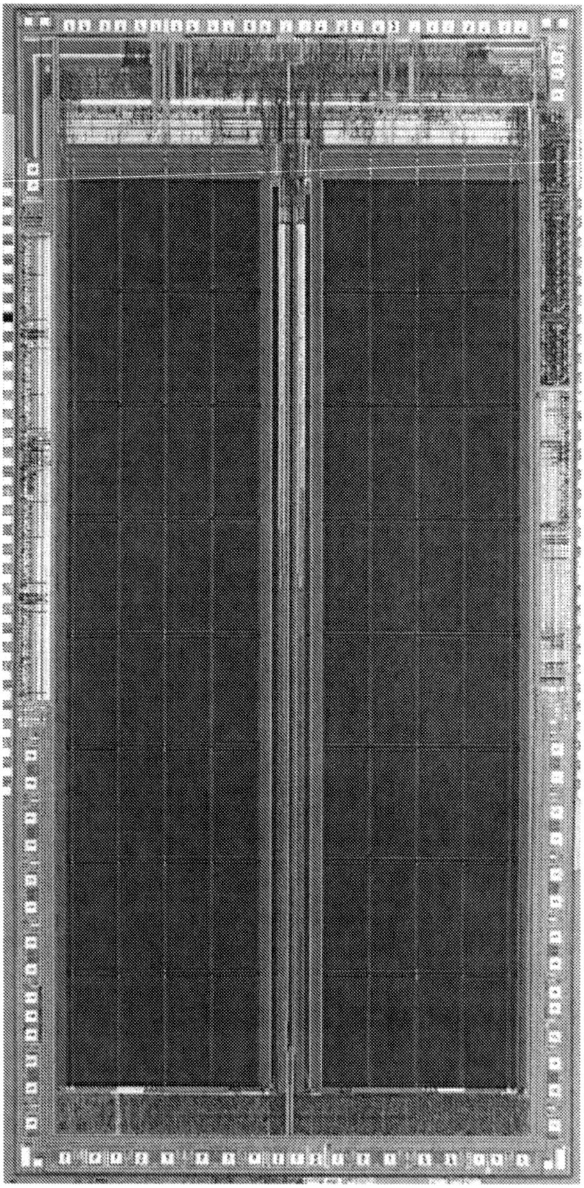

Fig. 23.33. Flash memory chip. A 64 Mbit multilevel Flash device in CMOS technology

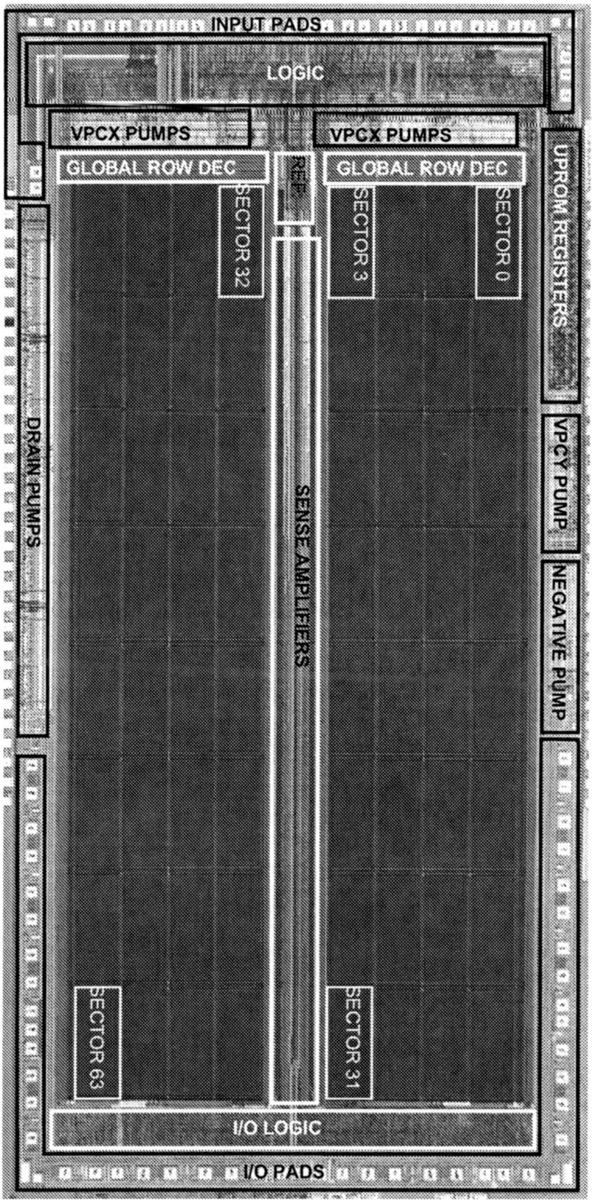

Fig. 23.34. Flash memory chip. The 64 memory sectors are shown with the necessary circuitry. Each sector is 1 Mbit from a logic point of view but is realized as half physical Mbit since the device is a multilevel 2bits/cell

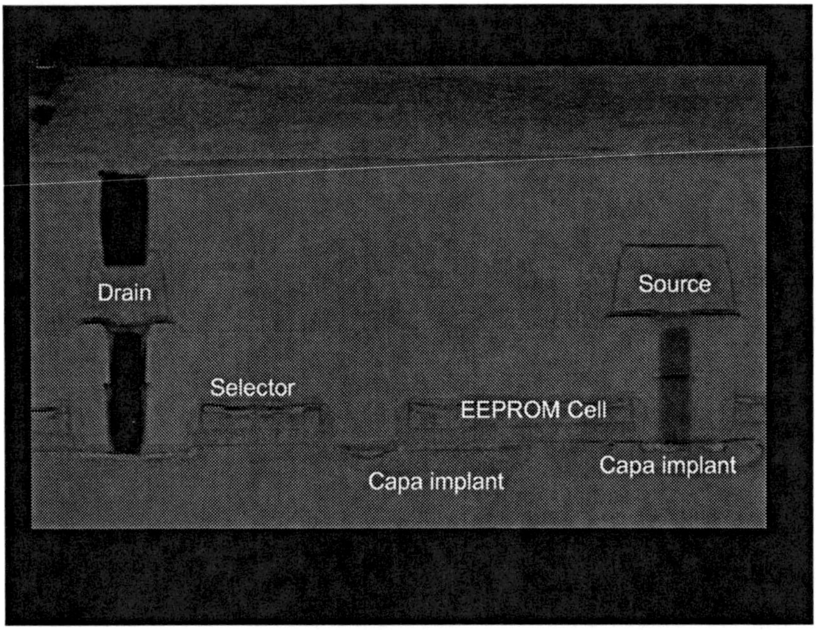

Fig. 23.35. Section of an EEPROM memory cell. The selector and the memory cell with the source and drain contacts are visible. The Capa implant is the implat realized to hold the high operating voltages during write and erase operations

Subject Index

Printed in the United Kingdom
by Lightning Source UK Ltd.
131152UK00001B/25-120/A

9 783540 201984